Implementation and Applications of

DSL

Technology

edited by
Philip Golden
Hervé Dedieu
Krista S. Jacobsen

Auerbach Publications
Taylor & Francis Group
Boca Raton   New York

Auerbach Publications is an imprint of the
Taylor & Francis Group, an **informa** business

CRC Press
Taylor & Francis Group
6000 Broken Sound Parkway NW, Suite 300
Boca Raton, FL 33487-2742

© 2008 by Taylor & Francis Group, LLC
CRC Press is an imprint of Taylor & Francis Group, an Informa business

No claim to original U.S. Government works
Printed in the United States of America on acid-free paper
10 9 8 7 6 5 4 3 2 1

International Standard Book Number-13: 978-0-8493-3423-8 (Hardcover)

Library of Congress Cataloging-in-Publication Data

Implementation and applications of DSL technology / edited by Philip Golden,
     Herve Dedieu, Krista S. Jacobsen.
          p. cm.
     Includes bibliographical references and index.
     ISBN-13: 978-0-8493-3423-8
     ISBN-10: 0-8493-3423-3
     1. Digital subscriber lines. I. Golden, Philip. II. Dedieu, Hervé. III. Jacobsen,
Krista. IV. Title.

     TK5103.78.I57 2007
     621.387'8--dc22   2007020715

Visit the Taylor & Francis Web site at
http://www.taylorandfrancis.com

and the CRC Press Web site at
http://www.crcpress.com

Dedication

Philip Golden dedicates this book to Hannah and Caroline.
Hervé Dedieu dedicates this book to Claire.
Krista S. Jacobsen dedicates this book to Jim.

Contents

Foreword

Although a success may have many parents, true lasting success spawns many offspring, and indeed this second book, *Implementation and Applications of DSL Technology,* is the well-endowed offspring of Golden, Dedieu, and Jacobsen's *Fundamentals of DSL Technology.* Each chapter of this second book peers into blossoming subindustries, all born of the once-mistermed "transitional" Digital Subscriber Line (DSL) industry (that will soon clear $100 billion in total annual service revenue). The trio of DSL editors has once again solicited, and then actually collected, contributions from the world's leading DSL experts to outline the expanding industry.

Once considered an afterthought to a DSL design, splitter circuits and microfilters have become an industry unto themselves. The lead editors set the tone for this second book by commencing with their speciality in this area and writing the very first chapter. While inevitably voice service will go digital, splitter circuits and their derivatives may well find use even when analog voice eventually disappears.

DSL components have become instrumental to the consumer electronics gateways of all serious semiconductor suppliers. In particular, as these components shrink in size, their use in libraries and opportunities to mix and match them with various other components become enabling to a connected world. Chapter 2 by Damien Macq investigates this emerging industry and the capabilities required of digital chipsets. The designers of the very first Very High Bit-Rate DSL (VDSL) analog front-ends, Dr. Nick Sands and Damien Macq, combine in Chapter 3 to explore analog design for DSL. Analog designers can profit from this exposé on an area that is still fraught with peril for those who may not appreciate the complexity of the DSL design, particularly the analog linearity requirements.

DSL testing has become its own industry, and laboratories around the world run thousands of different types of tests to qualify DSL equipment for deployment to hundreds of millions of customers. Chapter 4 is written by Jim Eyres and Alexander Stefanescu, and it shares their understanding of this growing area. Chapter 4 leads naturally to the various types of testing necessary for service qualification and deployment presented in the next two chapters. Chapter 5, which addresses the evolution of testing and provisioning of services from plain old telephone service (POTS) to DSL, is contributed by Roger Faulkner, Dr. Ken Kerpez (Telcordia's renowned

DSL expert), and editor Phil Golden. Chapter 6, by BT's Nigel Evans and Mark Fletcher, addresses loop qualification. BT contributions continue with Rob Kirkby joining Ken Kerpez to discuss the regulator's decree of spectrum management in Chapter 7. This area is particularly crucial to European and Asian deployment because in these regions service competition for the same pair is healthy, and the heavy footsteps of regulators raise fear among even the most emboldened telcos.

In the future, no other area will have a greater impact on DSL growth and health than dynamic spectrum management (DSM), which is now being issued in standardized form from the American National Standards Institute (ANSI). DSM paves the way to 100 Mbit/s DSLs and more bandwidth than consumers' current dreams. Dr. Taek Chung, formerly of ASSIA, the pioneering company in the DSM arena, describes DSM in Chapter 8.

Gavin Young of Cable & Wireless and Michael Brusca of incumbent provider Verizon pool their expertise in Chapter 9 to describe networks for multiple applications in the video, telephony, and Internet data areas and the associated network architectures. The story of the evolution of the DSL access multiplexer (DSLAM) architecture continues with the viewpoint of Netgear's Michael Clegg in Chapter 10. It is refreshing and significant that this chapter comes from an equipment provider not traditionally viewed as a telecommunications equipment vendor. Perhaps it signals more to come in the telecommunications industry, in which innovation typically has been sourced by upstarts, converts, and start-ups on the rise, while mammoth telco vendors sometimes try to quash the new ideas before they realize how to profit from them. Chapter 11 presents the basics of big telco operations and administration as viewed by Marko Löeffelholz, Thomas Haag, and Markus Freudenberger, who are from one of the biggest telcos of all, Deutsche Telekom. This chapter might be a good introduction to Chapter 8 for some readers, where the more adaptive additions to existing operational systems are described.

It seems no aspect of telecommunications today can be discussed without a thorough consideration of security. In Chapter 12, Amalfi's Randy Turner discusses security for DSL. Chapter 13, by Infineon's Vladimir Oksman, describes the packet transfer mechanism used in DSL, and Infineon's Ingo Volkening addresses voice over DSL in Chapter 14. Bonding of DSLs to multiplex and demultiplex a single bit stream onto and from several DSLs is reviewed by Alcatel's Lane Moss and Hatteras's Matt Squire in Chapter 15. Whereas bonding is a higher-layer technique to harness the bit rates of several loops for a single transceiver, vectoring offers a physical layer approach to increase bit rates beyond those of a bonded system. Chapter 16 considers the vector channel created by several lines in a binder and the various mechansims to capture considerably higher bandwidth than is possible with simple mutiplexing. The world experts at

Aktino, Michail Tsatsanis and Thorkell Gudmundson, join force to explain this rapidly growing area.

Finally, editor and barrister-to-be Dr. Krista S. Jacobsen is joined by standards icon Les Brown, standards-escapee and history buff Angus Carrick, Dr. Ragnar Jonsson, and Dr. Sigurd Schelstraete to review the continuing saga of standardization in DSL. Still vastly underappreciated by many, the standards area has fundamentally determined present (and future) telecommunications profitability and opportunity. Rumor has it this last chapter was written on napkins at the U.N. cafeteria in Geneva between the afternoon Q4 meeting session determining the embedded operations channel (EOC) mechanism and the evening's contentious ad hoc to decide whether the proper acronym is "eoc" or "EOC."

This book is sure to be on the shelf of every self-respecting telecommunications engineer in the world and completes the job initiated by the first book, at least for now. The communications industry continues to morph into every aspect of daily life in all corners of the planet, and those wired connections are prerequisite to all the other connections. (Indeed, wireless does not exist without first wires to at least one terminal.) This book helps further that ubiquitous penetration and the pursuant technology foundations. Special thanks to the many authors, and long live DSL!

John M. Cioffi

Acknowledgments

The editors are grateful for the participation of so many skilled engineers in the creation of *Implementation and Applications of DSL Technology*. As was the case for *Fundamentals of DSL Technology*, the quality of this volume reflects the talent and dedication of its chapter authors. The editors sincerely thank the authors for their outstanding contributions. The high quality of this volume is also due in part to the efforts of the excellent team of reviewers. The reviewers are DSL experts whose careful reviews of the material helped to ensure technical accuracy and clarity. The editors would like to thank the reviewers, some of whom did double duty as authors and reviewers, for their role in crafting this volume: Mike Agah, Hugh Barrass, Michael Brusca, Raphael Cendrillon, Andy Chattell, John Cook, Andrew Deczky, Kevin Foster, George Ginis, Barry Harvey, Rob Kirkby, Marko Löeffelholz, Cory Modlin, Barry O'Mahony, Peter Reusens, Nick Sands, Ken Schneider, Peter Silvermann, Massimo Sorbara, Tom Starr, Frank van der Putten, Gary Tennyson, Roxana Trofin, Gavin Young, and Wei Yu.

Phil Golden would like to thank his coeditors Hervé and Krista for their tremendous work and effort. Phil also wants to give special mention to both John Cioffi and Bruce Wooley, whose advice was invaluable in choosing to study at Stanford. While there the guidance and advice provided by Roxana Trofin, Bruce Wooley, Tom Lee, Boris Murmann, Stephen O'Driscoll, and Aaron Gibby were instrumental to his academic experience. Finally, Phil would like to thank Hannah for her love and patience. Although she considers the latter to be among the least of her qualities, being with Phil proves that it is one of her finest.

Hervé Dedieu would like to thank Krista and Phil for having taken most of the burden when he was heavily loaded with academic tasks. He would also like to thank Dominique and Claire for their patience and comprehension during the last summer.

Krista S. Jacobsen would like to thank Phil and Hervé for their efforts on this second and final volume of the set. Working with them both has been a pleasure. Krista thanks the good people of 2Wire for the consulting gig that allowed her to eat and pay the mortgage while still leaving ample time to figure out her next career move. Much love to Jim S. for his moral support, for his cooking, for putting up with countless weekends of writing, rewriting, and editing, and, of course, for checking the math. Finally, Krista thanks John M. Cioffi for his significant contributions to her education and to her life. By founding Amati, John not only planted the seeds of ADSL, but he also gave Krista the most fun job she has ever had in an environment that fostered innovation, learning, and friendships. No job in Krista's future could ever be as special as Amati was.

Chapter 1

POTS Protection and Voice-Data Multiplexing and Separation—xDSL Splitters

Philip Golden and Herve Dedieu

CONTENTS

2 ■ *Implementation and Applications of DSL Technology*

Abstract Splitter design has been a somewhat unexpected issue in overall Digital Subscriber Line (DSL) system design. Although apparently a quite simple and indeed somewhat peripheral function, there are in reality significant design issues. Furthermore splitters are fundamental to DSL's core advantage of utilizing the telephony network infrastructure. This chapter attempts to explain the various issues that a splitter designer needs to consider, and subsequently outlines a proven design technique that can be easily adapted to any DSL splitter design.

1.1   Introduction to the Splitter Function

Since the inception of the design and standardization of DSL systems, one of the least studied aspects of DSL has been the splitter function. Apart from a few notable exceptions such as [Cook 1995], very little has been published on splitter functionality or design. The reasons for this apparent lack of interest are diverse. One can consider that the splitter is almost always a purely analog function that is part of a predominantly digital system, and hence requires design techniques that might be of little use in other aspects of DSL. In addition, an inherent knowledge of telephony i.e., (plain old telephone service [POTS]) can be important in splitter design as will be seen later in this chapter; however, technical literature on telephony is not so widely available. Moreover, the basic functionality of the splitter is quite easily understood, and this can lead to the belief that splitter design is a trivial problem. Finally, some would suggest that the concept of a POTS splitter is merely a temporary solution that will be used only until all voice transmission is digitized from end-to-end.

One of the main goals of this chapter is to address some of the above issues. The proceeding subsections of Section 1.1 will discuss splitter functionality, both from a fundamental and a somewhat more detailed perspective. Section 1.2 will then explain how practical splitter configurations fit into a DSL system. The different filters that can be implemented in a practical splitter are discussed in Sections 1.3 through 1.7 from the perspective of requirements, and finally design techniques are explained in Section 1.8. Before all of that however, the final point in the previous paragraph is now addressed. Despite the fact that there is an undeniable trend toward digitized voice in the future, at present most of the deployments of services such as Voice-over-IP (VoIP) (see Chapter 14) are in quite early stages. The vast majority of existing DSL deployments use splitters, and this will continue to be the case for some time. Furthermore, in the short term, the significance of the splitter function is likely to increase due to at least two factors:

1. DSL chipsets have achieved consistent increases in density over time that have not been matched by the splitter function. In newer chipsets a traditionally implemented splitter function will dominate the size of the Asymmetric Digital Subscriber Line (ADSL) physical layer footprint,* which will be crucial in many local exchange deployments and also in remote terminals (see Section 1.2). Significant advances in so-called active splitter technology are addressing this issue (see Section 1.8).

---

* The entire physical layer may not be on a single board (see Section 1.2).

2. At present much commercial DSL is based upon a "best effort Internet" service. As this evolves (see Chapter 9) the effect of poorly designed splitters on DSL transmission will become critical. This will be accentuated by the deployment of home network technology etc. This will be further discussed in Section 1.4.

1.1.1 Coexistence of DSL with Low-Frequency Services

One of the key advantages of DSL technology is that it generally utilizes an existing network infrastructure (see chapter 5 of [Golden 2006]), i.e., the copper network. Rather than trying to immediately replace the services previously provided on the network in question, most deployments of DSL to date involve a coexistence of the DSL service and the legacy service on the same line. In practical terms, the most widespread deployment of DSL to date is an "ADSL over POTS,"* although in many countries an "ADSL over Integrated Services Digital Network (ISDN)" service is also offered. Very High Bit-Rate Digital Subscriber Line (VDSL) services are being offered over POTS and ISDN, much like ADSL. A feature of almost all of the new flavours of DSL that are proposed in the standardization bodies is that they have an option to coexist with POTS. The fundamental reason that DSL can coexist with some legacy services has to do with the fact that they can be transmitted using different frequency bands. For any given service, the maximum power that can be used over any given frequency band is usually specified by a power spectral density (PSD) template (see chapter 5 of [Golden 2006]). In the case of POTS, for instance, the transmission will generally be up to around 4 kHz. The most common version of ADSL however (as defined by Annex A of [G.992.1]), does not transmit significant energy below 25 kHz. Hence, one can consider the PSD diagram shown in Figure 1.1. In the discussion that follows, an ADSL over POTS splitter is used as an example; however, the principles involved are applicable to all DSL splitters (except where explicitly mentioned). Because of the fact that there is a significant† transition band between the upper frequency of the POTS band and the lower frequency of the ADSL band, it is possible to design a practical low pass filter (LPF) that will pass the POTS with very little attenuation, but will heavily attenuate the DSL signal. It is of course also possible to design a high pass filter (HPF) to perform the converse operation. The DSL splitter function is made up of these two filters. The splitter functionality can be thought of in terms of Figure 1.2. It should be

---

* This means that ADSL and POTS coexist on the same line. It is not to be confused with "Voice over DSL," which indicates that digitized voice is being sent using DSL technology.

† Significant here means that the width of the transition band is at least comparable (in terms of order of magnitude) to the upper frequency of the POTS band.

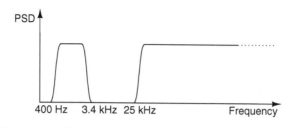

Figure 1.1 **Example of frequency spectra used for POTS and ADSL.**

noted that this block diagram represents the splitter function rather than any physical equipment configuration.

1.1.2 *Functionality of the Splitter*

As is evident from Figure 1.2, the basic function of the splitter is to prevent the DSL signal from interacting with the POTS terminal equipment (and wiring), while ensuring that the DSL equipment is electrically isolated from the POTS signals. This function is necessary due to the potential inter-ference between the two services. Some sources of this interference (for both the case of a POTS and an ISDN splitter) are detailed in the following subsections. Early attempts to deploy a form of "splitterless"* ADSL were unsuccessful due in large part to these issues.

1.1.2.1 *Protection of DSL from POTS or ISDN Interference*

One fundamental problem here is that POTS equipment was generally not designed with high-frequency performance in mind (see chapter 1 of [Golden 2006]). Typical specifications for the impedance of POTS

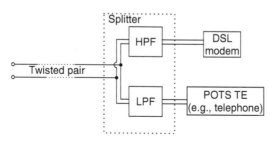

Figure 1.2 **Splitter functionality at either end of the loop.**

---

* Although the concept here was to completely do away with the splitter function, revised pro-posals included "microfilters" (see Section 1.2) that effectively implement the splitter function anyway. For this reason, some microfilter deployments are still referred to as "splitterless ADSL."

equipment are only for the "off-hook" state, and do not mention frequencies above the voiceband.* As detailed in section 1.1.5.7 of [Golden 2006], one consequence of this can be that POTS terminals present a very low impedance at frequencies used for DSL transmission. If the splitter were not present, this would effectively short circuit the line at these frequencies and thus destroy the DSL transmission. In addition, a POTS terminal could also send signals at frequencies higher than 200 kHz that could interfere with DSL; however, in practice this is somewhat less likely. Nevertheless the steady-state behavior of splitters is quite well studied, and is typically well specified by the standards bodies.†

In addition, POTS phenomena such as ring trip and on- or off-hook transitions (see chapter 1 of [Golden 2006]) can result in transient signals that can cause errors in DSL transmission. This issue is discussed in Section 1.4. In the case of ISDN, the potential impact on DSL is due principally to its static spectral characteristic. Because of the fact that most ISDN systems were developed and specified before the widespread popularization of DSL, the out of band PSD requirements for ISDN are potentially "unfriendly" to DSL transmission. Hence, the splitter is needed to provide the required isolation. Some newer ISDN systems effectively integrate the splitter low pass filter into the ISDN terminal, in what are known as "DSL friendly ISDN systems."

1.1.2.2   Protection of POTS or ISDN from DSL Interference

Again a major issue here is that POTS equipment was generally not designed with high-frequency performance in mind. Most modern POTS terminals employ solid-state devices that in some cases could potentially demodulate high-frequency DSL signals resulting in unwanted signals in the audible band (this is generally up to around 20 kHz). For example, one common requirement on POTS equipment is that they be polarity independent, virtually ensuring that a diode bridge (and hence a potential demodulator) is employed in the line interface. It should be noted that some EMC filtering is usually incorporated in POTS terminals for protection against radio signals, and this will provide some measure of protection from DSL.

1.1.2.3   Protection of DSL from Common Mode Interference

As is detailed in chapter 13 of [Golden 2006], common mode interference can cause issues for DSL performance. Many splitter designs will implement some form of common mode protection (usually a common mode choke)

---

* Revising these specifications is worthwhile; however, a huge amount of legacy POTS terminals are still in use.

† There are some notable exceptions for this, such as an occasional lack of a balance specification for the splitter in the DSL band.

that seeks to prevent common mode signals originating from the POTS terminal equipment or the customer premise wiring from affecting the DSL transmission. Although this is not part of the splitter function in terms of the block diagram of Figure 1.2, it can be extremely useful. It is particularly beneficial when the customer premise wiring is poorly balanced, and the benefit is accentuated when high-frequency services (such as HomePNA) are present on the home wiring.

1.2 Physical Location of the Splitter Function

Although Figure 1.2 shows each splitter as a high pass and a low pass filter that are separate from any terminal equipment, in reality this may not be the case. In particular, certain parts of the splitter can be implemented within terminal equipment (e.g., DSL modems will have a high pass filter at the input), and furthermore functions appearing as blocks in Figure 1.2 may be practically implemented as multiple different filters (see Section 1.2.2.2). The following subsections will review the practical splitter configurations that are (and will be) deployed.

1.2.1 Network Side

The network side splitter will always be physically "close"* to the Digital Subscriber Line Access Multiplexer (DSLAM). There are three basic categories to consider, as shown in Figure 1.3. One common point that should be noted about network side splitter implementations in general is that issues of size and density tend to be extremely important.

1.2.1.1 Local Exchange Splitters

In a large number of existing DSL deployments, the splitter and DSLAM are physically situated in a telephony exchange. This is a particularly elegant solution if the same service provider is supplying the POTS and the DSL, as then all of the relevant terminal equipment on the network side is in the same building. For some smaller local exchanges, however, the DSLAM (with or without the splitter) may be located in a building "close"† to the local exchange due to space constraints. This is often true for exchanges that terminate lines being used by more than one service provider (i.e., line sharing). A local exchange splitter can be located in a "splitter rack," i.e., metallic housing whose main function is to hold the splitter boards and provide the appropriate connectivity. It can also be located in the same

---

* This generally means within about 200 m.
† This generally means within a few hundred meters.

8 ■ *Implementation and Applications of DSL Technology*

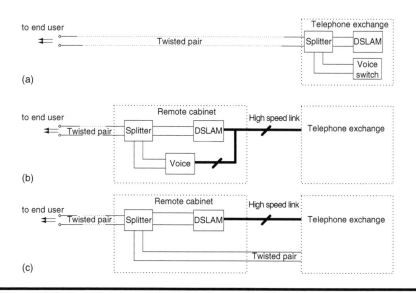

Figure 1.3   Physical locations of network side splitters.

housing as the DSLAM, insuring that the length of wire between the splitter and the DSL card is very short (a long line can cause performance issues for DSL). A third possible location for the splitter is within the main distribution frame (see chapter 1 of [Golden 2006]). This is especially popular in line sharing deployments, at least in part because it facilitates the routing of cables.

1.2.1.2   Remote Terminal Splitter, Colocated with the POTS Line Card

As mentioned in chapter 1 of [Golden 2006], there are certain deployments where the POTS customer is "fed" from a remote exchange that is usually connected to a local exchange via a high speed link. The most common example of this is the digital loop carrier (DLC). This is generally used where there is significant geographical dispersion of customers, as it can significantly increase the effective range of the POTS. In this case, a compact DSLAM and splitter combination will reside very close to the remote exchange, either in the same physical structure or within a few meters. For some remote exchanges, the existing POTS line card can be replaced with a so-called DSL ready POTS line card. This enables a much more controlled* POTS environment than would be typical with legacy POTS equipment and hence can greatly simplify the splitter requirements.

---

* For example, the ringing generator would not produce harmonics at DSL frequencies, a relatively common issue with traditional ringing generators used in POTS systems.

If this is not the case, i.e., the POTS line card is similar to that used in a traditional local exchange, then from a splitter perspective this is electrically identical to the case where all the network equipment is in the local exchange.*

1.2.1.3 Remote Terminal Splitter, Not Colocated with the POTS Line Card

This is the configuration shown in Figure 1.3c, and is generally used for shorter range DSL services such as VDSL. From an electrical perspective, the important point to note here is that there is a significant length of line between the network side splitter and the POTS line card. As will be seen in Section 1.3, this can have some impacts on the specification of the splitter, especially from the perspective of loss requirements.

1.2.2 Customer Side

In a broad sense, one can categorize all customer side splitters by means of two configurations:

1. Master splitter configuration
2. Distributed filter† configuration

1.2.2.1 Master Splitters

The master (or "central") splitter configuration is shown in Figure 1.4, and was used for almost all early ADSL deployments. The key point here is that the master splitter resides beside the demarcation point (as defined in chapter 1 of [Golden 2006]) of the premise, and is always a three-port device. A master splitter will usually be on the network side of the demarcation point, meaning that the splitter should be installed, owned, and maintained by the network operator. The actual physical location of the demarcation point can vary; however, it is worth noting that in many networks it may not be readily identifiable or accessible from the customer's perspective. The fact that the network operator will in general‡ have to install a master splitter at the customer premise is a very significant disadvantage of this configuration. With the popularization of "plug and play" DSL modems, installation of the master splitter becomes the predominant

---

* Because the "local loop" effectively starts from the remote exchange.

† Also known variously as "in-line" filters or "microfilters."

‡ This is not always the case. Certain networks have readily identifiable demarcation points at which the customer can install the master splitter.

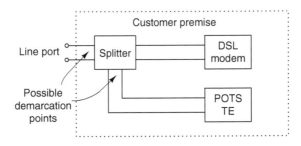

Figure 1.4   Master splitter configuration.

need for "truck roll," i.e., the deployment of service-provider-employed technicians to a customer premise to aid with DSL installation. With the growing numbers of DSLs, it is evident that truck roll could prove extremely expensive and problematic (in terms of resources) for a successful DSL provider. The desire to eliminate the need for truck roll was among the primary motivations for the attempted standardization of "splitterless DSL" mentioned in Section 1.1.2. The actual physical housing and form of master splitters will in general be very much network dependent, because of factors explained in chapter 1 of [Golden 2006].

1.2.2.2 Distributed Filters

The distributed filter configuration is shown in Figure 1.5. As implied above, the initial motivation in having this type of configuration came from the desire to have a "self-install" splitter configuration. In late 1997, a group of

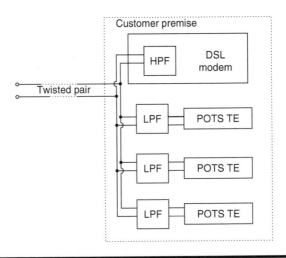

Figure 1.5   Distributed filter configuration.

computer manufacturers, ADSL vendors, and service providers formed the Universal ADSL Working Group (UAWG) to develop a specification for a low rate version ADSL known as "ADSL-lite." This was to use less power than standard ADSL* (as defined in [G.992.1]), and would enable a lower rate service to be deployed without any splitter at the customer side.† It quickly became clear however that this was unrealistic, due in large part to the technical issues described in Section 1.1.2.1. The fundamental problem was that the potential behavior of POTS terminal equipment at DSL frequencies means that some form of isolation is required for DSL to work well in the majority of installations. Hence, the UAWG defined a topology very similar to that shown in Figure 1.5, where each POTS terminal equipment is effectively isolated from the DSL by a low pass filter. The work of the UAWG became the basis of the International Telecommunications Union (ITU) G.lite standard [G.992.2], and these low pass filters became known as "microfilters," so-called because they were simple (typically second order) and hence relatively small filters. The apparent choice for service providers was thus between a full rate ADSL service that necessitated a master splitter and hence truck roll, or a lower rate ADSL-lite service with self-install filters. In reality, the majority of operators combined the two: deployment of full rate ADSL with self-install filters. It was considered that by increasing the order (and hence the complexity) of the so-called microfilters, they could provide enough isolation to allow full rate ADSL to be deployed with a self-install configuration. For this reason, standardization groups such as T1E1.4 and European Telecommunications Standards Institute (ETSI) have produced [T1.421-2001] and [TS101 952-1-5] respectively, to specify filters that can be deployed with full rate ADSL in the configuration shown in Figure 1.5. These are known as "distributed filters" in Europe and "in-line filters" in the United States. In physical terms, these filters will usually be two-port devices (a line and a phone port); however, they can also include a DSL port to allow direct connection of a DSL modem at various locations around the house. Their housing can be network dependent as they will often fit into the phone sockets in the premise (the form of these often being country dependent). In addition, the so-called wall mount filters are available, as well as "dual line" filters for POTS terminal equipment that uses two lines. Despite the fact that the configuration shown in Figure 1.5 is used in the vast majority of existing ADSL deployments, it does have the disadvantage that the DSL transmission is directly connected to the customer premise wiring. As explained in chapter 1 of [Golden 2006], this

---

* This was the main motivation of the computer chip manufacturers. Decreasing the bandwidth (and hence the power), and also not using trellis coding, significantly reduces the number of MIPS necessary in the processor.
† It should be noted that "splitterless" ADSL is a misnomer even in its original definition, as there was always a splitter defined at the network side.

can cause significant issues for DSL performance. This is likely to become more pronounced with the deployment of higher frequency DSLs (such as VDSL) and home network systems. Further issues with this configuration relating to impedance loading are explained in Section 1.9.4.

1.3 Low Pass Filter for POTS—Steady-State Behavior

The low pass filter used for full rate ADSL will typically be between fifth and eighth order to meet the necessary electrical standards. The performance of the low pass filter is hugely significant because of the fact that the POTS passes through it. As mentioned earlier, one of the key advantages to DSL is the ability to coexist peacefully with the low-frequency services that are already deployed on twisted pair lines. A good low pass filter design allows this to be the case.

1.3.1 DC Behavior

The DC (direct current) behavior of the low pass filter is important for a number of reasons. As has been mentioned in chapter 1 of [Golden 2006], DC signalling is used quite extensively in POTS systems [EN 300 0001]. The following POTS features are relevant to the DC behavior of splitters:

- Distinction between a "live" POTS line and a "dormant" line is the amount of DC in the loop (the former will have a current value above a certain threshold).
- All POTS terminal equipment need a minimum amount of DC and voltage in the off-hook case to function correctly.
- Voice transceiver sensitivities in many POTS terminal equipment are related to the DC in the loop, i.e., a low DC value can indicate a long loop causing the POTS equipment to increase the gain of the speech amplifiers to compensate for loop attenuation.
- Much of the POTS loop testing is carried out at DC, and will generally be done through the splitter (see Chapter 5).

For these reasons, it is clear that the DC characteristic of the splitter must be kept below a certain limit. In many of the existing splitter specifications, this limit is set to match the characteristic of a resistor. This is unnecessarily restrictive, however, as it can be shown that other characteristics are fully compatible with POTSs. In crude terms, the fact that any positive character- istic is acceptable is due to the fact that the "reach" of POTS will in general be far greater than that of ADSL. The difference in line length is then used as a margin that allows for some DC loss in the splitter low pass filter. A similar concept is used when setting voiceband insertion loss limits for the splitter (see next section). An interesting point to note here concerns the case of a splitter at a remote terminal, where the POTS line card is in a local exchange (as detailed in Section 1.2.1.3). In this situation, one could consider a very long POTS line with the splitter effectively inserted at some point, a few kilometers from the local exchange. Of course, here there would no longer be a margin, and any further loss introduced by the splitter could cause issues with the POTS. At the time of writing, this scenario has not been envisaged in any of the major splitter standards. In addition, the DC behavior of any low pass filter must be such that it will continue to operate in the presence of the large DC steady-state currents (generally up to about 100 mA) that can be present on POTS lines.

1.3.2 Requirements for Voice Signals

1.3.2.1 Voice-Band Insertion Loss

Voice transmission in telephony usually takes place between 200 Hz and 4 kHz. The most obvious requirement for the low pass filter here is a limit in the amount of voice signal that it can attenuate. For ease of measurement, and also due to the fact that many networks use complex impedance models for lines and terminal equipment at voice frequencies (the terminating impedances of the low pass filter), an insertion loss requirement is generally used. Typically, around 2 dB of insertion loss per splitter is allowed. As well as limiting the insertion loss, it is also desirable to limit the insertion loss distortion. This is to prevent a splitter that would say have 0.1 dB insertion loss at 1 kHz and 2 dB at 3 kHz, which would mean that some parts of the voice would be attenuated significantly* more than others. Furthermore, because of the fact that the filter will generally have complex terminating impedances, a negative limit on the insertion loss (i.e., a gain limit) must also be set.[†] Hence a typical specification for insertion loss is that the value cannot exceed ±1 dB at 1 kHz, with a maximum distortion of 1 dB over the 200 Hz–4 kHz band.

1.3.2.2 Voice-Band Return Loss

As explained in chapter 2 of [Golden 2006], for a distributed (rather than a lumped) medium, impedance mismatches in the transmission path of

---

* Significantly is not well defined here, because POTS performance is inherently subjective. It is related to whether or not a typical user would notice the effect.

[†] The default case for measuring insertion loss, i.e., without the filter present, will not be that of maximum power transfer when complex impedances are used (see chapter 2 of [Golden 2006]). Hence, a passive device can cause an "insertion gain."

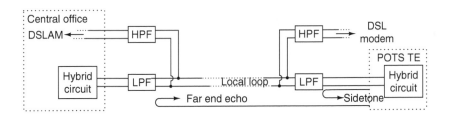

Figure 1.6   Impedance mismatches.

a signal will cause reflection of signals, as shown in Figure 1.6. Furthermore, the magnitude of the reflections is directly related to the difference between the incident impedance and the characteristic impedance of the transmission medium before the mismatch. In terms of voice transmission, these reflections will be heard as audible feedback. It is clear that insertion of low pass filters into the voice transmission path could potentially add more significant impedance mismatches. The effect of these mismatches will either cause sidetone or echo, depending on the round trip delay of the signal. To limit the effect of audible feedback, it is thus necessary to specify the input impedance of the low pass filter at both ports. The ideal situation would be that the filter would not change the impedance seen at any point of the loop. The return loss of the input impedance of the filter is specified against a reference impedance, and is shown in Figure 1.7, this reference impedance should be the expected impedance at the other port of the filter. Splitter return loss in Europe is generally specified as an un-weighted piecewise linear (in decibels) requirement over the voiceband. In the United States, a weighted requirement is used, with the weightings corresponding to the expected sensitivities of the human ear to different frequencies.

1.3.2.3 Group Delay

POTS transmission is evidently sensitive to delay, and so the group delay (and more importantly the group delay distortion) should be specified.

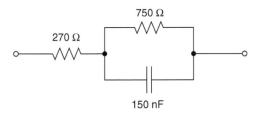

Figure 1.7   Example of reference impedance used for return loss measurement.

In practical splitter designs, however, it is evident that the complexity of the filter will almost always be limited by the attenuation rather than the group delay. Indeed, for any minimum phase transfer function, the attenuation of the filter with real impedance terminations is directly related to the phase angle (and henceforth the group delay) by a Hilbert transform.

1.3.3   Requirements for On-Hook Signalling

The voice related requirements described in Section 1.3.2 are valid for the off-hook case, when a POTS transmission is taking place.

As highlighted in chapter 1 of [Golden 2006], however, POTS signalling can take place in the on-hook case also. The most common example of this is calling line identification (CLI), e.g. [SIN 227] [SIN 242]. This will generally take place at frequencies between 200 Hz and 2.8 kHz, and involves TE impedance models significantly different from those used in off-hook. Indeed, one of the practical issues with some early deployments of ADSL was the number of customers who complained that their caller ID service stopped working after their broadband service was installed.*

1.3.4   Higher Frequency Behavior

The most obvious requirement of the low pass filter at higher frequencies is that it sufficiently isolates the DSL signal from the POTS terminal equipment. As for voice-band loss, this requirement is generally given as an insertion loss. In this case, the insertion loss is specified to be greater than a certain curve over the range of the DSL transmission. The DSL band insertion loss is usually termed the "isolation" of the low pass filter, and 55 dB is a typical requirement across the band. This requirement is of course valid in both the off-hook and the on-hook cases.

1.3.5   Balance

The balance of the low pass filter should also be specified throughout the frequency range of any signals that are likely to be incident upon it, to prevent conversion of common mode interference to differential mode signals that could be processed by the terminal equipment. One possible technique for measuring balance is shown in Figure 1.8. Here a common mode signal is injected at one port of the splitter, and the differential mode signal that results is measured. In a crude sense, the balance of the filter can

---

* This could be either due to a blocking of the CLI signal by the filter or due to interference to the CLI receiver caused by DSL signals leaking through the filter (see Section 1.3.4). The performance of the CLI receiver is generally not specified at DSL frequencies.

16 ■ *Implementation and Applications of DSL Technology*

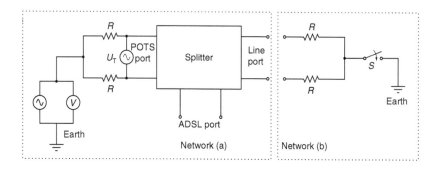

Figure 1.8   This shows a test setup for an LCL (longitudinal conversion loss) measurement at the POTS port. The other ports are in turn terminated with network (b) with switch *S* both open and closed.

be thought of as a measure of how closely the splitter "arms" are matched (see Figure 1.8).

1.3.6   *Loading Effects*

These are applicable only to customer side filters, and then only for the distributed filter configuration of Figure 1.5. It is clear from Figure 1.5 that a number of parallel low pass filters can be present in the distributed filter configuration. It is evident from Figure 1.5 that a signal incident at the line port of any of these low pass filters will also "see" the input impedance of each of the others in parallel. It is thus evident that a low impedance in the voice-band presented by one of the parallel filters (assuming it is connected to an on-hook device) could degrade the voice transmission to an off-hook POTS device that is connected to another filter. Hence, it is important that the input impedance of distributed filters in the voice-band remain high when the POTS device connected in is the on-hook state.

1.3.7   *Other Steady-State POTS Requirements*

These can include the following:

- Low pass filter must meet requirements on insulation resistance (i.e., the resistance between the tip and the ring at one port with the other open circuited) and resistance to earth to be compatible with POTS systems that include terminal equipment meeting similar requirements.

- Steady-state behavior at ringing signal frequencies can be specified so that ringing signals are passed without significant attenuation or distortion.
- Low pass filter must not generate excessive noise or inter-modulation distortion at either POTS or DSL frequencies.

1.4   Low Pass Filter for POTS—Transient Behavior

As mentioned in Section 1.1.2.1, the transient behavior of splitter is in general very poorly specified. The following POTS features make the transient behavior of the low pass filter important:

- Reversals in the polarity of the POTS feeding voltage are sometimes used to signal various events to POTS terminal equipment in chapter 1 of [Golden 2006].
- Use of relays to generate ringing signals cause large transient signals, in particular when the operation of the ring relay is not synchronized to the phase of the ringing voltage.
- When a ringing phone is answered (a process known as "ring trip"), it effectively puts a near short circuit across the pair. The shorting is done by a mechanical switch, so transients similar to those caused by ringing generation (as above) can be present. The difference is that the number of transients is much lower, and the source of the transients is the customer end of the loop.
- On- or off-hook transitions other than ring trip also cause transients; however, these are in general less severe than those caused by ring trip.
- Pulse (loop disconnect) dialing can cause large transient voltage spikes.

Specifying fixed requirements based upon transient performance is difficult, as the effects of POTS transient signals on DSL performance can vary significantly from modem to modem (both in terms of how sensitive the receivers are to these transient signals and also how quickly the modem can recover). One first step in setting effective transient requirements on the splitter low pass filter would be to deduce circuit models for the POTS phenomena described above. As mentioned earlier in this chapter, the lack of specifications of splitter transient behavior may be tolerable when DSL is offered to provide best effort Internet as a competitor to traditional voice-band modems. As DSL provision evolves to other applications however, which of course it must compete with other technologies, the lack of specification of transition splitter behavior will become more critical if not addressed. One can imagine a situation where the lifting of a phone handset causes some errors in DSL transmission. If the DSL is providing Internet service, then it is quite likely that these errors will be well tolerated (unless they are severe enough to take down the connection). For an application such as real-time video, however, these errors would likely be seen by the customer as visual interference. Practical Internet Protocol (IP) video systems are usually protected by forward error correction systems at the application layer (way beyond that implemented in the DSL physical layer) that can mitigate these effects.

1.5 Low Pass Filter for ISDN

The low pass filter for ISDN differs most significantly from that used in POTS by its passband. Typically, ISDN transmission has significant energy up to about 80 kHz, and so the passband of the filter will extend to at least a few kilohertz above this. The nature of the requirements is similar, if not as complicated, as those for the POTS low pass filter. The principal reason for this is the fact that ISDN systems use real impedance to model the line and terminal equipment. DC must still be passed, however; the main issue here is that the filter does not saturate in the presence of high DCs. Again a passband insertion loss is specified, and also a passband return loss (however, the latter is not so critical as for POTS owing to the echo canceling capability of ISDN transceivers). Group delay requirements are more strict for an ISDN filter than for its POTS counterpart, as would be expected for a digital transmission; however, they still do not tend to restrict the design any more than the attenuation requirements already have. Balance and isolation requirements are similar to those for POTS filters (the latter will of course start at a higher frequency). Of course, none of the POTS phenomena such as ringing and pulse dialing are present; however, switching at ISDN terminals can produce voltage spikes whose effect on DSL has yet to be determined. Although the majority of currently deployed ISDN splitters perform their functions at least adequately, there have been a small amount of instances reported of older ISDN terminals ceasing to operate in certain configurations when splitters are installed. This can be a significant issue because it typically requires replacement of the ISDN terminal equipment at the customer side. In addition, some detrimental influence from ADSL into ISDN in the presence of splitters has been reported in the standards bodies. This was as a result of laboratory testing, during which short scratches in established ISDN voice connections were evident during switching on or off of the ADSL modem. These two issues would indicate that there is still some work to be done in the standardization of ISDN splitters.

1.6 Low Pass Combi-Filter for POTS and ISDN

For certain deployment scenarios (notably in Germany), low pass filters that are intended to pass both POTS and ISDN are used. This basically involves a combination of the requirements described in Sections 1.3, 1.4, and 1.5, usually with some relaxation of POTS requirements to enable a more cost effective design.

1.7 High Pass Filter

As mentioned at the start of Section 1.3, there will always be a high pass filter at the input stage of a DSL modem. The high pass filter in the splitter device is thus usually either first order (i.e., a blocking capacitor on each line) or else will be nonexistent.

1.7.1 *Blocking Capacitors*

The issue of blocking capacitors has proven to be somewhat contentious in splitter standardization. At the time of writing, the U.S. splitter standard in Annex E of [T1.413-1998] mandates blocking capacitors for the network side splitter in general, but states that they may be "optional on splitters integrated with the equipment closely associated with the ATU-C (DSL modem)." The customer side splitter is defined without any high pass (i.e., a direct electrical connection between the line and the ADSL port). In addition, working group T1E1.4 has specified an additional network side splitter for the line sharing application. This has options both with and without the blocking capacitors. In Europe, the situation is equally complicated. At the time of writing, blocking capacitors are mandated for all ISDN splitters; however, they are optional for all POTS splitters. In addition, for the case of POTS splitters, a higher order high pass filter is optional; however, this is not widely deployed in practice.

1.7.1.1 Advantages of Blocking Capacitors

The blocking capacitors provide DC isolation between services, preventing the DC from the POTS or IDSN service flowing into the DSL modem. Generally, the modem will have its own blocking capacitors anyway; however, this is not always assured. If no capacitors were present (either in the splitter or in the modem), then the DC could cause significant damage to the DSL modem and loss of functionality of the POTS or the ISDN service. The capacitors also provide a low degree of AC isolation between services at low frequencies. Again this can prevent interference; however, for network side splitters it also has advantages in a line sharing deployment, where different operators provide the POTS (or ISDN) and DSL service respectively. In this case, if no blocking capacitors were present in the splitter, the DSL operator would effectively have full low-frequency access to the line. This situation has potential issues with privacy, as the DSL operator could effectively "listen" to the POTS transmission. It should be stated though that the level of isolation provided by blocking capacitors would not typically be sufficient to prevent this anyway. In the case of a customer premises equipment (CPE) filter, the role of the series high pass filter is less important. One reason for including some form of high pass filter is to present a more controlled impedance to the low pass filter at the ADSL port (e.g., it potentially enables the POTS to continue functioning in the case of a user short circuiting the ADSL port of the splitter).

1.7.1.2   Disadvantages of Blocking Capacitors

The most publicized disadvantage of blocking capacitors doubtlessly relates to test access. The presence of blocking capacitors prevents low-frequency access to the line from the DSL port of the splitter. As explained in Chapter 5, much of the line testing (e.g., fault detection) typically carried out by a service provider uses signals at low frequencies. Hence, the presence of blocking capacitors can be considered to prevent a service provider from performing necessary testing on a line. The other main disadvantage to blocking capacitors is their potential impact on the performance of DSL. The capacitors (if present) will obviously be in the path of the DSL signal, and in particular, they can cause attenuation and phase distortion that necessitates more equalization in the modem, if bit rate losses are to be avoided. In addition, the blocking capacitors will influence the impedance "seen" by the modem, which can potentially enhance echo signals. The potential degradation to DSL performance can be greatly increased if a higher order high pass filter is present in the splitter. It should be stated though that in terms of performance degradation, the DSL modem design can compensate for the presence of blocking capacitors. Crudely speaking, the DSL modem will usually have blocking capacitors of its own at its input. Boosting the value of these (usually by a factor of about two) should compensate to a large extent for the fact that the blocking capacitors are present. The situation is somewhat complicated by the potentially significant length of line between the modem and the blocking capacitors, but nevertheless adequate compensation should in general be possible. The practical problem then is that the modem designer must know whether the splitter to be used will contain blocking capacitors or not. In the case where they are optional in the relevant standard, this can cause confusion.

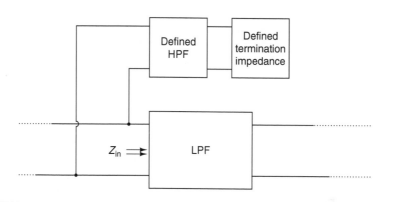

Figure 1.9   **Insertion loss of the high pass filter.**

1.7.1.3   Requirements for the High Pass Filter

Assuming that only blocking capacitors of a specified value (and tolerance) are used, it is clear that minimal electrical requirements are needed for the high pass filter. In practice, an insertion loss requirement will generally be set; however, reference to Figure 1.9 shows that this is in reality a require-ment on impedance at the line side of the low pass filter. Again the potential use of a higher order filter is problematic, as a whole range of electrical requirements would now be potentially needed.

1.8   Passive Design Techniques

1.8.1   Introduction

As highlighted in the previous sections, a great number of constraints have to be considered to design a splitter in accordance to its standard of relevance. Although it can be argued that a splitter is mainly a passive filter (the theory of which has been completely established three or four decades ago), it turns out that off-the-shelf standard techniques [Sedra 1978], [Williams 1988] which have been developed principally for lossless two-ports terminated on resistive impedances cannot be applied directly. The main reasons which make the splitter design a delicate issue are as follows:

- Splitter is a three-port, which is not equally terminated on its three ports.
- POTS and line ports are in general terminated on complex impeda-nces, which makes the "ideal" three-port not lossless.

- Even for a two-port terminated on complex impedances there is no available theory that gives the designer the best ladder RLC structure.
- Even by supposing that by optimization the three-port transfer function which fulfills all the constraints can be defined, component structure extraction is an open problem for a lossy three-port.
- Many specifications are defined with the ADSL port either opened or closed on its standard impedance.
- On-hook and off-hook requirements impose constraints when the POTS is terminated with different impedances ranging from high-impedance to a complex impedance close to 600 Ω.
- In some cases, specific constraints on the required splitter behavior in on-hook and off-hook conditions make the design unrealizable with standard components. The splitter design becomes not only a theoretical issue but also a technological issue in which some inductive components have to present values in henries depending on the presence or absence of current; the ratio of these two values can be as high as 5–10. Because these kinds of components are not offered by traditional magnetic vendors, expertise in magnetic component design is therefore required.
- Not all the constraints are listed in standard documents, in particular, some DSLAM manufacturers test the equipments of splitter vendors during the abrupt transient behaviors associated with telephony. Quality of inductive components (absence of saturation during high DC induced by ringing) and damping of attenuation poles (to have smooth phase transition) play a major role which must not be underestimated by the designer.
- Fierce competition between splitter vendors and drastic rationalization of the DSL market have pushed the splitter designers to develop their solution with minimum coils possible.
- When dealing with active design, stability is a very delicate issue in the presence of many possible termination conditions.

1.8.2 A Simple Example Showing Why the Ideal Splitter Is in General Lossy

Let us suppose that we have to design a first-order splitter that has for the moment its ADSL port open, simplifying the three-port design into a two-port one. Let us suppose that this splitter has to be designed when the loads at POTS and line ports are at the moment 600 Ω. Although very poor in performance, this elementary splitter will give some intuitive insight. Let us suppose also that we want a 3 dB cut-off frequency at 9 kHz. Starting from a Butterworth design, which is a simple inductance of 1 H for a pulsation of

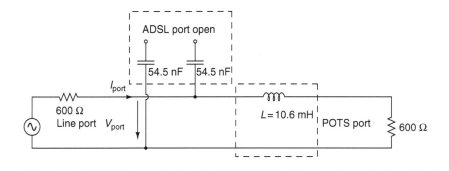

Figure 1.10   Elementary first-order splitter when ADSL port is open.

1 rad/s and a load of 1 Ω, we find by de-normalization both in impedance and frequency that

$$L = \frac{600}{2\pi \times 9000} = 10.6 \text{ mH}$$

This first-order splitter is shown in Figure 1.10.

The return loss is computed according to

$$RL = 20 \log 10 \left(\left| \frac{(V_{\text{port}}/I_{\text{port}}) + 600}{(V_{\text{port}}/I_{\text{port}}) - 600} \right| \right) \tag{1.1}$$

The return loss of elementary first-order splitter is shown in Figure 1.11.

Let us suppose now that the splitter has to be terminated on both line port and POTS port with the so-called European harmonized impedance Z_R, which is represented in Figure 1.12.

The terminating impedance Z_R can be expressed as

$$Z_R(s) = R_2 + \frac{R_1}{1 + R_1 C_1 s} \tag{1.2}$$

How to design a two-port terminated on Z_R which has exactly both the same return loss and insertion loss than its counterpart of Figure 1.11 is a trivial problem that is solved by exchanging L with $L \times \dfrac{Z_R(s)}{R}$ for $R = 600$ Ω. This is equivalent to replace the impedance $L s$ by a new one of the form

$$L s \longrightarrow \frac{R_2 L s}{R} + \frac{R_1}{R} \times \frac{L s}{1 + R_1 C_1 s} \tag{1.3}$$

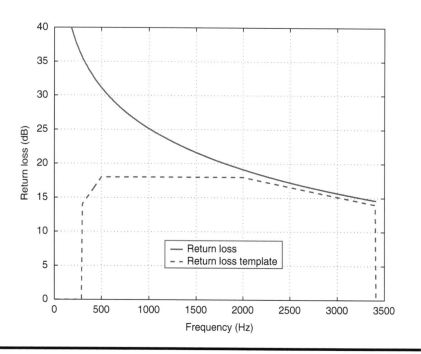

Figure 1.11   **Return loss of elementary first-order splitter of Figure 1.10.**

The right side of Equation 1.3 can be rewritten as

$$\frac{R_2 L\, s}{R} + \frac{R_1}{R} \times \frac{L\, s}{1 + R_1 C_1 s} = L_s\, s + \frac{L_p\, s}{1 + \frac{L_p}{R_p} s} \qquad (1.4)$$

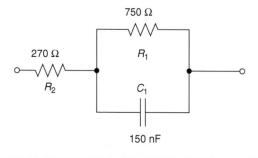

Figure 1.12   **European harmonized impedance Z_R.**

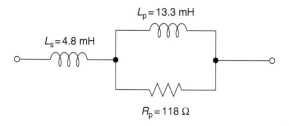

Figure 1.13   **New circuit replacing the inductance of Figure 1.10.**

where

$$L_s = \frac{R_2}{R}L$$

$$L_p = \frac{R_1}{R}L$$

$$R_p = \frac{L}{RC_1} \tag{1.5}$$

The new circuit replacing the inductance L is easily obtained by elementary circuit synthesis; it is made of a series of an inductance L_s with another inductance L_p having in parallel a resistance R_p as shown in Figure 1.13.

It is worth considering what happens in terms of adaptation of impedance when we connect the new lossy circuit with a circuit modeling the ADSL port. The new circuit with its terminations is shown in Figure 1.14. The return loss is now computed using

$$RL = 20 \log 10 \left(\left\| \frac{(V_{port}/I_{port}) + Z_R}{(V_{port}/I_{port}) - Z_R} \right\| \right) \tag{1.6}$$

and has to be measured both from the line and the POTS port with the ADSL port being connected or not. The return loss setup of the three-port is shown in Figure 1.15.

One can see from this example that the ideal first lossless splitter becomes lossy by exchanging the real terminations with the complex one. In our peculiar case, the connection of the ADSL port does not seem to complicate the design because the margin in return loss has been improved (this is not the case in general). This example seems also to suggest that it is

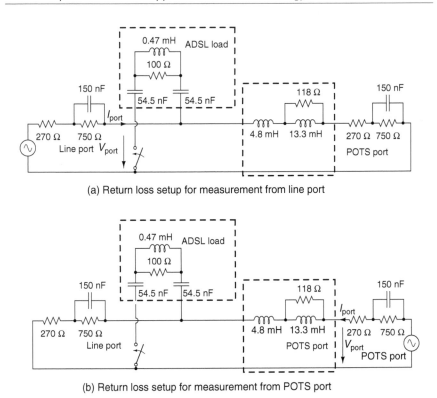

(a) Return loss setup for measurement from line port

(b) Return loss setup for measurement from POTS port

Figure 1.14 **New elementary splitter associated with complex terminations.**

relatively simple to design a splitter terminated with complex impedances; just design a splitter terminated on 600 Ω then apply a de-normalization in impedance by multiplying each impedance by $\frac{Z_R(s)}{R}$. This method has two drawbacks. First, it generates designs having twice as many coils with respect to the real terminated splitter; second, the transformation of capacitances leads to circuits which are no longer passive and would have to be synthesized with active circuitry. To confirm this passive realizability problem, let us suppose that we exchange $\frac{1}{Cs}$ with $\frac{1}{Cs}\frac{Z_R(s)}{R}$.

Starting from Equation 1.2, we find that

$$\frac{1}{Cs}\frac{Z_R(s)}{R} = \left(\frac{R_1+R_2}{R}\right)\frac{1}{Cs}\left(\frac{1+\dfrac{R_1}{R_1+R_2}R_2C_2s}{1+R_2C_2s}\right) = k\frac{1}{Cs}\frac{1+\tau_1 s}{1+\tau_2 s} \qquad (1.7)$$

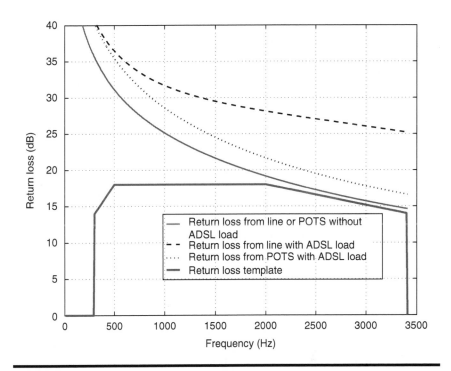

Figure 1.15 **Return loss of circuit of Figure 1.14 when the ADSL port is connected or not.**

with

$$k = \frac{R_1}{R_1 + R_2} \qquad \tau_2 = R_2 C_2 \qquad \tau_1 = \frac{R_1}{R_1 + R_2} \tau_2 \qquad (1.8)$$

From Equation 1.8, it results that $\tau_1 < \tau_2$. Therefore, the poles and zeroes of $\frac{1}{Cs} \frac{Z_R(s)}{R}$, which are sketched in Figure 1.16, do not alternate on the $j\omega$-axis making the passive synthesis impossible, as well known from elementary circuit theory.

The conclusions that can be drawn from this simple example are as follows:

- Impedance matching as required in DSL standards where the splitters need to become "transparent" imposes the presence of lossy circuits when the terminations are complex.
- Simple transformations of real impedance terminated splitters into complex impedance terminated splitters by impedance

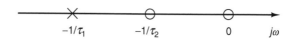

Figure 1.16 **Poles and zeroes of** $\dfrac{1}{Cs}\dfrac{Z_R(s)}{R}$.

re-normalization does not work both for economical and for passive synthesis realizability problems.
- Splitter design has to rely on specific optimization techniques owing to the large number of practical constraints.

1.8.3 Structure and Complexity for Splitter Design

In terms of area and cost, the inductive components are very significant considerations in splitter design. In general, good quality coils are chosen at the central office (CO) side (for reasons to be explained later), and therefore their price cannot be reduced below a significant limit. For instance, coils that are implemented at the CO splitter side undergo large ringing voltages; to avoid saturation of their magnetic core, these coils are usually ferrite pot cores which are of relatively big size (RM4 to RM6 format). Even for millions of pieces, the unit price of these POTS core after bobbin winding ranges between 30 and 60 cents. The budget of a CO splitter is in the range of 2–3 dollars per port, so minimization of the number of coils is of paramount importance. This is directly related to the splitter order.

To begin an optimization process, the following procedure can be useful in practice:

- First evaluate the complexity (order) as if the splitter were a two-port terminated on real impedances. Then design a first splitter architecture as a Cauer filter terminated on 600 Ω impedances.
- For complex impedance terminations, adapt the previous splitter structure by allowing only an additional extra coil which is damped by a resistor as explained for the one order splitter, i.e., by transformation of the line port inductance by de-normalization in impedance of the form $L\,s\dfrac{Z_R}{600}$.
- Use a dedicated optimization technique to adjust the components of the modified splitter to fulfill all the constraints relative to the required standard.

From this rudimentary technique of synthesis, one can see that the price paid for complex impedance adaptation is simply one extra coil. Although there is no proof that this technique is optimal in complexity, many empirical attempts to decrease this complexity have failed. Comparisons of splitter manufacturer solutions show that manufacturers implement filters of the same complexity for the same level of performance. Decrease of complexity is at the expense of the relaxation of some requirement, in general, the level of rejection at the beginning of the ADSL band.

1.8.3.1 First Assessment of the Splitter as a Real Impedance Terminated Two-Port

The ratio between the beginning of the ADSL band and the end of the POTS band is a good starting point. If 12 kHz is taken as the end of the POTS bandwidth (in order not to reject possible pulse metering signals) and 32 kHz for the beginning of the ADSL band, this ratio is close to $\Omega_s = 2.7$. The level of rejection needed in the ADSL band is also important. It can be initially assumed that the splitter has a real termination impedance, and that the sharpest transition band is required. It is well known from filter theory that the lowest order structure is elliptic (Cauer's structure). Assuming a reflection coefficient ρ about 10 percent, which would induce a return loss of about 20 dB in the POTS bandwidth, the smallest even order needed to provide a rejection of 55 dB at the normalized stop band pulsation $\Omega_s = 2.7$ can be found from elliptic filter tables. Cauer filters are tabulated according to the convention

$$C \; n \; \rho \; \theta$$

where C stands for Cauer, n is the order, ρ is the reflection coefficient which fixes the ripple in the passband as $R_{dB} = -10\log 10(1 - \rho^2)$ (or equivalently the return loss in the passband as $RL_{dB} = -10\log 10(\rho^2)$), and θ is the modular angle linked to Ω_s by the relation

$$\sin(\theta) = \frac{1}{\Omega_s}$$

For $\Omega_s = 2.7$, $\theta = 22°$, the rejection capabilities of even-order Cauer filters C04 10 22, C06 10 22, C08 10 22, etc. from Cauer's tables can be compared. C04 10 22 exhibits a rejection of 49 dB, while C06 10 22 allows 90 dB of rejection. A sixth order is therefore enough to design the splitter if it were terminated with real impedances. It can be argued that it is certainly enough to implement the odd order C05 10 22 filter which according to the Zverev tables [Zverev 1967] provides a sufficient 69 dB of rejection. Although of lower order, this filter does not consume less inductors than its sixth-order

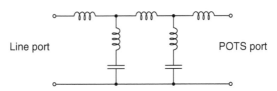

(a) Fifth-order Cauer structure with high impedance at high frequencies at line port

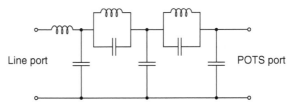

(b) Sixth-order Cauer structure with high impedance at high frequencies at line port

Figure 1.17   Comparison of consumption of coils for odd- and even-order Cauer structures with the constraint of high impedance at infinity for the line port.

counterpart. Because the line port has to be high impedance at infinite frequencies, the line has to begin with a series inductor, therefore the C05 10 22 filter of interest consumes two more coils* than its C06 10 22 counterpart, as shown in Figure 1.17.

1.8.3.2   *Modification of the Structure for Adaptation to Complex Impedance Terminations*

After computation of the real impedance terminated splitter, the line port series inductor L_1 is transformed as explained in Figure 1.18 by following the rule explained for the first-order splitter, i.e.,

$$L_{s1} = \frac{R_2}{R}L_1 \qquad L_{p1} = \frac{R_1}{R}L_1 \qquad R_{p1} = \frac{L_1}{R_1 C_1} \qquad (1.9)$$

where R_1, R_2, C_1 are relative to the impedance Z_R shown in Figure 1.12. To add a degree of freedom a capacitance C_{p1} can be put in parallel with L_{p1} and C_{p1}, this can improve the return loss performance of the splitter. The parallel of R_{p1}, L_{p1}, and C_{p1} will be considered in this chapter as a phase

---

* By applying special coupling techniques, the six coils of the CO5 10 22 filter can be reduced to four coils; however, two coils would require extra winding, making the design no more economical than that of the C06 10 22.

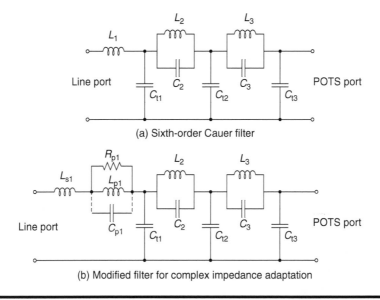

(a) Sixth-order Cauer filter

(b) Modified filter for complex impedance adaptation

Figure 1.18   Transformation of Cauer structure to allow impedance matching.

corrector circuit or corrector. Of course, the proposed circuit is not unique; the corrector can be placed at different places in the ladder.

1.8.3.3   *Optimization of the Circuit*

Let θ be the vector, the components of which are the set of capacitances, resistances, and inductances that describe the splitter architecture. Let n be the dimension of θ. The problem to be solved is to find θ_*, which minimizes cost function $C(\theta) : R^n \longrightarrow R$ subject to m constraints $g_i(\theta) \leq 0$, $i = 1, \ldots, m$ where $g : R^n \longrightarrow R$. The m constraints are about a tenth relative to the return loss and insertion loss functions measured from each of the three ports for different load conditions for different bands of frequencies. The cost function can be for instance a penalty function related to the splitter level of rejection in the ADSL band while all the other return loss and insertion functions are constrained through their admissible maximum. The algorithm of optimization (if nontrapped in a local minima) will give the best level of rejection in the ADSL band (compliant or not to the standard), while all the other functions will be constrained such that their performances are standard compliant. This multiconstraint can be posed in different ways and an arsenal of well-known constraint optimization techniques can be used. Among them is the method of feasible directions, which is an optimization algorithm that improves the cost function without violating the constraints. The main problem is to avoid being trapped in a local minima. Another problem for the designer is the amount of code that must be developed; most methods use gradient techniques or second-order derivatives when using quasi-Newton methods.

One of the simplest methods of optimization that offers a good trade-off between extremely good capabilities to avoid local minima and simplicity of implementation is the simulated annealing method, which has been successfully used in many phases of circuit design. As its name suggests, the simulated annealing exploits an analogy between the way in which a metal cools and freezes into a minimum energy crystalline structure (the annealing process) and the search for parameters which offer a minimum for a cost function. The major advantage of this method over other methods lies in its ability to avoid being trapped in local minima. The algorithm employs a random search which not only accepts changes that decrease the cost function but also accepts, if some conditions hold, an increase of the cost function. This peculiarity is essential and gives the algorithm its ability to pass "above" saddle points of the cost function. Given a temperature T which is high at the beginning of the "cooling process" and which follows a certain decreasing profile to be defined, the algorithm accepts a change in parameters $\delta\theta$, which increase the cost function $C(\theta + \Delta\theta) = C(\theta) + |\Delta C|$ by a positive value $|\Delta C|$ with a probability

$$p = \exp\left(\frac{-|\Delta C|}{T}\right) \tag{1.10}$$

In other words, the bigger the increase of the cost function, the weaker the probability of parameter change acceptance.

The implementation of the simulated annealing is remarkably easy in contrast to all the other methods. Compared to other methods it could be said that the simplicity of computation per iteration is paid by the number of iterations needed for convergence. The structure of the simulated annealing algorithm is given in Figure 1.19. A case study is presented in Section 1.8.4.

1.8.4   Case Study for the Design of an ETSI CO Splitter

The goal of this subsection is to detail part of the design of a CO splitter in which a great number of constraints have to be fulfilled. To exhibit a great number of constraints we have chosen to design a "universal" ETSI splitter, the goal of which is to be compliant either with option A or option B of the ETSI TS 101-1952-1-1 technical specification. The method can be used for any kind of splitter terminated with complex impedances.

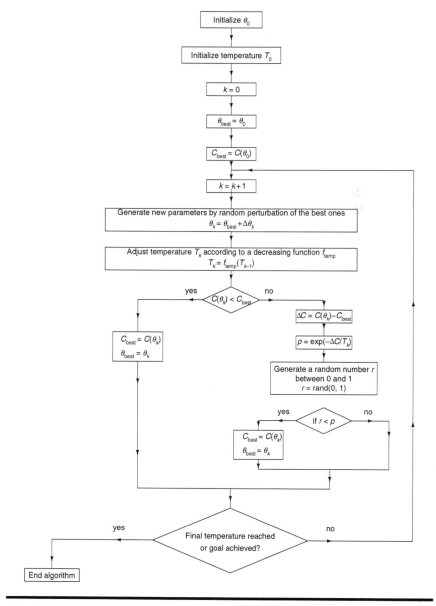

Figure 1.19 **Organigramme of the simulated annealing algorithm.**

Basically, a splitter compliant either with option A or option B has to exhibit a return loss of about 18 dB on the POTS band and an insertion of 55 dB all along the ADSL band. As explained in the previous subsection, this would correspond to a Cauer filter C06 10 22 if the splitter would be terminated with resistive terminations.

Table 1.1   Initial Guess for Filter between Resistive Terminations of 600 Ω. The Components Are Relative to the Circuit of Figure 1.18a

Component	Normalized Value	De-normalized Value
L_1	0.8180 H	6.5 mH
L_2	1.429 H	11.4 mH
L_3	1.352 H	10.8 mH
C_{t1}	1.654 F	36.5 nF
C_{t2}	1.792 F	39.6 nF
C_{t3}	0.9540 F	21.1 nF
C_2	0.08586 F	1.9 nF
C_3	0.0486 F	1.1 nF

For the normalized C06 10 22 filter, the values of the components are shown in Figure 1.18a. These values are listed in Table 1.1 before and after de-normalization. The de-normalization is both in impedance and in frequency. The de-normalization impedance is 600 Ω and the de-normalization frequency is 12 kHz.

By using Equation 1.9, the splitter can be modified for complex impedance terminations Z_R. These initial values are given in Table 1.2. To add more freedom, a capacitor is inserted in parallel with the corrector with a random value of 70 nF.

Table 1.2   Initial Guess for Filter between Complex Terminations Z_R. The Components Are Relative to the Circuit of Figure 1.18b

Component	Value
L_s	2.9 mH
L_p	8.1 mH
R_p	72 Ω
C_p	70 nF
L_2	11.4 mH
L_3	10.8 mH
C_{t1}	36.5 nF
C_{t2}	39.6 nF
C_{t3}	21.1 nF
C_2	1.9 nF
C_3	1.1 nF

Nine constraints have been taken into account. They are not all encompassing, but being the most severe they include most of the other constraints. These constraints are as follows:

- Four constraints are linked with the return loss function measured from both POTS and line ports, the ADSL being opened or closed on the ADSL modem impedance (e.g., $Z_{\text{ADSL}-1}$ of ETSI TS 101 952-1-1). The reference setup is shown in Figure 1.20 as well as the return loss template.
- Two constraints are linked to the insertion loss measured in off-hook mode at 1000 Hz. The insertion loss should be less than 1 dB for both POTS and line port terminated either with 600 Ω or Z_R, the ADSL being opened or closed on its ADSL modem impedance. The reference setup is shown in Figure 1.21.
- Two additional constraints are related to the insertion loss deviation in off-hook mode in the POTS band that should be less than ±1 dB with respect to the insertion loss measured at 1000 Hz for both POTS and line port terminated either with 600 Ω or Z_R, the ADSL being opened or closed on its ADSL modem impedance. The POTS band considered is in the range 200 Hz–4 kHz. The reference setup is shown in Figure 1.21.
- One additional constraint is the level of rejection in the ADSL band that in option A should be above 55 dB between 200 Hz and 4 kHz. The reference setup is shown in Figure 1.22.

Once the constraints are defined, the cost function that is to be mini-mized remains to be defined. There is a slight difficulty here with respect to the general constrained optimization framework that has been presented previously; the simulated annealing is more suited for nonconstrained optimizations. Therefore, the constraints have to be cast into the cost function. The nine constraints can be expressed as

$$g_i(\theta) \leq 0, \quad i = 1, \ldots, 9$$

The four return loss functions, which are constrained as shown in Figure 1.20, are defined as $RL_i(f)$, $i = 1, \ldots, 4$. Additionally, $TPL(f)$ is defined as the return loss template function. The four functions $g_i(\theta)$, $i = 1, \ldots, 4$ can be defined as

$$g_i(\theta) = \max_{f \in S} \left(TPL(f) - RL_i(\theta, f) \right), \quad i = 1, \ldots, 4$$

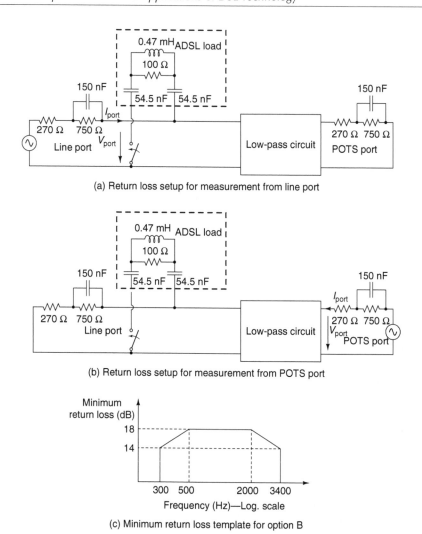

(a) Return loss setup for measurement from line port

(b) Return loss setup for measurement from POTS port

(c) Minimum return loss template for option B

Figure 1.20   **Return loss setup for ETSI TS 101 952-1-1 option B splitter.**

where

$$S = [300 \text{ Hz}, \ 3400 \text{ Hz}]$$

The two insertion functions to be computed as shown in the reference setup of Figure 1.21 are defined as $IL_i(f)$, $i = 5, \ldots, 6$. The two constraints at 1000 Hz can be expressed as

$$g_i(\theta) = \Big(IL_i(\theta, \ 1000 \text{ Hz}) - 1 \Big), \quad i = 5, \ldots, 6$$

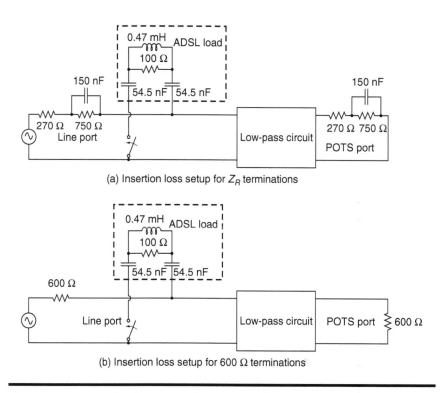

(a) Insertion loss setup for Z_R terminations

(b) Insertion loss setup for 600 Ω terminations

Figure 1.21 **Insertion loss setup for ETSI TS 101 952-1-1 option B splitter. The insertion loss measured at 1000 Hz should be less than 1 dB in configuration (a) or (b). The insertion loss deviation in the POTS band 200 Hz—4 kHz should be in the range ±1 dB with respect to insertion loss at 1000 Hz.**

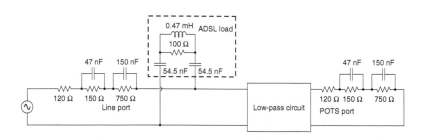

Figure 1.22 **Insertion loss in ADSL band. For ETSI TS 101 952-1-1 option B the insertion loss has to be greater than 55 dB on the band 32–1100 kHz.**

The two constraints on the deviation of the insertion loss in passband can be expressed as

$$g_i(\theta) = \max_{f \in V} \left(|IL_{i-2}(\theta, f) - IL_{i-2}(\theta, 1000 \text{ Hz})| - 1 \right), \qquad i = 7, \ldots, 8$$

where

$$V = [200 \text{ Hz}, 4000 \text{ Hz}]$$

The insertion loss in the ADSL band to be measured with the reference setup of Figure 1.22 is defined as $IL_9(\theta, f)$. The ninth constraint can be expressed as

$$g_9(\theta) = \max_{f \in W} \left(55 - IL_9(\theta, f) \right), \qquad i = 1, \ldots, 4$$

where

$$W = [32 \text{ kHz}, 1100 \text{ kHz}]$$

One possible solution for the implementation of the simulated annealing algorithm is to introduce a cost function $C(\theta)$ suited for the minimax optimization of the form

$$minimize \ C(\theta) = \max_{1 \leq i \leq 9} w_i g_i(\theta)$$

where w_i are weights which compensate for the disparities of the different g_i ranges.

Other variants are possible. In a variant implemented below, the g_i functions are set to zero as soon as the constraints are fulfilled with a predefinite margin and the cost function is such that the minimization problem becomes

$$minimize \ C(\theta) = \sum w_i^2 f_i^2(\theta)$$

where

$$f_i = g_i + |threshold_i| \quad \text{if} \quad \left(g_i + |threshold_i| \right) > 0$$

and

$$f_i = 0 \quad \text{if} \quad \left(g_i + |threshold_i| \right) \leq 0$$

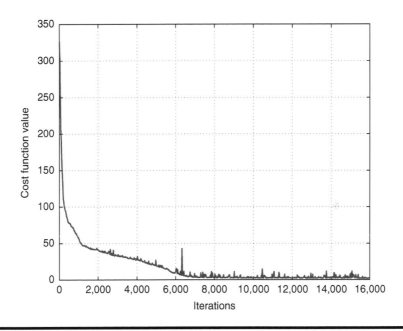

Figure 1.23   Cost function behavior in function of iterations of the simulated algorithm.

From the initial guess defined in Table 1.2,* Figure 1.22 shows the behavior of the cost function during the 16,000 first iterations of the simulated annealing algorithm. After these 16,000 iterations, all the constraints are fulfilled with margin. The initial temperature was set to 1000 and a linear decreasing profile was chosen in such a way that at the end of the process, the temperature decreases to 20 percent.

The results after optimization are given in Figures 1.23 through 1.28. Parasitic resistance has been introduced in series with the inductances. As a rule of thumb which has been shown to be quite accurate in practice, each inductance has been modelled with a series resistance in ohms which is half of the inductance value in millihenries. The parasitic capacitances have been neglected; however, care has to be taken when winding the bobbins, depending on the technique used these parasitic capacitances can vary from a tenth of a picofarad on a L_s up to one hundred of picofarads and impact the rejection at the end of the ADSL band. The component values after convergence are shown in Table 1.3, and a proposed balanced realization of the splitter is shown in Figure 1.29.

---

* We add three resistances of 30 kΩ in parallel with L_s, L_2, and L_3 to avoid abrupt phase transition.

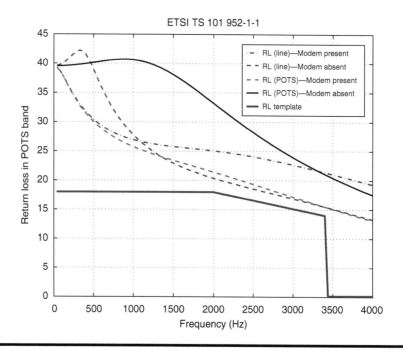

Figure 1.24   **Option B return loss margin is 1.5 dB after convergence.**

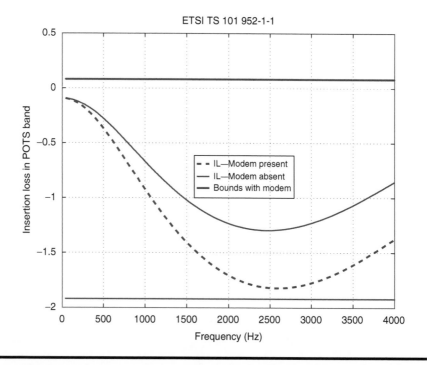

Figure 1.25   **Insertion loss in POTS band after convergence.**

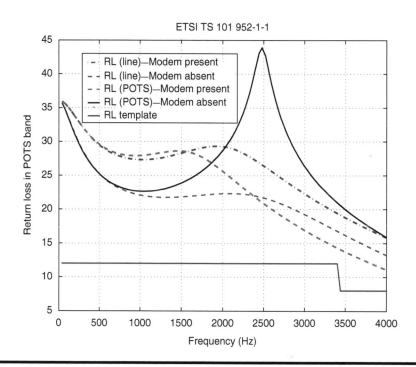

Figure 1.26 **Option A return loss margin is 2.3 dB after convergence.**

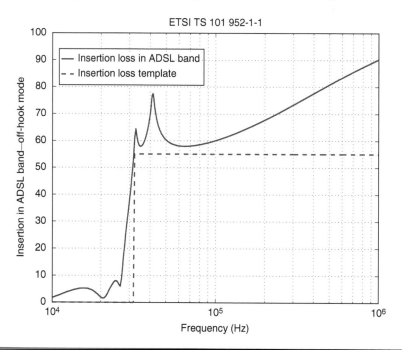

Figure 1.27 **Isolation in ADSL band (option B template).**

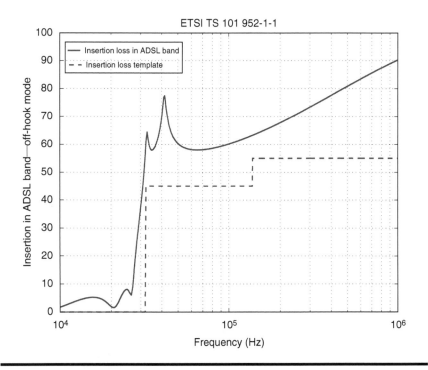

Figure 1.28   Isolation in ADSL band (option A template).

Table 1.3   Final Values of Components of Figure 1.18b after Convergence of the Simulated Annealing Algorithm

Component	Value
L_s	5.2641 mH
L_p	33.9831 mH
R_p	184.1662 Ω
C_p	192.8722 nF
L_2	3.4466 mH
L_3	3.6595 mH
C_{t1}	14.0315 nF
C_{t2}	15.0036 nF
C_{t3}	8.3661 nF
C_2	6.9112 nF
C_3	4.0202 nF

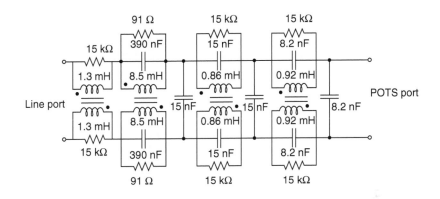

Figure 1.29   Balanced circuit version.

1.9   Active Design

1.9.1   Introduction

Although passive circuits have been shown capable of solving the splitter design problem, they suffer from some inherent drawbacks: they are bulky, heavy, expensive, and can be subject to relatively quick aging. They suffer also from some limitations of performance when both return loss and insertion loss have to be optimized with some margin with respect to standards dealing with complex impedance terminated networks. Nominal passive designs are very close to the specifications with poor margins, meaning that many times the specifications have to be slightly relaxed so that an entire manufacturer splitter batch can pass the specifications when magnetic component tolerance of ±7 percent applies. It can be even stated that splitter specifications (ETSI or American National Standards Institute [ANSI] documents) have evolved in such a way that the required performance is close to the boundaries of what is feasible with a state-of-the-art passive design state rather than what would be necessary to preserve the quality of service of the POTS network.

To improve the splitter design with respect to filter performance, cost, and size, some attempts have been made to develop active circuitry. Early attempts [Cook 1995] were essentially focused on filter performance improvement when the isolation requirement in ADSL band was higher than the present requirement. These early attempts were made using discrete active electronics. Resulting active splitters were shown to provide isolation better than 60 dB in the ADSL band and return loss better than 20 dB in the POTS band, outperforming passive designs. They were also shown to outperform their passive counterpart in terms of comfort of listening especially in the presence of room noise [Cook 1995]. However, these active designs were still bulky and much more expensive than their passive counterparts. The extra constraint of external power requirement discouraged the market to trade performance against cost, size, and power constraint. As it could be expected, most of the discrete electronics–based active designs stayed at the prototype level. However, there is now a tendency to develop solid-state specific circuit with the aim of decreasing the size of the splitter in such a way that it becomes part of the DSLAM. Recent research and development has been reported in [Dedieu 2004] and [Sackinger 2006].

Although the domain of filter design is a very mature domain, it is far from obvious with regard to the development of a fully active circuit. The main reasons that prevent an easy integration are as follows:

- Most active design documented techniques rely on the gyrator principle where currents are exchanged with voltages and vice versa making possible the exchange of an inductor by a capacitor [Chen 1995] [Dedieu 1991]. An inductor is said to be emulated or simulated by a couple made of a gyrator plus a capacitor. Unfortunately, as the AC and DC are coupled on the same twisted pair it is impossible to implement the gyrator principle directly because a capacitor across the DC path would block the signalling DC during off-hook. As a result, an adaptation of the gyrator principle has to be found to find a way to decouple the DC path from the AC path. The result is that most of the solutions found until now are "unconventional."

- There is a considerable heterogeneity of AC and DC signals flowing through the twisted pair. As a result, different ranges of voltages and currents have to be foreseen across the active elements. For instance, AC transients between tip and ring can be greater than 200 V peak-to-peak during ringing. These very low frequency signals can induce high currents close to DC as high as 200 mA. There also exists a considerable heterogeneity in the line impedance conditions which can range from high impedance to very low impedance. Table 1.4 gives a first idea of these different signalling and impedance conditions during on-hook and off-hook modes.

- Until now it has not been possible to provide a fully integrated splitter owing to the range of AC and DC signals that would have to be covered by a full solid-state device. As a result, implementations have been made with a mix of passive and active circuits with a minimal number of coils (one). State-of-the-art placement of the passive circuits allows a monitoring of AC and DC voltage and current ranges in such a way that the active circuits are never saturated during quiescent or transient behavior. Optimization techniques

Table 1.4 Most Common Signals and Line Termination in Both On-Hook and Off-Hook Mode

		Off-Hook State		
DC	Depends on line length	20 mA typ.	100 mA max.	
DC termination	Country dependant	600 Ω at CO typ.	300–600 Ω at CPE typ.	
POTS signals	Dyn. is country dependant	3 V peak-to-peak max.	Freq. range:	200 Hz– 4 kHz
AC POTS termination	US 900 Ω	Europe (ETSI Harm. Imp.)	Germany	UK
	CO, 600 Ω CPE	270 Ω + (750 Ω // 150 nF)	220 Ω + (820 Ω // 115 nF)	320 Ω + (1050 Ω // 230 nF)
ADSL signals	20 V peak-to-peak max.	Freq. range:	Upstream: 32–132 kHz	Downstream: 132 kHz– 1.1 MHz
ADSL termination	100 Ω typ.			
		On-Hook State		
DC	Max. authorized value is country dependant	100–500 μA max.	1–10 μA typ.	
DC termination	Country dependant	600 Ω at CO typ.	Receiver model at CPE:	1 MΩ //5 nF
Ringing signals	Country dependant	200 V peak-to-peak max.	Freq. range:	25–60 Hz
ADSL signals	20 V peak-to-peak max.	Freq. range:	Upstream: 32–132 kHz	Downstream: 132 kHz– 1.1 MHz
ADSL termination	100 Ω typ.			

which are used to tune the passive and active component parameters have also to ensure that the signals never exceed their nominal range of operation for any condition off-hook or on-hook, or for abrupt passages between on-hook and off-hook.

- A potential danger with active circuits is the possible triggering of oscillations for some specific abrupt change of line conditions. For instance, if the design is not unconditionally stable, oscillations can occur [Chen 1991] when passing from an off-hook mode to an on-hook mode or vice versa. With integrated circuits, it can easily happen that a part of their batch is slightly de-tuned and that a number of circuits are not free of oscillation for some line conditions. For instance, for a very long line the active system can oscillate although this was never the case for a short line. In practice, this oscillation problem can be a nightmare for the designer.

- Because of problems of integration precision, the circuits need to embed precise calibration techniques using, for instance, digital triggers that allow the circuit to be maintained in its nominal zone. Calibration is also necessary to ensure good LCL properties in particular when active circuits have to be balanced on tip and ring. Splitter calibration is not a problem that should be underestimated because it has a direct impact on cost.

The goal of this section is to present the few active splitter developments which have been documented. More academic research and innovative solutions are certainly needed on this subject which would profit from the collaborative talent of several designers.

1.9.2 Cook and Sheppard Gyrator

The solution proposed by Cook and Sheppard [Cook 1995] is an elegant solution, which provides an unconventional impedance converter. Cook and Sheppard were able to find a structure of two-port which converts a real impedance R_0 into a complex one $Z(s)$ without loading the DC path with an ohmic or capacitive load in contrast of what a conventional impedance converter would do. To succeed, Cook and Sheppard used a transformer and decoupled the active elements of the DC path by placing them into the transformer secondary path. This avoids any loading of the DC path.

Before investigating the Cook and Sheppard gyrator, let us define a floating impedance converter as the two-port of Figure 1.30. The impedance converter is such that

$$\frac{V_1}{I_1} = \alpha(s)\frac{V_2}{I_2} \tag{1.11}$$

The direction of the arrow in the schematic of Figure 1.30 indicates that if the right port is loaded with R_0, the left port will appear as a loading impedance $\alpha(s) \cdot R_0$.

The basic idea followed by Cook and Sheppard was therefore to design a filter with real input and output impedance R_0 using, for instance, Cauer's

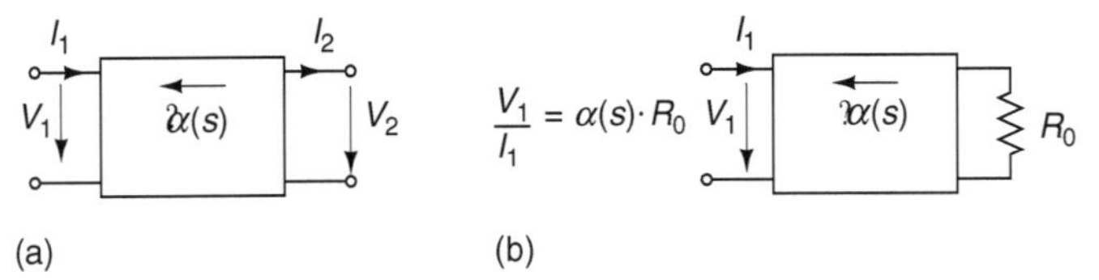

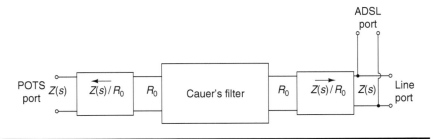

Figure 1.30 **(a) Impedance converter schematic. (b) Equivalent input impedance when the right port is loaded with R_0.**

structures providing rejection in the ADSL band greater than 80 dB with low order while giving a return loss against R_0 above 20 dB in the POTS band. By placing specific active impedance converters at both sides of their two-port, they were able to obtain the same return loss above 20 dB against the reference complex impedance $Z(s)$ used in the United Kingdom at both sides of their splitter. In other words, they were able to decouple the two problems of obtaining both a good rejection in the ADSL band and a good return loss on the POTS band against complex impedance by using two impedance converters, as described in Figure 1.31.

The Cook and Sheppard impedance converter in its grounded version is shown in Figure 1.32. The capacitor C_d is a decoupling capacitor, the role of which is to block the DC. Its impedance being considerably lower than R in the POTS band, the impedance of this blocking capacitor will be neglected in the sequel. The following equations govern the behavior of the circuit:

$$U_1 = L_1 \cdot s \cdot I_1 - M \cdot s \cdot I_e \qquad U_2 = L_2 \cdot s \cdot I_e - M \cdot s \cdot I_1 \qquad (1.12)$$

where $L_1 = L_2 = L$. Supposing that perfect coupling holds $M = \sqrt{L_1 L_2} = L$.

Figure 1.31 Cook and Sheppard proposed splitter system with impedance converters.

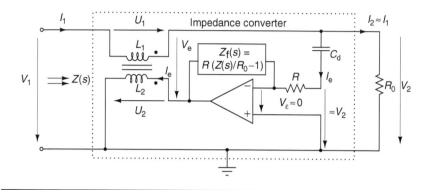

Figure 1.32   Grounded version of Cook and Sheppard impedance converter.

As $R \gg R_0$, it can be considered that $I_e \ll I_1$ and therefore

$$I_2 = \frac{V_2}{R_0} \approx I_1 \tag{1.13}$$

As $\frac{1}{C_d s} \ll R$ for frequencies in the POTS band and beyond, and $V_\epsilon \approx 0$, it can be stated that

$$R \cdot I_e \approx V_2 \tag{1.14}$$

Elementary circuit investigation shows that

$$V_e = \left(R \left(\frac{Z(s)}{R_0} - 1 \right) \right) I_e + R I_e = R \cdot \frac{Z(s)}{R_0} \cdot I_e \tag{1.15}$$

Injecting the results of Equation 1.14 into Equation 1.15, it can be seen that

$$V_e \approx \frac{Z(s)}{R_0} \cdot V_2 \tag{1.16}$$

Furthermore

$$V_1 = U_1 + V_e + U_2 \tag{1.17}$$

By introducing Equation 1.12 into Equation 1.17 with $L_1 = L_2 = L$ and $M = L$, it can be shown that

$$V_1 = U_1 + V_e + U_2 = L \, sI_1 - L \, sI_e + V_e + L \, sI_e - L \, sI_1 = V_e \tag{1.18}$$

By using Equation 1.16, Equation 1.18 becomes

$$V_1 = V_e \approx \frac{Z(s)}{R_0} \cdot V_2 \tag{1.19}$$

By dividing both members of Equation 1.19 by I_1 and by taking account that $I_2 \approx I_1$, it can be shown that

$$\frac{V_1}{I_1} \approx \frac{Z(s)}{R_0} \cdot \frac{V_2}{I_2} = \frac{Z(s)}{R_0} \cdot R_0 = Z(s) \tag{1.20}$$

Therefore, the two-port of Figure 1.20 has the desired property of transforming a real impedance R_0, loading the right port, into a complex impedance $Z(s)$ which is the input impedance of the left port.

As an example of Cook and Sheppard impedance converter, let $R_0 = 300\ \Omega$ and let us consider the United Kingdom reference impedance $Z(s)$, the schematic of which is given in Figure 1.33.

Hence

$$Z(s) = R_2 + \frac{R_1}{1 + R_1 C_1\,s} = 320 + \frac{1050}{1 + 2.415 \times 10^{-4}\,s}$$

and therefore the impedance converter parameter $\alpha(s)$ is such that

$$\alpha(s) = \frac{Z(s)}{R_0} = 1.067 + \frac{3.5}{1 + 2.415 \times 10^{-4}\,s}$$

Taking $R = 100\ \text{k}\Omega$, the network to be synthesized for the feedback circuit of the operational amplifier of Figure 1.33 is such that its impedance obeys

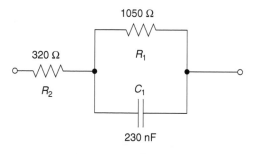

Figure 1.33 **United Kingdom reference impedance $Z(s)$.**

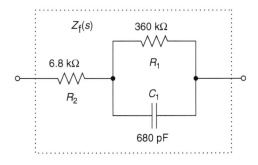

Figure 1.34 $Z_f(s)$ **feedback impedance.**

$$Z_f(s) = \left(R \left(\frac{Z(s)}{R_0} - 1 \right) \right) = 6.7 \times 10^4 + \frac{350 \times 10^3}{1 + 2.415 \times 10^{-4} \, s}$$

By taking components in the normalized series E24, $Z_f(s)$ can be modeled with a precision close to ± 3 percent by the circuit of Figure 1.34.

Taking $C_d = 150$ nF, the complete impedance converter circuit is shown in Figure 1.35.

The floating version of the impedance converter of Figure 1.35 has been proposed by Cook and Sheppard [Cook 1995]. The structure of this floating impedance converter is shown in Figure 1.36.

An example of filter transformation is given in [Cook 1997]. The passive filter of Figure 1.37 which has a reference impedance of 300 Ω is transformed into the splitter of Figure 1.38 which has a reference impedance of $Z(s)$. Two main modifications were made with respect to the general scheme of Figure 1.31. To reduce the amount of high-frequency ADSL signal received by the impedance converter close to the line port, the impedance converter was shifted on the left side of the transformer made of L_3 and L_{3p}. This had the consequence to lower the return loss on the line side and

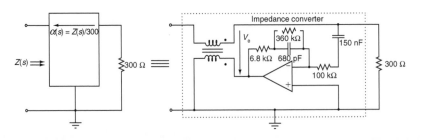

Figure 1.35 Grounded version of Cook and Sheppard impedance converter with component values.

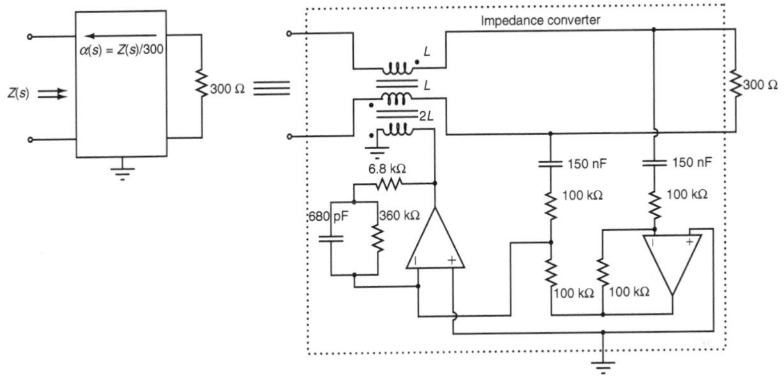

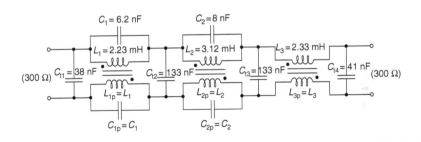

Figure 1.36   **Floating version of Cook and Sheppard impedance converter.**

Figure 1.37   **Low-pass filter of cut-off frequency $f_c = 6$ kHz and stop-band frequency $f_s = 21.4$ kHz. Level of rejection is 100 dB in the ADSL band. Return Loss is greater than 27 dB on 200–3400 Hz.**

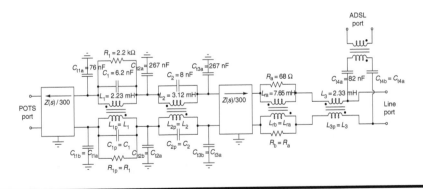

Figure 1.38   **Splitter realized from the low pass filter of Figure 1.37.**

imposed the presence of the transformer made of relative big coupled L_{ra} and L_{rb} inductors in parallel with 68 Ω resistances to enhance the return loss performance. The introduction of the damping resistors R_1 and R_2 is the second main modification. R_1 and R_2 are included to damp the parallel resonance of the transformer made of L_1 and L_{1p}, and C_1 and C_{1p}. This damping ensures stability of the active splitter in the stopband when the filter is "adversely terminated" (including short or open-circuits) [Cook 1997]. It should be noted that the capacitor C_{t4} has been split into two capacitors C_{t4a} and C_{t4b}. They appear in series with an ADSL transformer, which has an impedance close to zero in the POTS band.

In [Cook 1995], it is shown that the introduction of the active filter such as described in Figure 1.38 not only improves measurable quantities such as return loss, rejection in the ADSL band, and sidetone masking rating but also increases drastically comfort of listening. In particular, a customer opinion study was carried out according to a computed aided telephone network assessment program described in [CCITT Blue Book]. It is worth noting that the subjective opinion was such that it was quite difficult for a listener to make the difference between a situation where the ADSL path was on or off when the active splitter was connected. On the contrary, the subjective opinion was that the presence of a passive splitter lead to a deterioration of the listening comfort especially when the room noise was high in presence of the ADSL path. Despite their inherent advantages, the splitters developed by Cook and Sheppard were not deployed in mass of the extra cost of the active devices.

1.9.3 Silicon Integrated Splitters

As an example of integrated approach, the main principle which guided the designers of a mixed active–passive splitter remote powered with the line current [Krummenacher 2002] and [Dedieu 2004] is now presented. The architecture of the mixed passive–active circuit is shown in Figure 1.39. An active circuit made of two application-specific integrated circuits (ASICs) is embedded into a passive circuit reduced to one transformer and two capacitors. Each ASIC consists of a programmable active admittance which was designed to enhance the passive circuit capabilities. The frequency-selective active admittance is achieved by driving a voltage-controlled current source with a transconductor–capacitor circuit implementing a fourth-order Laplace transfer function (four poles and four zeroes). The device is self-powered through the DC line current in off-hook mode. A biasing current of 10 μA only allows the transmission of class signals in on-hook mode. Precise calibration of all relevant DC and AC parameters of

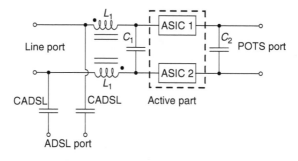

Figure 1.39   Mixed active–passive splitter.

the circuit is performed during testing and stored in the on-chip Electronically Erasable Programmable Read-Only Memory (EEPROM). The general features of the circuit are as follows:

- No external supply required
- Programmable AC response covering European and U.S. line impedance
- Embedded on-hook off-hook detection circuit
- Low-voltage low-power EEPROM Complementary Metal–Oxide–Semiconductor (CMOS) integration
- Low cost

During on-hook, the active circuits are switched off and are equivalent to resistors of about 50 Ω. During off-hook, the active circuits are switched-on and are equivalent to a fourth-order impedance such that

$$Z(s) = R_{\text{rect}} + R_{\text{oh}} \frac{1 + B_1 s + B_2 s^2 + B_3 s^3 + B_4 s^4}{1 + A_1 s + A_2 s^2 + A_3 s^3 + A_4 s^4} = \frac{N(s)}{D(s)} \qquad (1.21)$$

The term R_{rect} is a resistor because of a rectifier, the role of which is to power the active circuit with the right voltage direction according to the incoming direction of DC. Typically, R_{rect} is around 10 Ω. The term R_{oh} is a small resistor, which is about 2.5 Ω. The parameters A_i's and B_i's in Equation 1.21 are tuned in such a way that $|Z(j\omega)|$ reaches its maximum at the beginning of the ADSL band where a good rejection is needed. The A_i's and B_i's are also chosen in such a way that the two-port embedding the two ASICs exhibits passivity. The behavior of $|Z(j\omega)|$ is shown in Figure 1.40; the maximum is reached around 34 kHz. The global transfer function for both on-hook and off-hook modes is shown in Figure 1.41. The on-hook

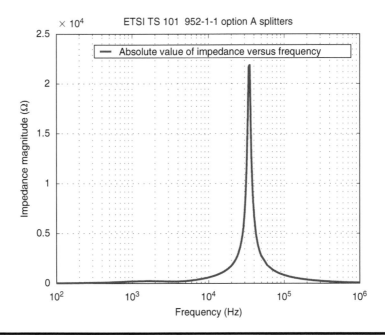

Figure 1.40   AC impedance of ASIC when switched on.

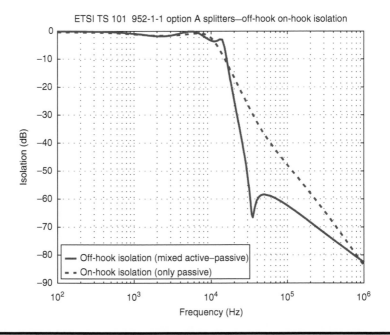

Figure 1.41   Transfer function in on-hook and off-hook modes.

transfer function is due to the passive circuit while the off-hook one results from the enhancing of the passive circuit by the active circuitry.

The heart of each ASIC is a fourth-order impedance, the model of which is depicted in Figure 1.42 [Krummenacher 2002].

The fourth-order admittance is a voltage-controlled current source. The voltage which controls the source is the voltage V, which is the voltage across each ASIC. The four amplifiers on the left of Figure 1.42 are Operational Transconductance Amplifiers (OTA). Each OTA has a differential input voltage $\Delta V(t)$ and converts this differential input voltage into an output current which obeys the following equation:

$$i(t) = g_m \Delta V(t) \tag{1.22}$$

The quantity g_m is referred to the transconductance of the amplifier. It will be shown that this transconductance can be controlled to program the so-called OTA. As R_{oh} is small and I_y is considerably greater than the currents entering the capacitors C_{11}, C_{21}, C_{31}, and C_{41}, it can be easily shown by elementary circuit analysis that

$$I_y = \frac{1}{R_{oh}} \frac{1 + A_1 s + A_2 s^2 + A_3 s^3 + A_4 s^4}{1 + B_1 s + B_2 s^2 + B_3 s^3 + B_4 s^4} \tag{1.23}$$

where the A_i's and B_i's are linked to the g_{mi}'s $i = 1, \ldots, 4$, C_{ij}'s $i = 1, \ldots, 4$, $j = 1, \ldots, 2$, R_{oh} of Figure 1.42 according to

$$C_{11} = g_1 A_1 \tag{1.24}$$

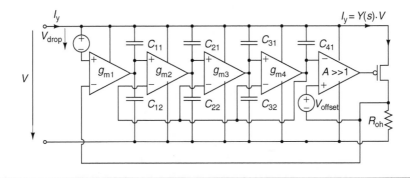

Figure 1.42   **Realization of fourth order admittance.**

$$C_{12} = g_1 B_1 - C_{11} \tag{1.25}$$

$$C_{21} = \frac{A_2 g_1 g_2}{(C_{11} + C_{12})} \tag{1.26}$$

$$C_{22} = \frac{B_2 g_1 g_2}{(C_{11} + C_{12})} - C_{21} \tag{1.27}$$

$$C_{31} = \frac{A_3 g_1 g_2 g_3}{(C_{11} + C_{12})(C_{21} + C_{22})} \tag{1.28}$$

$$C_{32} = \frac{B_3 g_1 g_2 g_3}{(C_{11} + C_{12})(C_{21} + C_{22})} - C_{31} \tag{1.29}$$

$$C_{41} = \frac{A_4 g_1 g_2 g_3 g_4}{(C_{11} + C_{12})(C_{21} + C_{22})(C_{31} + C_{32})} \tag{1.30}$$

$$C_{42} = \frac{B_4 g_1 g_2 g_3 g_4}{(C_{11} + C_{12})(C_{21} + C_{22})(C_{31} + C_{32})} - C_{41} \tag{1.31}$$

Precise calibration of all relevant DC and AC parameters of the circuit is performed during testing and stored into an on-chip EEPROM. An internal transient suppressor provides smooth transition from on-hook to off-hook mode and reduces cyclic redundancy check (CRC) errors on the ADSL traffic during POTS transients. Four programming modes are available that correspond to four transfer functions covering the main splitter standards. Mode 0 covers the POTS "complex" European impedance so-called ETSI-A; mode 1 covers the ETSI option-B impedance. Modes 2 and 3 cover respectively the ANSI ITU-T G992.1 E2 and ANSI ITU-T G992.1 E4 requirements. Figures 1.43 and 1.44 show good agreement between model and measurements for a specific ASIC. Circuit layout is shown in Figure 1.45.

1.9.4 CPE-Specific Design Problems

Although it is in general easier to design CPE splitters because they require less isolation, it is worth mentioning some specific problems. Very simple

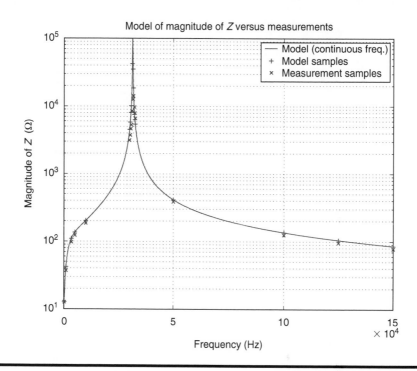

Figure 1.43   Comparison of magnitude of $Z(j\omega)$—model versus measurements.

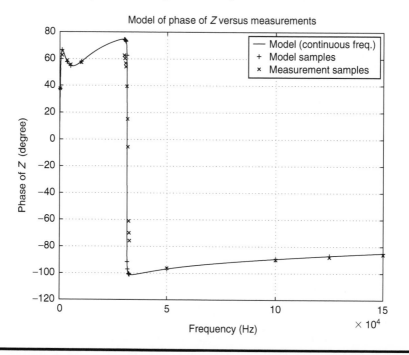

Figure 1.44   Comparison of phase of $Z(j\omega)$—model versus measurements.

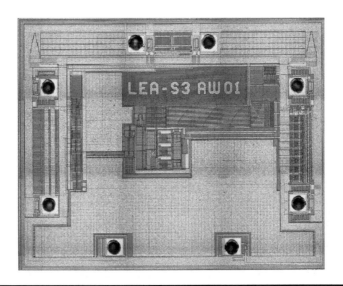

Figure 1.45   Circuit layout.

splitters with low isolation have been deployed worldwide with very inexpensive passive technology. As previously mentioned, this very simple technology can cause degradation of the POTS when several telephones are connected to the same line. As a second consequence, "one-dollar" splitters having poor isolation in the ADSL band can be responsible for poor impedance matching with the ADSL termination, which in turn degrades the ADSL rate. With triple-play applications requiring more bandwidth, low-cost CPE splitters can degrade the effectiveness of the ADSL link. In short, if minimal care is not taken with the design of the CPE splitters, the quality of both POTS and ADSL service can be deteriorated because of impedance matching problems both for the POTS and ADSL terminations. Both ETSI and ANSI standards have not been too demanding with the requirements of CPE splitters, and some national telcos have strengthened the requirements for their ADSL CPE splitters. This is the case for British Telecom, France Telecom, and Belgacom for instance. The [SIN 346] Suppliers' Information Note from British Telecom is one good example of extra requirements, which are asked of splitter manufacturers. It turns out that it is far from obvious to design splitters which are fully compliant with these new national standards. It is worth mentioning here some efforts that have been made by splitter manufacturers to improve their CPE splitters. The solutions presented use either passive or active circuits.

The impedance matching problem with POTS terminations when several telephones are connected to the same line can be easily explained using Figure 1.46. In this figure, it has been supposed that a cheap 600 Ω splitter is simply made of a second-order filter (one transformer plus a

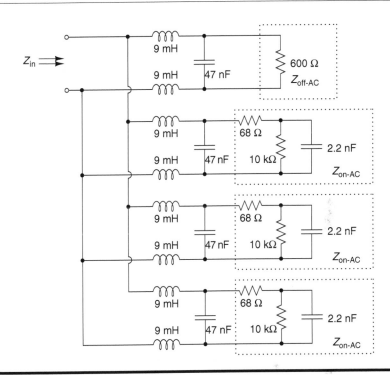

Figure 1.46 Model of one splitter associated with one telephone off-hook. In parallel are three splitters associated with on-hook telephones. The ADSL path has not been represented for simplification.

capacitor). It has been assumed that one telephone is off-hook and that three telephones are on-hook. The off-hook telephone exhibits an AC impedance of 600 Ω, the other telephones which are on-hook are supposed to be modeled with an AC impedance made of two resistors and one capacitor such as proposed in the [TS 101 952-1-1]. The splitter connected to the off-hook telephone has been designed to match a 600 Ω impedance. With only that splitter connected to the line, its return loss would be fine. However, the 600 Ω input impedance "sees" three other splitters in parallel, which are connected to roughly 10 kΩ impedances. If the input impedance of each on-hook telephone-associated splitter is assumed to be about 10 kΩ, the three telephones would present a $\frac{10}{3}$ kΩ impedance in parallel with 600 Ω and the return loss (although degraded) would still be fine. However, what actually happens is that each on-hook telephone- associated splitter presents an input impedance which is lower than 10 kΩ due to the presence of the splitter input capacitor seen in parallel with the 10 kΩ impedance. As a result, the return loss is completely out of specifications as shown in Figure 1.47.

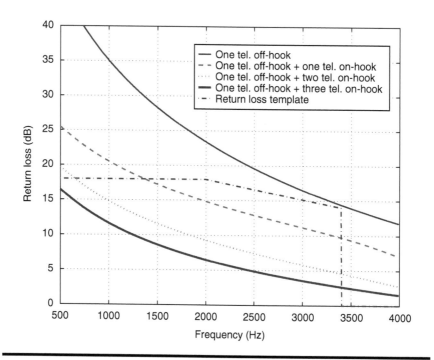

**Figure 1.47   Example of return loss degradation when up to three splitters asso-
ciated to on-hook telephones are connected in parallel to one splitter associated
with one telephone off-hook.**

To maintain a good return loss when several splitters are protecting telephones connected to the same line, a specific isolation mechanism needs
to be implemented. Two basic mechanisms using passive or active switches
are generally used. A first mechanism introduced by [Kiko 1999] [Kiko 2001]
consists in switching off the capacitors between tip and ring. The switch
can be either active or passive (Reed relay triggered with line current). The
mechanism is illustrated in Figure 1.48. During on-hook mode, the line current is weak and considerably below a certain current threshold i_{th}. The
input impedance during on-hook is therefore mainly fixed by L_1, L'_1, and
the small capacitor C_f. When the line current i_c is above an i_{th}, the Reed
relays close and the splitter takes its nominal shape.

Advantages of the solution are as follows:

1. During on-hook the caller ID signals in the POTS band are sent with
    only small attenuation to the TE.
2. When an active switch is used, its electronic noise during on-hook
    is partially filtered by the capacitor C_f in DSL band.

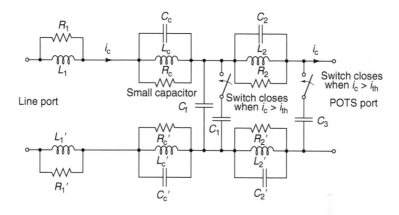

Figure 1.48   Switch between tip and ring [Kiko 1999]. The splitter architecture is suited for complex impedance network.

3. Threshold between on-hook and off-hook need not be very precise. A threshold of 14 mA, for instance, ensures to be above caller ID DC. The off-hook DC being above 20 mA, there is a wide range for the threshold.

4. Low cost Reed relays solutions can be cast into L_1 as shown in Figure 1.49.

Drawbacks of the solution are as follows:

1. Rejection in on-hook mode is very poor because of the following:

    (a) Filter total capacitance between tip and ring has been decreased from $C_1 + C_3 + C_f$ to C_f (from off-hook to on-hook)

    (b) High impedance of the TE

2. As a consequence there is possible injection of ADSL noise during ringing

3. Poor rejection in on-hook mode has forced splitter manufacturers to find a solution for a new DC-sensitive inductor which exhibits a large value during on-hook (typ. 40 mH) and a small value during off-hook (typ. 5 mH). This special inductor has an extra cost [Kiko 2002].

A second principle of operation has already been proposed by France Telecom [Rayer 2000]. It consists in putting an active switch in series with the inductance L_1 as shown on Figure 1.50.

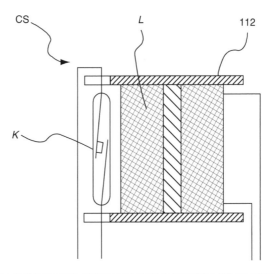

Figure 1.49 Example of Reed relay cast into a drum core L as shown in Excelsus Tech. patent [Kiko 1999]. When the DC exceeds i_{th}, the two blades of the Reed relay are in contact and allow the switching-on of one of the capacitors of Figure 1.48.

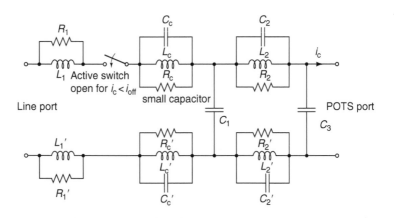

Figure 1.50 Active switch in series with inductor L_1 as proposed in [Rayer 2000]. The switch can be as simple as two diodes in parallel having opposite directions. The switch is usually balanced on tip and ring (two active switches).

Figure 1.50 shows a switch which is high impedance (infinite impedance) for $i_c < i_{off}$ where i_{off} is the level of off-hook current (typ. 20 μA).

Advantages of the solution are as follows:

1. Isolation of off-hook phone is excellent (provided that the switch is high impedance enough during on-hook).
2. L_1 does not need special crafting to exhibit a large value during off-hook.
3. Architecture is simpler than the previous one and does not require Reed relay material inside the inductors.

Drawbacks of the solution are as follows:

1. Value of the DC during off-hook is not so well known in advance. For instance in the United Kingdom, the insulation resistance between leads is not specified, hence a value of resistance is not specified. Instead, loop insulation resistance as a function of bleed current, defined in terms of the ringer equivalence number (REN) of the TE, is specified.

   (a) When REN $= 0$, the current drawn by the TE shall not be greater than 5 μA.
   (b) When $0 < $ REN ≤ 4, the current drawn by the TE shall not be greater than $30 \times$ REN μA.

   Because various RENs can be used, the level of DC in on-hook can be theoretically in the range of 5–120 μA, therefore the switch has to be designed in such a way that its impedances are high enough in this range. The switch has to be designed with a monotonic variation of its impedance with respect to the DC.
2. Passing Caller ID (CLI) signals are inherently more problematic with this architecture than that described in Figure 1.48. In particular, if the termination during CLI mode is such that the range of DC is not well above the DC off-hook current, the switch remains open and the CLI current cannot reach the TE.
3. Active elements in series as shown in Figure 1.50 can demodulate ADSL signals because of nonlinear effects. The impact of the filter on the number of CRC errors in the ADSL path has to be carefully checked.

1.10 Conclusion

Although an overall trend in telecommunications is clearly toward the digitization of voice, the splitter function will remain in a huge number of existing and new deployments of DSL. This chapter has explained the role of the splitter, and detailed the various contraints that a splitter designer should be aware of. Furthermore, it has presented a technique that can be applied to design splitters meeting the major specifications. Additional issues that have not been fully addressed by the standards bodies are discussed along with their potential effects on next generation DSL services.

References

[Chen 1991] W. K. Chen, *Active Network Analysis*. Advanced Series in Electrical Engineering—Vol. 2, World Scientific, River Edge, New Jersey 1991, p. 641.

[Chen 1995] W. K. Chen, Ed., *The Circuits and Filters Handbook*. CRC Handbook published in collaboration with IEEE, 1995, p. 2601.

[Cook 1995] J. Cook and P. Sheppard, *ADSL and VADSL Splitter Design and Telephony Performance*. IEEE Journal on Selected Areas in Communications, Vol. 13, No. 9, December 1995.

[Cook 1997] J. Cook, *Two Port Signalling Voltages Filter Arrangement*. US Patent 5623543, April 22, 1997.

[Dedieu 1996] H. Dedieu, C. Dehollain, M. Hasler, and J. Neirynck, *Filtres Électriques*. Presses Polytechniques et Universitaires Romandes, Lausanne, 1996, p. 434.

[Dedieu 2004] H. Dedieu, T. Fernandez, P. Golden, G. Nallatamby, J. Sevenhans, P. Reusens, J. P. Bardyn, E. Moons, and F. Krummenacher, *Seventh Order Low Voltage CMOS POTS/ADSL Splitter for DSLAM Size Reduction*. ISSCC (International Solid-State Circuits Conference), February 2004, San Francisco, paper 22.4, pp. 408–409.

[Golden 2006] P. Golden, H. Dedieu, and K. S. Jacobsen, Eds., *Fundamentals of DSL Technology*. Auerbach Publications, Boca Raton, Florida, 2006, p. 457.

[Kiko 1999] J. F. Kiko, *Impedance Blocking Filter Circuit*. Patent US 6,187,177, Filed on August 9, 1999.

[Kiko 2001] J. F. Kiko, *High Performance Micro-Filter and Splitter Apparatus*. Int. PCT WO 03/045043 A1, Filed on November 14, 2001.

[Kiko 2002] J. F. Kiko and C. Watts, *Controlled Inductance Device and Method*. Int. Patent US 7,009,482, Filed on September 17, 2002.

[Krummenacher 2002] F. Krummenacher, C. De Raad Iseli, G. Nallatamby, and H. Dedieu, *Application-Specific Integrated Circuit for a Low-Pass Filtering Device used for Decoupling xDSL Channels*. Patent WO03085829.

[Rayer 2000] A. Rayer, A. Bencivengo, and V. Durel, *Low-Pass Filtering Device with Integrated Insulated and Private Installation Comprising Same*. Patent US 6,980,645, Filed on April 21, 2000.

[Sackinger 2006]   E. Sackinger, A. Tennen, D. Shulman, B. Wani, M. Rambaud, D. Lim, F. Larsen, and G. S. Moschytz, *A 5V AC-Powered CMOS Filter-Selectivity Booster for POTS/ADSL Splitter Size Reduction*. ISSCC (International Solid-State Circuits Conference), February 2006, San Francisco, paper 28.1, pp. 512–513.

[Sedra 1978]   A. S. Sedra and P. O. Bracket, *Filter Theory and Design: Active and Passive*. Matrix Publishers, Inc., Beaverton, Oregon, 1978, p. 785.

[Williams 1988]   A. B. Williams and F. J. Taylor, *Electronic Filter Design Handbook—LC, Active, and Digital Filters*. McGraw-Hill Publishing Company, 1988.

[Zverev 1967]   A. I. Zverev, *Handbook of Filter Synthesis*. John Wiley & Sons, New York, 1967.

[CCITT Blue Book]   *Models for Predicting Transmission Quality from Objective Measurements*. CCITT Blue Book, Vol. V, P-Series, Supplement 3, 1988.

[EN 300 0001]   European Standard (Telecommunications Series) *Attachments to Public Switched Telephone Network (PSTN); General Technical Requirements for Equipment Connected to an Analogue Subscriber Interface PSTN*. V1.5.1 (1998-10).

[G.992.1]   ITU Standard G.992.1, *Asymmetrical Digital Subscriber Line (ADSL) Transceivers*, Annex E—*POTS and ISDN-BA Splitters* (pp. 198–219), June 1999.

[G992.2]   ITU-T Recommendation G.992.2, *Splitterless Asymmetric Digital Subscriber Line (ADSL) Transceivers*, June 1999.

[SIN 346]   Suppliers' Information Note for the BT Network, *BT ADSL Interface Description*, SIN 346, Issue 2.5, August 2004.

[SIN 227]   Suppliers' Information Note for the BT Network, *CDS^{TM} Calling Line Identification Service—Service Description*, SIN 227, Issue 3.4, June 2004.

[SIN 242]   Suppliers' Information Note for the BT Network, *CDS^{TM} Calling Line Identification Service—Part 1, Idle State, Down Stream Signalling—Part 2, Loop State Signalling*, SIN 242, Issue 2.3, January 2004.

[TR102 139]   ETSI TR 102 139 v1.1.1 (2000–2006), *Compatibility of POTS Terminal Equipment with xDSL Systems*.

[TS 101 952-1-1]   ETSI TS Technical Document 101 952-1-1, V1.1.5 (2003–2004) *Access Network xDSL Transmission Filters; Part 1: DSL Splitters for European Development; Sub-part 1: Specification of the Low Pass Part of the DSL over POTS Solution*.

[TS101 952-1-5]   ETSI TS 101 952-1-5 v1.1.1 (2003–2005), *Access Network xDSL Transmission Filters; Part 1: ADSL Splitters for European Deployment; Sub-part 5: Specification for ADSL over POTS Distributed Filters*.

[T1.413-1998]   ANSI T1.413-1998, *Network and Customer Installation Interfaces—Asymmetric Digital Subscriber Line (ADSL) Metallic Interface*.

[T1.421-2001]   ANSI T1.421-2001, *In Line Filter for Use with Voiceband Terminal Equipment Operating on the Same Wire Pair with High Frequency (up to 12 MHz) Devices*.

Chapter 2

DSL Specific Integrated Circuits: A Silicon Perspective

Damien Macq

CONTENTS

Abstract   This chapter is developing the challenges associated with the design of Asymmetric Digital Subscriber Line (ADSL) integrated circuits (ICs). After a brief overview of the last fifteen years of silicon developments, the key design challenges are highlighted by extracting system architecture and building blocks requirements from the high-level system specifications and the ADSL Standards. A compilation of implementations options published in the literature is provided. Chipset roadmaps are developed, demonstrating the leverage of the design and process advances on the product performance, power, cost, and features. Both ends of the line as well as other product derivatives like remote digital loop carrier (DLC) and other Multiunit product derivatives (MxU) are considered.

2.1   Fifteen Bumpy Years of DSL IC Developments

Using Digital Subscriber Line (DSL) technology to exploit the bandwidth of the telephone line well beyond a few hundred kilohertz was already a very hot R&D topic by the end of the 1980s, with the very aggressive but not yet fully articulated target to break the barrier of a few megabits per second on the local loop. This would suddenly transform the already installed telephone twisted pairs into a global broadband access network, paving the way for a completely new panoply of consumer businesses in areas such as video-on-demand, remote games, and video-conferencing. Many people predicted that this would result in a second life with a golden future for the lucky owners of this old but broadly deployed copper network. Others predicted that this would result in another failure following the unequal worldwide success of ISDN. At that time, it was obvious that whichever DSL standard or application would eventually emerge, a massive worldwide deployment of the technology would largely rely on the successful design of a brand-new family of custom integrated circuits (ICs). These custom ICs would have to explore new design techniques and take the maximum benefits from leading edge processes, to meet all cost, horsepower, and power consumption targets of this new product.

Looking back to the very first integrated solutions available from the mid-1990s [Amrany 1992], [Kuczynski 1993], [Chang 1995], [Macq 1998], and [Reusens 2001], they appear today quite ugly, in most cases implementing limited and proprietary flavors of the standard. The square centimeters and watts required for the digital chips were forcing designers to look for unusual and expensive chipset partitioning using custom packages, while the analog chips were pushed to operate too close or beyond the limits of what was feasible in the best process available at that time in terms of linearity, resolution, and power dissipation. These early solutions, quickly optimized in terms of cost, power, and standard compliance, were mostly developed by rather big incumbent telco providers and produced in relatively small quantities. Their biggest merit was probably to help to demonstrate the intrinsic capabilities and the relative robustness of the new technology in various field trials during the mid-1990s, mostly in the United States and Asia. Fortunately, for these young DSL product precursors, a killer application rapidly emerged at the same time; the unprecedented success of Internet suddenly boosted the demand for higher bandwidth up to the last mile, thus fueling capital for massive R&D efforts from the mid-1990s. From 1997, new chipset families appeared on the market, offering full interoperability and good performance at a reasonable cost and within the very tight power budget [Kiss 1999]. New IC players entered the market at that time with standard compliant ADSL discrete multitone (DMT) products. The following years were truly glory days for some ADSL IC players, with spectacularly successful IPOs, vertiginous stock valuations for some fabless semiconductor companies, and revenues multiplied by 600 percent or even 1000 percent on a year-to-year basis. No one could accurately predict where and how the huge DSL chipset orders were supposed to be deployed, and even how many technicians would be required in the field to support this deployment. Any demand forecast was systematically beaten quarter after quarter. The bubble finally burst at the end of 2000. From that time, a rather drastic consolidation occurred for the numerous IC companies still alive on the DSL playground. Nonetheless, more recent times have seen emerging DSL markets in Asia (first in Japan, Korea, and later in China) and new product requirements leading toward higher speed (from ADSL2+ to VDSL) and longer reach. These disruptive market evolutions are opportunities for a few top DSL IC players to prosper again.

This chapter presents some highlights and issues associated with the design of ADSL ICs. In the first part, some orders of magnitude in terms of horsepower, linearity, and signal-to-noise ratio are extracted from the system specifications, leading to a few considerations regarding the chipset requirements, mostly for the digital subsystem (the analog is covered in Chapter 3). In the second part of this chapter, various implementation options available in the open literature are briefly described and compared. Finally, the major trends in the DSL chipset integration are discussed, with emphasis on product, voltage, and technology roadmaps for the ICs. Both ends of the line will be considered: the central office (CO) side (covering as well other product derivatives like remote digital loop carrier (DLC) and other Multiunit product derivatives (MxU)) and the customer premises equipment (CPE) side.

2.2 A Preliminary Chipset Complexity Assessment from System Specifications

A quick evaluation of the main challenges associated with the design of a DMT ADSL modem can be extracted from a first overview on the high-level system specifications. The Shannon criterion dictates that the bit-rate increase, necessary to break the megabits per second barrier, requires extending the signal bandwidth well beyond the 4 kHz of a conventional voice band modem. The first DMT ADSL modem utilizes bandwidth up to 1.1 MHz. For illustrative purposes, a DMT ADSL modem could be viewed as the parallel combination of 256 narrowband modems regularly spread along the frequency axis between DC (direct current) and 1.1 MHz, each one centered around a carrier frequency of $n \times f_{BW}$ with f_{BW} equal to 4.3125 kHz and n an integer taking values between 0 and 255. Each carrier is quadrature amplitude modulation QAM-modulated with a maximum of 15 bits per symbol, i.e., 60 kbps/carrier for a 4 kHz symbol rate. This frequency multiplexing of 256 narrowband modems could lead theoretically to an aggregate upstream and downstream bit rate of 256×60 kbps, i.e., close to 15 Mbps.

Figure 2.1 illustrates the major building blocks of the digital portion of an ADSL DMT modem; further details on the functions of each of these blocks

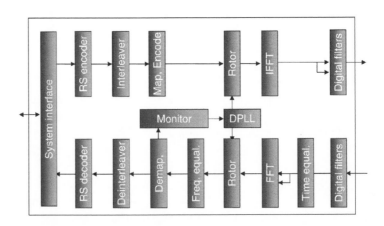

Figure 2.1   ADSL DMT block diagram.

may be found in chapter 7 of [Golden 2006]. In the transmit direction, the data bytes coming from the system interface are first scrambled, then encoded by the Reed–Solomon engine, interleaved and mapped through a 16-states Wei trellis encoder onto the appropriate QAM constellation point. The different QAM-encoded tones are then multiplied by the gain scaling coefficient, before being sent to the IFFT block. A cyclic extension is added to the symbol in the time domain. This signal is then processed by a combination of interpolators and filter stages to satisfy out-of-band power spectral density requirements in the POTS band (digital high-pass filter) and in the receive band (digital high-pass filter or low-pass filter), before being sent to the Digital to Analog (D/A) converter.

On the receiver side, after Analog to Digital (A/D) conversion the signal is first decimated and filtered to eliminate the quantization noise as well as any other undesired out-of-band component. In the case of echo-cancelled systems or systems with a reduced guard band between the transmit and the receive channels, a part of the downstream signal overlaps within the upstream channel. An echo cancellation block is then required to eliminate the residual echo of the transmit signal, which is already partially attenuated in the analog block by the hybrid and some other analog echo cancellation mechanisms. This echo cancellation block is often split into two components implemented respectively in the time and in the frequency domains. A time domain equalizer shortens the channel impulse response, and then the signal goes through the Fast Fourier Transformer (FFT). After FFT conversion and elimination of the cyclic prefix, a frequency domain equalizer (FEQ) tunes the different carriers in phase and amplitude, compensating for the dispersion owing to the telephone line and the analog components. A rotor may compensate as well for residual frequency mismatch between the clocks references of the modems connected at both sides of the line. The demapper, combined with a trellis decoder, makes a decision on the constellation points, and the retrieved data bytes are then sent to the deinterleaver, following which the Reed–Solomon decoder decodes the codeword and the resulting bytes are finally descrambled before being sent back to the system interface.

2.2.1   *Digital Signal Processing Requirements*

A rough analysis of the digital signal processing (DSP) processing requirements to support the ADSL datapump operations can be estimated by looking at the multiplication operations and evaluating the aggregate number of Million Multiplications–Accumulations Per Second (MMACS) required for each submodule [Cioffi 1993], [Wiese 2000], and [Naveh 1999]. Relying on optimized hardware design and efficient software coding style, a simplistic assumption is that there is enough parallel processing, pipelining, and auxiliary resources to keep each hardwired multiplier fully loaded, in such a way that each clock cycle allows a complete multiplication operation. This simplification has obviously its own limitations. For example, it will not provide a very accurate estimation of highly parallel operations required for a Reed–Solomon encoder and decoder or for the Viterbi algorithm. Also, straightforward architecture optimizations will often be applied to relax the MACS requirement of some fixed and dedicated blocs. For example, interpolation or decimation filters with limited programmability requirements will preferably be implemented without multipliers, by selecting properly the coefficients in such a way that only add and shift operations are needed.

That being said, analyzing the remote side can approximate a worst-case computation requirement. Indeed, from the system asymmetry the remote receiver handles eight times more traffic than the CO receiver, and the bulk of the transceiver complexity resides essentially on the receiver, because it has to handle difficult tasks like channel equalization and echo cancellation. We will look at a generic echo-cancelling ADSL system, operating at 8 Mbps downstream and 800 kbps upstream, on a 1.1 MHz bandwidth and with a 4 kHz symbol rate. The same method could be applied to any other DSL flavor (ADSL2+, SHDSL, VDSL, etc.) to compare the intrinsic DSP complexity of these systems.

2.2.1.1 Filters, Decimation, and Interpolation

The data processing required between the A/D–D/A converters and the DMT engine is very dependant on the overall transceiver architecture, especially the type of converter used, and the amount of filtering already provided by the analog front-end (AFE). We will assume that the system uses a Nyquist A/D converter, relying on an oversampling rate of four to relax the analog filtering requirements, leading to a sampling rate of 8.8 MHz. The filters and decimation by a factor of four could be implemented as in [Kiss 1999] with two sinx/x finite impulse response (FIR) filters, with 15 and 59 coefficients, respectively. Each filter also decimates the signal sampling frequency by two, leading to a total requirement of around $15 \times 8.8 \times 10^6 + 59 \times 4.4 \times 10^6$, or 391 million operations per second. In our case, it can be assumed that each operation leads to a single 16-bit shift-add operation, and we can calculate the equivalent 16-bits MACS operation as 391,000,000/16 or around 25 MMACS. The data processing on the transmit direction of the remote side involves interpolation and some filtering to reduce the Inverse Fast Fourier Transform (IFFT) sidelobes out of the upstream band. Because of the reduced bandwidth on the upstream channel, the amount of processing required is obviously lower; a rough estimate is 5–10 MMACS.

2.2.1.2 TEQ

A time domain equalizer (TEQ) is required in a DMT ADSL system to attenuate the intersymbol interference (ISI) resulting from channel dispersion. Channel dispersion in the ADSL band results in an impulse response generally much longer than the cyclic prefix inserted between the DMT symbols. The TEQ will shorten the effective impulse response in such a way that most of the interference is included within the cyclic prefix. At the remote side of the DSL, a typical TEQ is implemented by an adaptive 32-tap FIR filter running at the Nyquist frequency of 2.2 MHz, and will therefore require up to 2.2 × 32, or approximately 70 MMACS.

2.2.1.3 FFT and IFFT

The 2N-point real to complex FFT required for an N-carrier DMT symbol can be computed from an N-point complex to a complex FFT, followed by a post-processing stage, leading to $N/2 \times \log_2(N)$ complex multiplications for the FFT and $N/2$ other complex multiplications for the post-processing stage per DMT symbol. Note that the multiplier size is often larger than 16 bits, typically it is 20 bits for the data and 16–18 bits for the twiddle coefficients. For the FFT in the CPE receiver, this leads to a total of $2N \times [\log_2(N) + 1]$ multiplications per symbol, or 18 MMACS for a 4 kHz symbol rate and 256 carriers. The IFFT will process a reduced number of carriers and will only require from 2 to 4 MMACS.

2.2.1.4 FEQ and Rotor

An FEQ is used to align the received carriers (representing constellation points) on the X and Y axes of the complex plane, compensating for carrier-specific channel distortion. N complex multiplications are required, or a total of 4 MMACS (250 carriers, 4 real multiplication per carrier). A rotor can be used as well to compensate for the misalignments between the received signal frequency and the local oscillator. This subsystem consists of a feedback loop that includes the FFT and the demapper, and will closely monitor the phase of a dedicated tone selected as pilot. A measured phase error is calculated, and fed into an integrator filter whose output is used as a measure of the phase error to apply to each tone. This operation will again require N complex operations, or an extra 4 MMACS. On the transmit side, gain scaling is provided, requiring the multiplication of the 30 tones by a real number. Because of the lower bandwidth in the upstream direction this last component can be neglected.

2.2.1.5   Echo Cancelling

A digital echo canceller, combined with an analog hybrid and some high-pass filtering in the case of partial frequency division multiplexing (FDM), can provide an overall echo attenuation beyond 60 dB. This function may be split between the frequency domain and the time domain, as demonstrated in [Naveh 1999]. The frequency component requires N complex multiplications per symbol, or 4 MMACS for 250 carriers, while the circular echo synthesizer [Naveh 1999] is implemented as an FIR filter and requires around 7000 MAC per symbol or 28 MMACS.

2.2.1.6   Transmit and Receive Forward Error Correction
Blocks, Viterbi Decoder, and Adaptation of the Filters

The modem is completed with the scrambler and descrambler, interleaver and deinterleaver, mapper with a Wei–Treillis encoder and demapper combined with a Viterbi decoder. The aggregated million instructions per second (MIPS) or MACS required for these operations is much more implementation dependant and difficult to estimate because they do not rely principally on multipliers as was the case for the previous blocks. Because a deeper architectural analysis of these blocks would go far beyond the scope of this simple complexity evaluation, we will simply double the aggregate horsepower requirement to cover these mandatory features of the modem, as suggested in [Wiese 2000].

2.2.1.7   Discussion

Table 2.1 summarizes the different "equivalent MMACS" estimates for the remote modem. A total of about 320–400 MMACS is required for the entire modem, from the A/D–D/A converter up to the system interface. A similar estimation for the CO modem would lead to a lower estimate of 250–300 MMACS. These numbers should only be used to get a first indication of the intrinsic capabilities of a given hardware platform.

Table 2.1   Computing Requirements ADSL Modem, in Million Multiplications–Accumulations per Second (MMACS)

	ADSL CPE	ADSL CO
Digital filters	35	35
Time equalizer	70	35
FFT/IFFT	25	25
Frequency equalizer	10	5
Echo cancellation	30	20
Viterbi, Reed–Solomon, TC-layer	~150–230	~130–180
Total	**~320–400**	**~250–300**

At the time of the first ADSL technology developments in the early 1990s, there were no DSP cores powerful enough to provide that level of performance. Also, leading-edge cell-based application-specific integrated circuits (ASICs) implemented with latest Complementary Metal–Oxide–Semiconductor (CMOS) technology with 1.2–0.7 µm gate lengths would only run at frequencies close to 50 MHz, requiring a high level of parallelism to provide the processing power required per DSL channel. A three to four chip solution was therefore typically envisaged from the AFE interface up to the system interface. These early ADSL implementations were often derived from complex and expensive hardware systems, built with a mixture of standard off-the-shelf components, field programmable gate-arrays (FPGAs) and standalone DSPs. This evolutionary process strongly influenced the initial chipset partitioning and architecture choices. A certain degree of programmability was obviously mandatory to track the ADSL standard evolutions, or (even more tediously) the various standards "flavors." The programmability was also often welcome at a time when verification methodologies and tools for this kind of complex system were generally much less mature than today, and not yet efficiently embedded in the computer-aided design (CAD) tool flows. A first family of products was assembled from a set of custom engines running in parallel and handling the various subsystems of the complete DSL modem, without relying on a conventional DSP core [Kuczynski 1993] and [Kiss 1999]. A quite representative example of this first approach is detailed in Section 2.3.1.2. Other products would be developed around efficient DSP cores that were available in-house. Dedicated hardware accelerators provided support [Amrany 1992] as illustrated in Section 2.3.1.1. One of the major challenges for the IC design teams in charge of this architectural migration is to continuously meet or surpass the next product generation requirements in terms of cost, power consumption, and performances and features. They must also preserve interoperability with the existing products and all previous generations already deployed in the field. This sometimes leads to tricky trade-offs on the level of reuse for existing blocs, architectures, and algorithms; to continue to guarantee interoperability with legacy products, their various firmware versions, and any other software component already developed by the customer on older versions of the product. The barrier for any new player entering this very competitive market thus becomes higher and higher. The fickleness of existing product acceptance in the field necessitates programmable architectures, without any sacrifice in cost or power to quickly adapt the solution for interoperability with all deployed legacy solutions. It is also necessary to be able to present a convincing roadmap with new features and product differentiations in terms of reach and aggregate bit rate. To illustrate this, the solution proposed by a more recent DSL player is briefly discussed in Section 2.3.1.3 [Clarke 2001] and [Wilson 2002].

As of today, numerous solutions for the implementation of the digital part of the ADSL system are deployed in the field. A fair comparison between all of these architectures is close to impossible and would be highly controversial. The success of a given solution or supplier is no longer measured only in terms of the technical performance of a modem. It also strongly depends on the overall business strategy of the company and its capability to react appropriately to a rapidly evolving and demanding market. The DSL modem or datapump is becoming more and more of a subset, a feature embedded in a complex system combining voice, data, cells or packet processing, control, and management. Leading DSL IC players are required to be able to propose global solution (IC, hardware, and software) for a DSLAM at the CO or an all-in-one home gateway on the CPE side (e.g., DSL in voice, wireless, and data networking out).

2.2.2 Analog System Requirements

A detailed analysis of the technical challenges for a DSL AFE is presented in Chapter 3. The goal of the designer is to build a flexible and adaptive engine capable of digitizing (in both transmit and receive directions) the quite unfriendly DMT signal in a totally transparent way. Specifically, this entails having intrinsic noise and distortion levels significantly below the telephone line noise on nearly all possible line conditions for bandwidths on the order of megahertz. As illustrated in [Chang 1995], the DMT-based ADSL standard provided a very tough challenge for the first integrated AFE (Figure 2.2), designed back in the early 1990s. On the receive side, the combination of the dynamic range requirements for the very high crest factor of the DMT signal, the high signal-to-noise ratio requirements per DMT carrier, and the amount of echo still present at the A/D input translate into a very stringent requirement for the A/D and D/A converters. A resolution of 12–14 effective number of bits is required, this is an accuracy usually achievable only with sigma–delta converters. A sampling frequency above 4.4 or 8.8 MHz and a the signal bandwidth of 1.1 MHz is more in line with the speed of pipeline or subranging architectures, usually reaching 10–11 bits in production (without calibration). On the transmit side, regardless of which multiplexing method is selected by the system engineer, highly linear filters with up to 60–70 dB are required. Spurious free dynamic range (SFDR) is required to eliminate out-of-band signal or noise components. Additionally, these filters are constrained not to add excessive internal thermal noise components. Continuous-time approaches have been preferred so far over switched-type filters owing to potentially extended signal bandwidth. Finally, a low-noise highly linear line driver is needed, capable of

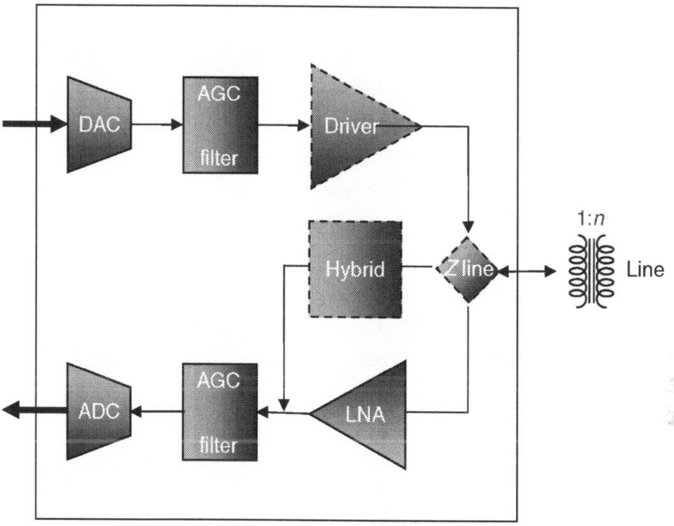

Figure 2.2   ADSL analog front-end generic block diagram.

efficiently transmitting up to 100 mW of power at the CO side of the link, while accommodating the impressive DMT peaks standing 15 dB above the root mean squared signal.

2.3   Chipset Implementations

These next sections present and try to compare different solutions proposed in the literature, with some circuit details, and advantages and drawbacks perceived from these architectures for implementations at both ends of the line. General trends in the design of broadband communication products were highlighted by [Samueli 1999] and [Cloetens 2001]. At the early deployment stage of a technology, parallel dedicated architectures (custom hardwired) are often preferred over centralized programmable solutions (DSP core-based solutions). This is due to the very large computational requirements of most of these systems. The amount of parallelism and programmability planned for a given chipset generation often results from a delicate trade-off between the additional power required for highly centralized and programmable engines and the cost penalty for the larger area expected from more parallel systems. This trade-off evolves from one process to the next, tending over time toward more and more programmable architectures and relying on a reduced set of hardware coprocessors. This is a very welcomed consequence of Moore's law (Figures 2.8 and 2.9) that mainly impacts the digital parts of the DSL modem. As a side effect, the system cost, area, and power become dominated by the analog part of the chipset or other analog discrete components; the digital cost and power eventually becomes insignificant in the complete solution. Another dimension to explore during the definition of an optimal chipset partitioning is the opportunity for a complete integration on a single die. In the case of DSL, high resolution analog functions have to be integrated with complex digital circuitry that is switching at very high-clock frequencies. For cost reasons, a pure digital CMOS process will often be preferred for this mixed-signal solution at the expense of options like double poly for capacitors. For the same cost reason, the design of the analog components integrated with a complex digital subsystem will have to be extremely robust as any parametric failure owing to analog performance variation across temperature, voltage, and process range will impact negatively on the production yield. The analog characteristics do not scale very well with the process; for the high sensitivity requirements of DSL products, the analog components size is ultimately determined by the amount of thermal noise tolerated in the system or in some cases by the capacitor matching. Also, the lower power supply required by the smaller geometries of advanced digital process reduces the overall dynamic range of the sensitive analog portion of the chip, and impacts negatively on the power consumption. The lower power supply also jeopardizes the future integration of higher voltage components such as the line driver if no additional process options can be considered (e.g., dual gate oxide). Finally, there are potential silicon re-spins owing to tedious second-order analog problems that are sometimes difficult to apprehend, simulate, or even to validate. These may end up penalizing heavily the overall product strategy success by delaying the start of production and dramatically increasing the development cost. The latter is largely owing to the prohibitive mask set price below 0.18 μm CMOS geometries. All these considerations will often justify the decision to keep the analog portion of the ADSL system on a separate die. This allows the selection of the most appropriate analog process in terms of cost, performance, and stability. This will also normally offer mature noise, distortion, and matching models. A System In Package (SiP) approach that consists of a combination of multiple dies and discrete components on a single package can often offer an attractive alternative in the quest for a single chip DSL solution.

2.3.1 Digital Signal Processing Examples

The first category of DSL IC solutions was the result of the long maturing of the very early solutions that took place over more than five successive chipset generations. A second category of more recent chipsets tries to take the maximum benefits from the latest DSP architecture evolutions toward very long instruction word/single instruction multiple data (VLIW/SIMD) engines to offer the higher flexibility required from a new entrant. This entails supporting any legacy product already deployed and also offering an attractive migration path toward enhanced product in terms of reach, speed, or multistandard support. The chipsets from the first category are based on either custom DSP engines [Amrany 1992] that are capable of handling one or two DSL channels, or on a set of dedicated hardwired processors [Kuczynski 1993] and [Kiss 1999] that are being set up during initialization through status registers and controlled globally by a medium-performance microcontroller. These two examples illustrate how massively parallel architectures can lead to power-efficient solutions, driven at fairly low-clock speed (lower than or close to 100 MHz), while preserving enough programmability to support the necessary platform evolutions. More recent solutions built around centralized powerful DSP engines will run at much higher speed (higher than 500 MHz in 0.13 or 0.09 μ CMOS) and are capable of handling multiple DSLs on a unique hardware platform.

2.3.1.1   A Flexible and Distributed DSP Approach

As early as 1992, Amrany et al. [Amrany 1992] presented quite an innovative way to reach the required horsepower of a DSL modem without sacrificing the programmability of the solution. The idea is to build a "single fully programmable application-specific fast-signal processor" based on a hierarchy of custom programmable processors. This hierarchy of processors, the so-called General Adaptive FIR Filter (GAFF), is capable of handling the full duplex operation for early carrierless amplitude and phase (CAP) versions of ADSL, including echo cancellation, decision feedback equalization, and Tomlinson precoding. For more complex processing, multiple GAFF processors are connected through a time division multiplexed bus allowing complete system and interconnect reconfiguration. At the bottom of the hierarchical engine, a set of three LIW/SIMD engines, called the FIR engines (FE), perform highly parallel operations such as complex multiplication or coefficient updates. Above the FEs, an FIR processor defines the sequence of operation handled by all FEs. A binary processor feeds the data in and out of the FEs, performs binary operations such as data slicing, and defines the connectivity to the time division multiplexed (TDM) bus. Finally, an external processor interface is used to manage the real-time control, code downloads, and any type of monitoring. The total transistor count was 240 K. In a CMOS 0.9 μ process, the chip ran at 33 MHz and the set of three FE executed 100 MMACS, with 30 million additional MACS supplied by the binary processor. This early system supported, for example, a 190-tap echo canceller with 100 percent update rate at 257 Kbaud symbol rate.

2.3.1.2   Custom Programmable Processor

In [Kiss 1999], Kiss et al. presented "SACHEM," a single digital component capable of handling the complete digital processing of the DMT modulation scheme and the transmission convergence layer. The same engine is used for both CO and remote applications. Overall, SACHEM is assembled from a set of dedicated hardwired engines. It has some level of programmability to handle all different modes of operations required at the CO or the customer premise side, and enough observability with access to local memories and test loops around multiple engines. The massively parallel component requires a rather low-clock frequency to run the entire ADSL modem, and the distributed structure with local memories and local processing limits the power that would be wasted driving the longer buses of a more centralized architecture. Finally, lowering, for example, the supply voltage below the nominal values of the process can further reduce power consumption.

Connected to the AFE, a first engine called a "DSP interface" handles decimation, time equalization, and interpolation. Most of the filters in this block are built from fixed structures with possible by-pass and programmable coefficients to handle different sampling rate options, supporting CO or customer premises operation. The DSP interface handles sampling rates up to 8.8 MHz in the AFE.

The core of the modem is built around a second engine, handling the FFT, rotor, frequency equalization, and fine gain tuning functions. This block is a programmable machine with its own instruction set, and a dedicated pipeline multiplier-accumulator arithmetic logic unit, with two 18×20-bit fixed point multipliers. Rotors are added to compensate for the misalignments between the received signal frequency and the local oscillator on the receive side, and in the transmit direction to adjust the desired transmit frequency. Other dedicated engines handle the constellation decoding with a Viterbi decoder, the mapper and demapper, some monitor functions to calculate coefficient updates for the frequency and time equalizers, and a digital phase locked loop (PLL). Dedicated engines also handle the TC (transmission convergence)-layer after the demapper with rectangular interleaving, the Reed–Solomon decoder, some hardwired scramblers and descramblers, and finally, all ATM-TC operations (with provisions for bit error rate (BER) measurement). To design such a custom, dedicated, and optimized engine with enough confidence to get it right the first time, advanced high-level modelling approaches were defined that combined C++, behavioral VHDL (Very high speed integrated circuit Hardware Description Language), and gate-level simulations. These models were used within a complete simulation environment, relying on behavioral, gate-level, or the mapped database on a hardware emulator. The same simulation environment also allowed the software engineers to be involved very early with the system validation and the overall software development, well before the chip was sent to the silicon foundry.

2.3.1.3 A Centralized Architecture: Single DSP Approach

In [Wilson 2002], Wilson presents the architecture of FIREPATH^{TM}, a custom DSP processor used at the core of a 12-channel DSL digital transceiver for CO applications. This DSP core was originally targeted at bulk data processing, for applications where large amount of data is acted upon repeatedly by the same algorithm. The overall structure is based on two identical 64-bit data paths running in parallel, with 64-bit general purpose register sets and associated 8-bits predicate registers. The engine is a LIW/SIMD machine capable of processing 128-bit instructions. Operations in the SIMD data paths are predicated on a byte-by-byte basis, allowing the machine to conditionally execute operations on individual bytes without branching. The two paths share a set of general-purpose registers and each path has a dedicated set of multiply-accumulate registers. The DSP could run up to 8 × 16-bit MAC operation per clock tick (7 cycle multiply), while a deep pipeline (7 stages) allowed the issuing of MAC instructions back-to-back. Specific instructions were defined for communications algorithms such as Galois field arithmetic. The orthogonal simple two-way long instruction simplifies the task of the compiler, and a C-code compiler was developed for this engine. At 500 MHz, the DSP provides up to 4 million 16-bit MACS. A 256-point complex FFT can be performed in 1290 cycles. The design is fully captured in register transfer level (RTL), 100 percent cell-based, assembled with an in-house standard cell library, and has partial pre-placement and routing.

2.3.1.4 Comments

The previous examples illustrate the wide diversity in digital implementations, from fully programmable [Amrany 1992] and [Wilson 2002] to more customized [Kuczynski 1993] and [Kiss 1999], or from low frequency distributed [Amrany 1992], [Kuczynski 1993], and [Kiss 1999] to high-speed centralized [Wilson 2002] solutions. Once an architecture is developed and validated in the field, there is a strong reluctance to further platform evolutions, as minor hardware changes may sometime heavily impact the software and firmware. There is less variety on the analog side, eventually leading to more similarity amongst the AFE architectures from different vendors after two or three generations.

 Some quite generic comments can nevertheless be made about the three very different digital architecture examples presented previously. Obviously, programmability is a must: chips already deployed in the field have to interoperate with new generation products from numerous vendors that offer extra features in terms of reach and maximum bit rate. Hence, sufficient flexibility for firmware upgrades is always valuable and necessary. For CO applications, a highly centralized architecture built around a single DSP (as presented in [Wilson 2002]) is supposed to be cost effective compared to a parallel solution using distributed DSP or other custom engines, because the same hardware and program code can be shared across multiple channels. Additionally, bigger memory blocks can be used, leading overall to area-efficient memory sizes. In these architectures, the level of integration expressed in terms of number of DSL channels per chip is directly proportional to the number of MACS delivered by the DSP core. As discussed earlier, because around 300 MACS are required per DSL at the CO, a single DSP core such as FIREPATH^{TM} clocked at 500 MHz could handle from 5 to 10 channels in parallel, depending on the overhead required for the various breaks in the pipeline, the operation and management of the modem, and the higher level protocols. This race for the higher speed on a single platform drives a rapid design migration to the latest CMOS technology, which unfortunately is both expensive and immature for early productions. Power consumption may become a problem because faster versions of advanced processes may present excessive leakage currents, especially in the high-temperature environment of a DSLAM. Fortunately, the digital power becomes negligible in the aggregate power consumption per channel. As discussed later in Section 2.4.1.1, digital power today represents close to 15 percent of the overall power dissipation of a DSL channel. Another potential limitation of a centralized architecture comes from reduced flexibility in cases when the same core has to run different DSL flavors on a per channel basis (e.g., a combination of "plain vanilla" ADSL, long reach ADSL, and higher speed ADSL2+ for three distinct channels on the same DSP, or even totally different flavors of DSL, like HDSL2 [High Bit-Rate Digital Subscriber Line with 6 db noise margin] or SHDSL [Single-pair HDSL]), as this would require the core to simultaneously load and manage different code images and any unexpected event on a given channel, impacting somewhat the pipeline loading and the overall DSP performance. A distributed system on the other end is by constructing a more straightforward approach to treat each channel completely independent of the others. For example, a robust reset strategy, with a hierarchy of independent reset signals per DSL channel, can be easily defined. Finally, centralized systems are by definition less easy to scale: the building of a CPE product, requiring only a fraction of the horsepower of the corresponding multiline CO chip, may lead to a totally different architecture. This limits reusability in terms of code, macro blocks, development tools, and product maturity.

2.3.1.5   Verification, Validation, and Testability

As pointed out in different papers [Kiss 1999] and [Jahner 2001], the system complexity of a DSL modem often requires the development of in-house custom verification environments covering the different levels of the design, from behavioral down to gate level, also including software co-design tools to allow other teams to be also involved with the complex verification process. The complexity of the verification engineer's job will increase dramatically when the system is assembled with different modules reused from previous designs and third parties. These can be developed under different environments with a wide variety of design methodologies, RTL languages (e.g., VHDL models in Verilog flows), and abstraction level (synthesizable netlist mixed with hard macro). A hierarchical approach may be adopted, allowing the abstraction of the different reused blocks as a set of interconnected black-boxes and focusing the system verification process on the higher level interconnections. The number of test case scenarios will still explode at system level: up to 1000 test cases to be covered are reported in [Jahner 2001]. Hardware emulation tools, although quite expensive and sometimes difficult to set up, are used to help reduce the verification time by a factor of up to 100 compared to behavioral simulations, shortening a complete verification cycle from four months to less than four days [Jahner 2001].

To allow the automatic test of such a complex system on a chip at a reasonable cost, a combination of design-for-testability methodologies can be considered. These include conventional scan inserted with automatic test pattern generation (ATPG) tools, boundary scan, and also a set of functional simulations for third party macros or any speed-sensitive part of the design not completely covered by a scan path (because the scan insertion could negatively affect the performance). The availability of DSP core and on-chip microcontrollers may be exploited to develop fairly complex test cases to cover on-chip memories or other macros.

Interaction between the system architects, who are expected to provide complete and crystal clear specifications, and the logic designers, who are supposed to generate 100 percent error-free blocks and interfaces, is always critical and potentially subject to a lot of misunderstanding. For difficult parts of the design, loop backs, backup modes of operation, backdoor mechanisms to access the different memories or processors, and by-pass modes should always be considered. With advanced digital CMOS processes, the cost of the extra few gates required for most of these hidden hardware capabilities is totally negligible compared to the potential advantages they can provide. This can include the provision of a number of "mission-critical" new features that are sometimes required to be added to a product already deployed in the field. These capabilities will also increase the overall system observability, softening the validation job performed by the system, test, and firmware engineers.

2.4   Chipset Roadmap

2.4.1   CO and CPE Products

Despite the built-in system asymmetry, early ADSL chipsets were often designed in such a way that at least the most complex parts of the analog or digital chips could be reused at both sides of the link, with configuration registers defining the mode of operation [Chang 1995]. However, divergent product roadmaps were rapidly envisaged to support the different product architecture of CPE and CO equipment. The ADSL technology and other DSL flavors are now more and more integrated within feature-rich products combining data and voice traffic on ATM or IP, and are capable of interconnecting multiple access technologies inside and outside the house. Furthermore, they are frequently associated with complex multiplexing and aggregation functions in COs, remote digital loop carrier equipment, or multidwelling units for business or consumer applications. The roadmap complexity reflects the large variety of products relying today on broadband DSL technologies.

2.4.1.1   Central Office

On the CO side, the challenge is to concentrate on as many DSLs as possible in a confined environment, with typically less than 100 W available on a single line card. Depending on the architecture of the DSLAM, DLC, or MxU, various levels of system software features will also be required. These handle, mix, multiplex, and aggregate cells, packets, or any proprietary raw data traffic from all the DSLs. In some cases, POTS traffic could also be handled on the same line card. Aggregated data will be sent to the backplane or the uplink interfaces through high-speed serial interfaces. A line card processor with a dedicated or embedded control interface will manage the boot functionalities, monitor the DSL status, and report various line and traffic parameters (as well as different alarms and errors status) to the upper software layers. A line card today (Figure 2.3) may integrate from 48 up to 96 DSL channels and be combined with a multiplexing and aggregation subsystem interfacing with the uplink. A quite natural integration path is to combine all line card functions in a single chip: the highest number of DSLs, a cell or packet aggregator, and a line card controller. This chip should have the ability to access a sizeable amount of external memory to handle all

Figure 2.3   Portion of a CO reference design. (Courtesy GlobespanVirata.)

necessary packet or cell buffering requirements and would be connected to the AFE with an efficient proprietary interface that is optimized in terms of pin-count. The digital component will quickly migrate to the next leading edge CMOS process to get greater channel densities, reducing cost, and power. Integration between 8 and 24 DSL channels on a single chip was recently reported by major DSL chipset suppliers (Figure 2.5).

The focus is also on power and density on the analog side: to integrate as many low-power DSL channels as practical in a single device, permitting the sharing of some common functions (references, PLL, etc.). Present solutions may integrate all analog functions, with the exceptions of highly dissipative line driver, line transformer, and line protection. In a multiline, DSL codec between 4 and 12 analog channels are integrated on a single device. For particular systems mixing POTS and DSL on the line card, a mix between DSL and POTS subsystems could be investigated. This could lead to a totally different partitioning, with a combined POTS plus DSL AFE in a low-voltage analog CMOS process and a combined subscriber line interface circuit (SLIC) plus DSL driver using a high-voltage bipolar process [Zojer 2000].

CO chipsets may also offer multistandard capabilities, and more generally, firmware upgradeability to cope with the changing nature of the different ADSL standards, which are moving toward higher speeds and longer reach (ADSL, ADSL2+, migrations to SHDSL, VDSL, etc.). This sometimes requires significant levels of programmability in the AFE as well, to adjust filter bandwidths or sampling frequencies.

As of today, power dissipation is the barrier limiting the level of integration in most DSLAM architectures. This justifies special attention to different power reduction techniques. This is not only at the chipset level (especially around the line driver), but also at system level such as, for example, some advanced power management techniques for optimizing the aggregate line card power. Leading edge solutions today are consuming less than a watt per ADSL channel, with close to 75 percent of this power being consumed by the line driver (Figure 2.9).

Special attention should be paid as well to the reduction of the supply voltages: each different supply requires a dedicated supply brick on the line card, dissipating more power, and further limiting the achievable density. Leading edge chipsets will typically require only three supplies, one for the digital core (1.8· · ·1.2 V), one for the analog front-end and the digital interfaces (3.3 V), and a symmetrical supply for the line driver (±12 V). Figure 2.4 illustrates the line card density achievable today. Less than one

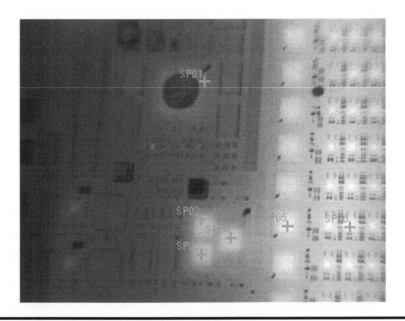

Figure 2.4   Hot spots on a DSL CO card: Infrared picture of a DSLAM board, clear spots indicate hot area, mostly around the line driver and to a lesser extent around the analog chips area (right side of the picture). (Courtesy GlobespanVirata.)

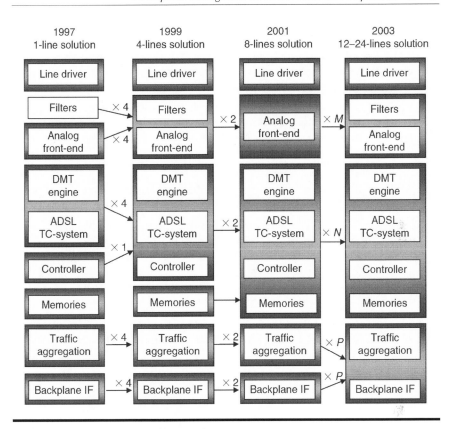

Figure 2.5   Generic ADSL CO roadmap.

square inch is required to hold a complete DSL, from line protection up to the ATM interface.

2.4.1.2 *Customer Premises Equipment*

On the CPE side, cost is driving the roadmap, leading to higher levels of integration and broad product portfolios to address the numerous consumer and professional markets. ADSL was once envisioned as a modem upgrade that was to be directly connected to (or even inside) the PC, an incremental (big) step after voiceband modems. A number of solutions combining ADSL and a 56 Kbps modem on the same peripheral component interface (PCI) board were even developed from 1998, and PC products pre-equipped with combo DSL/V90 modems are still sold today. ADSL chipsets with integrated PCI or universal serial bus (USB) interfaces are designed to address that particular market, and V90 capability may or may not be provided on the same platform. The success of this concept is somewhat limited, though. This is probably due to line provisioning and interoperability problems associated with early ADSL deployments. Today different network configuration allows the sharing of a single DSL pipe between multiple users inside the house or the office.

To target standalone DSL products, another chipset is needed. This contains the necessary software and hardware components to offer an Ethernet interface, with on-chip bridging or even routing capabilities and generic interfaces to other access technologies (wireless LAN, HomePNA, POTS, etc.).

The CPE analog front-end integrates nearly all analog functions, all filters, and the line driver: the 13-dBm power required on the line can be provided by a single 12 V or even 5 V device integrated on a CMOS or BiCMOS (bipolar CMOS) process. Protection, line isolation, and voltage scaling require close to 20 discrete components on the board, including the line transformer.

The ADSL modem on the CPE side will eventually become a single or dual chip solution with aggregate power consumption well below 1.5 W. Figure 2.6 shows a typical integration roadmap at the CPE side, while Figure 2.7 shows a typical example of a high-end ADSL CPE product.

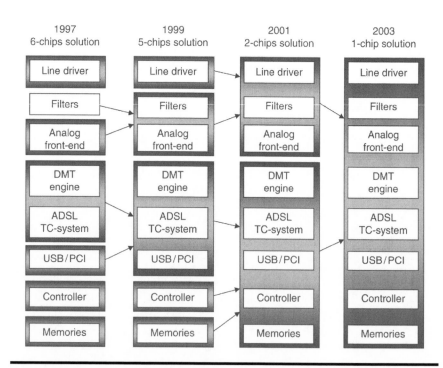

Figure 2.6   Generic ADSL CPE roadmap.

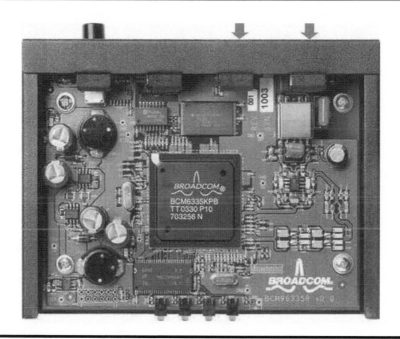

Figure 2.7 **Picture CPE products: Reference design for an ADSL router. (Courtesy Broadcom.)**

2.4.2 Power and Cost Evolution

As already discussed earlier, the power and the cost of DSL continues to drive further chipset enhancements. The very first solutions developed in the early 1990s typically relied on two impressive and power hungry digital chips implemented with a 0.7 μ CMOS process, an external controller, and an AFE. The system was counting on an impressive number of discrete components to complement the analog processing of the signal, especially in terms of filtering with quite expensive inductors and capacitors, and also with discrete low-noise amplifiers and line drivers. The overall power consumption was well above 5 W, and the aggregate silicon area was greater than 10 cm^2 for each DSL channel. Fortunately, DSL enjoyed the benefits of Moore's law, bringing impressive cost and power reductions for all digital components as shown in Figure 2.8. The analog parts were improved as discussed previously, though often in a less spectacular way. Currently, leading edge ADSL implementations require less than a watt on the CO side, with more than 70 percent of the power sunk by the line driver delivering 20 dBm to the line, and close to 1.5 W on the customer premise side, mostly dissipated by the integrated AFE because of the very high sensitivity required on the receiver side and the rather low efficiency achieved with integrated class AB line drivers.

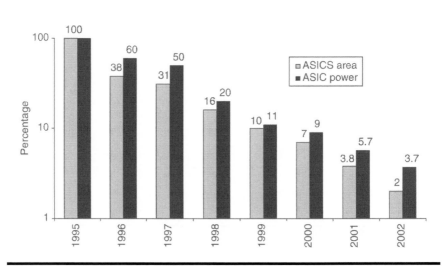

Figure 2.8 **ADSL IC cost and power evolution (compilation from various datasheets and publications).**

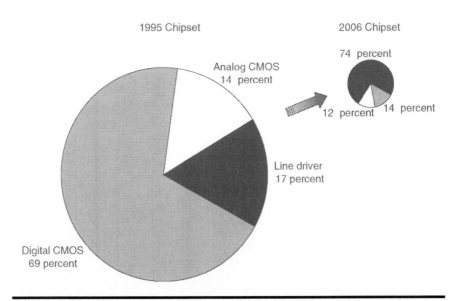

Figure 2.9 **Relative power consumption: Power for a single DSL channel, a 1995 generation and a 2006 generation (compilation from various datasheets and publications).**

2.5 Conclusions

Just at the end of the first decade of ADSL IC developments, designs have reached a great level of maturity in terms of cost, power, integration, and performance. A wide variety of DSL products is today deployed in the field, for both consumer or business applications, and with a large degree of interoperability between up to ten silicon vendors. Some growth is still being observed despite the long downturn in the telecom industry and the ADSL becoming a commodity market. This, of course, is not the end of the story. A new battlefield has emerged lately with the need for higher speed ADSL2+ and VDSL products. This is perceived for new entrants as an opportunity to disrupt the market and challenge current ADSL players with new chipset families, using sometimes different flavors of line coding. At the same time, incumbent ADSL players tend to displace the need for the new products by increasing the bit rate of existing solutions with firmware upgrades, offering "turbo" ADSL2+ solutions capable of delivering 12, 16, or 24 Mbps or even more, with products spectrally compliant and fully interoperable with deployed ADSL products (largely DMT-based). A second controversial area concerns longer reach, directly translatable in larger deployment. This is another opportunity for product differentiation, with systems capable of reaching CPE located up to 24 kft from the CO, and here again firmware upgrades are by far the preferred solution, based on currently deployed ADSL platforms. A third opportunity is in the provision of multistandard capabilities. Some chipset suppliers propose a unique platform capable of supporting many ADSL flavors described in the annexes of the standard, sometimes including symmetric capabilities, as long as each supported product variant remains competitive in terms of cost, power, and performance. This could be a nice value proposition for a customer with the nice flexibility to deploy a DSL unique platform worldwide. All these new features need to be duplicated at both sides of the link, and combined with the optimization of existing products still supporting all legacy flavors deployed in the field, explain the complexity of the current IC designs, typically handled today by impressive teams of a few hundred talented engineers.

2.6 Acknowledgments

The author would like to thank the reviewers for their extensive work on this chapter, the editors for their meticulous follow-up and organization of this gigantic task of compiling a book with such a broad panel of authors, and also all colleagues, competitors, and friends he had the chance to meet in the last decade while struggling on so many generations of ADSL products at Alcatel Bell, Amati, Alcatel Microelectronics, and GlobespanVirata, for their direct or indirect contribution to the content of this chapter.

References

[Amrany 1992]   Amrany, D., Shlomo, G., and Magid, D., *A programmable DSP engine for high-rate modems. Proceedings of IEEE ISSCC'92*, San-Francisco, 222, 1992.

[Chang 1995]   Chang, Z.Y., Macq, D., Haspeslagh, D., Spruyt, P.M.P., and Goffart, L.A.G., *A CMOS analog front-end circuit for an FDM-based ADSL system. IEEE Journal of Solid-State Circuits*, Vol. 30, Iss. 12, 1449, December 1995.

[Cioffi 1993]   Cioffi, J. and Tong, P., *VLSI DMT implementation for ADSL.* T1E1.4 contribution number 93-025, March 8, 1993.

[Cloetens 2001]   Cloetens, L., *Broadband access: The last mile. Proceedings of IEEE ISSCC'01*, San-Francisco, 18, 2001.

[Golden 2006]   Golden, P., Dedieu, H., Jacobsen, K.S., Eds. *Fundamentals of DSL Technology.* Auerbach Publications, Boca Raton, Florida, 2006, p. 457.

[Jahner 2001]   Jahner, M., *How to verify ADSL chips. EE Times*, December 13, 2001.

[Kiss 1999]   Kiss, L. et al., *SACHEM, a versatile DMT-based modem transceiver for ADSL. IEEE Journal of Solid-State Circuits*, Vol. 34, Iss. 7, 1001, July 1999.

[Kuczynski 1993]   Kuczynski, M.A. et al., *A 1 Mb/s digital subscriber line transceiver signal processor. Proceedings of IEEE ISSCC'93*, San-Francisco, 26, 1993.

[Macq 1998]   Macq, D., *DMT based ADSL circuits and systems. Proceedings of ISSCC: xDSL Broadband Interactive Communications*, San-Francisco, 476, 1998.

[Naveh 1999]   Naveh, G. et al., *SW implementation of an ADSL modem on the Carmel^{TM} DSP core, documentation on Carmel^{TM} core.* Siemens EZM Villach, Austria, 1999.

[Reusens 2001]   Reusens, P. et al., *A practical ADSL technology following a decade of efforts. IEEE Communications Magazine*, 145, October 2001.

[Samueli 1999]   Samueli, H., *Broadband communications ICs: Enabling high bandwidth connectivity in the home and office. Proceedings of IEEE ISSCC'99*, San-Francisco, 26, 1999.

[Wiese 2000]   Wiese, B.R. and Chow, J., *Programmable implementations of xDSL transceiver systems. IEEE Communications Magazine*, 114, May 2000.

[Wilson 2002]   Wilson, S., *Firepath^{TM} processor architecture and micro-architecture, Hot Chips Conference*, Stanford University, August 2002.

[Zojer 2000]   Zojer, B., Koban, R., Pichler, J., and Paoli, G., *A broadband high-voltage SLIC for a splitter and transformerless combined ADSL-Lite/POTS linecard. IEEE Journal of Solid-State Circuits*, Vol. 35, Iss. 12, December 2000.

Bibliography

[Clarke 2001] Clarke, P., *Broadcom firepath combines RISC DSP elements. EE Times*, June 14, 2001.

[Conroy 1999] Conroy, C. et al., *A CMOS analog front-end IC for DMT ADSL. Proceedings of IEEE ISSCC'99*, San-Francisco, 240, 1999.

[Cornil 1999] Cornil, J.P. et al., *A 0.5 μm CMOS ADSL analog front-end IC. Proceedings of IEEE ISSCC'99*, San-Francisco, 238, 1999.

[Egerer 1998] Egerer, J. et al., *A low distortion linear-tunable continuous-time 488 kHz fifth-order Bessel filter. Proceedings Eleventh Annual IEEE International ASIC Conference 1998*, 47, September 1998.

[Guangming 1994] Guangming, Y. and Sansen, W., *A high-frequency and high-resolution fourth-order SD A/D converter in BiCMOS technology. IEEE Journal of Solid-State Circuits*, Vol. 29, Iss. 8, 857, August 1994.

[Gupta 2002] Gupta, S.K. and Fong, V., *A 64-MHz clock rate SD ADC with 88-dB SNDR and ⊆105-dB distortion at 1.5-MHz signal frequency. IEEE Journal of Solid-State Circuits*, Vol. 37, Iss. 12, 1653, December 2002.

[Hester 1999] Hester, R. et al., *CODEC for echo-cancelling full-rate ADSL modems. IEEE Journal of Solid-State Circuits*, Vol. 34, Iss. 12, 1973, December 1999.

[Ingels 2002] Ingels, M., Bojja, S. and Wouters, P., *A 0.5 m CMOS low-distortion low-power line driver with embedded digital adaptive bias algorithm for integrated ADSL analog front-ends. Proceedings of IEEE ISSCC'02*, San-Francisco, 258, 2002.

[Lee 2002] Lee, S.S., *Integration and system design trends of ADSL analog front-ends and hybrid line interfaces. Proceedings of IEEE 2002 CICC*, Orlando, 37, 2002.

[Matsumoto 2001] Matsumoto, C., *Legerity ADSL driver pushes integration. EE Times Magazine*, June 11, 2001.

[Mestdagh 1993] Mestdagh, D.J.G., Spruyt, P.M.P., and Biran, B., *Effect of amplitude clipping in DMT-ADSL transceivers. Electronics Letters*, Vol. 29, Iss. 15, 1354, 22, July 1993.

[Moon 1993] Moon, U.K. and Song, B.S., *Design of a low distortion 22-kHz fifth-order Bessel filter. IEEE Journal of Solid-State Circuits*, Vol. 28, Iss. 12, 1254, December 1993.

[Morizio 2000] Morizio, J.C. et al., *14-bit 2.2-MS/s sigma-delta ADC's. IEEE Journal of Solid-State Circuits*, Vol. 35, Iss. 7, 968, July 2000.

[Moyal 1999] Moyal, M., Groepl, M., and Blon, T., *A 25-kft, 768-kb/s CMOS analog front-end for multiple-bit-rate DSL transceiver. IEEE Journal of Solid-State Circuits*, Vol. 34, Iss. 12, 1961, December 1999.

[Pécourt 1999] Pécourt, F., Hauptmann, J., and Tenen, A., *An integrated adaptive analog balancing hybrid for use in (A)DSL modems. Proceedings of IEEE ISSCC'99*, San-Francisco, 252, 1999.

[Pierdomenico 2002] Pierdomenico, J., Wurcer, S., and Day, B., *A 744mW adaptive supply full-rate ADSL CO driver. Proceedings of IEEE ISSCC'02*, San-Francisco, 320, 2002.

[Piessens 2001]   Piessens, T. and Steyaert, M., *SOPA: A high-efficiency line driver in 0.35-μm CMOS using a self-oscillating power amplifier*. Proceedings of IEEE ISSCC'01, San-Francisco, 306, 2001.

[Piessens 2003]   Piessens, T. and Steyaert, M., *Highly efficient xDSL line drivers in 0.35-μm CMOS using a self-oscillating power amplifier*. IEEE *Journal of Solid-State Circuits*, Vol. 38, Iss. 1, 22, January 2003.

[Shorb 2002]   Shorb, J., Allstot, D.J., and Roze, R., *Class AB-D-G line driver for central office asymmetric digital subscriber line systems*. *International Symposium on Circuits and Systems*, ISCAS 2002, V-405, 2002.

[Siniscalchi 2001] Siniscalchi, P.P., *A CMOS ADSL codec for central office applications*. *IEEE Journal of Solid-State Circuits*, Vol. 36, Iss. 3, 356, March 2001.

[Van der Plas 1999] Van der Plas, et al., *A 14-bit intrinsic accuracy Q^2 random walk CMOS DAC*. IEEE *Journal of Solid-State Circuits*, Vol. 34, Iss. 12, 1708, December 1999.

[Vecci 2003]   Vecci, D. and Morandi, C., *A 750 mW class G ADSL line driver with offset-controlled amplifier hand-over*. *Southwest Symposium on Mixed-Signal Design*, 253, 2003.

[Weinberger 2002]   Weinberger, H. et al., *An ADSL-RT full-rate analog front-end IC with integrated line driver*. IEEE *Journal of Solid-State Circuits*, Vol. 37, Iss. 7, 857, July 2002.

[Wurcer 2003]   Wurcer, S., *Carefully evaluate your ADSL line driver efficiency*. *Communication Systems Design*, February 21, 2003.

Chapter 3

Analog Front-End

Nicholas P. Sands and Damien Macq

CONTENTS

Abstract   The functions in the Digital Subscriber Line (DSL) analog front-end (AFE) can be grouped into three general areas: transmitter, receiver, and hybrid. The transmitter transforms the digital representation of the transmit signal into a form that can be launched into the twisted-pair cable. The receiver senses the signal being transmitted from the far end of the loop and digitizes it such that it can be processed, i.e., demodulated, by the digital section. The hybrid couples the receiver and transmitter to the line to minimize the amount of transmit signal superimposed on the receive signal.

3.1   Introduction

The transmit signal is presented to the analog transmitter in the form of a stream of words of 14–16 bits, updated at the sample rate, which will typically be in the range of 2–70 MHz as shown in Figure 3.1. The digital-to-analog converter (DAC) turns this digital stream into an analog signal. Often, some additional spectrum shaping is needed, if for no other reason than to suppress the image spectrum that is a consequence of the digital-to-analog conversion. This function is performed by an analog filter after the DAC; the filter order is typically anywhere between four and seven. The final active stage of the transmitter is the line driver that amplifies the generally low-level (1–2 V peak-to-peak) signal to the correct amplitude—and hence power spectral density—suitable for launching onto the twisted pair. The function of the hybrid with respect to the transmit function is to pass as much of the transmit signal as possible onto the transformer and line. Because the line driver must create an output voltage range (up to 33 V peak-to-peak) in an integrated circuit (IC) process that might be

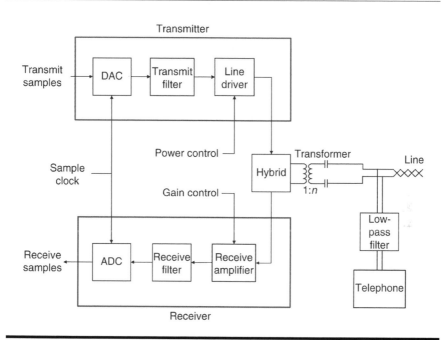

Figure 3.1 **Block diagram of DSL analog front-end, showing main functional units.**

incompatible with the rest of the analog system, this function might be implemented in a different process technologies. Bipolar is the usual choice, although much work has gone into integrated complementary metal-oxide semiconductor (CMOS) line drivers for some DSL applications. The twisted-pair cable is designed for differential operation; the line driver must create a balanced differential signal. A transformer enables this process while also providing isolation for the receiver from noise sources that couple onto the line principally in the common mode. The transformer offers the added advantage in that it has almost no response at DC (direct current). (An ideal transformer has a zero at DC.) This feature assists with the plain old telephone service (POTS) splitter function, forming in essence a high-pass filter that blocks the baseband telephony signal.

As the signal propagates along the twisted pair, it is attenuated and distorted by the nonidealities of the cable. Because the received signal might be quite small, some amplification is needed to bring the signal up to a size compatible with the 2–4 V input range of the analog-to-digital converter (ADC), otherwise the desired signal will be dwarfed (or rivaled) by the inherent quantization noise of the ADC. The system has to operate on a wide range of loop lengths, and hence the amount of gain needed depends on the particular loop on which the system operates. Therefore, this amplifier must have variable gain that can be adjusted under software control during the initialization procedure. Gain values between −10 and +40 dB, in steps of about 1 dB or sometimes less, are typical in many DSL applications. An anti-alias filter removes signal and noise components above the Nyquist rate to prevent folding of high-frequency noise and interference down into the signal band. The ADC samples the analog input signal in time (to provide a transition to discrete time) with a sample-and-hold unit, and it quantizes the signal amplitude, i.e., the continuous analog input is compared to a number of thresholds and a digital output word is produced based on this comparison. The more thresholds that are built into the ADC—the more dense they are—the more accurate a representation of the input as the digital word, and hence the error between the two is smaller. ADC designers refer to this error as quantization noise. Because it is related deterministically to the input signal, it is not really noise, but it behaves in most important ways like noise. Most ADCs used in DSL transceivers have between 12- and 14-bits of resolution, i.e., 4,096–16,384 thresholds. Having more thresholds available reduces the amount of quantization noise; however, this leads to an increasing burden of circuit design complexity, die area, and power consumption.

Nyquist's sampling theorem dictates that if a signal is band-limited up to some frequency f, then sampling at a rate of $2f$ or higher is sufficient to capture all significant features of the signal. All DSL systems have a defined (by standard, see Chapter 17) maximum frequency, which imposes an upper limit on the bandwidth used for transmission. Generally, the receiver will need to sample at least twice this frequency, and in some applications it may be a good decision to sample yet faster. Sampling at a rate higher than $2f$ is known as oversampling, to be described later.

The transmit signal developed by the line driver is necessarily large, and the receive signal is usually small because of the attenuation caused by the loop. This disparity in amplitudes leads to some significant challenges in transceiver design. If the output impedance of the modem were identical to the impedance of its load—the line and everything connected to it—then a simple hybrid circuit would suffice to couple the transmitter, line, and receiver together, and the transmit signal would be eliminated from the receiver. This is rarely the case in practice. The complex impedance of a lossy transmission line has a complicated dependence on frequency that is not easily modeled. If there are gauge changes in the loop, such as at the junction between the drop wire and the binder, reflections are created that return to the transceiver; most inconvenient are the effects of bridged taps, which are unterminated stubs that can lead to quite large echoes that return to the transceiver with unknown delay. (See chapter 2 of [Golden 2006] for more details of bridged taps.)

The extent to which the driving and load impedances differ causes reflection of the transmitted wave, which returns to the receiver as an echo, i.e., an attenuated and distorted version of the transmitted signal. In its mild form, this reflection causes interference that the transceiver must somehow mitigate, either by confining the spectrum of the transmitted signal using aggressive transmit filtering (such as used in pure frequency-division duplexed systems), or by using some kind of adaptive canceling technique in the receiver. Use of cancelation is required if the transmit and receive spectra overlap. If the echo is so large as to saturate the receiver, then digital cancelation techniques are at a loss; analog techniques must be used.

Note that the seriousness of the echo problem is related to the transmit power being used, the frequency band usage (duplexing method), and the presence and details of the bridged taps and other discontinuities in the loop plant. It has been reported that bridged taps are more likely to be closer to the subscriber than the central office (CO). (See chapter 1 of [Golden 2006].) For this reason, echo canceling will be more appropriate for implementation in the customer premises equipment (CPE), although if transmit levels, frequency band usage, and loop plant permit, it may not be necessary.

The need for duplexing arises because one wire pair is used to send information in both directions. Some way must be found to separate the transmitted and received signals. Time-division duplexing (TDD) is in some ways the simplest approach. The downstream and upstream transmitters are allocated disjoint time durations in which they can use the channel; this means that a hybrid is not needed in either transceiver because its transmitter is dormant while its receiver is active, and vice versa. Furthermore, power is saved because only the transmitter or receiver is turned on at any time. Another way to separate transmit and receive signals is to allocate a disjoint frequency band to each. This method is known as frequency-division duplexing (FDD) and is the most widespread method of duplexing used in DSL. A hybrid is needed, and analog filters in the transmit and receive paths help to ensure that significant amounts of the transmit signal do not leak into the receiver. If the plan of frequency usage requires several upstream and downstream bands (such as in VDSL), the analog filtering requirements become quite onerous. Another approach to FDD that obviates the need for analog filters is known as "digital duplexing" [Cioffi 1999]. In digitally duplexed systems, orthogonality between the transmitted and received signals is obtained by maintaining time alignment at both ends of the loop [Cioffi 1999]. (See chapter 7 of [Golden 2006] for details of digital duplexing.) The lack of analog filtering, however, means that the receiver is completely exposed to nonlinear distortion arising in the transmitter that leaks into the receiver via imperfect hybrid return loss. In such systems, transmitter linearity and hybrid matching are of utmost importance to achieve good receiver performance.

To get the most use out of the twisted-pair channel, transmit and receive spectra can be partially or completely overlapped, as is done in one variant of ADSL. In this case, additional measures are needed to remove the transmit signal, known as an echo canceler (EC). An approximate replica of the echo signal is generated locally using the transmit signal and a model of the echo channel. The approximation of the echo is then subtracted from the received signal to yield, ideally, just the part containing the signal from the far end of the loop. In practice, of course, there is almost always some residual, uncanceled echo. Echo-canceled systems hold the promise of the highest capacity, because the available bandwidth is used more efficiently. However, it is quite challenging to realize this promise: the EC itself has significant implementation cost, and it must also be trained to learn the echo channel during initialization. The result is that the noise floor of the part of the spectrum occupied by the transmit signal is often significantly higher than the remainder, thereby reducing the gain in data rate that was expected.

3.2 Implementation

When implementing the AFE, the general objective is to integrate as much of the analog circuitry as possible into one monolithic package. Integration is the way to reduce costs (because only one package is needed) and minimize board area, both of which are much desired by system designers. Further, power consumption is reduced because there is less interchip communication. The "single-chip transceiver" has long been the ultimate goal of DSL modem designers. Much progress has been made since the early days of DMT modems. Amati's ADSL prototype, called "Prelude," occupied a case about the size of a conventional desktop computer, whereas today's DSL modems are typically composed of only three small IC chips and some discrete components. The challenges in system integration usually involve the functional density (transistors, etc.) and compatibility (memory versus logic, digital versus analog, large analog versus small). The amazing improvements in functional density in CMOS technology suggest that it should be possible to create the much sought after monolithic transceiver. However, the two problems of signal compatibility remain, as alluded to above. The first is that analog circuits are susceptible to noise generated by a large active digital section on the same chip. For this reason, AFEs have so far remained self-contained, i.e., isolated from the rest of the system. The second conflict, within the analog system itself, was touched on in the introduction of the line driver. For many DSL systems, up to 15–30 V has to be developed at the output of the line driver. Voltages of this magnitude are not compatible with the current high-density technology of choice, which is CMOS. Indeed, in the past, bipolar transistors were favored for analog signal processing; more recently, CMOS has gained rapid acceptance due to its high-density, low-power dissipation, simple manufacturing process, and hence low cost, partly driven, it must be pointed out, by its massive success for digital circuits. Analog designers have overcome their earlier distaste for CMOS motivated largely by its low cost, but it is at a price in terms of voltage tolerance, linearity, and the achievable noise floor. For this reason, some designers have partitioned the system in a way that appears to have the best of both worlds. The high-density, logic-rich, low-voltage parts of the analog system (the ADC, DAC, and active filters) are implemented in CMOS. A separate bipolar IC is devoted to the line driver (which requires high-voltage tolerance and excellent linearity) and the receive amplifier (once again, good linearity is needed with low noise).

It should be noted here that this issue is less significant with regard to some of the DSL variants that use lower transmit power, such as VDSL1 and VDSL2 (which may have maximum transmitted power levels as low as $+11.5$ dBm). Similarly, the ADSL transmitter at the customer end of the line (the CPE) uses only $+13.5$ dBm of power, compared to the $+20.4$ dBm used by the ADSL transmitter located at the CO. Line drivers integrated with the AFE have been reported for both ADSL CPE [?], [Weinberger 2002], and [Oswal 2004] and VDSL [Moyal 2003] transceivers.

3.3 DSL System Parameters That Drive Analog Design

The AFE linearity and noise performance are critical for overall system performance in terms of maximum rate and reach, determining ultimately the percentage of telephone lines over which a reliable DSL service can be provided by an operator. The amplitude variations of the received discrete multitone (DMT) signal and the strong echo superimposed on it place requirements of more than 100 dB of dynamic range and close to 75 dB of linearity on the CPE receive path. The signal amplitude at the input of the receiver is a complex function of the line characteristics, including the line impedance (which affects the performance of the hybrid), the line length (which can vary from 0 to 20 kft and, as a result, can cause signal attenuation of more than 60 dB), and the presence of impedance discontinuities such as bridged taps and wire gauge variations. The noise level at the line interface is typically assumed to be -140 dBm/Hz, which corresponds to 31.6 nV/rtHz into the (assumed) 100 Ω line impedance, although some lines in the field can present much lower levels such as −145 dBm/Hz or even lower in the downstream band. The noise generated by the AFE referred to the line is expected to be 3–5 dB below these levels, while maintaining reasonable power dissipation levels and a high level of integration with a small silicon area. This is a very severe trade-off, especially for the customer premises side.

To minimize interference, it is typical for regulators to apply limits to the transmitted power spectral density (see Chapter 7). These limits vary with frequency to protect certain services such as amateur radio. Each DSL standard (ADSL, VDSL, etc.) specifies the frequency usage; for example, certain bands are used for downstream, some bands are shared, and so on (see Chapter 17). Each standard also specifies a maximum allowed transmit power, which is usually expressed in dBm (dB relative to 1 mW). Typical values for DSL systems range from +13.5 dBm for ADSL upstream transmissions to +20.4 dBm for ADSL downstream transmissions and one variant of VDSL2. This value can be related to the root-mean-square (rms) voltage and current required via the characteristic impedance of the twisted-pair, usually assumed to be 100 Ω nominal, with a 10 percent tolerance allowed. A key parameter of interest to the analog designer is the maximum voltage swing that the line driver should produce. This is related to the mean-square voltage by the crest factor (CF), which is defined as the ratio of the peak voltage to rms voltage. An alternative term is peak-to-average ratio (PAR), which is just the CF expressed as a power ratio, i.e., $PAR = 20 \log_{10}(CF)$ dB. It is one of the characteristics of DMT that this ratio is rather high compared to other modulation schemes. (See chapter 7 of [Golden 2006].) It is related to the large number (in the sense of the central limit theorem of statistics) of subcarrier signals that are combined by the inverse discrete Fourier transform (IDFT) to form the transmit signal; usually more than 200 subcarriers are used. An extreme example would be if all the subcarriers happened to have the same phase; the resulting transmit waveform would have a peak of (in the case of 200 subcarriers) 200 times the average value. This is an extremely unlikely situation, and in practice, a statistical approach is used, whereby the maximum peak voltage relative to the mean is constrained to some reasonable value (usually 14–15 dB) and larger voltages, when they occur, are clipped by the DAC. Because the number of carriers being added together is large, the voltage distribution is assumed to be approximately Gaussian, and the frequency of the clip events can be computed. It is typical for clip events to be constrained to be less than one in 10^7 samples transmitted. The result is that for ADSL, for example, the line driver must (occasionally) produce around 35 V peak-to-peak swing, compared with an average power of 100 mW, which is equivalent to 3.3 V_{rms}.

This requirement poses a challenge for the designer, because the combination of large-signal voltages with low distortion and noise is a conflict. It also means that the supply voltage will have to be large because the output voltage of conventional amplifiers usually swings only between, but not beyond, the power supplies. One approach is to use a step-up transformer. The transformer is required anyway, to generate the balanced signal suitable for the twisted-pair line, and by using an unequal turns ratio (for example, 1:2 or 1:3), the required output voltage can be created on the line by a smaller line driver output, with corresponding larger output current, of course. The drawback is that signals on the line side of the transformer (i.e., being received from the far end) are stepped down by the same ratio. This may cause problems with the receiver design, because all receiver noise sources must now be smaller by that ratio to comply with the receiver noise budget, which is considered in more detail later. The best choice will depend on the system details, and specifically on the compromise between transmitter performance, line driver architecture, and receiver noise budget.

There is one other consequence of the possible large supply voltage, which has been touched on above in connection with the voltage tolerance of the various circuit technologies. In addition, line drivers are usually characterized by some power efficiency. The efficiency describes how much power is drawn from the supply to deliver a given power to a load. For conventional op-amp line drivers, the efficiency might be as low as 10 percent. Much research has been directed in the ADSL world, as well as elsewhere, on how to improve this situation [Pierdomenico 2002] and [Vecci 2003]. If the power efficiency is fixed, (for example, at 20 percent), the requirement for a larger power supply voltage results in more power dissipated by the system as a whole.

The relatively high peak-to-average ratio has consequences for the data converters. The more detailed discussion on noise budgets for the transmitter and receiver will show that the PAR comes into the noise budgets in the general sense of requiring more dynamic range than, say, a sinusoidal signal (which has a PAR of 3 dB). Dynamic range is defined, rather vaguely, as the ratio between the amplitudes of the largest signal that must be accommodated and the smallest signal that must be resolved (detected). Obviously, the noise in the system must be less than the latter. The larger the PAR, for the same maximum amplitude, the closer the average power will be to the noise floor, thus limiting the signal-to-noise ratio (SNR). In other words, providing a given SNR with a signal of higher PAR (than a sinusoid) requires more dynamic range (specifically, PAR − 3 dB) than to provide the same SNR with a sinusoid.

The maximum signal bandwidth for a DSL service will be defined by the relevant standard. For example, the upstream signal in annex A of

ADSL1 occupies bandwidth from about 26 to 138 kHz, and the downstream signal occupies up to 1.104 MHz. The other extreme is VDSL2, optional service bandwidth up to 30 MHz has been proposed. The available bandwidth imposes a lower bound on the sampling rate of the data converters involved, the sampling rate must be at least twice the maximum bandwidth of the signal. It may be advantageous to sample faster than this, i.e., to oversample. For the purely analog parts of the system the analog filters, amplifiers, etc., the maximum signal bandwidth has implications for the bandwidth and slew rate that the individual circuits must support. For ADSL line drivers, bandwidth of 40 MHz is typical.

Having discussed in general terms the influences on analog component specifications, how does the designer arrive at more detailed design objectives? Obviously, more converter resolution, higher bandwidth, and lower noise will lead to better analog system performance, but in the real world, the goal is to design a system that is just good enough to meet the overall performance target, with good power and die size efficiency. One approach is to construct an SNR template, or envelope, for the analog system such that the effects of impairments arising from the analog circuits are negligible in almost all reasonable conditions of operation. The objective is to hide the analog noise and impairments as much as possible below the inevitable effects of the channel and noise environment, thus minimizing the effects of the analog system on the attainable data rate or detection margin.

3.4 Transmitter Noise Budget

The SNR required for reliable communication with a given constellation size 2^b, where $b \geq 4$, can be estimated using the gap approximation (see chapter 6 of [Golden 2006])

$$SNR(b) = 3(b-2) + \gamma_m - \gamma_c + 14.5 \, dB \tag{3.1}$$

where
b is the number of bits, γ_m is the margin, and γ_c is the coding gain.

A margin of 6 dB and coding gain of 3 dB are typical for DSL systems. For the purposes of analog noise budget computation, an extra 9 dB margin is assumed, yielding an overall transmitter SNR objective of

$$SNR_{tx}(b) = 3b + 20.5 \, dB$$

For most DSL systems, the largest constellations are those with 32,768 points (i.e., $b = 15$), which results in a transmitter SNR objective of 65.5 dB.

Note that because of the very strong dependence of loop attenuation on frequency, it will not be necessary to support this SNR over the entire band; line and receiver noise will be more significant at frequencies where attenuation is most severe. This SNR level should be regarded as a target that should be met for some lower-frequency portion of the transmit band.

Some of the impairments inherent in the transmitter have already been mentioned, including DAC quantization error, nonlinear distortion, circuit noise, etc. For the purposes of computing a simplified transmitter noise budget, four sources are considered:

1. Line driver noise
2. Line driver nonlinear distortion
3. DAC quantization noise (including the effects of dynamic nonlinearity)
4. DAC output noise

Obviously, if active filtering or other stages are also included in the system, then their linearity and noise must share the noise budget with these four main sources. To simplify the analysis, it will be assumed that each source's contribution to the total noise at the output of the transmitter may be equal, although in practice the relative contributions may have to be adjusted based on the details of the circuits involved. If the transmit power spectral density (PSD) is -40 dBm/Hz, then for an SNR objective of 65.5 dB, the total noise may be as high as -105.5 dBm/Hz. For all four sources to combine equally to result in -105.5 dBm/Hz, each must be no greater than -111.5 dBm/Hz when referred to the transmitter output on the line.

3.4.1 Line Driver Noise

A power level of -111.5 dBm/Hz referred to a line impedance of 100 Ω is equivalent to a voltage spectral density of 841 nV/rtHz. A line driver might have gain of 20 dB, so the input-referred voltage noise of the line driver must be less than 84 nV/rtHz.

3.4.2 Line Driver Nonlinear Distortion

Amplifier distortion comes in many forms, but most relevant to line drivers will be odd-order harmonic distortion arising from soft clipping associated with the limits of the driver output swing. The most direct way to characterize nonlinear distortion appropriate to DMT signals is the missing tone power ratio (MTPR) [ADSL2]. Recall the transmitter SNR objective of 65.5 dB, the MTPR for line driver linearity must therefore be greater than 71.5 dB.

3.4.3 DAC Output Noise

This noise will be amplified by the line driver gain. Assuming a line driver gain of 20 dB, the DAC output noise must be held below 84 nV/rtHz.

3.4.4 DAC Quantization Error

The contribution allowed from this source is 84 nV/rtHz referred to the DAC output. Ideal DAC quantization noise can be expressed as a voltage spectral density

$$v_{rms} = \left(\frac{V_{pp}}{2^n} \right) \sqrt{\frac{1}{12f_s}} V/\sqrt{Hz} \tag{3.2}$$

where
n is the number of bits, and f_s is the sample rate.

V_{pp} is the output voltage range. From this equation, the minimum n can be computed as

$$n = \log_2 \left[\frac{V_{pp}}{84nV\sqrt{12f_s}} \right] \tag{3.3}$$

If it is assumed that the DAC output range is 2 V peak-to-peak at sample rate of 2.2 MHz (which would be appropriate for downstream ADSL transmission), the allowance for ideal DAC quantization noise can be computed as 12.2 bits. Increasing the sample rate above the minimum distributes the quantization noise over a larger bandwidth, thereby lowering the minimum n. This technique is known as oversampling and is widely used.

In practice, the levels of a real DAC are not quite evenly distributed, and the output stage suffers from large-signal distortion. If the actual noise density is measured in the presence of a realistic signal (usually using the MTPR), the effective number of bits (ENOB) can be computed from the equation

$$ENOB = \log_2 \left[\frac{V_{pp}}{v_{rms,\ measured}\sqrt{12f_s}} \right] \tag{3.4}$$

Typical DACs in this range of resolution and sample rate achieve an ENOB of about 0.5–1.5 worse than ideal, so for the simple example of an output range of 2 V peak-to-peak at sample rate of 2.2 MHz, a DAC with 13 or 14 bits would be suitable.

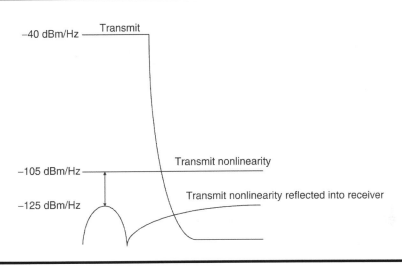

Figure 3.2   Transmit PSD with nonlinear distortion and its effect on receiver performance.

Systems with overlapping spectra or those without analog filters to separate transmit and receive bands (for example, systems that use digital duplexing) are subject to an additional constraint with regard to transmitter linearity. The nonlinear distortion creates a wideband spectrum (not spectrally confined to the transmit band) that leaks into the receiver as a consequence of imperfect hybrid matching. This leakage will be most serious on loops with bridged taps or wire gauge discontinuities. Echo return-loss (ERL) is a measure of how much signal is reflected back from the transmitter into the receiver. It may be severe as only 20 dB below the transmitted signal. The implication for receiver performance is illustrated in Figure 3.2. The uppermost curve represents the transmit PSD, which is at −40 dBm/Hz. If transmitter linearity is characterized by an MTPR of 65 dB, the distortion in the transmitter causes the nonlinearity to appear at −105 dBm/Hz. A worst-case ERL of 20 dB will cause this nonlinear interference to appear in the receiver with a PSD of −125 dBm/Hz, which may be a serious obstacle to system performance. For this reason, transmitter linearity and hybrid matching will be of great significance when designing such systems.

3.5   Receiver Noise Budget

It is typical for system designers to assume that there is noise on the line at −140 dBm/Hz, so the budget for receiver noise should be somewhat lower than this. How much lower has to be a compromise between the increasing power consumption and complexity of designing low-noise circuits, and in particular the receive amplifier (a programmable gain amplifier [PGA] or low-noise amplifier [LNA]) and the inevitable impairments of the ADC? Noise in the receiver comes from the transmitter via the attenuation of the line, thermal noise in the high-gain PGA, the ADC, and from jitter in the sampling clock. All of these sources will be referred to the signal level at the termination of the line, and thus it will be the equivalent noise at the receiver input caused by sampling clock jitter that is combined with input-referred ADC impairments, etc.

If a sinusoid of frequency f is sampled with jitter σ, the SNR is given by

$$\text{SNR}(f) = \frac{1}{(2\pi f \sigma)^2}$$

The equivalent PSD at the receiver input can then be calculated as

$$S_{\text{jitter}}(f) = (2\pi f \sigma)^2 \left|H(f)\right|^2 S_{\text{tx}} \tag{3.5}$$

Because of the steep roll-off in $H(f)$ with increasing frequency, the jitter PSD will be worst at low frequencies. More precisely, the effect of jitter would likely be most severe at the frequencies at which the channel attenuation is least, notwithstanding the linear increasing factor in Equation 3.5.

As an example, consider the downstream band of ADSL, which is shown diagrammatically in Figure 3.3. Channel attenuation might range from 30 to 80 dB. At the low end of the band, around 140 kHz, the channel attenuation could be 30 dB, yielding a receive PSD of −70 dBm/Hz. The transmitter noise floor is 65 dB below this level, at −135.5 dBm/Hz. Close to the maximum bit loading of 15 bits/subcarrier may be obtained as long as the effects of jitter and any other receiver effects can be kept below −142 dBm/Hz. To achieve this level, Equation 3.5 is used to solve for the maximum jitter that can be tolerated:

$$20 \log_{10}(2\pi f \sigma) = -142 - (-40) + 30 = -72\,\text{dB}$$

The limit on sampling jitter is obtained with $f = 140$ kHz as

$$\sigma = \frac{10^{-3.6}}{2\pi f} = 285\,\text{ps}$$

The effects of PGA noise and ADC impairments may be assessed by considering higher frequencies, at which the channel attenuation means

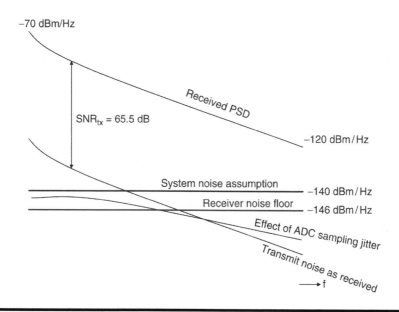

Figure 3.3 **Schematic representation of received signal power spectral density, with associated noise spectra.**

that sampling jitter will be less significant. If the total receiver noise bud-geted for these sources is −146 dBm/Hz, then each may be as high as −149 dBm/Hz. This level is equivalent to 10 nV/rtHz input-referred noise. Further, assume that the total gain between the receiver input and the ADC is 20 dB, and the ADC input range is 2 V, and the sample rate f_s is 2.2 MHz. Equation 3.3 can be used again, this time with the required noise PSD at the ADC input of 100 nV/rtHz

$$n = \log_2 \left[\frac{V_{pp}}{100nV\sqrt{12f_s}} \right] = 11.9 \, \text{bits}$$

As with the DAC, the actual resolution of the ADC will have to be somewhat higher to allow for circuit imperfections. It is typical to allow 1–2 extra bits, so the actual ADC resolution will have to be 13 or 14 bits with performance (as characterized by the ENOB) to 12 bits.

As with the transmitter, other impairments such as nonlinear distortion or circuit noise of possible active filter stages can be included to fit within this noise budget of −146 dBm/Hz.

3.6 Digital-to-Analog Converter

A DAC is designed for a certain resolution, i.e., the number of output levels it can produce. Even an ideal DAC will introduce an error owing to finite quantization. The difference between the unquantized input and the perfectly quantized output is referred to as quantization error, or somewhat misleadingly as quantization noise.* If the DAC has resolution of n bits, then it is capable of producing any of 2^n levels, and the quantization error has a uniform distribution $U(0, \text{lsb})$, which has variance (power) of $\text{lsb}^2/12$. The DAC has a defined full-scale voltage V_{pp}, so that the power of the DMT signal, correctly scaled to minimize clipping is $V_{pp}^2/4CF^2$. Therefore, the signal-to-quantization noise ratio (SQR) in decibels of an ideal DAC with n-bit resolution (for a DMT signal) is

$$SQR = 10\log_{10}\left(\frac{12V_{pp}^2}{4CF^2\text{lsb}^2}\right)$$

$$= 10\log_{10}\left(3\frac{2^{2n}}{CF^2}\right)$$

$$= 10\log_{10}(3) + 2n\log_{10}(2) - 2\log_{10}(CF)\,\text{dB}$$

Thus, if $CF = 5$, the familiar approximation results:

$$SQR = 6n - 9.2\,\text{dB} \tag{3.6}$$

To achieve an SQR of 65 dB, at least 12.3-bit resolution is required, assuming a perfect DAC. The forthcoming analysis will show that impairments caused by component tolerances and mismatch lead to DAC performance somewhat poorer than this, and it is common practice to provide 1–2 bits (extra bits of resolution sometimes known as "safety bits") to allow for these effects.

Transmit power control is an important feature for DSL systems to reduce crosstalk (see chapter 3 of [Golden 2006]), particularly, in situations where adjacent pairs serve CPEs with great disparity in range. The transmitted PSD is reduced on those loops with relatively less attenuation (i.e., those that are shorter). If DAC resolution is high, the PSD reduction may be

---

* The quantization error is correlated with the input signal, and so in the strict sense it is not noise, but "quantization noise" has become the conventional term.

done digitally by scaling the transmit signal before conversion to the analog domain, most efficiently in the frequency domain, prior to the IDFT. This provides for accurate control of the transmit power and further allows shaped (nonuniform) PSD.

For high levels of cutback, however, the resulting postconversion PSD might approach the DAC noise floor. In this situation, it is better to do the large, uniform cutback in the analog domain, to maintain acceptable SNR. This can be done conveniently in the current-steering DAC (see Section 3.6.4.1) by scaling the reference current back by the appropriate amount. In the case of the capacitor DAC (see Section 3.6.4.2), scaled copies of the voltage references must be provided. An analog cutback range of up to 20 dB, in steps of 3 dB, is a typical configuration.

3.6.1 DAC Impairments

If the DAC was perfect, it would produce evenly spaced output levels. In practice, component mismatch causes the levels to be slightly uneven, an effect known as differential nonlinearity (DNL). Further, output stage limitations lead to increasing divergence between the output voltage and the ideal as the extremes of the range are approached. This effect is called integral nonlinearity (INL). At 14-bit resolution, worst-case INL of about 1.5–2 LSB and DNL of 0.25 LSB are typical in DACs for DSL applications.

In addition to these static nonlinearity parameters, the DAC has also dynamic nonlinearity, for example, when the output stage slew rate is limited such that the settling time to reproduce a full-scale step differs from that for a smaller step. This kind of impairment will be seen as degradation of the DAC linearity performance as the input frequency is changed.

Thermal noise afflicts all electronic circuits. Careful attention must be paid to the design of the output stage so that thermal noise does not overwhelm the other noise sources. For DSL transceivers, this puts the requirement at or below approximately 50 pA/rtHz.

3.6.2 Functional Performance Parameters

If the SQR of a DAC is measured, Equation 3.6 can be used in reverse to derive the ENOB, which represents the noise-equivalent bit resolution of a perfect DAC. Typical DACs used in DSL applications have an ENOB that is 0.5–2 bits less than their actual resolution.

For systems employing DMT, the most useful measurement of DAC performance is the MTPR. It aims to take into account the CF and wideband nature of multicarrier signals. Definitions vary; some use a multiplet of four to eight orthogonal sinusoids, although others use higher number of carriers, generated in a similar way to the DMT signal, via the IDFT. The essence of the method is that some predefined carriers are left out of the signal so that spectral notches are created, which, in the absence of distortion and noise, would be infinitely deep. The degree to which the notches are filled in indicates amount distortion and noise, expressed as a power ratio relative to neighboring carriers. Under the assumption that the noise and distortion floor is approximately constant from one carrier to the next, it is directly analogous to the noise floor seen by the DMT receiver.

3.6.3   Requirements

A tight requirement comes from the out-of-band noise and distortion components, which are reflected into the receive band after some attenuation through the hybrid. To a lesser extent, another constraint comes from the maximum amount of noise tolerated on the line within the voice band to protect the existing POTS, as specified by the applicable standard. Consequently, expensive on-chip or even off-chip high-pass and low-pass filters were added in early implementations to compensate for the limited DAC performance. System simulations reported in [Siniscalchi 2001] show, for example, that for a minimal hybrid rejection of 12 dB (generally, accepted as a worst case on the CO side, whereas the customer premises side could present even worse hybrid rejection performance), given a 22 dB line driver gain, a noise level below −137 dBm/Hz is needed at the DAC output, leading to 13.2 effective number of bits. Other sources [Moyal 1999] similarly suggest to design the overall transmit path in such a way that the echo signal does not degrade a line with −140 dBm/Hz noise, leading to an overall transmit signal-to-noise-and-distortion (SND) of at least 75 dB for all components, and close to 84 dB SND for the DAC, corresponding to 20 percent of the total transmit noise contribution. At the time of writing, 13–14 bit DACs, some with up to four times oversampling, are commonly adopted for CO DACs. On the customer premises side, more than a 14-bit DAC may be needed, given the very low noise conditions observed on that side of the line, though on a smaller signal bandwidth.

3.6.4   DAC Architectures

Two types of DACs are usually adopted for DSL transceivers. The first category uses different flavors of current-steering architecture, offering excellent linearity with the improved matching performance of advanced CMOS processes. A second category uses high-order sigma–delta modulators followed by switched-capacitor and continuous-time filters. In either case, an additional low-pass filter is required to eliminate the image spectrum created by aliasing above the Nyquist frequency.

3.6.4.1 Current-Steering

The current-steering DAC consists of a reference generator, current source array, and the switch array. The array of current sources develops scaled versions of the reference current that are connected to the load via the switch array controlled by the digital input word. Full-scale current of 20 mA into a 50 Ω load, providing 1 Vpp output are typical parameters. Figure 3.4 shows a simplified depiction of this arrangement. Not shown is an additional array of complementary current sources that provides cancelation of common-mode errors.

DACs of this type with resolution up to 14 bits run comfortably above 100 Msamples/s, with power dissipation in the range 100–150 mW and MTPR values in the mid-to-high 60s [AD9744]. Extremely careful layout of the current sources combined with advanced switching algorithms allows compensation for first- and second-order spatial errors that create systematic current source mismatch in the matrix, and some designs include dynamic techniques to improve unary source matching. An overall linearity in the 14-bit range is currently reported by advanced designs (see, for example, [Van der Plas 1999]). Dynamic performance is limited mostly by the imperfect synchronization between all of the control signals at the current switches. The dynamic performance can be enhanced by careful layout definition and the careful placement of data latches within the current source matrix. In [Cornil 1999], a 12-bit DAC is reported in a CMOS 0.5 μ process, with 8 MSB unary current sources and a weighted current matrix for the 4 LSBs. The sampling clock is 8.8 MHz, and the device is

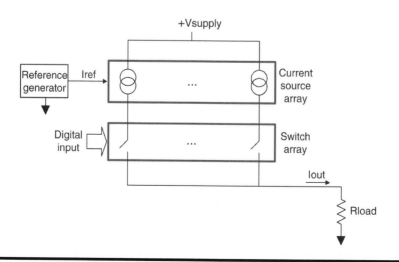

Figure 3.4   **Current-steering DAC—simplified architecture.**

based on a 39 µA unary source. Another CMOS 0.6 µ implementation pro-posed in [Hester 1999] reports 14-bit linearity, based on a calibrated 7 MSB unary current source array, and another 7 LSB weighted current source array.

3.6.4.2 Capacitor DAC

CMOS technology naturally lends itself to capacitors, in which the area of the capacitor on the die determines the capacitance. If an array of capaci-tors with binary weighted values is created, with switches that selectively connect one plate to a voltage reference or ground, and the other plates are all connected to the inverting terminal of an amplifier with capacitive feed-back, then a capacitor DAC results [Sands 1999], as shown in Figure 3.5. For the higher DAC resolution demanded in DSL applications, the ratio of the largest capacitor to the smallest is rather large, so segmentation is used. With segmentation, a smaller scaled replica of the voltage reference is developed for the least significant bits, in this way, the disparity in the capacitor sizes can be controlled.

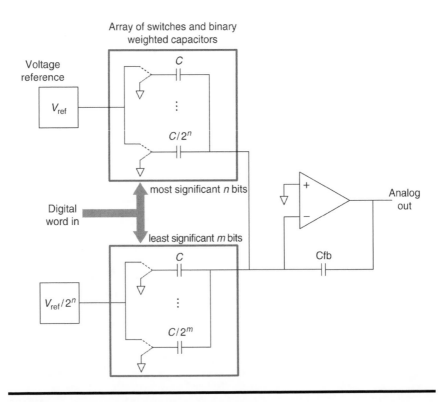

Figure 3.5　**Capacitor DAC with segmentation.**

3.6.4.3 Oversampled DACs and Noise Shaping—the Sigma–Delta Converter

Oversampling and noise shaping are techniques that can be used with either the current-steering or capacitor DAC architectures, in combination with additional digital and analog filters, to increase the SQR in the signal band at the expense of the SQR in other parts of the spectrum.

The previous discussion of DAC quantization noise indicated that the noise has a fixed power determined only by the quantization step size relative to the DAC output range. If the signal occupies only a fraction of the DAC Nyquist band, then oversampling can be used. Oversampling disperses the fixed quantization noise power over a wider bandwidth, thus lowering its spectral density. Noise shaping moves the noise PSD into unused spectrum, so that the in-band SNR can significantly enhance. An analog postfilter is used to reduce the out-of-band quantization noise.

In the first step of oversampling and noise shaping, the high-precision digital signal is applied to a feedback loop with a loop filter $H(z)$ and a quantizer that matches the resolution of the DAC. It is typical to treat quantization error as noise that is uncorrelated with the signal. The model used for system analysis is shown in the lower portion of Figure 3.6 [Jantzi 1993]. Ignoring the postfilter for the moment, the output sequence $Y(z)$ can be written in terms of the input sequence $X(z)$, the quantization noise $N(z)$, and the filter transfer function

$$Y(z) = \frac{N(z)}{1 + H(z)} + \frac{H(z)X(z)}{1 + H(z)}$$

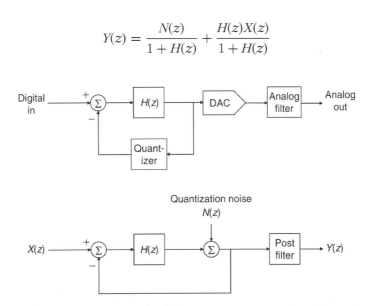

Figure 3.6   **Oversampled DAC, with noise-equivalent system for analysis.**

The filter $H(z)$ will be chosen so that its gain is large within the signal band and small outside it. In this way, the noise and signal experience different transfer functions: the signal gain is

$$\frac{Y(z)}{X(z)} = \frac{H(z)}{1 + H(z)}$$

whereas the noise gain is

$$\frac{Y(z)}{N(z)} = \frac{1}{1 + H(z)}$$

In the signal band, $H(z)$ is large, so the signal gain $Y(z)/X(z)$ is close to unity and $Y(z)/N(z)$ is small. Out of band, $H(z)$ is small, so the noise gain is large. The effect of the filter is to redistribute the noise out of the signal band, into unused spectrum.

A simple example illustrates how this procedure works. Assume the signal bandwidth is concentrated in the lowest fraction of the DAC Nyquist band. A common choice for the filter is a simple integrator (i.e., an accumulator)

$$H(z) = \frac{z^{-1}}{1 - z^{-1}}$$

so that

$$\frac{Y(z)}{X(z)} = z^{-1}$$

and

$$\frac{Y(z)}{N(z)} = 1 - z^{-1}$$

Thus, the noise gain has a null at DC, meaning that if only the very small portion of the band near DC is used by the signal, then, in principle, the SQR can be arbitrarily high. Finally, the importance of the analog postfilter can be seen as the noise gain approaches unity at the Nyquist frequency of the DAC. In most cases, a simple low-pass filter will do.

The use of oversampling and noise shaping can be considered a trade-off between the sample rate and the effective number of bits. To add one more bit-equivalent of DAC SQR performance requires increasing the sample rate by a factor of four, just by virtue of dispersing the quantization noise over a wider bandwidth.

An early implementation [Chang 1995] was based on a single-bit DAC, built from a sixth-order modulator with multiple feedback loops and an oversampling rate of 32, leading to a theoretical 88 dB SQR after the modulator. Limitations from kT/C and amplifier noise in the analog switch-capacitor filtering reduced overall accuracy to 12 bits. This limitation required the addition of significant filtering on the transmit side to eliminate all remaining out-of-band noise components. Another problem with this type of 1-bit DAC is the presence of very high amplitude limit cycles at close to half the sampling frequency. These strong tones are susceptible to folding within the signal band by self-mixing or any mixing with the master clock, owing to any nonlinearity in the continuous-time filter. Dithering can be applied to reduce this effect, at the expense of increased background noise. Another approach is to work with higher resolution DACs, as illustrated in [Moyal 1999], which provides some details on a 6-bit DAC sampling at a 20.48 MHz with 15 dB/octave noise shaping. The converter reaches an 84 dB SQR. The linearity of the current-steering 6-bit DAC is improved by dynamic element matching techniques, which are designed to shift the distortion caused by current mismatches beyond the useful signal bandwidth.

Recent DSL implementations using either oversampling, current-steering, or a combination of both principles generally fulfill the 14-bit SND requirements in such a way that the high-pass filter, necessary to eliminate out-of-band components at the CO side, can even become optional. This high-pass filter was originally partly integrated on-chip, but to improve linearity and accommodate noise constraints, an extra high-pass filter was often required using expensive discrete components (capacitors and inductors). The significant performance improvement of integrated DACs has led to impressive area and power improvements for recent CO designs, allowing very aggressive integration levels for these front-ends. For example, recent products can combine up to 8 or 12 ADSL AFEs on the same die, while the number of external discrete components is reduced to a minimum. This is probably one of the most significant cost and area gains observed in the analog portion of the latest generations of ADSL CO products.

3.7 Filtering

Despite the general preference for digital signal processing, analog filters are needed in both the transmitter and the receiver to provide additional spectrum shaping, to correct undesirable features of the signal created by the digital system, or to protect the receiver from external interference.

One of the defining properties of DSL is its coexistence on the same pair with baseband services such as POTS or integrated services digital network (ISDN). Coexistence implies the need for a high-pass filter to protect these baseband services from the DSL signals, and vice versa. The isolation is usually accomplished using a passive network, incorporating the line-coupling transformer as an inductive element and series capacitors that also serve to block the DC that is used in telephony to signal off-hook and other conditions. This simple network forms a second-order filter, more attenuation may be obtained by using two more capacitors on the modem side of the transformer. If a dual winding transformer is to be used, the number of capacitors can be cut in half by inserting them in series between the windings on each side. To obtain the best stop-band attenuation, the corner frequency is set as close as possible to the DSL band, however, attention must be paid to group delay distortion that occurs in the transition band of this filter. The delay distortion can increase the length of the overall impulse response of the combined channel formed by the transmitter, loop, and receiver. See chapters 7 and 11 of [Golden 2006] for discussions of the channel impulse response and time-domain equalizer issues.

Specific filter prototypes are determined by a compromise between in-band flatness (typically 0.5–1 dB), steepness of roll-off, and complexity. For the low-pass filter, phase linearity in the transition band is not normally a significant concern, pass-band flatness and steepness of roll-off are the two most important parameters. The Chebyshev prototype is a good choice for this application.

Some level of filter tuning is desirable, not only to compensate for the drift of the filter characteristics with temperature, supply voltage, or process, but also to provide some amount of system-level adaptivity to specific line conditions.

3.7.1 Transmit Filter

DSL standards incorporate PSD masks, which are profiles within which the emitted power spectral density must fall (see Chapter 17). Outside the transmit bands, the transmitted PSD must comply with much lower limits. These limits are imposed to protect other services, such as other DSLs or telephony, and to control crosstalk. The transmit gain scaling parameters in the DMT transmitter provide gross control over the PSD in-band; however, there are several mechanisms that cause spectral leakage out of the transmit band. This unintended emission must be controlled to comply with the PSD masks, and analog filtering may be the only way to adequately suppress the transmitted PSD outside of the transmit band(s). Further, in situations of poor return loss (such as on loops with bridged taps or nonuniform line impedance), this out-of-band PSD falls into the receive band, and it will thus degrade receiver performance because it appears as noise.

There are three specific mechanisms by which the transmit signal leaks out of its desired band. The DAC output response to an impulse at its discrete-time input is an one-sample-width pulse of unit amplitude (in other words, a zero-order-hold). In the frequency domain, this pulse has a spectrum described by $\mathrm{sinc}(f/f_s)$, and the conversion from discrete time to continuous time means that this spectrum as defined in discrete time will be aliased (images will be created) around frequencies nf_s, with n as any integer. Most DSL systems deployed to date have used a rectangular IDFT symbol window (see chapter 7 of [Golden 2006]), whereby the cyclic prefix is simply appended (prepended) to the start of the symbol. This procedure causes spectral leakage for each tone at f_i according to $\mathrm{sinc}((f-f_i)/f_{\mathrm{symbol}})$.*
Nonlinearities such as DAC quantization and amplifier distortion cause signal energy to leak from the intended band at a relative level indicated by the MTPR specification.

In the case of the conventional rectangular IDFT window, the PSD of the emitted spectrum can be written as

$$S_{\mathrm{tx}}(f) = \mathrm{sinc}^2\left(\frac{f}{f_s}\right) \sum_{<i>} X_i \, \mathrm{sinc}^2\left(\frac{f-f_i}{f_{\mathrm{symbol}}}\right) + \frac{\bar{X}_i}{\mathrm{MTPR}} \qquad (3.7)$$

where X_i is the PSD of a tone used in the transmit direction and $\bar{X}_i$ is the nominal transmit PSD. The last term of the expression represents the generally flat PSD of transmitter noise and nonlinearity.

As an example, consider the upper portion of the ADSL upstream transmit spectrum shown in Figure 3.7.

The spectral mask specified in ITU-T Recommendation G.992.1 [ADSL1] is also shown. To comply with this mask, approximately 20 dB more attenuation than that provided by the rectangular window is required at 307 kHz. Use of a filter design program such as [FilterPro] shows that a third-order Chebyshev filter will provide this level of attenuation. Because bridged taps and gauge discontinuities are more likely at the CPE, the effect of transmitter nonlinearity on receiver performance must also be considered. If the transmitter MTPR is assumed to be 65 dB, this places the unfiltered nonlinear floor at a level as high as $-34.5\,\mathrm{dBm/Hz} - 65\,\mathrm{dB} = -100\,\mathrm{dBm/Hz}$. Assuming a worst-case return loss of 14 dB, if no transmit filter is present, the nonlinear effects appear in the receiver at $-114\,\mathrm{dBm/Hz}$. The third-order transmit filter reduces this effect to $-134\,\mathrm{dBm/Hz}$ at 307 kHz.

---

* More recent systems (notably, VDSL and VDSL2) employ more sophisticated transmit windowing, in which the beginning and end of the symbol have "soft" edges. See chapter 13 of [Golden 2006] for more details on transmit windowing.

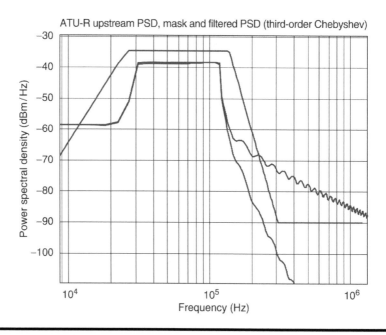

Figure 3.7   ADSL1 upstream spectral mask, with unfiltered and filtered PSD.

3.7.2   Receive Filter

The primary task of the receive filter is to reduce the effect of aliasing caused by the ADC. The faster the ADC is able to sample, the more relaxed the requirements are on the receiver's anti-alias filter. One particular concern is relevant to DSL systems that operate at frequencies near amateur radio bands. Transmissions in these bands may cause strong interference (up to 0 dBm is assumed in the CPE). If the DSL pass-band extends near one of these bands, the anti-alias filter must have sufficient attenuation to reduce the radio interference to more manageable levels, such as −20 dBm.

A subsidiary role of the receive filter is to protect the receiver from out-of-band noise, which is a role shared with the transmit filter. Once again, challenges arise when operating in the presence of bridged taps or other causes of poor return loss. The echo signal that appears in the receiver may be large enough to cause the receiver to saturate or force the use of lower receive gain than would be optimal. A more aggressive receive filter reduces the amplitude of the echo in the receiver, thereby enabling the use of a higher gain to amplify the desired signal.

As an example, consider the situation at an ATU-R with a total transmit power of +12.5 dBm and overall echo attenuation of 20 dB, which leads to an echo amplitude of 1.3 Vpp. It is typical in FDD systems to allocate a guard band of several tones between the upstream and the downstream bands; in this case, assume tones 6–25 are used in the upstream direction, and tones 33–255 are used in the downstream direction. If the ADC input range is taken to be 2 Vpp and the receive gain is assumed to be 20 dB, then ADC saturation at this receiver gain setting occurs for a receiver input amplitude of 200 mV. The receive filter must reduce the echo amplitude of 1.3 V to somewhat less than 200 mV to enable the receiver gain setting of 20 dB to be used without ADC saturation. (20 dB is assumed to be a design objective.)

The amplitude of the filtered echo remaining in the receiver can be computed as

$$V_{echo} = \int_0^\infty \left| H_{hpf}\left(f\right) \right|^2 H_{RL}^2 S_{tx}\left(f\right) df$$

Equation 3.7 can be used to compute $S_{tx}(f)$, neglecting distortion in this case. A fourth-order, 1 dB ripple Chebyshev high-pass filter with a corner frequency at 138 kHz in the receive channel would attenuate the residual echo in the receiver by just 20 dB.

3.7.3 Filter Implementation

Filters can be implemented with passive components up to about five orders. Standard design techniques, such as those in [Filter Free] and [FilterPro], can be used to synthesize filters with the desired input and output impedance and cut-off frequency. Component tolerance, and the cost and board area required, for inductors in particular, will all be of concern with the passive filter, except for the simplest of designs. An example of a fifth-order passive filter with cut-off frequency at 11 MHz is shown in Figure 3.8.

Active filters offer the advantage of integration: the board area and cost of the discrete components are moved into the AFE chip. Possible

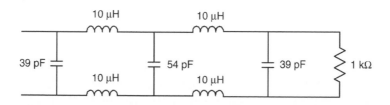

Figure 3.8 **Example passive fifth-order Chebyshev low-pass with cut-off at 11 MHz, passband ripple 1 dB.**

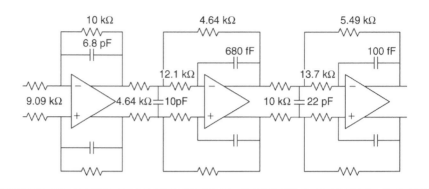

Figure 3.9   **Active transmit filter–fully differential, fifth-order, 8.8 MHz low-pass Chebyshev with 1 dB pass-band ripple.**

disadvantages of active filters are noise, distortion, and, once again, component tolerances that limit the accuracy of the frequency response. Noise budgets must be carefully examined so that the op-amps used in this part of the circuit do not contribute significant noise to the system. For those applications in which frequency tolerance is critical, it may be necessary to employ frequency self-calibration or tuning to mitigate the effects of component tolerance. Figure 3.9 illustrates an example active filter circuit with cut-off frequency at 8 MHz. In this fully differential configuration, three op-amps are needed to implement the fifth-order filter.

3.7.3.1 Continuous-Time RC Filters

Excellent dynamic range and linearity performance can be achieved with straightforward active RC implementations, either off-chip, using discrete resistors and capacitors around, for example, a receive low-noise amplifier or the line driver, or on-chip, using high-resistivity polysilicon and metal–metal or poly–poly capacitances. In most cases, on-chip active RC filters will require some tuning to compensate for filter characteristic deterioration because of process dispersion, or supply voltage and temperature drift. A tuning range in excess of ±40 percent was implemented in [Chang 1995] using capacitor arrays with 16 levels, leading to 5 percent accuracy for the cut-off frequency. A drawback of this first approach is the necessity to freeze the tuning after the modem initializes, because significant voltage glitches could be produced if the tuning engine switches were to unload extra capacitors. As a consequence, only the process variations are efficiently compensated. A second drawback of this approach is the significant area necessary to implement the 40 percent tuning capacitors, at places where the extremely low noise level specification dictates a maximum limit on the resistor size and therefore a minimum capacitor size. For example, the total capacitance for a single fourth-order low-pass filter reported in [Chang 1995] was 760 pF. Nevertheless, because of the extremely high linearity and dynamic range requirements, RC-based Rauch or Sallen–Key structures are typical solutions for low-order filters. In integrated implementations, differential circuits are preferred because of their immunity to noise. This precludes the use of the Sallen–Key topology; multiloop feedback (MLF) circuits are often used in this case.

3.7.3.2 Continuous-Time gm-R-C Filters

Another filtering approach was also developed in [Chang 1995] to deal with higher-order filtering requirements of a frequency-division multiplex system, while still maintaining area and power within affordable limits. In this paper, the authors selected gm-C structures for their low-power and high-speed capabilities, although the linearity of these filters was reputed to be relatively poor, in the 40–50 dB range. A new transconductance scheme was described, using a polysilicon resistance with local feedback to boost the linearity above 60 dB. Based on this concept, two sets of tunable 14th-order band pass filters were assembled in a CMOS 0.7 µ process, with third-order harmonic distortion close to 63 dB at 100 kHz and with more than 1 percent frequency accuracy on a wide tuning range. The filter dissipated 70 mW from a 5 V supply voltage. Drawbacks of the architecture were the somewhat excessive level of noise ($300 \text{ nV}/(\text{Hz})^2$) associated with some noise peaking effects, and the difficulty of scaling this concept to more advanced processes, because the large number of stacked transistors within the integrator necessitates a rather high supply voltage.

3.7.3.3 R-MOS-C Filters

A quite popular structure [Moon 1993] has been used for DSL applications, as described in [Egerer 1998]. Tunable resistors are built from the combination of a passive polysilicon resistor and an active transistor that is preferably operating in its linear region. By carefully selecting the relative contribution of each device to the total resistance and adding two extra current-steering MOS transistors, it is possible to trade the linearity of the global resistance for its tuning range. Higher-order filters can be built using this combined resistor element within conventional Rauch, Sallen–Key, or Tow–Thomas active RC filter structures. Linearity beyond 75 dB (with a 150 kHz input signal) is reported for the fifth-order low-pass filter in [Egerer 1998] with a 3.3 V CMOS 0.5 µ process, which offers a tuning range close to ±50 percent around a nominal frequency of 438 kHz. The tuning range in this case is essentially limited by the second-order components

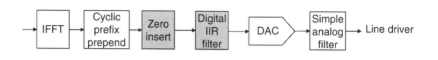

Figure 3.10 **Oversampled transmitter with zero insertion and digital IIR (low-pass) filter.**

created from the intrinsically nonsymmetrical mode of operation of the tuning current-steering transistors.

3.7.3.4 Oversampling

The use of oversampling and interpolation is a digital approach to solving the problem of aliasing. This approach can reduce the complexity of the analog filter, with a concomitant expense in digital circuitry and, of course, a DAC that is capable of running faster. Two possible approaches are shown in Figures 3.10 and 3.11. The first method uses interpolation that is accomplished by the combination of zero-insertion and a digital filter that runs at the oversampled rate.

An alternative way of achieving oversampling and interpolation that exploits the inherent computational efficiency of the fast Fourier transform (FFT)* is shown in Figure 3.11. In this case, the IFFT is several times larger than is required for data transmission, the extra unused tones it generates are set to zero.

3.8 Line-Coupling Transformer

Twisted-pair cable is used in balanced mode: the signal on each conductor is a replica of the other with opposite polarity. The line-coupling transformer aids in creating the balanced signals, as well as rejecting common-mode interference that originates either from crosstalk between pairs or

Figure 3.11 **Oversampled transmitter with extra large size IFFT.**

---

* The FFT is typically used to implement the IDFT required in DMT transmitters and the DFT required in the DMT receiver.

outside the binder. The galvanic isolation it provides is also critical to meet all system safety requirements (including the impact of lightning).

An additional benefit of the transformer is its ability to transform voltages via unequal turns ratio. This can be advantageous in those applications where large transmit voltages must be created, because a step-up transformer reduces the voltage range that must be produced by the line driver and hence reduces the supply voltage. The inductance of the transformer combined with the series coupling capacitors (see Figure 3.1) form a low-frequency pole that filters out interference from base-band telephony service. This pole must be low enough in frequency not to attenuate the DSL signals too much, but care must be taken so that the resulting impulse response does not significantly increase intersymbol interference.

An ideal transformer would provide perfect coupling between the two lossless windings, have no parasitic capacitance (and hence no insertion loss), all over infinite bandwidth and with infinite return loss. Practical transformers, unfortunately, fall short of these objectives. To understand the behavior of real transformers, it is useful to review the equivalent circuit in Figure 3.12, which illustrates a practical transformer as an ideal 1:n transformer surrounded by fictitious components that represent impairments. L_1 represents the open-circuit inductance, i.e., the desired inductance of the windings and core. L_2 is the leakage inductance that accounts for the flux produced in the core that is not linked by the other winding. C_w and R_p embody the capacitance and equivalent parallel resistance of the windings. Finally, R_0 accounts for the ohmic resistance of the windings.

In addition to undesired insertion loss, transformer impairments cause impedance mismatch between the line and the line driver, which in turn causes signal reflections that can have a deleterious effect on receiver performance. Therefore, control of return loss is also of great significance in transformer design.

The goal in transformer design is to minimize the insertion loss and maximize the return loss over sufficient bandwidth to support both received and transmitted signals, while maintaining adequate common-mode rejection

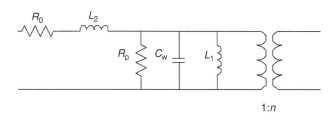

Figure 3.12   Line-coupling transformer—equivalent circuit.

and DC isolation. Pass-band insertion loss is governed by the ratio of the leakage inductance to the open-circuit inductance. Primary inductance of around 1 mH is typical for ADSL, with leakage inductance usually in the range 5–15 μH, so an insertion loss of less than 0.5 dB in-band is practical. High-frequency bandwidth is limited by the shunting of the signal by the winding capacitance and the resonant circuit formed by the winding capacitance in combination with the leakage inductance. For ADSL, winding capacitance of a few tens of picofarads will give sufficient performance with leakage inductance of around 10 μH. In contrast, VDSL presents a special challenge owing to its wide bandwidth requirements: from tens of kHz up to 30 MHz. Insertion loss at the low end of the band is determined by the open-circuit inductance combined with the line impedance. A compromise between the open-circuit inductance, winding capacitance, and the leakage inductance must be found to achieve adequate bandwidth.

3.9 Line Driver

The function of the line driver is to launch the transmit signal on the line with the correct PSD, while also providing a source termination to match the characteristic impedance of the line. The maximum output power that is allowed to be delivered to the line is specified in the relevant standards (see Chapter 17). For the downstream direction of ADSL2 over POTS, for example, the ATU-C transmit power is 20.4 dBm when overlapped spectra are used, or 19.9 dBm when nonoverlapped spectra are used. In the upstream direction, the maximum total power is 12.5 dBm for ADSL over POTS. VDSL is characterized by a generally lower PSD to minimize interference, however, its high bandwidth (up to 30 MHz) presents a design challenge, particularly when combined with the typically aggressive power consumption target for network-deployed equipment. As with many applications, the power consumption is a critical figure of merit of the complete transceiver. The power consumption affects how densely transceivers can be placed in access multiplexer at the CO. The advances of CMOS technology that have enabled smaller geometry have also enabled significant reductions in the power consumption of digital circuits. These power savings have become so large that the line driver has become (relatively) one of the most power hungry functions. For example, first generation ATU-C line drivers dissipated as much as 2.4 W [Wurcer 2003]. Today, 700 mW is typical. For this reason, the main design challenge in the line driver is one of optimizing the power efficiency while maintaining acceptable linearity performance.

The aggregation of transceivers at the CO, combined with the higher power needed for downstream ADSL transmissions, means that the most attention will be devoted to the specific problem of the ATU-C transmitter,

Table 3.1 Comparison of Transmit Signal Parameters for Various DSLs

	ADSL1 or 2 over POTS Downstream (Overlapped)	ADSL2+ over POTS Downstream	ADSL1 or 2 over POTS Upstream	VDSL2 Downstream, Profile 8d	Units
Signal bandwidth	1.104	2.208	138	30,00	kHz
Max total power	20.4	20.4	12.5	14.5	dBm
Crest Factor assumption	5.3	5.3	5.3	5.3	
Reference impedance	100	100	100	100	Ohms
rms Voltage	3.31	3.31	1.33	1.68	V
Voltage swing on line (peak-to-peak)	35.1	35.1	14.1	17.8	Vpp
Peak current	175.5	175.5	70.7	89.0	mA

for which 20.4 dBm of power is needed over a 1.1 MHz bandwidth. However, the discussion will be equally relevant to the newer DSL variants such as ADSL2+ or VDSL, (see Chapter 17). Table 3.1 compares transmit signal characteristics for various DSLs.

The stringent linearity requirements on the transmitter, in excess of 65 dB MTPR, have so far limited the options in terms of driver architectures: linear class-AB drivers or derivatives (class G and H) are the most widely deployed solution in CO or customer premises chipsets.

An interesting approach was proposed in [Matsumoto 2001], wherein the driver could be supplied directly from the 48 V typically available on the DSLAM backplane, the objective being to eliminate from the line card the components necessary for the generation of a differential 12 V supply. The driver would be combined with a line transformer presenting an unusual turns ratio smaller than one. Aside from the cost, power, and area savings that result from the elimination of a DC–DC converter, some extra power savings can be achieved because the higher impedance presented at the driver output will permit the reduction of the peak current requirements and lower somewhat the losses in the parasitic passives.

3.9.1 Class-AB

The defining characteristic of the class-AB amplifier is that the current drawn from the supply I_s is given by the sum of the quiescent current I_q and the absolute value of the output current I_{out}. If the amplifier is operating from supply voltage V_s, it will dissipate instantaneous power

$$P(t) = V_s I_s = V_s \left(I_q + \left| I_{out}(t) \right| \right)$$

The current needed will be scaled up by the transformer step-up ratio n, yielding

$$I_{out}(t) = \frac{n V_{tx}(t)}{Z_0}$$

The average power $\bar{P}$ can be obtained by taking the average of $P(t)$ and noting that for an approximately Gaussian signal such as that produced by a DMT transmitter,

$$E\left[\left| V_{tx} \right| \right] = \sqrt{\frac{2}{\pi}} V_{rms}$$

so that

$$\bar{I}_s = I_q + \frac{n}{Z_0} \sqrt{\frac{2}{\pi}} V_{rms}$$

For an ideal amplifier, the supply voltage needed is equal to the peak output voltage, this is determined by the peak voltage to be transmitted $CF V_{rms}$, scaled down by the transformer ratio n, and scaled up by the ratio of the termination resistor R_m to the transformed line impedance Z_0/n^2. Thus,

$$V_{out, peak} = \frac{CF V_{rms}}{n} \frac{(R_m + Z_0/n^2)}{Z_0/n^2}$$

First-generation line drivers used dual high-current op-amps configured in a balanced circuit with series resistors in the outputs to provide back-termination, i.e., source resistance R_m matched to the line characteristic impedance. An overview of this configuration is shown in Figure 3.13.

3.9.2 Active Termination

With passive back-termination, the output impedance of the driver is set by the series resistor R_m, which is chosen to be equal to the real part of the

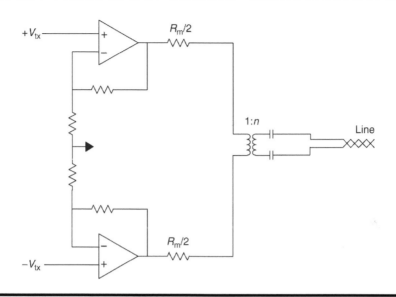

Figure 3.13 Conventional passively terminated line driver configuration.

transformed line impedance Z_0/n^2. In this situation, the power consumed
by the back-termination resistor is equal to the power delivered to the load,
so this naturally presents an attractive target for power saving. More recent
implementations have exploited active termination [Cresi 2001] in which
positive feedback around the amplifier is used to scale up the apparent
output impedance, allowing the use of smaller termination resistance
$R_m = kZ_0/n^2$. Figure 3.14 shows the general idea. The power savings
come at the expense of linearity performance and stability that deterio-
rate as a consequence of positive feedback, so this degradation must be
carefully controlled to preserve receiver performance.

A scaling factor k in the range of 0.1–0.2 is typical, and thus the amplifier
load becomes $(1 + k)Z_0/n^2$, where $k = 1$ represents passive termination.
The use of active termination reduces the voltage swing required of the
line driver to

$$V_{\text{out, peak}} = \frac{\text{CF}(1 + k)V_{\text{rms}}}{n}$$

To maintain good linearity, real amplifiers require a supply voltage with
some volts of headroom V_h in excess of $V_{\text{out, peak}}$. Thus, the minimum

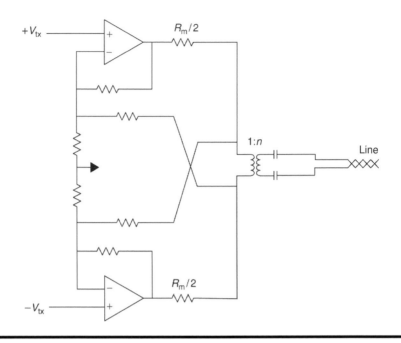

Figure 3.14 **Active termination with impedance synthesis created by positive feedback.**

supply voltage required can be written as

$$V_s = V_h + \frac{CF(1 + k)V_{rms}}{n}$$

yielding the average power consumption

$$\bar{P} = \left[V_h + \frac{CF(1 + k)V_{rms}}{n} \right] \left[I_q + \frac{n}{Z_0}\sqrt{\frac{2}{\pi}} V_{rms} \right]$$

As mentioned in [Wurcer 2003], early ATU-C drivers operated with $V_s = \pm 15$ V and $I_q = 15$ mA to produce an output power of 20.4 dBm ($V_{tx,rms} = 3.31$ V). If a crest factor $CF = 5.3$ is assumed along with a transformer step-up ratio of two, the power consumption would have been 2.0 W. The headroom voltage was typically around 2.5 V, amounting to 17 percent loss for the headroom alone. More modern designs using a 12 V supply have reduced the headroom to 1.2 V, which is still a 10 percent loss. Rail-to-rail circuits can be used, with the caveat that they have inferior bandwidth (for the same quiescent current) relative to conventional emitter–follower output stages.

The influence of the transformer ratio depends on the details of I_q and V_h, its effect can only be assessed once these are known.

If the amplifier is ideal, i.e., if $V_h = I_q = 0$,

$$\bar{P}_{ideal} = \sqrt{\frac{2}{\pi}} \frac{CF(1+k)V_{rms}^2}{Z_0} = \sqrt{\frac{2}{\pi}} CF(1+k)P_{tx} \tag{3.8}$$

then the theoretical efficiency P_{tx}/P_{ideal} depends only on the characteristics of the signal (via the CF and the $\sqrt{\frac{2}{\pi}}$ factor), and the active termination scaling factor k. With a CF of 5.3 and assuming $k = 0.1$, the power consumption of the ATU-C will be 510 mW. Implementations using this approach have achieved 20.4 dBm output with power consumption of 740 mW [Sabouri 2002].

Equation 3.8 highlights the dominance of the high-crest-factor DMT signal on the line driver efficiency; with a class-AB line driver, the supply voltage must accommodate the highest peaks, regardless of their relative infrequency. This observation has motivated two enhancements to the class-AB architecture. They exploit the infrequency of large-signal peaks to improve the line driver efficiency by reducing the main supply voltage to a level that supports the signal output most of the time, invoking special circuitry to support signal peaks only when necessary.

3.9.3 Class-G

The class-G driver [Maclean 2003] and [Vecci 2003] incorporates an additional supply V_{s1} that provides current most of the time, i.e., while the amplifier output voltage can be accommodated by V_{s1}. The larger voltage rail $V_{s2} = CF(1+k)V_{rms}/n$ supplies current only when the demanded voltage exceeds V_{s1}. The power consumed is found by integrating over two regions: from zero to V_{s1} and from V_{s1} to V_{s2} yielding the result

$$P = \sqrt{\frac{2}{\pi}} \frac{V_{rms}^2}{Z_0}(1+k)\left[a\left(1 - e^{-a^2/2}\right) + CFe^{-a^2/2}\right] \tag{3.9}$$

where a is the lower supply voltage scaled in terms of standard deviations of the transmit voltage so that

$$V_{s1} = \frac{a\left(1 + k\right) V_{rms}}{n}$$

The power consumption is minimized by differentiating Equation 3.9 with respect to a and solving the resulting equation numerically, arriving at, for $a = 2.0153$,

$$P = 2.45\sqrt{\frac{2}{\pi}\frac{V_{rms}^2}{Z_0}}(1 + k)$$

The efficiency improvement over the actively terminated ideal class-AB is CF/2.45, or about a factor of two.

3.9.4 Class-H

Another approach, known as class-H, obviates the extra power supply by creating the higher voltage supply temporarily, using a charge-pump and capacitor circuit that charges while the output voltage is less than V_{s1} and then discharges for signal peaks. The power consumption is from [Pierdomenico 2002]:

$$P = \sqrt{\frac{2}{\pi}\frac{V_{rms}^2}{Z_0}}(1 + k)a\left(1 + e^{-a^2/2}\right)$$

So with $a = 2.0153$ as before,

$$P = 2.28\sqrt{\frac{2}{\pi}\frac{V_{rms}^2}{Z_0}}(1 + k)$$

which indicates that the performance is slightly better than with class-G, although the main appeal of class-H is that the extra supply has been removed.

Table 3.2 shows a comparison of power consumption and efficiency for the various line driver architectures for the 20.4 dBm ATU-C, assuming a CF of 5.3, and neglecting the quiescent current and headroom voltage. However, the implementation details have a significant impact on achievable efficiency, therefore, a survey of practical ATU-C 20.4 dBm driver power consumption values is included in Table 3.3.

Table 3.2 Comparison of Power Consumption and Efficiency for Ideal Amplifiers

	P_{tx} (dBm)	V_{rms}	k	n	a	V_s	V_{s1}	ϵ	P (mW)
Ideal class-AB, passive term	20.4	3.31	1	2		17.55		0.118	927
Ideal class-AB, active term	20.4	3.31	0.1	2		9.65		0.215	510
Ideal class-G	20.4	3.31	0.1	2	2.02	9.65	3.67	0.466	235
Ideal class-H	20.4	3.31	0.1	2	2.02	9.65	3.67	0.500	219

An efficient implementation for an ATU-C published in [Maclean 2003] consumes 610 mW, which is a great improvement over early designs. However, static power dominates, and much focus is now placed on reducing the implementation overhead factors such as quiescent current and headroom voltage.

3.9.5 Switching Drivers, SOPA

Recent developments have led to more aggressive line driver architectures, based on successful results obtained with switching voiceband drivers. Although the maximum efficiency can be expected from these devices (because an ideal switch does not dissipate power), many issues have yet to be resolved to reach the linearity targets for DSL and satisfy the strong

Table 3.3 Survey of 20.4 dBm ATU-C Line Drivers

Author	Amplifier Type	Supplies (V)	Technology	Power (mW)	Remarks
[Shorb 2002]	Class-AB-D-G	+3.3	0.35 µm/3.3 V CMOS	1300	1:12.4 transf.
[Sabouri 2002]	Class-AB	+/−12	26 V 3 GHz Comp. Bipolar	710	Iq:4 mA
[Pierdomenico 2002]	Class-H	12	HV Bipolar SOI	684	1:1 transf.
[Bicakci 2003]	Class-H	6	0.25 µ CMOS	700	1:2.4 transf.
[Maclean 2003]	Class-G	+/−4, +/−8	0.7 µm BiCMOS	610	
[Vecci 2003]	Class-G	+/−5, +/−12	0.8 µm BCD technology	750	1:1.4 transf.

PSD limitations outside the signal bandwidth. In [Shorb 2002], Shorb et al. introduced the concept of the class-ABDG driver. The device combines a class-D driver to provide most of the power to the load and a highly linear class-AB driver responsible for the linearization of the output signal. An aggressive implementation of this concept was simulated in a 3.3 V standard CMOS 0.35 μ process. This driver needed a 1:12.4 transformer turns ratio, leading to a quite challenging peak current greater than 2.3 A and termination resistances of 0.6 Ω. The total power dissipation was 1.3 W, with significant power wasted to drive the class-D switches (83 percent of the total power). Different optimizations are discussed in [Shorb 2002] to reach efficiencies in line with current leading edge implementations.

The self-oscillating power amplifiers (SOPA) concept was first presented by Piessens and Steyaert at ISSCC in 2001 [Piessens 2001]. The principle is to apply the signal to be amplified to a self-oscillating device, built from positive feedback around a comparator, a driver, and a filter. The device is not clocked, and the loop is entirely analog. When no signal is applied, the system oscillates at a given frequency, the limit cycle, determined by the loop gain and the filter bandwidth. The output of the comparator is a square wave signal, and the power dissipation is mostly due to charging and discharging of the output node capacitance. When a signal is applied to the system, the output is still a square wave, combining now the limit cycle and the forced signal. By combining two self-oscillators and connecting them in a differential way, the system rejects the square wave limit cycle, which ideally appears as a common mode on the primary of the transformer, and only the forced input signal is present on the secondary of the transformer. In [Piessens 2001] and further described in [Piessens 2002], efficiencies close to 61 percent were reported, although the distortion was still quite high (an MTPR of only 41 dB was measured for an ADSL signal with a crest factor of 5). An improved version of the first design is reported in [Piessens 2003], using a third-order filter and providing MTPR of over 55 dB with efficiency of nearly 50 percent.

Although, self-oscillating and other switching drivers are probably not yet mature enough to be accepted for mass deployment because of their current poor linearity, they could present an interesting alternative in the future to more conventional "linear" schemes. Such devices may be invaluable in the realization of cheap and highly efficient CMOS line drivers, as they have been already in the audio world.

3.10 Hybrid

The large signal developed by the line driver must be launched into the line with the minimum of disturbance to the small receive signal arriving from the far end, while enabling full duplex operation. This is a similar function to that required in baseband telephony, whereby the upstream and downstream signals must be separated at the ends of the subscriber loop for amplification and subsequent processing, and is known as a hybrid. Its effectiveness can have a significant effect on the performance of FDD and overlapped spectrum systems because leakage from the transmitter (known as trans-hybrid loss, or ERL) appears as noise to the receiver. At its most severe extent, the remnant of the transmitted signal may be large enough to cause receiver saturation, for which there is no remedy in the receiver, other than to reduce receiver gain [Chen 1992].

A passive hybrid is a common way to separate the received signal from the transmitted signal. The ideal hybrid relies on perfect matching between the line impedance and a locally synthesized impedance to generate two analog electrical signals, which are subtracted further on to eliminate the transmitted signal, with its own noise and distortion components, from the received signal. Different hybrid schemes have been used, based on either transformers, resistor bridges, or active components, with some level of shaping to model, as closely as possible, the phone line frequency response [Lee 2002].

Conventional hybrids use variants of the Wheatstone bridge circuit. The voltage divider formed by the load and the source impedance is paired with another divider with impedances chosen so that the transmit voltage is duplicated. The receive signal is derived from the difference between the two arms of the divider. This arrangement is shown schematically in Figure 3.15 in a single-ended configuration for clarity. Z_2 is a network of passive components that approximates the input impedance of the load that is formed by the line-coupling high-pass filter and line. If the condition for balance is met, i.e., if $Z_1/Z_2 = R_m/Z_{load}$, then none of the transmit signal appears in the receiver, and the hybrid does a perfect job.

In reality, perfect matching is difficult to achieve. First, the line-coupling high-pass filter consisting of the transformer and series capacitors will

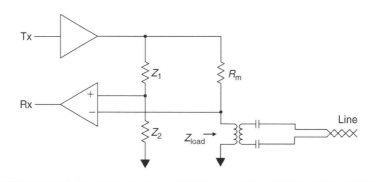

Figure 3.15   Conventional fixed hybrid.

136 ■ *Implementation and Applications of DSL Technology*

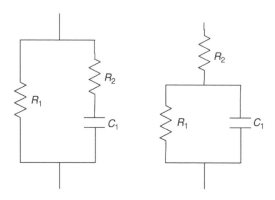

Figure 3.16   Example of passive hybrid matching networks. (From Lee, S.S., *Proceedings of IEEE 2002 CICC*, 37, 2002.)

have some mismatch with Z_1 and Z_2. Second, the impedance of the twisted-pair has a complicated dependence on frequency that is difficult to duplicate with passive components (see chapter 2 of [Golden 2006]). Lastly, impedance discontinuities, such as line gauge changes or bridged taps, cause reflections with large delay that are also difficult to synthesize with analog circuits.

The simplest approach to hybrid design, which is often used in practice, is to design a compromise matching network for Z_2 that provides adequate return loss over the expected component tolerances and loop conditions and hope for the best. Figure 3.16 shows example networks from [Lee 2002]. This approach may be sufficient for FDD systems with generous guard bands and with aggressive analog transmit filters deployed at the CO, where the effect of bridged taps and gauge changes are minimal. In [Hellums 2003], an active filter implements a fourth-order transfer function (four poles, four zeros) to compensate for the third-order line-coupling high-pass filter. This design achieves better than 25 dB of return loss over the ADSL band, however, behavior on bridged tap loops is not quoted.

To better cope with component tolerances and drift, adaptive hybrids [Dotter 1980] have been developed, with switchable impedance networks that are selected based on the conditions observed during modem initialization. In [Oswal 2004], three selectable hybrid settings are provided. Pecourt's approach [Pécourt 1999] is actually an analog EC with tuning of five parameters in 32 steps to implement a second-order transfer function. Return loss of better than 25 dB on the ANSI ADSL test loops is reported. (See chapter 2 of [Golden 2006] and [Starr 1999] for details of the ANSI test loops.) Challenges arise with bounding the range of switchable components to cover the expected variation in operating conditions, and the development of stable tuning algorithms [Wen-Kuang Su 1994] that provide confidence in the convergence to a global optimum. The authors of [Pécourt 1999] and [Oswal 2004] do not provide any details on how the settings are selected.

In the most demanding applications, such as those with overlapped spectra and digital duplexing [Cioffi 1999] (see Chapter 17), additional methods may be called for that are capable of synthesizing large time delays. This requirement suggests the use of digital processing with an extra DAC to inject a cancelation signal at the receiver input. This observation naturally leads into the next section on analog echo cancelation.

3.11 Echo Cancelation

As already mentioned (see Sections 3.10 and 3.7.1), the residual analog echo signal received after hybrid attenuation can dramatically affect the receiver sensitivity and consume a significant portion of the total receiver dynamic range. Echo cancelation falls into the general category of interference cancelation [Widrow 1985], in which a reference input is used to derive an estimate of the interference and thereby cancel it from the observations, leaving the desired signal. The discussion in this section will concentrate on analog echo cancelation, in which the cancelation is done in the analog domain. However, the signal that drives the canceler may in fact be derived from the digital system. This approach is in contrast to purely digital echo cancelation [Ho 1996], which is incapable of dealing with the issue of front-end saturation; rather, it is motivated by the need to remove the residual echo signal in systems with overlapped spectra.

ECs associated with telephony have a long history in communications engineering. Early implementations were analog [White 1972], however, advances in circuit technology allowed digital techniques to become prevalent. For telephony, the bandwidth of interest is approximately 400 Hz–4.3 kHz. The EC must be tuned, or tune itself, to model the trans-hybrid loss over this bandwidth. Analog approaches focused on the adjustment of only two or three parameters, because the input impedance of loops could be adequately characterized by simple resistor–capacitor lumped element models. Gazioglu's adaptive hybrid of [Gazioglu 1979] yielded good performance on a 1 km mixed gauge loop by adjusting only three parameters, although its performance degrades on longer loops.

The situation in DSL is quite different. To illustrate the extent of the problem, consider the echo responses shown in Figure 3.17, which are from a simulated test loop known as CSA4, which is defined for ADSL. CSA4 is a particularly difficult loop, consisting of 550 ft of 26 AWG line, followed by a 1200 ft bridged tap, followed by a gauge change to 24 AWG

138 ■ *Implementation and Applications of DSL Technology*

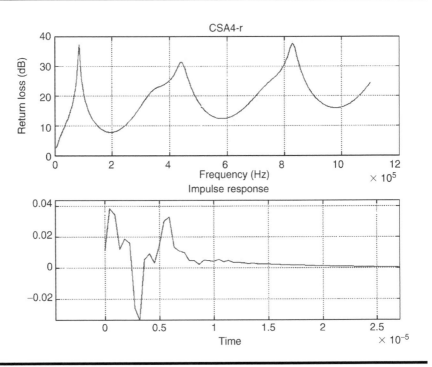

Figure 3.17   CSA4 at subscriber end.

and then back to 26 AWG, with another 650 ft bridged tap only 300 ft from the CPE. (See chapter 2 of [Golden 2006] for details of CSA4 and other DSL test loops.) It is the proximity of the bridged tap to the CPE in this case that causes the problem: there is insufficient attenuation of the echo before it returns to the CPE.

The upper part of Figure 3.17 shows the simulated return loss as function of frequency, and the lower part depicts the echo impulse response that is computed from its inverse transform. An important feature of this impulse response is the presence of two significant echoes with delays of approximately 3 and 6 μs. Any EC that seeks to be effective on loops like this one must encompass at least 6 μs of delay, with sufficient flexibility to cope with variations in the precise configuration of bridged taps. This requirement will have serious implications on the required complexity of ECs, as will be seen later.

Figure 3.18 shows a natural way to implement the EC with an analog filter that uses a copy of the transmit output as developed by the line driver. The analog filter $H_{ec}(j\omega)$ is designed to approximate the echo response $H_r(j\omega)$. As long as $H_{ec}(j\omega)$ forms a good approximation to $H_r(j\omega)$, the structure yields good echo suppression, with the advantage that nonlinearity

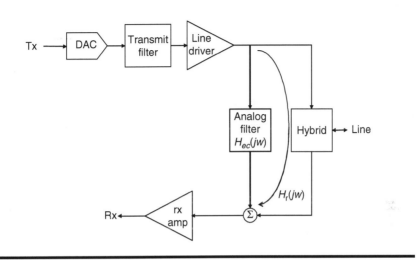

Figure 3.18   Analog echo canceler.

and noise developed in the transmit signal path is also canceled at the
receiver input, provided that the analog echo filter is very linear and contributes little noise. Examples of this architecture have been described in
[Pécourt 1999] and [Oswal 2004]. An active filter with possibly adjustable
poles and zeroes duplicates the echo response. There are two shortcomings to this approach. The first has already been alluded to above: if the
analog filter structure is unable to adequately model the echo response,
then there will be significant residual echo. The other difficulty is associated with adaptivity. Adaptive hybrids and ECs are definitely of benefit
because they are able to adjust to conditions that are unknown when the
system is designed, including component tolerances or mismatch, and the
unknown loop configuration on which the system will be operating. Pole
or zero filters are notoriously difficult to train. Furthermore, the range of
component values that will be needed and the resolution of the adjustment
steps must be assessed a priori.

A combination of digital signal processing with analog cancelation
would therefore have major advantages, in that the adaptivity could be
digitally controlled via an echo model embodied as a digital FIR filter. Such
a system is described in [Conroy 1999], [Hester 1999], and [Pal 2003]. An
example of this architecture is shown in Figure 3.19. The digital form of
the transmit signal is used as the input to a digital filter that synthesizes the
echo. An extra DAC converts this signal to analog form, and the signal then
cancels the echo at the receiver input. Training of the digital echo model is
done under the control of the digital system using well-known algorithms
such as least mean-squares (LMS) [Widrow 1985]. The fixed analog filter

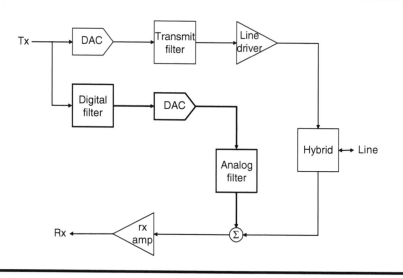

Figure 3.19   Digitally driven analog echo canceler.

at the DAC output is needed only for mild spectrum shaping and is not complex. The disadvantage of this approach is that the receiver is exposed to any transmitter noise and nonlinear distortion via the trans-hybrid loss $H_r(j\omega)$.

In the literature, only [Pécourt 1999] has aimed analog echo cancelation directly at the problem of bridged taps. It reports a potential reduction in the ADC resolution by more than 2 bits as well as a significant relaxation in out-of-band linearity and noise requirements for the transmit path. The adaptive echo canceling path is built around three op-amp biquad structures, using eight resistors and two capacitors to implement a generic second-order transfer function, in a 0.6 μ BiCMOS process. Impressive hybrid rejections are reported between 42 dB (on straight loop ANSI #7) and 26 dB (for the more complex ANSI #13, which has two bridged taps close to the customer premises). The main challenge is to limit the noise contribution of this active hybrid path to a level below the line noise without significantly increasing power or area, which can be difficult at the customer end of the line because the line noise can be below −145 dBm/Hz.

In the discussion of the transmitter noise budget, the specifications on transmitter linearity were driven by concerns about poor ERL. If the nonlinearity can be canceled in the receiver, this allows some relaxation in the linearity requirement. If the transmit signal has values from a fairly small set, as in binary or low radix single-carrier modulation, then the nonlinear distortion can be expressed as a look-up table or random access memory (RAM), which is included in the cancelation path. This idea goes back to the papers of [Messerschmitt 1980], [Agazzi 1982], and [Michaelides 1988]. However, for multilevel and particularly for multicarrier signals such as DMT, the look-up table would become very complex. If it can be assumed that the majority of the distortion comes from one or two simple effects, such as low-order polynomial distortion; this can be incorporated in the design of the digital filter that forms the model for the echo. Success of these reduced complexity approaches will depend on careful characterization of the possible nonlinear effects that might be encountered.

3.12 Receive Amplifier—Programmable Gain Amplifier (PGA)

As the signal propagates along the loop, it suffers attenuation. The receive amplifier is needed to boost the signal up to a level suitable for processing by the ADC. Because the required gain depends on the details of the loop, a control loop is formed by the combination of the receive amplifier and digital signal processor via the ADC, whereby the received signal amplitude is measured and gain corrections are computed by the DSP, which then programs the desired gain into the PGA during the initialization procedure. After the gain has been adjusted to the correct level for the ADC, it is fixed. Adjustments to allow for small dynamic variations in the received signal amplitude are made by the frequency-domain equalizer (see chapter 7 of [Golden 2006]), which is under control of the DSP.

For example, downstream ADSL signals are transmitted at up to 20.4 dBm. Overall attenuation on CSA range loops (defined as a 26 AWG loop of 9 kft or the roughly equivalent 24 AWG loop of 12 kft) is about 40 dB. The ADC will have an input range of 1–4 Vpp, which is equivalent to an input power of about 0 dBm. Thus, a receive gain of about 20–30 dB will be needed. On shorter loops, the attenuation will be much less, and in fact, it is desirable (usually for testing purposes) that the modem is able to connect satisfactorily even on null-loops, i.e., loops that are just a few feet of cable. The consequence is that the amplifier gain must be programmable from a few decibels of attenuation (negative gain) to 20 or 30 dB, with a step size that makes a reasonable compromise between the onset of clipping (gain too high) and excess quantization noise (gain too low). The penalty in SNR of gain misadjustment on the low side is equal to the amount in decibels by which the gain is incorrect. The effect of clipping is discussed in [Mestdagh 1993]. Step-sizes of 0.25–1 dB are typical.

Figure 3.20 shows a typical architecture for the receive amplifier. FET switches are used to select various resistor combinations, which in turn set the gain of the amplifier. Resistors generate thermal noise. An alternative is

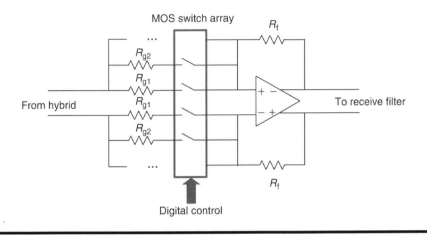

Figure 3.20   Programmable gain amplifier.

to use capacitors, which are noiseless, instead of resistors; however, there can be issues with settling and charge injection.

Note that receiver noise is usually specified as an input-referred, line-equivalent noise, i.e., as an equivalent voltage source on the line, so that the receiver gain and transformer turns ratio are divided out. At high-gain settings, the input-referred noise will be approximately constant for all loop lengths because it is dominated by the first gain stage of the amplifier. If a transformer with a step-up ratio from the modem to the line is used, the receive signal is attenuated, thereby magnifying the effect of noise when considered in line-equivalent terms. At low gain settings, other noise sources in the receiver become significant, particularly the quantization noise from the ADC. However, the low gain implies operation on a relatively short loop. Hence, attenuation is minimal, and noise magnitude is not so significant.

From the discussion of the receiver noise budget in Section 3.5, −150 dBm/Hz is a good target for PGA noise, which is equivalent to an input voltage noise of 10 nV/rtHz. This budget has to be met by the combination of op-amp noise, which is typically dominated by collector or drain current shot noise, and the thermal noise generated in the gain-setting resistors R_f and R_g, which is equivalent to the parallel combination resistance. Thus, the input-referred voltage noise of the amplifier with feedback is

$$v_n^2 = 2v_a^2 + 4kT\left(\frac{R_f R_g}{R_f + R_g}\right)$$

where the term $2v_a^2$ represents the amplifier noise, assuming the input stage is a differential pair. For both bipolar and MOS devices, there is generally an inverse relationship between input voltage noise and trans-conductance. The relationship for bipolar devices is [Gray 1977]

$$v_a^2 = 4kT\left(\frac{1}{2g_m} + r_b\right)$$

and for MOS devices the relationship is

$$v_a^2 = 4kT\left(\frac{2}{3g_m} + \frac{K'I_D^d}{g_m^2 f}\right)$$

If it can be assumed that the intrinsic device parameters r_b and the flicker noise term $\dfrac{K'I_D^d}{g_m^2 f}$ can be made insignificant by device design, then these equations imply that noise can be reduced by running high emitter or source current in the input devices, thus increasing g_m but consequently increasing the power consumption. A practical limit on the overall noise will be imposed by other design considerations on the feedback and gain setting resistors, resulting in a parallel resistance of perhaps 100–200 Ω. If these assumptions are combined with trans-conductance of 5 mA/V, the amplifier noise density is

$$v_n^2 = 4kT(200\ \Omega + 100\ \Omega) = \left(2.2\,\mathrm{nV}/\sqrt{\mathrm{Hz}}\right)^2$$

for bipolar devices and

$$v_n^2 = 4kT(267\ \Omega + 100\ \Omega) = \left(2.5\ \mathrm{nV}/\sqrt{\mathrm{Hz}}\right)^2$$

for MOS devices.

Op-amps for DSL application with input noise down to 2 nV/rtHz [Siniscalchi 2001] have been fabricated in CMOS running from a 3.3 V supply. Power consumption of receive amplifiers has been reported between 10 [Cornil 1999] and 35 mW [De Wilde 2002]. Other circuit design considerations relevant to low-noise amplifiers that have been reported are the use of low threshold voltage devices to reduce $1/f$ noise [Siniscalchi 2001], selection of PMOS devices as the first input pair (also with the aim of reducing $1/f$ noise) [Langford 1998], and the use of gain-dependent compensation to reduce the gain-bandwidth requirements [Cornil 1999].

3.13   Analog-to-Digital Converter

The simplest, and fastest, ADC is known as a flash converter. It consists of an array of comparators with evenly spaced thresholds generated by a resistor ladder network. Although fast in sampling rate and low in delay, the flash converter suffers from the drawback of exponentially increasing complexity as the number of bits is increased. Other converter architectures have been devised that strike a balance between sample rate, resolution latency, and complexity, where the targeted application dictates the best choice. For DSL transceivers, sample rate requirements range from 2 MHz to more than 70 MHz at resolution (as discussed previously) from 12 to 14 bits. Further, this application is not sensitive to conversion delay (latency) of a few clock cycles, because this delay appears to the transceiver as extra line delay (i.e., it is indistinguishable from line delay to the transceiver). This fact can be exploited in the choice of ADC architecture and has led to the use of two types of converters in this application: the multistage pipeline and the sigma–delta converter.

Any ADC consists of two main components: the sample-hold unit, which acquires the value of the input signal at the sampling instant and holds it steady, and the quantizer, which subsequently classifies the input relative to one of the digital output codes. The essence of the sample-hold is a circuit with several capacitors whose charging is controlled by CMOS transistor switches. During one half-cycle, the first capacitor is charged up to a voltage equal to the input signal, then this charge is transferred to the second capacitor, and so on. Fast settling is required (to within 1 LSB or so), which is aided by the use of small capacitors. A lower limit is imposed by the gate-drain and gate-source capacitance of the switch transistors, because if the capacitances are significant relative to the sample-hold capacitors, nonlinear charge feed-through effects will degrade ADC performance.

3.13.1   Pipeline ADC

The pipeline converter is a direct hardware implementation of a binary search algorithm, with each step of the algorithm being implemented in a different stage. For each stage of the pipeline, the signal is quantified in n bits through a sub-flash converter, and the residue is amplified before being sent to the next stage. The throughput of the converter is independent of the number of stages, each stage will be designed to operate at high sampling

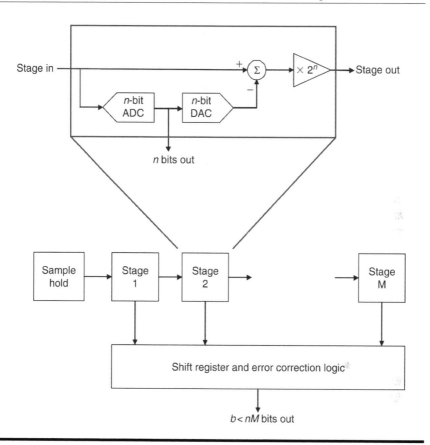

Figure 3.21   Multistage pipeline ADC.

rates. Therefore, the pipeline architecture is suitable for medium-speed (>1 Msamples/s), high-resolution converters (>10 bits). A basic block diagram of such a converter is shown in Figure 3.21. This type of converter uses redundancy combined with digital correction, which relaxes the comparator specifications and the specifications of the references used in the sub-flash ADCs. The main sources of nonlinearity are the gain error between stages, and the nonlinearity in the DAC subconverter calculating the residue, leading to tight requirements for capacitance matching and the amplifier gain. Although the matching is improving with process scaling for a given capacitance value, the design of high DC-gain amplifiers becomes a real challenge for deep sub-micron technologies, where the reduced supply voltage limits the stacking of cascode transistors. Calibration techniques may be used to improve the ADC linearity [Conroy 1999] and [Hester 1999] when linearity above 12–13 bits is required, at the expense of additional silicon area. Other accuracy limitations arise because of the thermal noise injected during the sampling process (kT/C noise) and clock jitter.

An example of a pipeline implementation is illustrated in [Cornil 1999]. Behavioral simulations led the authors to an optimal design of a six-stage structure, with each stage delivering 3 bits combined digitally. This structure is reported to offer the best power performance for the required sampling rate and linearity. True 12-bit performance is reported, using four-times oversampling at 8.8 MHz without trimming or calibration. Matching and kT/C noise considerations led to a minimal size for the sampling capacitors of 2.5 pF in the first sample-hold stage. Gain boosting techniques were used to guarantee a minimal DC gain for the amplifiers greater than 100 dB.

3.13.2 Sigma–Delta ADC

Sigma–delta techniques are commonly used for high-resolution, low-speed ADCs. The sigma–delta converter trades the high quantization noise of intrinsically linear single-bit or low-resolution multibit ADCs and DACs with high oversampling rates and high-order filters, for quantization noise shaping to reach the required signal-to-noise-plus-distortion ratio. This type of converter, illustrated in Figure 3.22, offers numerous advantages in the DSL space compared to other high-speed conversion schemes, which are usually limited to 10–11 bit accuracy. However, their first application ended up being quite challenging, as will be explained later. To achieve the required 13–14 bit accuracy on, for example, the ADSL or ADSL2 bandwidth with reasonable power consumption, third- to fourth-order noise shaping is implemented to limit the oversampling ratio

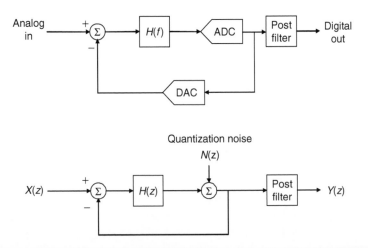

Figure 3.22   ADC with noise-shaping.

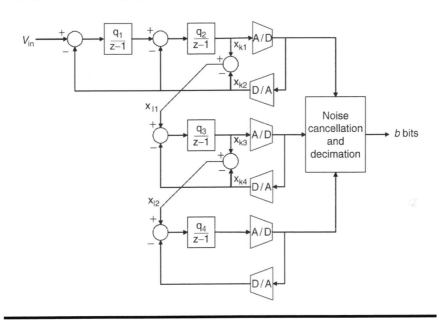

Figure 3.23   Sigma–Delta ADC—fourth-order MASH structure.

of the converter to 24 or 32, which leads to a 48–70 MHz sampling
clock speed.* Because single loop modulators exhibit stability problems
for orders higher than two, the converters are built from a cascade of a
second-order modulator followed by lower-order modulators, which is also
referred to as a MultistAge noise SHaping converter (MASH) structure, as
shown in Figure 3.23. A digital logic component combines the modulator
outputs to cancel the quantization noise from all intermediate modulator stages, which produces a higher-order quantization noise shaping.
A digital decimator filters out most of the out-of-band quantization noise
from this recombined signal. Numerous realizations for DSL applications
have been published in the last five years, based on various cascading
of modulators. For example, two second-order modulators in cascade
(also referred to as a "2-2" structure), "2-1-1" structures [Guangming 1994],
[Morizio 2000], and [Gupta 2002] or combinations of some of these structures are used with multibit quantizers in the last stages [Moyal 1999],
[Siniscalchi 2001], and [Gupta 2002] to reduce the sensitivity to uncanceled
quantization errors from the earlier stages. One of the major limitations
of all these architectures is the integrator gain error, which degrades the
accuracy of the digital noise cancellation operation. High accuracy requires
very low settling time for the amplifiers and good capacitor matching.

---

* To use this approach with VDSL2, which uses signal bandwidth up to 30 MHz, sample rate of
at least 720 MHz would be required.

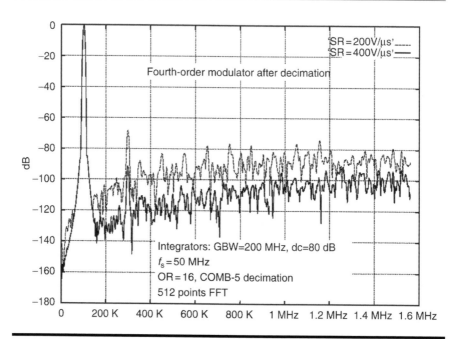

Figure 3.24 **Nonlinearity in multiloop Sigma–Delta ADCs (FFT after decimation, 100 kHz input; dotted line: op-amps with limited 200 V/μs slew-rate performance; plain line: op-amp with 400 V/μs slew-rate).**

A sigma–delta implementation was proposed for the first integrated ADSL AFE in 1995 [Chang 1995], at a time when the speed of the leading edge 0.7 μ CMOS process limited the signal bandwidth for this type of converter to a few hundred kilohertz. To limit the power consumption and relax the global design risks of this first AFE, the overall frequency plan of this first ADSL system was scaled down to 700 kHz rather than the 1.1 MHz required afterwards by the first version of the standard. The ADC dynamic range was 71 dB, most probably limited by the gain errors of the integrators and excessive substrate crosstalk from the digital components. Figure 3.24 illustrates the dramatic effects of slew-rate limitation for a 100 kHz input signal that generates nonlinearity (third harmonic distortion increased by 30 dB) and reduces the effectiveness of the noise cancelation logic (noise floor after decimation increased by 20 dB). Overall, this first design dissipated more than 350 mW for the analog modulator alone. The decimation filter was integrated on the same die and burned a similar amount of power. A significant power and area saving was provided later from the transfer of the digital portion of the converter to the digital chip, which was migrated to a more advanced digital CMOS processes. This led to a natural split between the analog and the digital integrated circuits at the modulator output, enabling a reduced set of high-speed pulse data modulated (PDM) signals between the two chips. Current sigma–delta based designs use a similar multistage approach with power consumption closer to 50 mW for the modulator itself. The area and power dissipation of the few thousands of gates required for the digital decimator are insignificant in advanced 0.13 μ process.

References

[ADSL1]   ITU-T Recommendation G.992.1: *Asymmetric digital subscriber line (ADSL) transceivers*, July 1999, available at http://www.itu.int/rec/T-REC-G.992.1-199907-I/en

[ADSL2]   ITU-T Recommendation G.992.3: *Asymmetric Digital Subscriber Line Transceivers 2 (ADSL2)*, January 2005, available at http://www.itu.int/rec/T-REC-G.992.3/en

[AD9744]   Analog Devices AD9744 Datasheet, available from www.analog.com.

[Agazzi 1982]   Agazzi, O., Messerschmitt, D., and Hodges, D., Nonlinear echo cancelation of data signals, *IEEE Transactions on Communications*, Vol. 30, Iss. 11, November 1982, pp. 2421–2433.

[Bicakci 2003]   Bicakci, A., Chun-Sup Kim, and Sang-Soo Lee, A CMOS line driver for ADSL central office applications, *IEEE Journal of Solid-State Circuits,* Vol. 38, Iss. 12, December 2003, pp. 2201–2208.

[Chang 1995]   Chang, Z.Y, Macq, D., Haspeslagh, D., Spruyt, P.M.P., and Goffart, L.A.G, A CMOS analog front-end circuit for an FDM-based ADSL system, *IEEE Journal of Solid-State Circuits*, Vol. 30, Iss. 12, December 1995, p. 1449.

[Chen 1992]   Chen, W.Y., The distribution of echo signal and required ADC precision, *Proceedings of IEEE ISCAS'92*, Vol. 2, May 10–13, 1992, pp. 581–584.

[Cioffi 1999]   Cioffi, J.M., G.vdsl: Digital duplexing: VDSL performance improvement by aversion of frequency guard bands, *ITU SG-15*, Question 4/15, NT-041, Nashville, Tennessee, November 1999.

[Conroy 1999]   Conroy, C., et al., A CMOS analog front-end IC for DMT ADSL, *IEEE International Solid-State Circuits Conference*, Digest of Technical Papers, Papers 14.2, 1999.

[Cornil 1999]   Cornil, J.P., et al., A 0.5 μm CMOS ADSL analog front-end IC, *IEEE International Solid-State Circuits Conference*, Digest of Technical Papers, Papers 14.1, 1999.

[Cresi 2001]   Cresi, M., Hester, R., Maclean, K., Agah, M., Quarfoot, J., Kozak, C., Gibson, N., and Hagen, T., An ADSL central office analog front-end integrating actively-terminated line driver, receiver and filters, *2001 IEEE International Solid-State Circuits Conference, 2001*. ISSCC Digest of Technical Papers, pp. 304–305, 2001.

[De Wilde 2002] De Wilde, W., Scantamburlo, N., Combe, M., Van Leeuwe, J., Doorakkers, K., Mazoyer, Y., Renous, C., Petigny, R., Bonin, A., Bayrackie, B., Belhi, B., Moons, E., and Sevenhans, J., Analog front-end for DMT-Based VDSL, *ISSCC 2002*, Paper 19.5.

150 ■ *Implementation and Applications of DSL Technology*

[Dotter 1980]   Dotter et al., Implementation of an adaptive balancing hybrid, *IEEE Transactions on Communications*, Vol. 28, August 1980, p. 1408.

[Egerer 1998]   Egerer, J., et al., A low distortion linear-tunable continuous-time 488 kHz fifth-order Bessel filter, *Proceedings of the 11th Annual IEEE International ASIC Conference 1998*, Vol. 47, September 1998.

[Filter Free]   Filter Free, available from www.nuhertz.com.

[FilterPro]   FilterPro, available at www.ti.com.

[Gazioglu 1979]   Gazioglu, H., Homer, D., and Sewell, J., An improved adaptive electronic circulator for telephone applications, *IEEE Transactions on Communications*, Vol. 27, Iss. 8, August 1979, pp. 1218–1224.

[Golden 2006]   Golden, P., Dedieu, H., Jacobsen, K.S., Eds., *Fundamentals of DSL Technology*, Auerbach Publications, Boca Raton, Florida, 2006, p. 457.

[Gray 1977]   Gray P.R., and Meyer, R.G., *Analysis and Design of Analog Integrated Circuits*, Chapter 11, John Wiley & Sons, New York, 1977.

[Guangming 1994]   Guangming Y. and Sansen, W., A high-frequency and high-resolution fourth-order SD ADC converter in BiCMOS technology, *IEEE Journal of Solid-State Circuits*, Vol. 29, Iss. 8, August 1994, p. 857.

[Gupta 2002]   Gupta, S.K. and Fong, V., A 64-MHz clock rate SD ADC with 88-dB SNDR and −105-dB distortion at 1.5-MHz signal frequency, *IEEE Journal of Solid-State Circuits*, Vol. 37, Iss. 12, December 2002, p. 1653.

[Hellums 2003]   Hellums, J., Hester, R., Corsi, M., Hagan, T., and Halbach, R. An ADSL integrated active hybrid circuit, *Proceedings of the 29th European Solid-State Circuits Conference, 2003, ESSCIRC'03*, September 16–18, 2003, pp. 517–520.

[Hester 1999]   Hester, R. et al., CODEC for echo-canceling full-rate ADSL modems, *IEEE Journal of Solid-State Circuits*, Vol. 34, Iss. 12, December 1999, p. 1973.

[Ho 1996]   Ho, M., Cioffi, J.M., and Bingham, J.A.C., Discrete multitone echo cancelation, *IEEE Transactions on Communications*, Vol. 44, Iss. 7, July 1996.

[Ingels 2002]   Ingels, M., Bojja, S., and Wouters, P., A 0.5 μm CMOS low-distortion low-power line driver with embedded digital adaptive bias algorithm for integrated ADSL analog front-ends, *IEEE International Solid-State Circuits Conference, Digest of Technical Papers,* Paper 19.3, 2002.

[Jantzi 1993]   Jantzi, S.A., Sielgrove, W.M., and Ferguson, P.F., A fourth-order bandpass sigma–delta modulator, *IEEE Journal of Solid-State Circuits*, Vol. 28, Iss. 3, March 1993.

[Langford 1998]   Langford, D.S., Tesch, B.J., Williams, B.E., and Nelson G.R., Jr., Harris Semiconductor, A BiCMOS analog front-end circuit for an FDM-based ADSL system, *JSSC 1998*, September, pp. 1383–1393.

[Lee 2002]   Lee, S.S., Integration and system design trends of ADSL analog front-ends and hybrid line interfaces, *Proceedings of IEEE 2002 CICC*, 37, 2002.

[Maclean 2003]   Maclean, K., Corsi, M., Hester, R.K., Quarfoot, J., Melsa, P., Halbach, R., Kozak, C., and Hagan, T., A 610-mW zero-overhead class-G full-rate ADSL CO line driver, *IEEE Journal of Solid-State Circuits*, Vol. 38, Iss. 12, December 2003, pp. 2191–2200.

[Matsumoto 2001]   Matsumoto, C., Legerity ADSL driver pushes integration, *EE Times magazine*, June 11, 2001.

[Messerschmitt 1980] Messerschmitt, D., An electronic hybrid with adaptive balancing for telephony, *IEEE Transactions on Communications*, Vol. 28, Iss. 8, August 1980, pp. 1399–1407.

[Mestdagh 1993]   Mestdagh, D.J.G., Spruyt, P.M.P., and Biran, B., Effect of amplitude clipping in DMT-ADSL transceivers, *Electronics Letters*, Vol. 29, Iss. 15, July 22, 1993, p. 1354.

[Michaelides 1988] Michaelides, J.F. and Kabal, P., Nonlinear adaptive filtering for echo cancellation, *Proceedings of IEEE International Conference on Communications* (Philadelphia, PA), June 1988, pp. 30.3.1–30.3.6.

[Moon 1993]   Moon, U.K. and Song, B.S., Design of a low distortion 22-kHz fifth-order Bessel filter, *IEEE Journal of Solid-State Circuits*, Vol. 28, Iss. 12, December 1993, p. 1254.

[Morizio 2000]   Morizio, J.C, et al., 14-bit 2.2-MS/s sigma–delta ADC's, *IEEE Journal of Solid-State Circuits*, Vol. 35, Iss. 7, July 2000, p. 968.

[Moyal 1999]   Moyal, M., Groepl, M., and Blon, T., A 25-kft, 768-kb/s CMOS analog front-end for multiple-bit-rate DSL transceiver, *IEEE Journal of Solid-State Circuits*, Vol. 34, Iss. 12, December 1999, p. 1961.

[Moyal 2003]   Moyal, M., Groepl, M., Werker, H., Mitteregger, G., and Schambacher, J., A 700/900 mW/channel CMOS dual analog front-end IC for VDSL with integrated 11.5/14.5dBm line drivers, *2003 IEEE International Solid-State Circuits Conference, Digest of Technical Papers*, pp. 416–504.

[Oswal 2004]   Oswal, S., Mujica, F., Prasad, S., Srinivasa, R., Sharma, B., Radychoudary, A., Khasnis, H., Sharma, A., Sriram, R., Vijayvardhan, B., Menon, R., Gireesh, R., Ahuja, N., Gambhir, M., and Sadafale, M., An integrated ADSL CPE analog front-end in 0.13 μm CMOS SoC, *ISSCC 2004*, Paper 22.2, *Digest of Technical Papers*, 2004.

[Pal 2003]   Pal, D., Chari, S., Hansen, C., and Lu, C.-L., Digitally-tunable echo-canceling analog front-end for wireline communications devices, US Patent 6,542,477. Issued April 2003.

[Pécourt 1999]   Pécourt, F., Hauptmann, J., and Tenen, A., An integrated adaptive analog balancing hybrid for use in (A)DSL modems, *Proceedings of IEEE ISSCC'99*, 252, 1999.

[Pierdomenico 2002] Pierdomenico, J., Wurcer, S., and Day, B., A 684mW Adaptive Supply Full-Rate ADSL CO Driver, *IEEE Journal of Solid State Circuits*, Vol. 37, Iss. 12, December 2002.

[Piessens 2001]   Piessens, T. and Steyaert, M., SOPA: A high-efficiency line driver in 0.35 μm CMOS using a self-oscillating power amplifier, *IEEE International Solid-State Circuits Conference*, Digest of Technical Papers, 2001, pp. 306–307.

[Piessens 2002]   Piessens, T. and Steyaert, M., A central office combined ADSL–VDSL line driver solution in 0.35μ/spl mu/m CMOS, *Custom Integrated Circuits Conference, 2002. Proceedings of IEEE 2002*, Vol., Iss., 2002, pp. 45–48.

[Piessens 2003]   Piessens, T. and Steyaert, M., Highly efficient xDSL line drivers in 0.35-/spl mu/m CMOS using a self-oscillating power amplifier, *IEEE Journal of Solid-State Circuits*, Vol. 38, Iss. 1, January 2003, pp. 22-29.

[Sabouri 2002]   Sabouri, F. and Shariatdoust, R., A 740 mW ADSL line driver for central office with 75 dB MTPR, *2002 IEEE International Solid-State Circuits Conference, ISSCC Digest of Technical Papers,* Vol. 1, Iss., 2002, pp. 322–470.

[Sands 1999]   Sands, N.P., Naviasky, E., Evans, W., Mengele, M., Faison, K., Frost, C., Casas, M., and Williams, M., An integrated analog front-end for VDSL, *1999 IEEE International Solid-State Circuits Conference, Digest of Technical Papers*, pp. 246–247.

[Shorb 2002]   Shorb, J., Allstot, D.J., and Roze, R., Class AB-D-G line driver for central office asymmetric digital subscriber line systems. *IEEE ISCAS 2002.* Vol. 5, May 26–29, 2002. pp. V-405–V-408.

[Siniscalchi 2001]   Siniscalchi, P.P., A CMOS ADSL codec for central office applications, *IEEE Journal of Solid-State Circuits*, Vol. 36, Iss. 3, March 2001, p. 356.

[Starr 1999]   T. Starr, Cioffi, J.M., and Silverman, P.J., *Understanding Digital Subscriber Line Technology*, Prentice-Hall, Upper Saddle River, New Jersey, 1999.

[Van der Plas 1999]   Van der Plas, et al., A 14-bit intrinsic accuracy Q^2 random walk CMOS DAC, *IEEE Journal of Solid-State Circuits*, Vol. 34, Iss. 12, December 1999, p. 1708.

[Vecci 2003]   Vecci, D. and Morandi, C., A 750 mW class G ADSL line driver with offset-controlled amplifier hand-over, *Southwest Symposium on Mixed-Signal Design*, 2003, p. 253.

[Weinberger 2002]   Weinberger, H., et al., An ADSL-RT full-rate analog front-end IC with integrated line driver, *IEEE Journal of Solid-State Circuits*, Vol. 37, Iss. 7, July 2002, p. 857.

[Wen-Kuang Su 1994]   Wen-Kuang Su, Yih-Rong Chen, and Lin, D.W., Optimization of hybrid circuits for echo cancelation in high-rate digital subscriber line transmission, *APCCAS'94., IEEE Asia-Pacific Conference on Circuits and Systems*, December 5–8, 1994, pp. 334–339.

[White 1972]   White, S., An adaptive electronic hybrid transformer, *IEEE Transactions on Communications*, Vol. 20, Iss. 6, December 1972, pp. 1184–1188.

[Widrow 1985]   Widrow, B. and Stearns, S.D., *Adaptive Signal Processing,* Prentice-Hall, Englewood Cliffs, New Jersey, 1985.

[Wurcer 2003]   Wurcer, S., Carefully evaluate your ADSL line driver efficiency, *Communication Systems Design*, February 2003.

Bibliography

[Bingham 2000]   Bingham, J.A.C., *ADSL, VDSL, and Multicarrier Modulation.* John Wiley & Sons, New York, 2000.

Chapter 4

Local Loop Simulation and Simulators

Jim Eyres and Alexander Stefanescu

CONTENTS

Abstract   This chapter will be of special interest to engineers who are testing performance of prototype or production Digital Subscriber Line (DSL) modems. It starts with a brief history of telephone line simulation, and continues by discussing DSL loop simulators in more detail. It examines the requirements typically needed to test communications over the local loop and the importance of various parameters. Impairments are also mentioned, in that they are needed in addition to the test loop to fully test a DSL modem's physical layer. Various different types of DSL loop simulators are discussed, as well as the main advantages and disadvantages of each. Repeatability and accuracy of loop simulators along with other nonelectrical characteristics such as remote control are discussed. Finally, terms used by standards bodies, such as DSL Forum, to measure the quality of the loop simulation are explained together with a description of accuracy expressed in "Mean Error" and "Mean Absolute Error."

4.1 Introduction

In the DSL industry, simulation has developed in parallel with new products. As new products are designed, simulation of the design or the environment must be used to predict how well the product will perform in practice. The need for simulation is present during several stages of the product development cycle, including initial research, synthesis, design testing, and testing at subsequent stages, such as during first production runs and final production manufacture.

In DSL, the initial research and synthesis of the design of a modem is normally done using mathematical modeling of some kind. Copper cable plant from the telephone company building, the central office (CO), to the customer premises is usually known as the local loop. This chapter is about simulation of the local loop, using physical simulators. It also discusses the addition to the test environment of impairments that make data transmission more difficult. These simulators enable designers of DSL modems, Digital Subscriber Line access multiplexers (DSLAMs), line drivers, and other DSL transmission equipment devices to predict and test how well the equipment will perform in the real world. Note that for the rest of this chapter, the word "modem" is used to denote any of these devices.

4.2 History and Background of Loop Simulators

For the first 90 years of the telephone, the principal objective was to transfer voice from one end of the line to the other with maximum clarity. It was not until the late 1960s that there was enough interest in sending data over the telephone line that a commercial loop simulator was developed. Before that, many companies who manufactured computer equipment also developed their own modems for the transmission of data over the telephone network, but test equipment for these modems was proprietary.

One of the first wireline simulators, said to be the first in existence, was developed by the Rome Air Development Center in 1968. Although details of the operation of this simulator are sparse, it is known that the simulator could function as an impaired channel or as a typical twisted pair wire link. It was used to compare different communications devices [Rome]. This research establishment, which is run by the United States Air Force, is located in Rome, New York, and is now called the Rome Laboratory.

Another early simulator was sold by Bradley. Named the 2A/2B, it was an all-analog telephone channel simulator built in an instrument case. The simulator had a handle, which made it portable and reasonably rugged. It was designed to be used in the laboratory but could easily be carried elsewhere if needed. The case opened to reveal two panels with knobs, each of which had its own designation. The simulator was designed to simulate the 200 Hz–4 kHz telephone channel, along with a large variety of impairments, such as attenuation at the high- and low-frequency ends of the channel, and delay distortion. These impairments could be changed by setting the knobs on the panels. There was no such thing as remote control or any digitization of the signals. The Bradley simulator was used to test four-wire voice band modems at the design stage. One of the problems with this unit was repeatability of results, because the settings were determined by the position of the knobs, which were set by hand and thus prone to variability. However, the simulator worked well for testing the voiceband modems of that time and was one of the better known and frequently used instruments.

Wandel and Goltermann (W&G) produced another device known as the PKN-1 and called it a "cable simulator." It was intended for testing T1 and equivalent alternate mark inversion (AMI) or high density bipolar three code (HDB3) digital services at line rates of 1.544 or 2.048 Mbit/s. The device provided adjustable group delay and the attenuation response for transmission in one direction only. The user was able to adjust it to simulate the attenuation and group delay of a variety of twisted-pair line types at a variety of distances. The simulator was AC-coupled and covered the frequency range from 1 kHz to 10 MHz. One major advantage was that it could be controlled remotely.

In the PKN-1 simulator, no attempt was made to simulate the delay of the line. The input impedance of the line was simulated by a fixed resistor. As well as being used to test modems, it could also be used to test repeaters, which normally are spaced at 6000 ft intervals for this type of transmission. The PKN-1 was an active device, and as such suffered from clipping when a signal much larger than expected was received. Like all active simulators, it carried the signal through its electronics, and the results were to some extent dependent on the quality of the output stage of the simulator rather than the devices under test. Although T1 services are still widely used, more sophisticated simulators have been developed, and the PKN-1 was taken off the market many years ago.

These early types of simulators were targeted at specific transmission specifications. The Bradley 2A/2B was designed to simulate the voice-frequency telephone channel provided by the telephone company, and the types of impairments often found on this channel; W&G's PKN-1 was intended for testing HDB3 modems and repeaters. These simulators, which shared some commonality, were used to test modems that were completely different. However, both were used for four-wire connections, or at least tested transmission in one direction at a time on any one pair of wires.

Simulation of the bi-directional flow of signals on a telephone line was initially done in the early 1970s using passive circuits. Lear–Siegler offered a test bed with a position for a series of plug-in modules that emulated local loop characteristics up to the edge of the voice band. AEA, later a division of consultronics, offered two instruments, one intended to simulate a loaded line, which has 88 mH loading coils at 6 kft intervals, and one to simulate the attenuation characteristics of nonloaded twisted pair line up to approximately 10 kHz.

One advantage of the Lear–Siegler test bed was its flexibility. The test bed consisted of a metal chassis with banana sockets at both ends, which were connected as a balanced interface to the modem's under test. The metal chassis had positions for plugging in a dozen modules. If a particular position was not used, there was an automatic bypass from one side to the other. As a result, in the initial empty state, with no modules plugged in, the test bed represented a zero-length loop. Modules represented a specific length of line and could be plugged in to the test bed to construct a model of almost any line desired. Loading coils could be added, if desired, to simulate those inserted on a normal voice circuit. The architecture could be extended by connecting test beds together to make up a limitless variety of configurations. One of the problems associated with this device, as with many simulators, was wear on the test bed because of the constant plugging-in and unplugging of modules.

The W&G TLN-1 (telephone channel simulator) and the AEA line simulator known as the S1 used switches to allow the user to change settings. Both devices simulated group delay and attenuation characteristics of a telephone channel. Each instrument was limited by its initial configuration; there was no way to expand either unit, so if additional functionality was needed, the only option was to buy a second unit. The S1 was entirely passive. It allowed the user to change the group delay at both the high and low ends of the telephone channel in a series of fixed steps, and it also provided independent attenuation distortion at the high and low ends of the channel, again in a series of fixed steps. The TLN-1 was similar in concept. It simulated group delay of loaded lines but had added features in that noise could be added to the signal passing through the device. The TLN-1 was an active device, and thus it also had a higher noise floor.

The S2 from AEA was designed to let the user test a typical modem of the day over nonloaded lines. In that era, it was possible to use bandwidth outside the 3 kHz telephone channel using a leased line with compatible units at both ends. Not all telephone companies allowed the use of these types of modems, and the power output from them was severely limited. The performance was limited as well. The modems were almost always four-wire. Sometimes, a two-wire half-duplex approach was also used. Echo cancelation was unheard of, although voice grade modems made use of frequency-division multiplexing (FDM) to achieve two-wire full-duplex capability.

A different type of simulator arrived in the early 1980s. This unit simulated the telephone channel from one user to another and was designed to test the performance of dial-up modems. It was an active type of simulator (see Section 4.6.3) that employed hybrid circuits at the interface to the modems and provided one channel in each direction. Early versions of this type of simulator were used to test a modem's immunity to such impairments as impulse noise, attenuation distortion, and varying levels of background noise. As the simulator developed, additional impairment types were added, such as satellite echo delay. Later still, it could also be used to test a modem's ability to dial and auto-answer calls. Both AEA and TAS manufactured this type of simulator until the mid 1990s. During that time, modems evolved rapidly, and the sophistication of the simulators changed in response. This type of simulator reproduced the end-to-end transmission channel and generated impairments and attenuation using digitally controlled analog circuitry built around transistors and op-amps. The output at each end could be attenuated, and the frequency response and group delay could be changed, so that this type of unit provided simulation of the complete channel from user to user.

One of the features of this type of simulator was its remote control capability, allowing the user to write a set of software tests and apply them under computer control to modem after modem. This functionality was very important for users testing large numbers of modems, often with little or no intervention from test personnel.

The first time that a commercial wideband* loop simulator became available was in the mid-1980s. This was AEA's DLS 100. It was targeted specifically at testing the new modems for Integrated services digital network (ISDN). The simulator was a passive simulator capable of carrying traffic in both directions on one pair of wires. It could also be configured to carry the North American four-wire Digital Data System (DDS-type transmission) [Bell]. This simulator used two types of modules. One type was specified from DC (direct current) to 100 kHz and the other from DC to 300 kHz. This range was a tremendous improvement in bandwidth compared to previous loop simulators. Unfortunately, the improvement came with a drawback: increased physical size. For example, the DLS 100A would fit into a 19-inch rack and was 10.5 inches high.

4.3   Applications of Wideband Loop Simulators

Chapter 2 of [Golden 2006] outlines some of the physical characteristics of a local loop and develops some of the theory behind differential mode transmission of signals over the loop. Loop simulators are used rather than real cables for a variety of reasons, including

- Ease of use
- Repeatability
- Ability to change length and configuration easily
- Standardization of tests and
- Greater accuracy of simulators to standard lines

Essentially, the local loop acts as a transmission line, sometimes composed of many different sections with different line characteristics for each part. Wideband loop simulators replicate the analog characteristics of this transmission line. More specifically, they emulate the attenuation, delay, and impedance characteristics of a local loop. Any simulator has limitations brought about by the compromises in its design between accuracy, size, and cost. A big question facing a potential user is how much the limitations of any particular simulator affect the intended application. There are several applications in which a simulator is used, often leading to distinctly different specifications that are relevant for different applications. These differences are discussed in Sections 4.3.1 through 4.3.4.

---

* Previous simulators were able to reproduce the response from DC to approximately 10 kHz on nonloaded lines. Here, "wideband" means far beyond the plane old telephone service (POTS) bandwidth.

4.3.1   To Help Establish Practical Modem Specifications

To establish modem specifications, designers generally perform computer simulation on the theory of the design. This helps to establish specifications for the modem, such as the required signal-to-noise ratio (SNR) and how well the modem handles interference of various types. Computer simulations help to establish the expected data rate under various loop lengths and configurations and with different types of crosstalk, impulse noise, and other impairments. Furthermore, computer simulation is needed to help designers understand and estimate the effects of implementation losses, such as analog-to-digital-converter (ADC) quantization error (see Chapter 3). The use of a loop simulator, together with a noise generator, can be regarded as an extension of simulation in the design process. They are used together to check that the design performs as expected, that there are no unforeseen design problems, and that the specifications for the modem can be met.

4.3.2   Testing of Prototypes

When a number of prototypes have been made, and the first prototype runs are being tested, the loop simulator is used to characterize modem performance under various levels of stress. Both designers and design verification testers may be involved with testing of the prototypes. At this stage, modem testers want to be able to change the line length and configuration as well as apply different types of noise, such as crosstalk, impulse noise, RFI (radio frequency interference) noise, and other ingress onto the line. This stage is generally the one at which the greatest versatility is required from the simulator.

Modem designers need to know how the prototypes perform under varying conditions. Tests may include

- Effects of remote powering.
- Performance of a DSL service and POTS on the same loop. Particularly important is the effect of ringing current and, just as importantly, the effect of interruption of ringing current and of loop current.
- Coexistence of a DSL service and ISDN on the same loop.
- Coupling of signals from neighbouring circuits and their influence on the performance of the modem.
- Effect of POTS splitters on modem performance.
- How well the modems work with repeaters.
- How the modems behave during start up (initialization) and following temporary line interruptions.
- Tests to determine if the line monitoring, performance, and quality indications given by the modem are accurate.

The loop simulator should allow full testing under as many of these conditions as the user needs for the type of modem being tested. Therefore, the loop simulator becomes an integral part of the test setup. Designers use this setup to determine that the modem meets a certain set of specifications. Furthermore, they can then specify a subset of these conditions that can repeatably and accurately ensure the integrity of the units being tested, but that also fully stress the design.

4.3.3 Testing Production Modems

Driven by economic considerations, testing carried out at the production level must be rapid but must still identify any deficiencies in the production unit. At this stage of the product cycle, the production test engineer assumes she has been given a solid design. She does not have time to do exhaustive testing. Instead, she is looking for abnormal performance owing to some deviation in a production model.

In this application, loop simulators need to be inexpensive. The test engineer determines simple tests to be carried out either on each unit manufactured or on some set of samples. This suite of tests determines the requirements of the loop simulating equipment. Generally, loop simulators used in this application are very different from simulators used elsewhere. The production test simulators often provide a small subset of the loops that are used in testing elsewhere. In addition, they may have wider tolerance limits on their specifications, may not simulate delay or group delay, and often provide multiple identical loops so that the user can test a number of modems all at once using the same loop simulator.

4.3.4 Conformance Testing

Conformance testing refers to the testing of modem equipment, often by an independent test laboratory (ITL) or sometimes in the laboratory of the service provider purchasing the modems. This testing determines whether modems of a particular type conform to the established specifications. To produce consistent results, loop simulators must accurately simulate loops based on models specified by the relevant standard. In general, real cables are far too variable from pair to pair inside a multi-pair binder and from cable to cable to give consistent results. Also, the characteristics of real cable vary depending upon how the cable is laid out. Furthermore, testing normally needs to be comprehensive, and, as such, it is typically very time consuming. Automation of the testing procedure is essential. The same modem pair should give the same results at different ITLs. Operators often use the results of tests by an ITL to help them determine if the modems tested are of high enough quality to be deployed in their networks.

4.4   Simulator Characteristics

At various stages of modem testing, the simulator may have to emulate a wide variety of different characteristics. For example, the simulator may have to transmit or emulate

- Differential and common mode signals
- Insertion loss of the transmitted signal
- Impedance (both magnitude and phase) presented by the line to the modems under test
- End-to-end signal delay
- Signal delay of one frequency relative to another, called group delay or envelope delay
- Impairments
- DC continuity
- DC resistance
- Maximum signal level

These parameters are discussed in more detail in the rest of this section.

4.4.1   Differential and Common Mode

Line simulators for the testing of DSL modems and systems concentrate on the correct simulation of differential mode signals. Because all DSL transmissions use the differential mode, there is little need to simulate common mode signal transmission characteristics in the loop simulator. Certain other specialized uses of wireline simulators would require simulation of common mode characteristics. Such uses include modeling the effects of lightning strikes and transmission over very long loops, such as overhead high-voltage power lines. Common mode simulation is also remarkably difficult and expensive to perform accurately, and for these and other reasons the common mode is never simulated in commercial line simulators for DSL testing.

However, the fact that common mode simulation is not currently embodied in DSL loop simulators does not mean that the common mode characteristics of modems can be ignored. On the contrary, many of the interfering signals that arrive at a modem's receiver in practice are unwanted common mode signals. The modem must reject these signals. Because there is always some conversion of longitudinal mode signals to differential mode signals within a modem, a longitudinal conversion loss (LCL) test is often specified. This test is performed on the modems when a simulator is not present.

Recently, and especially for DSL that uses transmission frequencies above 2 MHz, there has been recognition that RFI will be picked up on aerial cables, and that the unwanted RFI signals are converted to the differential mode at the cable ends by a variety of mechanisms to do with cable and termination unbalance. (See chapter 13 of [Golden 2006].) At this time, the effects can be reproduced in the test environment by applying some form of common mode impairment at the modem under test. The standards bodies are becoming increasingly aware of the need to test the effects of RFI on Very High Bit-Rate DSL (VDSL) transmissions as well as ADSL (Asymmetric DSL)2plus. In the future, standards bodies will likely specify common mode testing using impairments applied at the ends of the simulator.

4.4.2 Insertion Loss of the Transmitted Signal

Insertion loss (IL), which is often referred to as attenuation, is the most important characteristic of a loop simulator.* Chapter 4 of [Golden 2006] shows that the theoretical data rate achievable over a loop depends on the SNR at the receiver. Because the modem's maximum transmission power is very strictly regulated, the received power is dependent on the line attenuation. It is important that the power at the receiver is simulated accurately to get a good idea of the data rate achievable.

Loop simulators are a vital part of the test setup used to determine modem performance. Loop loss must be simulated accurately to allow the performance of modems from different vendors to be accurately assessed. In addition, it is just as important that the same modem pair give the same results at different times and at different places around the world. The need for consistent test results has placed an increased demand on loop simulators for repeatability. Prompted by industry demand, manufacturers of wireline simulators have made several improvements to help achieve repeatability for testing the latest DSL technologies. These improvements are discussed in detail in Section 4.8.

4.4.3 Impedance Presented by the Line to the Modems under Test

The impedance presented by the twisted pair to the modems has sometimes been overlooked by the technical standards bodies, but it is important. Modems must send and receive data on a line that has impedance characteristics that vary significantly with frequency. Moreover, this impedance is not

---

* It should be noted that the IL is normally measured in decibels (dB). IL figures are positive by definition so that a figure of, say, 10 dB represents attenuation or loss. However, common usage has muddied the waters. A network analyser shows a graph of gain versus frequency. Almost without exception, transmission engineers refer to this gain as "loss."

the same for every loop. It changes greatly, depending on the construction of the loop. Modems are typically designed to match a specific impedance, which is typically real-valued, 100 or 135 Ω, and independent of frequency. As a result, modems are generally mismatched to the loop at some frequencies, particularly if the loop is constructed of multiple, different-gauge segments or has bridged taps. The amount of mismatch significantly affects modem performance. Consequently, a simulator must simulate the actual impedance of any modeled loop to accurately test a modem's performance.

Loop impedance varies with frequency and is a complex function with both real and imaginary parts. Sometimes the loop impedance is expressed in terms of magnitude and phase. Typically, the impedance for a long, straight line looks like that shown in Figure 4.1.* This figure, which is for a long length of North American 26 AWG cable, shows both the real and imaginary parts. This is a typical impedance for a long line. Note that the variability of the real and imaginary components of the impedance is significant at low frequencies. At higher frequencies, the impedance becomes almost entirely real.

DSL modems also have to operate effectively on more complicated loops, such as those composed of different gauge wires, and possibly with

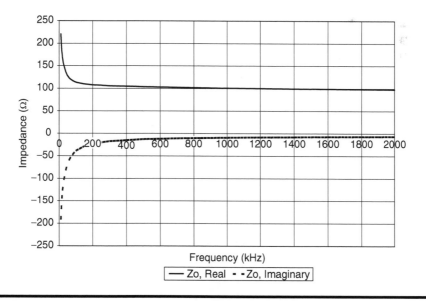

Figure 4.1   Characteristic impedance of 26 AWG line.

---

* All graphs are calculated from theoretical models of cables unless otherwise stated. Real cables vary a great deal over temperature and from batch to batch, and even from loop to loop within a binder.

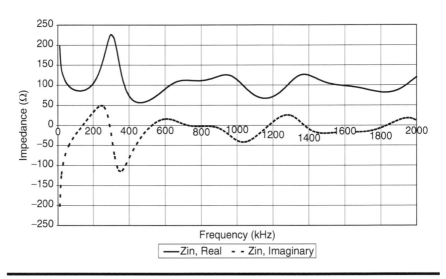

Figure 4.2   Impedance at CPE side of CSA4 test loop.

bridged taps. Figure 4.2 shows the complex impedance looking into the CPE end of a loop made up as shown in Figure 4.3, with two bridged (unterminated) taps and a relatively short in-line section between the second tap and the customer. This impedance is unusual, but nevertheless the modems must be able to handle it. The loop in Figure 4.3 is often known as customer service area loop 4 (CSA4). It is one of several loops that are commonly used for testing purposes in North America. The suite of test loops is shown in many publications, including Annex A of [G.991.2] and chapter 2 of [Golden 2006].

Chapter 2 of [Golden 2006] shows how the primary parameters R, L, G, and C can be used to calculate the impedance seen by the modem. It shows how to calculate the input impedance of a loop with two or more different types of cable, and also with bridged taps, in the loop. One of the more difficult loops for the modems may well be one where there is a bridged tap at the modem's end of the loop. The presence of a bridged tap so close to the modem approximately halves the impedance seen by the modem, and

		0.8 kft 26 AWG		0.4 kft 26 AWG	
CO	0.8 kft 26 AWG		6.25 kft 26 AWG2	0.55 kft 26 AWG	CPE

Figure 4.3   CSA4, with three in-line sections and two bridged taps, all 26 AWG line.

the impedance can vary widely because of the bridged tap. The modem must be able to drive this lower impedance and, just as importantly, be able to efficiently receive a signal from this sort of mismatched complex impedance.

The ability to simulate bridged taps as well as in-line gauge changes may be very important, depending on the application, and it can have large effects. For example, a signal source of $100\,\Omega$ feeding into a load of $100\,\Omega$ with no line in between has an insertion loss of zero by definition. When a bridged tap is added, there is some loss because of the insertion of the (bridged tap) line. Calculating the effect of, for example, a 400 ft bridged tap, using the standard parameters for 26 AWG cable results in the two graphs of Figures 4.4 and 4.5. Figure 4.4 shows the impedance seen when the line length is zero, and the transmitter is terminated with $100\,\Omega$ plus the bridged tap. Both the real and the imaginary parts of the impedance vary with frequency. At just under 400 kHz, the impedance seen by the modem's transmitter is in the region of $20\,\Omega$. Figure 4.5 shows the attenuation, or insertion loss because of the bridged tap, which rises to just under 10 dB.

For higher-frequency applications, the requirements for loop simulators are for short bridged taps on the order of a few tens of feet only. The accuracy of simulation in these applications is very important because a change of 1 percent in the length of a bridged tap changes the notches of the frequency response by a few tens of kilohertz.

As well as causing variability in the impedance presented to the modem, discontinuities in the line such as gauge changes, bridged taps, or a

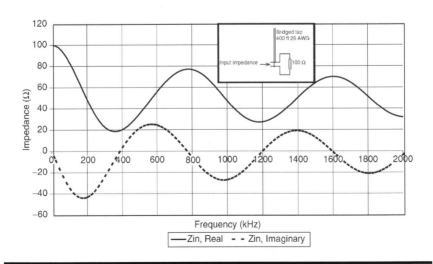

Figure 4.4   Input impedance of 400 ft bridged tap of 26 AWG line. The inset circuit shows the measurement setup.

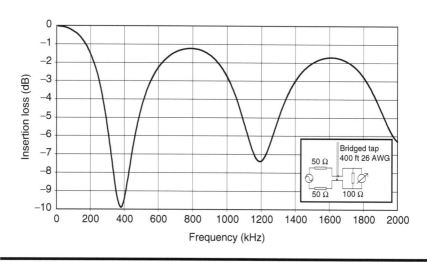

Figure 4.5   Insertion loss because of a bridged tap of 400 ft of 26 AWG line. The inset circuit shows the measurement setup.

mismatch between the wireline and its termination cause part of the signal transmitted by the modem to be reflected back to the modem. Chapter 2 of [Golden 2006] shows that the reflection of part of the signal causes the impedance of an input sine wave to change with frequency. This impedance function is often plotted against frequency to see how well a simulator is performing. It should be remembered that a modem is likely transmitting a constantly changing signal, and in this case it is the way that the modem handles and equalizes a reflected pulse that determines how well it performs. An ideal loop simulator provides the correct reflections from various parts of the line, and therefore will stress the modem's equalizers and echo cancelation circuits correctly.

4.4.4   End-to-End Signal Delay

A key difference between a high-quality simulator and one that is used for simple production-line performance verification testing is whether the signal delay through the various sections of the loop is simulated. End-to-end delay is easily overlooked, but for high-quality loop simulation, simulation of the correct delay is essential. End-to-end delay occurs because the signal propagates with a finite velocity along the loop. It is calculated as the ratio of the phase shift to the radian frequency.

The effects of signal delay are seen by a DSL modem in the reflections at its receiver caused by its own transmitter. The modem's receiver must reject these reflections to avoid degrading the SNR of the received signal. In addition, it is these reflections that cause the variations in impedance seen on a loop with bridged taps or multiple gauges.

During the testing of modems on the production line, it is often possible to dispense with delay in the loop simulator, making it much less expensive. For instance, if the objective is to test that under certain well-understood conditions, the modem behaves properly and similarly to the rest of the modems coming off the production line, it is likely that modeling of delay in the simulator is not needed. Almost certainly, it is necessary to simulate the changes in loop impedance with frequency presented to the modem by the simulator for any specific loop, but for any specific loop the modeling of impedance variations can be achieved with analog circuitry in the simulator.

4.4.5   Signal Delay of One Frequency Relative to Another (Group Delay or Envelope Delay)

Group delay can be calculated from the end-to-end phase shift. It is defined for any radian frequency as

$$-\frac{d\theta}{d\omega}$$

where θ is the phase shift and ω is the radian frequency. The group delay is often specified because it is easier to measure than end-to-end delay and in many cases leads to same answers. Furthermore, the group delay is much more sensitive to gauge changes and bridged taps than is end-to-end delay, and it can even be negative, whereas end-to-end delay is always positive.

4.4.6   Impairments

Impairments cover a class of interferences that disturb modem communication and cause transmission over the line to be less than ideal. Examples of impairments that are often used to test modems are

- Additive white Gaussian noise (AWGN)
- Crosstalk noises
- Impulse noise
- Very short loop interruptions known as micro-interruptions and
- RFI interference

The simulator should not produce any of these impairments. The main difficulty for the simulator is likely to be the low level of AWGN. Not to adversely affect the performance of a modem being tested, the simulator carrying the modem's signal should be quieter than any source

The noise generated by an ideal twisted-pair line is around −170 dBm/Hz at frequencies above 100 kHz. This value is the thermal noise at room temperature that is created by the characteristic impedance of the loop. In practice, the background noise in a CO and even at a remote end modem is not nearly as quiet as −170 dBm/Hz, and for test purposes AWGN having a PSD of −140 dBm/Hz is normally applied in addition to any other interference sources.

The simulator must be quiet enough that a background noise of −140 dBm/Hz plus any additional noises used during any test are not increased by the noise level generated inside the simulator. Also, the test environment should be quiet enough to allow proper testing of the modems. In some cases, this is difficult to achieve, and instead some form of allowance must be made for the nonideal environmental noise. For instance, local radio stations often produce interference that couples into the test setup at levels well above the background noise. This sort of effect must generally be accepted as unavoidable.

Although, passive simulators can achieve a noise floor significantly lower than −140 dBm/Hz, active simulators generally have had difficulty doing so. So far, any simulator with a noise floor above −140 dBm/Hz has limited usefulness when testing to international standards such as those produced by the ITU-T, ANSI, and ETSI.

4.4.7 DC Continuity

If the DSL modems under test are supposed to work in the presence of POTS, or if they are CPE and being powered from the CO end of the line, then the loop simulator must pass DC through its simulated loop. Also, if the network between the CO and the CPE uses repeaters, then the loop simulator must also pass DC. In these cases, an AC-coupled simulator just will not be sufficient. Some simulators deal with this problem by combining a high pass response with a filter and a separate DC path. As a result, there are two parallel paths within the unit. From a practical point of view, if the passbands supplied by the simulator meet the specifications of the signals the modems need, then such an approach is sufficient. But if one is concerned with what happens if some unexpected and out-of-band frequencies couple into the line, it is better if the circuit simulation itself extends as low as DC and up to some upper frequency without disruption, as is the case on real lines.

4.4.8 Unwanted Effects

In addition to the desired properties, a loop simulator may have some unwanted effects that affect the accuracy of testing. These effects arise owing to the actual implementation of the loop simulator. They are a consequence of the design realization and can be controlled to some extent during the design process of the loop simulator. Among these unwanted effects are

- Input-to-output crosstalk
- Unbalance to ground
- Common mode signal
- Harmonic distortion and
- Intermodulation distortion

Any of these effects can severely affect the performance of the modem under test. Often the presence of these effects is subtle and not easily detected before modems are actually tested.

Crosstalk or coupling from the input to the output of the line simulator is of great concern. Here, the attenuation of the simulator may be on the order of 95 dB. As an example, 80 dB attenuation corresponds to the voltage being attenuated by 10,000, and it is easy for this much voltage to "jump" from the input to the output. Sometimes, this occurs because of external coupling, particularly, if the leads going into the simulator at its two ends are close to each other at some point, and it can also happen inside the simulator. Because crosstalk increases with increasing frequency, simulator manufacturers will have to take even greater care to prevent crosstalk inside the simulator as DSL technologies use higher frequencies (for example, as deployments move from ADSL to ADSL2plus to VDSL2).

The effect of crosstalk, whether from input to output or from one part of the simulator to another, is very distinctive. It is shown in Figure 4.6 for a typical line of 10 kft length and 26 AWG with extra input-to-output coupling of one part in ten thousand. This coupling is often very difficult to prevent, and if it occurs in the frequency band, can severely affect throughput.

Unbalance to ground on either a real line or a simulator converts common mode signals to differential mode signals. In addition, common mode signals can be converted to differential mode signals by any unbalance in a modem.

Harmonic distortion and intermodulation distortion give false signals to the modem's receivers, which in turn reduce the SNR, thereby artificially reducing performance as a consequence of the simulator's shortcomings.

4.5   Derivation of Equations

Differential mode transmission on a loop can be viewed as a two-port circuit. Test loops can be defined by one or a cascade of loop sections.

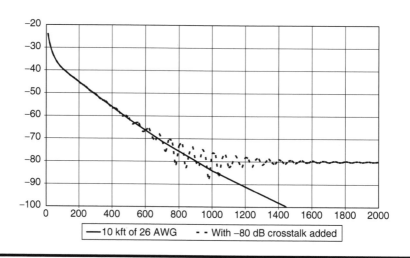

Figure 4.6 **Effect of −80 dB of crosstalk on the signal at the end of a 10 kft 26 AWG line.**

The characteristics of each section are specified by the line length and by the primary line parameters of resistance (R), inductance (L), conductance (G), and capacitance (C) per unit length. From these primary parameters, any number of different matrices can be derived to describe transmission over the loop. For more details see Chapter 2 of [Golden 2006].

4.6 Different Types of Loop Simulators

There are four main types of loop simulators used for DSL transmission testing. In each case, differential mode characteristics only are simulated. The goal in loop simulation is to develop an instrument with the same secondary characteristics as those of the line being simulated. The passive types of simulator attempt to simulate these characteristics by directly simulating distributed R, L, G, and C parameters. The same task can also be done to some extent by using wave equations in an active simulation.

4.6.1 Passive Analog Simulation

The oldest and most popular method of wideband simulation is passive analog simulation. Its advantages include

- Simultaneous simulation of all secondary characteristics, i.e., attenuation, propagation delay, and characteristic impedance
- Continuity in frequency, down to DC

- Inherently low noise
- Bidirectional transmission
- No clipping
- No distortion in a well-designed unit

A passive analog loop simulator is generally made up of a large number of sections of circuits, sometimes known as rungs, which is the name used here. Each rung represents a length of line. In general, the higher the frequency that a rung simulates, the shorter is the length of line that it simulates. Many rungs can be cascaded in series to construct the desired length of the line. Different rungs can be used to simulate different gauges of line, and the configuration can be changed to simulate specific loops selected by the user. The configuration is normally changed under some sort of computer control by relays, but it may also be changed by hand.

Equations for a transmission line are derived in chapter 2 of [Golden 2006]. In particular, the propagation constant is (see Equation 2.8 of [Golden 2006])

$$\gamma = \alpha + j\beta = \sqrt{(R + jL\omega)(G + jC\omega)} \tag{4.1}$$

and the characteristic impedance (see Equation (2.14) of [Golden 2006]) is given by

$$Z_0 = \sqrt{\frac{R + jL\omega}{G + jC\omega}} \tag{4.2}$$

The real and imaginary parts of the propagation constant, γ, are equivalent to the transmission line attenuation and delay, respectively. The attenuation and delay, as well as the real and imaginary parts of the characteristic impedance are all measurable quantities that are known as the four secondary characteristics of the line. The primary parameters (R, L, C, and G) cannot be measured and instead are calculated from the secondary characteristics. The simulator must reflect the correct values of these secondary characteristics. If it does the modeling well, then the way that differential mode DSL signals travel through the simulator is similar to the way they travel over the real loop.

One of the assumptions used for deriving Equations 4.1 and 4.2 above is that the line consists of an infinite number of infinitely small sections. But what happens if the sections are not infinitely small, and the number is finite? Suppose the section consists of some resistance (R) and inductance (L) in series and some conductance (G) and capacitance (C) in parallel. Figure 4.7 shows two possible models for a rung. The Pi-model is one of many models for the rung of an analog wireline simulator. Another model,

174 ■ *Implementation and Applications of DSL Technology*

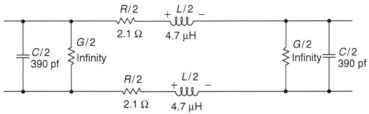

(a) Pi-model: Values are approximate for a 50 ft rung of 26 AWG at low frequencies.

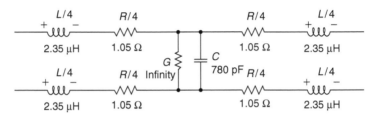

(b) T-model: Values are approximate for a 50 ft rung of 26 AWG at low frequencies.

Figure 4.7   Schematics of one rung of a typical wireline simulator.

the T-model, is also shown. It is possible to simulate the line using either one of these models, or a mixture of both, and there are many other models that can be used.

If all the rungs are the same, and there are nPi-model rungs used in series to simulate 1 kft of line, typical values of R, L, G, and C can be calculated from the loop resistance, loop inductance, loop conductance, and loop capacitance per unit length of line. Tables of these primary values at different frequencies are published in many places, including [ANSI T1.601-1992]. A more comprehensive set of tables for North American wire gauges at frequencies up to 5 MHz was published in [Bell], but this is now out of print.

Considering the Pi-model, $R/2$ and $L/2$ are used in the upper and lower arms. Values for $C/2$ and $G/2$ are placed at the end of each section or rung. Normally, C is held to be constant, or nearly so, at the frequencies used for DSL, and G is so small that in the simulation it is often ignored. Values of R and L change with frequency, and the analog circuit must be designed so that the values in the rung change to match. Values for R, L, C, and G per unit length of line vary according to the line gauge and other factors and are discussed in [Golden 2006]. The designer of a loop simulator must find a way to make both the resistance and the inductance match, as closely as possible, the corresponding characteristics of a real twisted-pair line in frequency. It may also be necessary for the rung to contain conductance. The number of rungs per unit length helps to determine the maximum frequency range for any design. Using the example values given in Figure 4.7 for the Pi-model of 50 ft length, there will be some frequency at which the inductance and capacitance in the rung resonates. This frequency is well above the frequency at which the rung provides good line simulation. For the example, the resonant frequency is

$$1/2\pi\sqrt{LC} = 3.7 \text{ MHz}$$

To make a simulator that is useful at higher frequencies, the rungs must contain lower values of inductance and capacitance, which means each rung must represent a shorter length. One other main consideration for this type of simulator is how to switch the rungs in and out of the simulated line. The switching is normally accomplished by relays, so that there can be remote control of the line setting. One advantage of this type of simulator is that very short bridged taps and segments of line can be simulated. This feature is important in high-bandwidth applications such as VDSL and VDSL2, where there are requirements for bridged taps that are only 10 or 20 ft long. One disadvantage of this type of simulation for use in DSL testing is that because of the many analog components used, not all simulators are identical; so secondary parameters may vary between production batches. Today this variability is being largely overcome by factory-provided calibration and compensation for each individual simulator and by superior methods of loop specification by use of mean error (ME) and mean average error (MAE). ME and MAE are discussed in Section 4.8.1.

4.6.2 Production Line Passive Analog Simulation

Because of the complexity and cost of the type of analog simulation described above, simulators intended for use on the production line normally do not simulate delay, and they may not simulate impedance very well. If the test engineer decides that impedance features of the loop are not important for testing, but he still wants to perform a test over a simulated line, he may choose not to simulate the delay characteristics of the line. For simplicity, these types of simulators generally contain only one segment of a homogenous line (i.e., a loop without bridged taps). This simplification reduces the simulation to that of a filter with lumped parameters and often only an approximation of the input impedance. The advantages include reduced cost and greatly reduced size. This sort of simulator is not generally used for modem design, standard conformance, or development purposes.

4.6.3 Active Simulation

Active simulation is a method in which simulation is done electronically by active circuits. Active simulation allows each of the four secondary line parameters to be simulated independently. This approach may result in more accurate simulation of certain effects, such as attenuation. Furthermore, the inclusion of active circuits allows extra features to be added to the simulator. Some of the advantages of active simulation include

- Ability to simulate a wide variety of different types of line with one simulator unit, and to change the line types almost instantaneously
- Ability to save and recall the characteristics of any line type and the make-up of any loop under user control
- Simulation of line characteristics at different temperatures
- Addition of some impairments to the signal passing through the simulator

Because of the DC and ringing tone put on to many DSL circuits, and the sensitive nature of active circuits used in this type of simulator, the simulation circuit must be AC-coupled to the line. Often these simulators handle DC using a separate circuit constructed in parallel to and filtered from the signal simulation circuit. A problem for the active simulator is providing the correct input impedance, and another problem is that active simulators always have a higher noise floor than passive simulators.

There are two main methods of providing active simulation. One uses op-amp circuits for the attenuation and some of the group delay shaping. End-to-end delay is simulated by digitizing the signal and delaying it through a first-in-first-out (FIFO) buffer or a memory buffer. The second type digitizes the received signal from the modem as soon as possible and uses all-digital processing to simulate the end-to-end delay and the frequency response. In both cases, analog circuits are situated close to the output of the simulator to simulate the input impedance. There is a great deal of commonality between the two approaches.

4.6.3.1 Active Analog Simulation with Op-Amps

This type of simulator is composed of one or more active modules. Each module provides a length of line of a known gauge. The length and gauge can be varied quickly and are both under user control. In most cases, a loop configuration module is also provided, either through the front panel or via remote control, to interconnect the modules for the user so that an entire loop can be reconfigured.

Because electronic circuits can normally handle signals traveling in one direction only, an active simulator has to overcome the problem of signals traveling simultaneously in both directions on the line. The requirement for accommodating bidirectional signals is generally met by using a hybrid

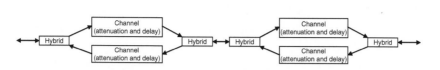

Figure 4.8   Block diagram of an active analog simulator, consisting of two modules with a node in the middle.

circuit, which is a derivative of the Wheatstone bridge (see Chapter 3). There are many different variants of hybrids. One hybrid circuit resides near each end of a module. The module has two sets of active circuits, one for each direction, connected between the hybrids. Figure 4.8 shows a block diagram of an active simulator with two modules in series.

Each module must present the correct impedance to whatever is attached at its end. In the case of a one module simulation, the modems under test would be connected at each end of that module. The impedance is provided by a passive network that presents a known impedance at both ends of the module. It is extremely likely that this passive network is tightly bound to the hybrid network, and for this reason it has not been shown separately in Figure 4.8. Each hybrid has an associated impedance network. Often the passive network in this module can be switched by relays to represent several different impedances that are designed to match the impedances of several different lines. However, the number of different line types available at any one time is limited by both cost and design constraints.

Once the two signal directions have been separated, the signals can be processed in two identical channels. The characteristics of attenuation and delay can be programmed in the channel. Attenuation is relatively easy to program using analog or digital circuitry, or a mixture of both. End-to-end delay is achieved by first digitizing the signal, then using a digital delay circuit, and then converting the signal back to the analog domain. A suitable digital delay could be a circular buffer or a FIFO adjusted to the correct end-to-end delay for the line length set for the module. In addition, the simulator must provide frequency-dependent corrections to this end-to-end delay to simulate the group delay of different frequencies.

One of the problems in this process is the dynamic range, not only of the analog interface to the line, but also of the analog-to-digital and digital-to-analog converters. For instance, on long loops, the modems being tested may send a signal that requires 95 dB of SNR over the test loop. In addition, the signal from DMT modems, which are used for ADSL, ADSL2, and ADSL2plus, typically has a crest factor of 5 times (15 dB). If the ADC is to process this signal with full integrity, it needs a 110 dB dynamic range! This dynamic range is not easily achieved at the sample rates needed for wideband simulation, and doing so may lead to unwanted compromises such as a higher noise floor or increased distortion.

Attenuation and delay of the module can be programmed based on the length and type of line that is being simulated. In simple loops that have just one type of line with no bridged taps, the attenuation and delay of the module is the same as the attenuation and delay of the loop. If there is more than one line type in the loop, or if there is a bridged tap, each module is programmed for the length and type of line it is simulating. Different loops can be simulated by connecting together a number of modules in the correct configurations. There may also be a physical limitation on the minimum length of a module because of the possibility of oscillation or swinging when a segment has only a very small attenuation.

Standards bodies such as the DSL Forum normally specify the loop configuration for any particular modem test. Loop CSA4, which is shown in Figure 4.3, is an example of one such loop. If simulation of a standardized test loop is required, the user may be able to set the loop by name. Sometimes simulator manufacturers sell libraries of wireline types and loops that not only provide all the line characteristics but also the configurations and line lengths used for these standardized test loops.

The active analog type of simulator has difficulty simulating short bridged taps and short segments of line. It is also subject to the problems associated with all active circuits, such as overload, distortion, dynamic range, and noise. Normally, the main DSL signal is AC-coupled to the outside of a module, giving it a high-pass response. DC response is assured by using a low-pass circuit in parallel with the main simulation. This configuration presents a potential problem in the frequency range between the low pass and the high pass cut-off frequencies.

4.6.3.2 Active Digital Simulation

At the time of writing, no commercially available simulator using active digital simulation was available. However, lab versions have been built and demonstrated. A digital processing simulator must separate the two signal paths using the same sort of circuits and block diagram as that of the active analog simulator. It is possible that an electronic echo canceler circuit may be used to improve the quality of the separated signals. After the two signal directions have been separated, the signals can be processed in two channels.

As shown in Figure 4.9, this type of simulator has the same sort of block diagram as the active analog type in that, conceptually, the simulator is composed of one or more active modules. However, in practice the simulator has a number of internal nodes. These nodes occur where the line lengths of different types of cables are connected together. The interactions of the different line types that compose a loop are handled by

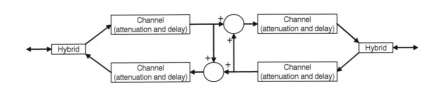

Figure 4.9   Block diagram of a digital active simulator with one internal node.

the digital signal processing that takes place in the node. As a result, only one module is needed to represent any loop, provided enough nodes for the loop being simulated are available within the simulator. Potentially, this approach yields a very powerful simulator in a very compact size. As in the active analog simulator, the digital simulator must present the correct impedance to the modems that are attached to it, and the impedance is simulated in the same way, by analog circuits. This means that the number of different line types available at any one time is limited as in the case of active analog simulators.

As far as the user is concerned, the loop topology can be constructed using a graphical user interface (GUI) on the unit. Bridged taps and in-line sections can be specified through the GUI. After the user has finished programming the loop, the unit can configure itself to provide the correct attenuation and delay. The optimum analog components for the impedance at each end of the line are chosen by the unit, and simulation can begin.

As with the active analog simulator, one of the problems with digital simulators is the very large dynamic range required by the analog interface to the line and the ADC. Another limitation is the number of nodes that can be accommodated by the simulator. The maximum number is built-in when the simulator is designed and can only be expanded by adding another simulator in series.

Other problems include the ability of the loop simulator to simulate short loops or short sections of a loop. Signals travel down a loop at a speed in the order of 200 m/μs (roughly 650 ft/μs). As an example, for a short loop of length 200 m, any reflection will return to the starting point after only 2 μs. There is some minimum processing time associated with much of the filtering being done in the unit. Clearly, if this processing cannot be completed within 2 μs, then the shortest loop that can be simulated is somewhat longer than 650 ft.

The digital simulator is also subject to the problems associated with all active circuits, such as overload, distortion, dynamic range, and noise. Normally, the main DSL signal is AC-coupled to the outside of a module, thereby giving it a high-pass response. The DC response is assured by using a low-pass circuit in parallel with the main simulation. This situation presents a potential problem in the frequency range between the low pass and the high pass cut-off frequencies.

4.7   Nonelectrical Characteristics of the Simulator

"Simulator Characteristics" discusses a range of parameters that a loop simulator may simulate. These are all electrical characteristics that must be considered when a loop simulator is used for testing, and the level to which each matter depends on what is being tested. This section considers the nonelectrical characteristics of a loop simulator. These are

- Repeatability
- Accuracy
- Remote control capability
- Scripting software

4.7.1   Repeatability

One of the key requirements when testing large numbers of units is the ability to repeat results from test to test and from location to location. This requirement means the test setup must be stable, and different units must be manufactured to the same specifications. At present, it is not unusual to find that if the modem pair, or the loop simulator, or even the impairments generator is changed to one made by a different manufacturer, the results are different.

Repeatability is a difficult subject. Because of intense competition among modem manufacturers, engineering design teams try to squeeze out the best possible performance, in terms of data rate, from their designs. As a result, relatively small differences in the test environment may lead to large differences in modem performance. In some cases, the results of data rate tests using the same type of simulator from the same manufacturer may lead to different results. The differences that cause this phenomenon are difficult to isolate.

For example, SHDSL (Symmetric High Bit-Rate Digital Subscriber Line) systems transmit and receive at the same time over the same frequency range. Transmit and receive signals are separated in the modem by a hybrid circuit and echo cancelers (see Chapter 17). Any distortion fed back from the transmitter into the receiver cannot be canceled and appears as noise at the receiver. On a long loop, at some frequencies, the received signal may be 90 dB less than the transmitted signal. Therefore, any distortion of the transmitted signal that is reflected back to the transmitter is at a level equivalent to noise in the receiver. Distortion this low is very difficult to measure.

4.7.2   Accuracy

The DSL Forum Test and Interoperability working group has produced test procedures for both ADSL and ADSL2plus so that modems and DSLAMs from different manufacturers may be tested against one another. In general, the accuracy requirement in both specifications for both the noise source and the line simulator is better than 0.5 dB.

But just how does one determine accuracy over a band of frequencies? If the line attenuation of the simulator is accurate to 0.05 dB over 500 kHz of the 700 kHz range, and then changes to 3 dB error over the other 200 kHz, is this better or worse than a simulator that is accurate to 0.3 dB over the whole range? If the simulator is specified as accurate to within 1 dB over a frequency range from 20 to 1000 kHz, is this better than one specified to within 4 percent of the attenuation in decibels over that range? This sort of question is not easy to answer.

One approach to specifying accuracy is to require that the accuracy of the attenuation must be "within X percent". In 2002, the ETSI TM6 working group specified in [ETSI TS 101 388] that for ADSL, the accuracy of a loop simulator should be within 3 percent on a decibel scale in the band from 20 kHz to 1.8 MHz. Unfortunately, this requirement tends to lead to surprisingly misleading specifications. The problem is connected with the data rate that one should be able to get. Using Shannon's "noisy coding theorem" [Shannon 1948], it is possible to calculate the theoretical capacity of a channel over a specific frequency range, when the transmitter is limited to a maximum power spectral density and the receiver has a specified noise floor as its lower limit.

Standards specify the maximum power a modem is allowed to transmit and how that power can be distributed across the available frequency band (see Chapter 17). For testing purposes, the noise at the receiver is also specified. Therefore, the SNR at the receiver depends on how accurately the loop simulator matches the theoretical loop characteristics. The purpose of the test is to see how well a modem performs under a specific set of repeatable conditions. The test conditions may or may not correspond to the conditions that confront the modem when it is connected to a real loop. What is important is that the total power received, and the way the received power varies with frequency, is within a certain target. Then the simulator accuracy can be specified in terms of how closely the received PSD matches the received PSD predicted by theory. This in turn leads to a specification of the simulator accuracy in terms of how well it attenuates the signal over the entire bandwidth, which is a measure of its ME, and how well the variation of attenuation with frequency matches the attenuation of the ideal loop. The DSL Forum has chosen to specify simulator attenuation accuracy in terms of ME and MAE, both of which are described in Section 4.8.1.

4.7.3 *Remote Control Capability and Programming*

Although the main purpose of a simulator is to simulate a local loop, there has to be some method of controlling the simulator. Especially, when doing tests repeatedly on many different units, or overnight on one unit, there is a need to be able to control the simulator remotely by sending commands from a computer. Often, this computer controls many aspects of the tests. Sometimes it controls the complete set of tests being performed on the modems. In addition, the ease with which a test engineer can program these commands and put them into a cohesive software application is vital.

It used to be that nearly everyone programming remote commands during the test phases of DSL modem production used the IEEE 488 (general purpose interface bus [GPIB]) bus. A serial port was also used, but the use was never very widespread. A higher level language, such as HP Vee or National Instruments' LabView, was often used in the test department of large modem manufacturers. Today, with more and more buildings being comprehensively wired for Ethernet applications and computer-to-computer interconnection, there is a general need to provide high level commands over Ethernet using TCP/IP (Transmission Control Protocol/Internet Protocol).

One of the ways to use Ethernet is to provide a software scripting language such as TCL (Tool command language), which can be used to program the simulator, the modems, and all the other test gear easily, quickly, and at a high level. Although providing remote control and scripting can be time consuming for the simulator manufacturer, it is of value to the user.

4.8 Recent Advances in Loop Simulator Design and Specifications

4.8.1 Attenuation Specifications

Starting in the year 2000, simulator manufacturers began working in conjunction with standards bodies to specify the ME and MAE requirements on the attenuation of loop simulators. The DSL Forum has provided definitions of ME and MAE for an ADSL simulator in Section 4.1 of [DSL For. TR-067]. The definitions for ADSL2plus simulators are specified in [DSL For. WT-100]. Readers should be aware that the frequency and attenuation ranges used to calculate ME and MAE vary according to the type of DSL signal being carried by the simulator.

Let us consider the PSD of the transmitting modem to be -40 dBm/Hz over the frequency range from 25.875 to 1104 kHz (as in ADSL, see Chapter 17). Figure 4.10a shows the theoretical attenuation of a 26 AWG loop 8 kft long. This sort of smooth attenuation curve is typical for a loop that consists of one type of line without bridged taps.

The received PSD is the transmitted PSD in decibels, plus the insertion loss in decibels, of the loop. The received PSD is shown in the graph of

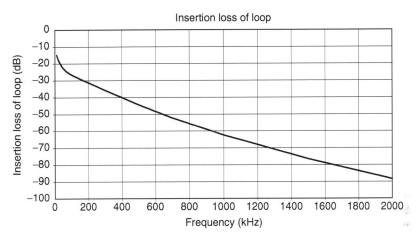

(a) Insertion loss of 8 kft of 26 AWG.

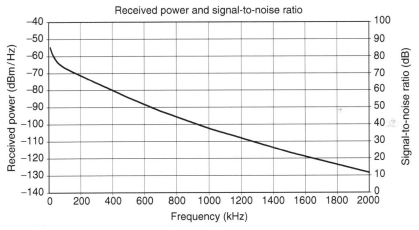

(b) Received power spectral density and SNR at the receiver.

Figure 4.10   Example loop attenuation, received PSD, and receiver SNR.

Figure 4.10b on the scale at the left hand side of the graph. Assuming that AWGN at a level of −140 dBm/Hz is added to the signal at the receiver, the SNR over the available frequency range is as shown on the scale at the right hand side of the bottom graph.

The area under the nominal received signal level and above the −140 dBm/Hz AWGN level can be calculated along with the average (mean) SNR over any specific frequency range. Any attenuation curve that gives the same mean SNR over the same bandwidth would provide an ideal modem with a communication channel capable of supporting the same data rate, provided the amplitude probability distribution of the noise is Gaussian. For

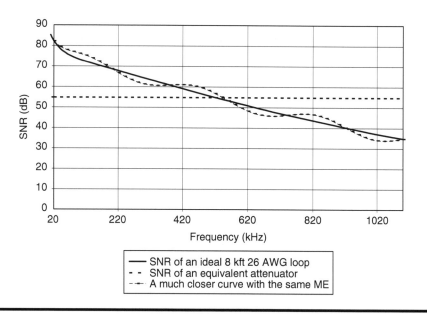

Figure 4.11   SNR of three different attenuation simulations.

the graph of Figure 4.10b, the area under SNR curve between the lowest frequency of 25.875 kHz and the highest frequency of 1104 kHz is an area of 58,563 dB-kHz, or an average of 54.1 dB over the entire bandwidth.

Figure 4.11 shows the received SNR as it varies with frequency for three cases. The solid line shows the SNR for a circuit that has the ideal attenuation of the 8 kft, 26 AWG loop shown in Figure 4.10a. The dashed horizontal line represents the extreme case of an attenuator that is constant over the entire frequency band of interest, and the dash-dot curve is an approximation to the 8 kft of 26 AWG of the first curve. This last curve represents the variation of attenuation with frequency that might result from a relatively poor simulation of the loop. All three of these curves have the same area under them.

Because the average receiver SNR for all three of these curves is the same, an ideal modem should be able to receive data at the same rate on any of them. Assuming that the first curve is the theoretical one, both the horizontal curve and the third, simulated curve are said to have zero mean error or ME. There is an infinite number of curves that would give a ME of zero. The horizontal line shown here is extreme, but it shows that although a ME of zero is desirable, by itself zero ME is not sufficient to enable a judgment of whether the simulated loop attenuation is accurate.

The MAE is a number that is used to determine whether the simulated curve fits well to the theoretical one. This calculation considers the magnitude or absolute value of the error of the measured curve in decibels at a number of frequency intervals across the transmission band.

For both ME and MAE, the attenuation of the simulator is measured up to some value of attenuation at equal frequency intervals in the frequency range of the modems. In the past, these measurements were taken at 10 kHz intervals. For ADSL and ADSL2plus modems, it is now more common to measure the attenuation at the center of each subchannel, where the subchannels are spaced at 4.3125 kHz.

To determine MAE, the value of attenuation at each frequency is measured, the difference between it and the calculated value of the ideal curve is calculated, and the absolute value of the difference at each frequency is summed. Then the average value is calculated by dividing this sum by the number of frequency points used. This figure cannot be better than the ME, but it can be significantly worse.

In the graph shown in Figure 4.11, the ME is zero but the MAE for the attenuator with the flat frequency response is 12.5 dB, and the MAE for the (poorly) simulated response is 1.95 dB. Detailed requirements for measuring and calculating ME and MAE are given in some of the test plans written by the DSL Forum. These requirements include the attenuation range and frequency range over which they shall be calculated. The DSL forum is addressing these requirements for ADSL2plus and VDSL modem testing at the time of writing.

It should be emphasized that both ME and MAE depend on the frequency range of interest. For instance, a simulator that carries frequency-division multiplexed ADSL2plus over POTS is required to meet the ME and MAE requirements over the range from 138 to 2208 kHz for downstream transmissions, and over the range from 25 to 138 kHz for upstream transmissions. The MAE values for these two ranges are almost certain to be different from each other and different from the MAE over the whole range.

Both ME and MAE are calculated for attenuation within a certain range. When the attenuation becomes too large, the receiver in the DSL modem cannot receive meaningful data, and the accuracy of the simulator becomes irrelevant.

It should be mentioned that after compensation, a high quality loop simulator is accurate to an MAE figure of better than 0.5 dB for any loop it simulates. This accuracy holds for both simple loops and more complicated loops that are composed of various sections of line and bridged taps.

An example of loop compensation using ME and MAE is shown in Figure 4.12. The plain curve of Figure 4.12a shows the theoretical attenuation of a loop of approximately 11.5 kft with a 250 ft bridged tap, and the curve with markers is the measured attenuation of a passive analog loop simulator. Figure 4.12b shows the theoretical attenuation as well as the measured attenuation of the same simulator when it had been fully compensated. It is clear that the compensated line is much closer to the

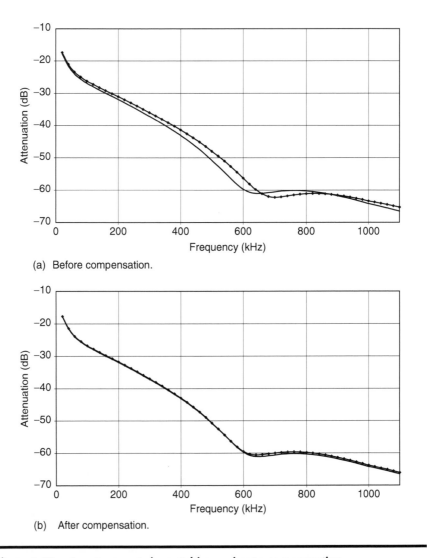

(a) Before compensation.

(b) After compensation.

Figure 4.12   Improvements in matching owing to compensation.

ideal curve. The ME and MAE figures of the uncompensated simulator are
−0.94 and 1.38 dB, and the same figures for the compensated simulator
are −0.31 and 0.32 dB.

Compensation is needed to optimize a simulator's performance. The
factory may supply compensation for each individual simulator, and there
may be some software supplied with the simulator that enables the user
to recompensate the simulator using a network analyzer at suitable time
intervals. Compensation software is also useful when it is not possible to

position the simulator close to the modems under test, and one or both of them have to be connected to the simulator using an extra length of wire. The software can be run to compensate, to some extent, for the extra length of line used. The effectiveness of the compensation is limited in high-frequency applications, especially when the extra line used is a different type than the line simulated.

There are a number of reasons why ME and MAE may not accurately predict the performance of modems deployed on real loops in the real world. For instance, the noise seen by the modem should have an amplitude distribution that is Gaussian in nature. In addition, if the noise seen by the receiver has crosstalk or other noise components that have a very different shape in the frequency domain than the white noise level assumed, the ME calculation could be distorted. Also, DMT modems assign a certain number of bits to each subchannel depending on the SNR of the subchannel, and to say that the only element of the simulator that affects data rate is the area under the curve, or ME, is somewhat of a simplification.

4.8.2   *Improvements in Internal Crosstalk*

In the early 1990s, there was little interest in a simulator with more attenuation than 70 dB. Modems just did not pick up signals attenuated by 70 dB. Now the advances in DSL receiver technology are such that some manufacturers are claiming that their front-ends are as quiet as −150 dBm/Hz, and their modems receive signals as low as 95 dB below the transmit level. This improved receiver performance has led to improvements in simulators so that the new advances can be tested. Recently, it has been announced that simulators have improved the internal crosstalk to less than −100 dB, and a response down to 95 dB of attenuation is achievable.

Bibliography

[ANSI T1.601-1992] American National Standard For Telecommunications, ANSI, T1.601-1992.

[Bell] Bell System Technical Reference Pub 62310, September 1983, *Digital Data System Channel Interface Specification.*

[DSL For. TR-067] DSL Forum Technical Report TR-067, *ADSL Interoperability Test Plan.*

[DSL For. WT-100] DSL Forum WT-100 ADSL2/ADSL2plus, *Performance Test Plan.*

[ETSI TS 101 388] ETSI TS 101 388 V1.3.1 (2002), *Asymmetric Digital Subscriber Line—European Requirements.*

[Golden 2006] P. Golden, H. Dedieu, and K.S. Jacobsen, Eds., *Fundamentals of DSL Technology.* Auerbach Publications, Boca Raton, Florida, 2006, p. 457.

188 ■ *Implementation and Applications of DSL Technology*

[G.991.2]   ITU-T Standard G.991.2, published December 2003, *Single-Pair High-Speed Subscriber Line (G.SHDSL) Transceivers.*

[Rome]   Wireline Simulator. Rome Air Development Center in Rome, New York. May be accessed at http://www.rf.af.mil/history/1960s/1968.html.

[Shannon 1948]   C.E. Shannon, A mathematical theory of communications, *The Bell System Technical Journal*, 27, 1948.

Chapter 5

Evolving Test and Provisioning from POTS to xDSL Services

Roger Faulkner, Ken Kerpez, and Philip Golden

CONTENTS

Abstract   We discuss methods of testing the electrical performance of copper pairs, including the detection and location of fault conditions.

5.1   Introduction

This chapter discusses methods of testing the electrical performance of copper pairs, including the detection and location of fault conditions. Line test has long been recognized as an essential component of running an efficient access network, and its technical evolution must progress in parallel with the technologies deployed. The chapter begins with a description of the legacy test systems used with traditional plain old telephone service (POTS) in Section 5.2. This is followed in Section 5.3 by an analysis of the relevancy of these systems for Digital Subscriber Line (DSL) related testing. Issues of test access for DSL systems are explained in Section 5.4, while Section 5.5 gives an introductory description of reflectometry techniques. Finally, Section 5.6 outlines aspects of test procedures being developed specifically for DSL services.

5.2   Line Test for POTS

The scope of this section concerns testing of the POTS access network, i.e., from the POTS line card (usually in the local exchange) to the customer.* A POTS line test system should accomplish the following effectively:

- Estimation of line characteristics (length, unbalance, termination)
- Detection of faults
- Assessment of whether the fault is affecting the quality of service
- Reconciliation of a detected fault with a customer complaint
- Location of fault and indication of initial activity required to clear the fault
- Interactive tests and tone generation

The focus of this chapter will be on the first two of these, as they are most relevant to DSL systems.

5.2.1   *Physical Connection of an External Test Head*

Although issues of test access for DSL can be quite complex (see Section 5.4), in the case of traditional POTS testing, the physical connection is usually straightforward. Virtually, all telephone exchanges are equipped with the ability to make parametric tests on the physical twisted pair lines that are connected to the line cards within the voice switch. Nonetheless, the bulk of line testing is performed by external line test systems that only make use of the test access of the switch to gain direct metallic access to the physical line. This metallic access allows testing at DC (direct current) to well above voice frequencies.

The external test head will be physically connected to the line via a test bus and the POTS line card, as shown in Figure 5.1. In addition, it will be connected to exchange earth and battery references for the lines that it may measure (this is detailed in Section 5.2.4). The primary reason for this move to external line test systems is the desire for a common line test platform within a multiple switch vendor environment as practiced by most telephone companies. Such systems perform the task of making measurements specifically for the purpose of providing diagnostics and fault dispatching recommendations.

---

* This "local loop" POTS testing is quite distinct from intra-exchange POTS testing, principally, because it is a less controlled environment that requires testing with a variety of possible terminations.

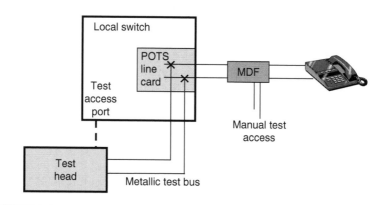

Figure 5.1   Single ended loop test access.

External POTS line test systems for voice switches request line test by directory number or equipment number through the various command or response interfaces provided by each switch type through a test access port, as shown in Figure 5.1. Once commanded to give test access, the switch will check the state of the line and either grant or refuse access. A refusal will usually be because of the line being busy,* however, there may be an alternative service-specific reason. Assuming that test access is granted, then typically a relay will operate within the POTS line card to give metallic access to the line on a test bus. One popular implementation is the so-called two-wire access. For this configuration, the default position of the relay is shown in Figure 5.2. The relay can be switched to

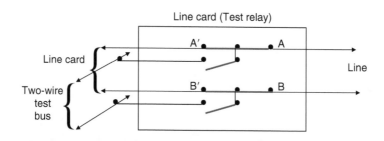

Figure 5.2   Two-wire metallic test access (default). Two-wire test bus, normal connection: The test relay which is bridging a connection A′ to A and B′ to B. Therefore, subscriber can signal to the switch as line feed is available through the relay. Test bus has no connection to the line.

---

* To avoid complaints from subscribers, it is generally accepted that testing should not intrude over voice calls.

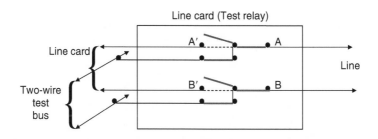

Figure 5.3 **Two-wire metallic test access (test head connected to line only). Two-wire test bus, line split, looking out: The test relay which was bridging a connection A′ to A and B′ to B is now open. The connection to the test bus is now closed so the test bus is now connected to the line.**

- Provide the connection between the test head and the line only (Figure 5.3)
- Provide the connection between the test head and the line card only (Figure 5.4)
- Provide the connection between the test head, the line, and the line card (Figure 5.5)

Alternatively, a "four-wire" metallic access is possible too, with two wires connected to the line and two wires connected to line card. In this scenario, the external test head must provide any bridging required. This configuration is shown in Figures 5.6 and 5.7. At the end of an externally requested test (or after a predetermined time out period), the switch will be commanded to take down the test connection. The normal POTS line feed (see chapter 1 of [Golden 2006]) will now be re-connected to the

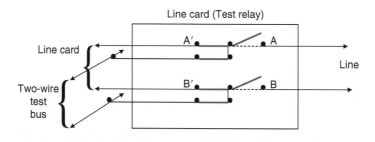

Figure 5.4 **Two-wire metallic test access (test head connected to line card only). Two-wire test bus, line split, looking in: The test relay which was bridging a connection A′ to A and B′ to B is now open again. The connection to the test bus is now closed so the test bus is now connected to the line card, typically to test that line feed is present and dial tone can be drawn.**

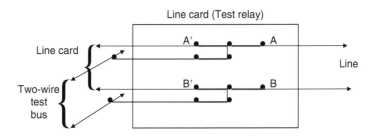

Figure 5.5 **Two-wire metallic test access (test head connected to line card and line simultaneously). Two-wire test bus, line bridged: The test relay which was bridging a connection A′ to A and B′ to B is now closed again. The connection to the test bus is now closed so the test bus is now connected to the line card and the line.**

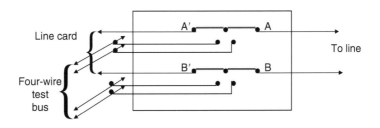

Figure 5.6 **Four-wire metallic test access (default). No test in progress, test relay has bridged A′ to A and B′ to B.**

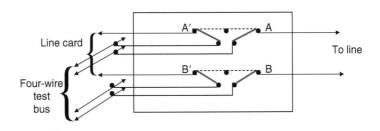

Figure 5.7 **Four-wire metallic test access (default). Test in progress, test relay has connected A′ and B′ to test bus and A and B to test bus. With this four-wire test bus, the test head must bridge the connection between line feed to customer under test if a bridged test is required.**

subscriber line. The metallic test bus connection to the line under test will be presented at a specific location within the switch. It is desirable that the path between this presentation of the test pair(s) and the pair exiting the exchange via the main distribution frame (MDF) be reasonably short so as to limit the effect of the exchange wiring on the measurement results.

In all of these cases, the test bus is usually common to either 128 or 256 line cards on the same shelf.

5.2.2 Parametric and Functional Tests

Once the test head has been granted access to the required line, it can then perform desired parametric and functional testing. The methods by which such tests are carried out are usually proprietary to the test system in question. However, typical tests that can be performed by the external test head are detailed below. (For the purpose of this description, the pair of wires are labeled "A"(tip) and "B"(ring), the earth reference is the exchange earth, and the battery reference is the exchange battery. Furthermore, it is assumed throughout this chapter that the A or tip wire is normally at earth potential, and the B or ring wire is normally at battery potential.)

The parametric measurements can be divided into the following categories:

- AC Voltage (A to Earth, B to Earth, A to B)
- DC Voltage (A to Earth, B to Earth, A to B)
- Capacitance (A to Earth, B to Earth, A to B)
- Resistance (A to Earth, B to Earth, A to B, B to A, A to Battery and B to Battery)
- Voice band noise

Additionally, the test head may perform a number of functional tests and tone generation that may include the following:

- Termination detection
- Signature recognition (specific equipment on a pair may sometimes be identified through consistent parametric results that are returned when tested)
- Interactive voice or monitor functions with the customer
- Interactive dial pad tests with the customer
- Interactive fault location with a field craft technician
- Trace tone generation
- Howler tones generation

"Trace tone" is an audible tone transmitted on the pair to enable a field technician to identify a particular pair in a cable. Generally, this tone will persist only when the input impedance of the line is high, i.e., telephone is in an "on-hook" state (see chapter 1 in [Golden 2006]). However, it may also be present if the line has a short circuit fault.

Howler is a tone or composite of many tones that may be transmitted on the pair when the line is low impedance, specifically when the telephone is off-hook and no call is in progress. This is done in an attempt to persuade the subscriber to replace the receiver.

The frequencies used in AC measurements for POTS are generally low (below 100 Hz), which is advantageous owing to the lower line attenuation at these frequencies. It should be noted that the test bus between the test head and the line under test in the exchange will usually be included in the measurements as previously mentioned, hence the desire to keep this wiring reasonably short. Compensation values may be applied in an attempt to offset the effect of the test bus wiring.

Inspection of Figure 5.1 shows a second potential test access point to the line at the MDF, sometimes called a "test shoe." This often involves a technician physically making the connection with a manual test probe to enable additional testing using multimeter techniques, Time domain reflectometry (TDR), or frequency domain reflectometry (FDR). These are explained in Section 5.4.

5.2.3 Line Length Estimation

As previously mentioned, one of the key results of POTS line testing is the estimation of the line length. This is typically done using a capacitance measurement. There are two types of capacitive measurement, that which is between the two wires, and that between wire and earth. Although they can be influenced by many factors, capacitive measurements can be a good indication of the length of a cable. Good quality telephone wires should have narrowly defined specifications for the characteristic capacitance per unit length of line, suggesting the assumption that measured capacitance increases uniformly with length. Some typical capacitance values for European and U.S. networks are given in Table 5.1. Much of Europe (with exceptions such as BT) uses quad cables, whilst cables comprising only pairs are favored by U.S. networks. Quads have a bigger separation between A and B wires which tends to reduce pair capacitance when compared with cables containing only pairs. The table shows the two different cable types (quad and pair) in use in two different telephone companies, note that despite there being different gauges and dielectrics each cable type has a similar capacitance per unit length of line. Measurements of capacitance to earth were made assuming that all surrounding pairs were in-service lines.

Table 5.1 Capacitance per Unit Length

Type	Gauge	Dielectric	Capacitance A-B (nF/km)	Capacitance A-Earth (nF/km)
Quad	0.65 mm ~ 22 AWG	Paper	45.6	86
Quad	0.4 mm ~ 26 AWG	Paper	45.8	86.6
Quad	0.65 mm ~ 22 AWG	Polyethylene	46.2	87.2
Quad	0.4 mm ~ 26 AWG	Polyethylene	45.7	86.3
Pair	26 AWG ~ 0.4 mm	Paper	51.6	56
Pair	24 AWG ~ 0.5 mm	Paper	51.8	56.3
Pair	26 AWG ~ 0.4 mm	Polyethylene	51.2	56.4
Pair	24 AWG ~ 0.5 mm	Polyethylene	51.7	56.7

5.2.3.1 *Capacitance between Wires*

In the case of capacitance measured between wires, the above assumption with regard to length of the customer premises can be somewhat tarnished by the capacitive effect of termination (e.g., telephone, modem) and inside wiring. Each of these can add to the effective capacitance between A and B wire, making the line appear to be longer than it is. In addition, the drop wire construction is generally very different to network cabling and may have very different pair capacitance per unit length values.

5.2.3.2 *Capacitance to Earth*

Figure 5.8 shows a few of the possible capacitive interactions that are possible in a bundle of wire pairs. The capacitance values arising from these multiple interactions are dependent on the geometry of the cable and the dielectric, the insulating material surrounding each wire core (see chapter 2 in [Golden 2006]). The number of pairs in use in the cable relative to its capacity (known as "cable fill," pairs not in use often called "spare") also has an effect on measured capacitance to earth, that the effect becoming very small as cable fill exceeds 60 percent, which commonly is exceeded in the access network. Generally, an AC measurement voltage is applied with respect to exchange ground on both A and B wires to perform this measurement. Thus, there is no potential difference between A and B and, therefore, unlike the case of capacitance between wires, the termination, drop, and inside wiring do not significantly affect capacitance measured to earth.

The capacitance then measured is highly indicative of cable length from the MDF to the final distribution point but no further in the case of single pair drop wires.

The most often calculated value for line length (assuming the loop is terminated) is based on measurement of capacitance to earth divided by

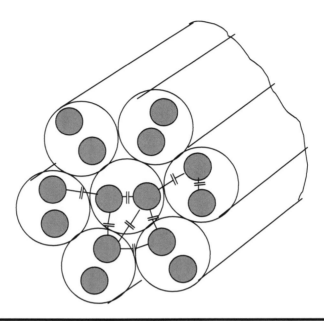

Figure 5.8   Capacitive coupling between wires.

some constant that best represents the average capacitance to earth for the cables deployed. This length value is taken to be representative of the distance between the line card and the final distribution point.

When the pair under test exits the cable, e.g., single pair in the drop wire to the customer premises, there is no more contribution to the earth capacitance measured, though pair capacitance measured between A and B continues to increase.

5.2.4   Fault Testing

POTS line testing may be carried out on a routine basis (often nightly intervals), for the purposes of pre-emptive or pro-active maintenance. These tests are particularly used to identify resistive faults which may cause problems on the affected line(s) but do not necessarily cause the POTS to fail completely. Testing is required on a demand basis too. It is quite normal for a POTS customer to report a problem before line test finds it, particularly true for disconnection type faults that cause instantaneous service failure. Generally, most common faults in POTS tend to be resistive contacts. These are DC faults that occur, for example, when the wires in a pair touch each other, or a wire in an adjacent pair. This can either be direct physical contact, or contact via a non-metallic path. In the latter case, the path is usually provided by a wet salt solution (see Section 5.3.2.1).

The relative percentage of disconnection type faults to resistive type faults tends to be seasonal, overall the quantity of disconnection type faults in the network tends to be stable, problems involving resistive contact increase during periods of wet weather.

Resistive faults can have a dramatic effect on the quality of the telephony service, affecting voice transmission by giving rise to noise and causing attenuation of POTS signals. In the worst case, the result can be complete loss of service.

As explained in chapter 1 of [Golden 2006], each telephone subscriber pair is wired from the POTS line card onto the MDF, and then into a cable containing multiple pairs in close proximity. The line card connection for each pair gives a so-called line feed (a DC potential difference across the pair) that is related to the battery voltage from which the exchange is powered. One wire of the pair presents a low impedance path to the exchange earth (hence the assumption of the A wire being grounded), while the other wire of the pair is connected to the exchange earth through the battery feed of the exchange. This is illustrated in Figure 5.9. As mentioned in chapter 1 of [Golden 2006], the latter wire will generally be at a negative potential relative to earth. Assuming that the exchange battery has a low source impedance, each wire in a cable thus presents a low impedance path to exchange earth. As mentioned in Section 5.2.1, the test head will also be connected to the exchange earth.

When a copper pair is switched for line testing, the line feed is removed from the pair under test. After discharging any residual voltage that may be present, the basic tests of Table 5.2 will be carried out. In the case where any of the measured results are not the desired results as given in Table 5.2, it is likely that a fault condition exists.

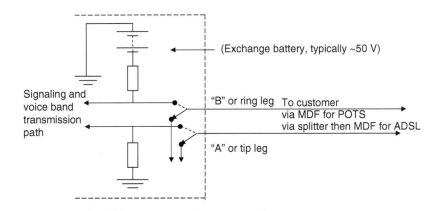

Figure 5.9 **Basic earth and battery references for a telephone pair (automated line test connection also shown).**

Table 5.2   Basic POTS Line Tests

Parameter	Test Points	Desired Result
AC voltage	A to Earth B to Earth	Very near zero
DC voltage	A to Earth B to Earth	Very near zero*
Resistance	A to Earth B to Earth	At least 1 MΩ
Resistance	A to Battery B to Battery	At least 1 MΩ
Resistance[†]	A to B	At least 1 MΩ
Capacitance	A to Earth B to Earth	A-eth = B-eth
Capacitance[‡]	A to B	Between limits

Of course, further information can be deduced from the measurements of Table 5.2 in the case where a fault is present. For example, an AC voltage of 220 V would indicate contact with an electrical line. Additionally, some information regarding fault location can be deduced particularly for open circuits when capacitance measurements can be used to give distance to disconnection. When accurate location information is needed for resistive type faults, then additional measurements may be made using hand-held instruments at the MDF and in the access network to locate such a fault. Regardless of fault type, many POTS line test systems will, on encountering a fault, make a dispatch recommendation which indicates the location of the fault at a high level typically:

1. Central office (CO)
2. Local network
3. Drop, customer premises equipment (CPE), or inside wiring

In this way, the correct work group can be identified to resolve the problem, and a decision is made to send a repairperson to the customer premises or not.

---

* Although DC voltage can indicate that there is a contact to another telephone pair, the severity of such a contact is best given by the resistance between the line under test to reference battery and to reference earth.

[†] Two measurements are often made by driving current in opposing directions, any rectification caused by corrosive faults can thus be identified (see Section 5.3.2.1).

[‡] Capacitance limits are highly dependent on the technique used to assess A–B capacitance limits, which are set above that expected for an open circuit and below the value that would indicate too many ringers attached.

5.3   Use of POTS Testing for DSL Services

As mentioned in the introduction to this chapter, using the access network for broadband data transmission necessitates new requirements for line testing. POTS line test systems were not designed for these requirements, however, their functionality can still be useful as detailed in the following sections. Even with the limitations mentioned, there is a good case for maximizing and evolving the use of existing line test systems in support of xDSL services. Leveraging existing POTS line test systems to perform an assessment of access network health is economically advantageous, requiring no costly manual intervention, especially before xDSL services are deployed.

5.3.1   Line Length and Insertion Loss

As detailed in Chapter 6, the insertion loss of a line at a single frequency such as 300 kHz is often a critical parameter for deployment rules. There is often a good correlation between the insertion loss and the line length, and hence it is clear that the length estimates determined by POTS measurements (see Section 5.2.3) could be of use in DSL systems.

The insertion loss of the cable can vary significantly with the diameter of conductor and the dielectric used. Hence some additional network information can increase the accuracy of the estimation of insertion loss from line length. For some access networks, the planning rules are such that up to some specific distance from the CO only one cable type is deployed, thereafter another cable type is deployed. For example, it may be that planning rules dictate that all pairs exiting the MDF must be copper of gauge 0.4 mm, thereafter 0.6 mm gauge copper will be used. In this case, knowledge of loop length (D) alone can give a reasonable estimate of loss (provided of course the line is free of faults). As a rule of thumb, using the above assumptions one can estimate* the insertion loss for lines over 2 km in length using

$$\text{Insertion loss at 300 kHz} = 10 \times (D - 2) + 28 \text{ dB}$$

And for lines under 2 km in length

$$\text{Insertion loss at 300 kHz} = D \times 14 \text{ dB}$$

It should be noted that an estimation of insertion loss at DSL frequencies can also be made by extrapolating direct AC measurements made at POTS frequencies and above (assuming that the test equipment has the ability to make these measurements). The accuracy of this method will principally

---

* Cable engineering rules vary widely between countries so specific equations would need to be constructed to reflect those rules.

Table 5.3   Example of Insertion Loss (dB) Variation with Frequency and Bridged Tap

Frequency (kHz)	1	5	10	50	100	200	500	700	800	1000
2 km	11.7	12.0	12.8	18.6	21.4	25.2	35.9	42.2	45.2	50.7
2 km with 200 m tap	11.7	12.1	12.8	19.1	22.9	32.7	37.4	48.5	48.6	52.9

be affected by load coils, metallic faults, termination types, and bridged taps. (One advantage of this technique is that it is fairly immune to noise, very little crosstalk is seen at low frequencies, and noise in-band for DSL does not enter into the measurements.) For example, the loss of a pair of wires 0.4 mm diameter copper with polythene dielectric terminated in 100 Ω was found. The loss of 2 km loop with no bridged tap is compared to the loss of a 2 km loop with a 200 m bridged tap (Table 5.3).

Interesting to note is that these extrapolation techniques, though being inaccurate compared with instruments at particular frequencies (e.g., 300 kHz with a 150 m tap) does give a good estimate of the Asymmetric Digital Subscriber Line (ADSL) performance. However, if a measurement happens to be taken at a frequency at which a null is caused by a bridged tap, then there is a pessimistic assessment of the ability of that line to support service. The extrapolation technique should give a value for loss on a 2 km line with a 200 m tap of a loss slightly above a line of 2.2 km long without a tap.

5.3.2   Fault Detection for DSL

The detection of resistive faults is as vital for DSL as it is for POTS, even though, the actual effect of the fault may be different. In particular, certain faults may allow that some level of reduced DSL performance to be maintained, but certainly the impairment should be recognized, located, and repaired.

As highlighted by chapter 2 of [Golden 2006], line balance is of particular importance to the performance of DSL systems. Poor balance often results in high noise levels. Comparatively, large line imbalances can be recognized by traditional line test systems by comparing capacitance of each wire to earth. Values that differ by a marked amount can indicate a split pair or a high resistive joint (see below). The reverse is not certain though, i.e., well-matched capacitance values to earth on each leg do not necessarily indicate that the line is balanced.

5.3.2.1   Faults at Splice Joints

A "split pair" occurs when a wire from one pair is accidentally spliced to the

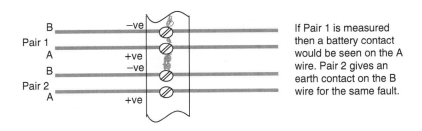

Figure 5.10   Two adjacent pairs terminated onto a connection strip.

that the twisting of the pair is obviously disturbed means that it is very susceptible to crosstalk. Split pairs can be found by analyzing impedance measurements to earth of each leg, with respect to impedance between A and B wires. For a split pair, there will be an unexpectedly large impedance between A and B compared to that which would be expected from the impedance to ground.

High resistance faults can be caused by a poor connection between two wires at the MDF or a subsequent connection point. If oxygen is present in a splice or connector, then an oxide layer will form on the conductors to provide a joint in which unexpectedly high resistance can be seen. Even worse, such an oxidized joint may be unstable mechanically or electrically such that the resistance associated with the joint may change rapidly with time if the joint is disturbed by movement or by changing currents through the joint, such as when the receiver goes off-hook.

Additionally, at joints and connection points with some exposed metallic conductor, it may be possible to form a low impedance path between two connectors that can become a problem, a potential a source of noise in both POTS and DSL transmission. Figure 5.10 shows two pairs terminated on a connection strip. Provided that the connection strip remains dry, no low impedance pathway can form. However, if there is moisture present on the nonconductive substrate then a fault can form. This kind of fault is inherently unstable, however, it is polarized. The latter quality greatly assists in its detection, as resistance measured with current flowing in one direction is likely to be measured differently if current is flowing in the other direction.

From a test point-of-view, testing that one wire of the pair has different electrical properties to the other is a starting point for recognizing imbalance faults. Refining this to see whether the fault is unstable is typically done by two means:

- Apply ramped DC voltage with respect to ground on either legs of the pair and measure any current that may flow. For a stable fault, current measured should be directly proportional to voltage applied. The faults that will be relatively more service affecting will be those where there is a very rapid change of current that is not proportional to the applied voltage, particularly where the current takes rapid excursions from the expected value.

- Measure the balance of the line with respect to the measured phase angle between each leg and earth. If the two values coincide at all frequencies, then the line could be expected to be reasonably balanced. However, if the two values are different, then one could reasonably expect that the pair is presenting an unbalanced signal to the modem. If the wires are now pulsed with a voltage to drive current through the fault and the phase angle re-measured, changes in that phase angle are likely because of the fault being unstable.

Corroded splices can behave somewhat like a diode and cause a nonlinear response. These could be identified as non-Ohmic by a nonlinear I-V curve, or by harmonics generated by a sine wave or sine waves.

5.3.2.2 DSL Response to Unstable Faults

Generally, xDSL communications are quite fault tolerant but unstable faults can be particularly detrimental to data transmission, therefore, it is highly desirable to find such unstable faults through testing. The reason why unstable faults affect data transmission more than stable faults (even some very severe stable faults) is because of how xDSL or other modems interface to the line, track carrier phase, adjust access gateway controller (AGC), and adapt their echo cancelers.

If the fault remains unstable whilst training proceeds, the modem may fail to train altogether.*

5.3.3 Loop Electronics

Sometimes, a loop loaded has a filter or other attached electronics which can unexpectedly disable DSL on a good POTS line, but which were not identified in loop records at the time of service provisioning. These are outlined and methods for identifying them are given in Table 5.4.

---

* Modems do contain dynamic phase adaptation circuitry, the agility of such circuitry will dictate performance in the presence of unstable faults.

Table 5.4   Loop Impairments

Impairment	Test Method
Load coils	Testing the line at multiple frequencies and analyzing how the impedance changes as frequency increases
Radio filters	Very difficult to detect by measurement except through end-to-end transmission testing with the telephone terminating device(s) removed
Shared line multiplexers	Single-ended testing using standard or frequency enhanced parametric values

5.3.4   *Enhanced POTS Test Equipment to Estimate Loop Loss*

Enhancing traditional line test to enable estimation of the primary line transmission parameters allows insertion loss to be calculated (see chapter 2 of [Golden 2006]). This offers an economical means of assessing lines for ADSL (other DSLs are generally not accessible through the voice switch, or perhaps only use a small portion of the overall loop, see Section 5.4). Critically, such an enhancement would need to use single-ended measurement techniques to return an estimate of insertion loss at higher frequencies, much higher than would normally be used by such line test systems for measurements. Estimates of primary line parameters can be found using this technique with advanced analysis methods.

Even though, measurements are only possible either below or at the lower end of the DSL spectrum, this technique can be an inexpensive enhancement to existing POTS test equipment and can give some assessment of loss in the presence of bridged taps. Lower frequency measurements are affected by the loss owing to the tap but will not experience destructive interference that could be present at specific reference frequencies as a function of bridged tap length.

Testing of lines using POTS test equipment has an important function in that tests may be made network wide, and give a channel for remote field tests to be automatically analyzed and interpreted for use by the operator receiving customer complaints to support their decision making. For lines that are ADSL enabled, the POTS based test equipment is able to look for repairable faults and is also able to recognize that a splitter(s) of a particular type(s) is installed and is wired correctly.

5.4   Access Issues for DSL Testing

At present, there are a number of xDSL services with fundamentally different automated test access criteria. Examples of these are the following:

- *Example* 1: ADSL broadband data is provided by a Digital Subscriber Line access multiplexer (DSLAM) over baseband POTS or Integrated Services Digital Network (ISDN) service. In this type of connection, the physical pair is connected to the voice switch and the DSLAM in the exchange. The subscriber typically has a single line for telephony and data services, hence a low-frequency metallic connection to the line is available via the switch test bus. This is one access point, however, it is clear that the bandwidth of the testing will be limited to below about 10 kHz owing to the POTS splitter. There will generally be a second access point via the DSLAM for broadband access. The location of the broadband access point in relation to the splitter is important. In the case where it is on the loop side of the splitter, full bandwidth metallic testing should be possible. If it is on the DSLAM side of the splitter, the DC blocking capacitors that may be present in the splitter (see Chapter 1) will block all DC and low frequency tests. It can also corrupt TDR tests.
- *Example* 2: Symmetric High Bit-Rate Digital Subscriber Line (SHDSL) broadband data is provided by a DSLAM located in the exchange but not over POTS or ISDN. In this case, the data occupies the entire frequency spectrum available on the line, therefore, no voice switch is required hence no access for testing via the voice switch.
- *Example* 3: Very High Bit-Rate Digital Subscriber Line (VDSL) broadband data is provided by a DSLAM (or equivalent) located within the vicinity of the subscriber. In this case, DSLAMs are often deployed remote of the exchange, and typically linked to the exchange by fiber. Metallic pairs run between the DSLAM and the subscribers in the served area, which for example may be a number of businesses with a business park. The voice may either be via POTS or may be digitized.

Hence, traditional test systems are not able to use the switch test bus method for SHDSL or VDSL (where voice is digitized), because there is no metallic connection between the exchange and the DSLAM providing such services. Even where there is a metallic connection between voice switch and the physical pair providing the data service (such as with most ADSL deployment), the subscriber pair is connected to a splitter such that high frequency access to the line is available to the ATU-C (or equivalent) and only the low frequency access to the line is available to the voice switch.

The economics of providing a large public network is generally driving change from a predominately circuit switched voice network to a packet switched data network, including packets containing voice data. In turn, this change spells the demise of traditional Class 5 switches and DSLAMs to be replaced by multi-service access nodes (MSANs) providing POTS and xDSL services using single, multi-service line cards for each connected pair. The additional desire to increase revenue from value added services that require higher bandwidth (such as Internet Protocol Television (IPTV)) will also encourage the placement of MSANs much closer to the customer. Fiber to the curb with MSANs placed at streetside will serve only several hundred customers. These developments, in turn, may be a stepping stone to "fiber-to-the-home" (FTTH). This latter development would of course spell the end to metallic testing of the subscriber loop.

The demise of "traditional" Class 5 switches also means of course that "traditional" access for metallic line test will also disappear over time. Telcos that wish to retain existing line test systems (particularly, those which have developed extensive operational practices and procedures for maintenance and repair based on metallic line testing) are investing in test access matrices (TAMs) that are able to provide a metallic test connection to any of the pairs wired to the MSAN. There are many issues that must be considered when implementing such equipment starting with the cost per port set against possible benefits such as remote configuration with no re-jumpering required. Then, given that the economics are such that there is positive benefit from retaining existing investment in a line test system driving maintenance and repair with TAMs, there are technical issues that then need to be considered:

- Transfer engineering—verification that lines moved from Class 5 switch to the MSAN TAM environment are connected to the correct ports.
- TAMs provide metallic access to the line but do not provide line status information (busy or free) prior to test, or back-busying of the subscriber line under test.
- Issue of TAM control is also not straightforward, typically there will be multiple uses of the TAM, connecting of other external equipment, remote line configuration changes as well as line testing. These user requests need to be prioritized and managed.

The upside of using TAMs is that typically they will provide four-wire broadband access to the line. This means that traditional line test systems are no longer restricted by the bandwidth available at the switch access or, even more restrictive, the low pass side of the CO splitter.

5.4.1 Intrusive Testing

In the case of DSL measurements, the POTS line state may not be known before the measurement is carried out. It would then be very important that any tests be nonintrusive to any possible voice call that may be present.

If tests are to be made through an access matrix, it is highly desirable that the TAM and line test work in a coordinated fashion to provide

- Line state before making any intrusive tests.
- Back-busying of the subscriber to incoming calls during any tests that are being made.

In general, low frequency testing through the voice switch is not intrusive either to voice calls or data calls as the switch provides the POTS line state, and low frequency measurements do not affect high frequency data transmissions. Other techniques that use higher frequency measurements and that do not assess POTS line state may be intrusive to active voice and data calls. Service interruption is less of an issue when the end customer has complained and is actively involved with the on-going resolution of the complaint, however, and pro-active or pre-emptive testing could become an issue and a source of complaints from customers.

5.5 Testing the Next Generation Network

Legacy analog line test systems with four-wire broadband access to the line can now be augmented to provide noise and PSD measurements—confirm spectral occupancy or "policing" the access network frequency plan (ANFP). In this unrestricted broadband access environment, other broadband tests can be added to the analog test heads such as FDR for fault location. Another narrowband and analog test type that can be added is that of perceptually motivated waveform analysis. There is both an active ITU standard P.862 (PESQ) and a passive standard P.563, for making measurements of a customers' opinion of voice quality being carried by a telephony network.

Legacy line test systems typically use draw dial tone and call completion to ensure that a call can be successfully connected across the POTS network. There is typically no standard test of voice quality made by the test head because it is assumed that a modern Class 5 digital switch network will always maintain "toll grade" quality from end-to-end. However, the movement to packet switched networks requires fundamental network change, see Figure 5.11. (In Figure 5.11, TAMs and LDUs are shown at each MSAN location, this may not always be the case, some MSANs will continue with fiber to the subscribers premises, some telcos may use whatever SELT or DELT facilities might be available through the line cards.)

The various components of a packet switched network are more than able to produce performance statistics such as packet loss and packet delay

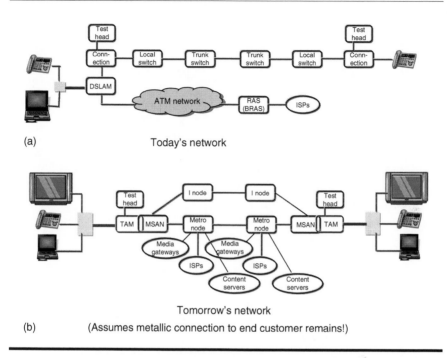

(a)   Today's network

(b)   (Assumes metallic connection to end customer remains!)

Figure 5.11   Fundamental network change owing to switched networks.

variations, alarms, etc. However, it is very difficult to use such statistics to predict the performance of a service such as Voice-over-Internet Protocol (VoIP).

Being based on a customer-oriented perception of the call quality, using perceptually motivated techniques to test the media pathway for the call gives, arguably, a better view of the service quality being delivered to the end customer than would network performance statistics alone.

Analog perceptual based measurements to evaluate call quality by analyzing the spectral content of the "speech" being monitored. The simplest implementation of such testing would involve interception of the customer's line, monitoring a conversation in progress (with permission granted by the customer of course) which would then be analyzed for the spectral content of the incoming voice tones. There is a standard for such passive tests (ITU Standard P.563), which will deliver a limited mean opinion score (MOS) for the call. There are several limitations of such testing. For instance, latency cannot be measured as there is no way in which to synchronize the incoming voice with the analysis being made.

Active PESQ testing (ITU Standard P.862) provides a much more powerful solution. An active PESQ test involves the generation of an analog speech reference signal, which is then transmitted across some part of the VoIP network. The receiving device performs a PESQ analysis by comparing received speech with the known reference signal. Again a MOS can be generated but there are additional diagnostics available, for example, jitter and "clipping" of the speech owing to incorrect operation of voice activity detection (VAD).

A simple active PESQ test could involve the simultaneous transmission and reception of the reference signal at a single test head. A call is placed by the test head that will be eventually switched back to itself via a Metro node. The time stamps for transmission and reception of the PESQ signal give the latency for the connection. This value, in addition to the MOS and diagnostics described above, is then available for fault finding in this limited part of the Internet Protocol (IP) network.

Given that test heads are able to access all customer pairs, and generate and analyze speech reference signals, a useful implementation of active PESQ testing would be to use one test head to generate a reference speech stimulus at one location at the edge of the network whilst another test head is set to analyze the incoming signal at another "edge" location. Multiple calls could be placed across the network to form a "mesh" of tests, analysis of PESQ results could then be used to highlight individual network nodes that may be causing problems to the media path used for VoIP calls. Such a scheme would require centralized marshaling of the test requests and analysis of results to reveal network "blackspots."

5.6   Introduction to Reflectometry Measurement

As mentioned in Section 5.2.1, there is usually a test access point at the MDF where measurements are typically made using hand held instruments. These instruments often use TDR, or FDR. In the case of POTS systems, these methods can be employed when a fault has been detected, and needs to be accurately located by a technician.

In the case of DSL measurements, reflectometry is used both with hand-held instruments and also with automated access and test equipment. The following gives the basic principles of TDR and FDR.

Single-ended tests based upon reflectometer techniques rely on the fact that a signal propagates through a medium but is reflected by discontinuities in that medium (see Chapter 2 of [Golden 2006]). These reflectometer or echometric techniques may be in the frequency domain (FDR) or in the time domain (TDR). Often a little "m" is used to make the distinction "metallic" time domain reflectometry (mTDR) to distinguish from optical methods. Reflectometry can identify difficult echo responses that may upset echo cancelation, it can identify loop length, and it can even be input to recently developed analysis routines to estimate a loop make-up including gauge changes and bridged taps.

Reflectometers are based on reflected signals which in turn are dependent on the termination impedance at the customer end of the loop. The "on-hook" telephone termination is approximately an open circuit [Galli 2001]. The accuracy of all reflectometry measurements are affected by noise and crosstalk on the loop, as well as by microreflections caused by slight imperfections along what is otherwise generally a good transmission line.

Converting reflectometry measurements to cable length requires knowledge of the propagation velocity. Temperature changes the propagation velocity by about 1 percent for each 10°C, but this can be compensated for if the measurement temperature is known. Propagation velocity also varies by up to about 6 percent with wire gauge. Propagation velocity can change unpredictably with age and manufacturing run. Older cables with less stable dielectrics can modify the propagation velocity as the dielectric degrades over time, and every new cable can vary by as much as ±3 percent from expected value.

5.6.1   FDR

FDR involves the transmission of a broad range of frequencies into the medium under test, i.e., the pair of wires. FDR is generally performed by a network analyzer that reports return loss, or equivalently, input impedance, as a function of frequency. Assuming that the pair of wires is reasonably homogenous throughout its length and that there are no significant faults that cause other reflections, then a major discontinuity at the end of the line (such as an open circuit) will cause the significant reflections. The broad spectrum of frequencies that are transmitted will be reflected and arrive back at the source but will be altered in phase. A vector network analyzer can record the phase as well as the amplitude of this reflection. Superposition of the transmitted and received spectrum will reveal frequencies at which constructive and destructive interference has taken place to give a standing wave at much lower frequency than those transmitted down the line, as seen in Figure 5.12. The usefulness of such a test lies in the fact that the peak-to-peak amplitude of the superposition is dependent upon cable loss and the magnitude of the impedance mismatch occurring at the end of the line under test, while the frequency of the modulation is a function of the line length to that impedance mismatch.

Bridged taps create nulls in the returned spectrum. The propagation velocity on twisted pair is approximately $v = 6.0 \times 10^8$ ft/s. At a given frequency, a bridged tap one-quarter wavelength long reflects a 180°

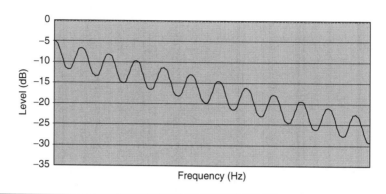

Figure 5.12   Chart to show Idealized FDR response.

phase-shifted signal, which cancels part of the signal (not all, because it is attenuated by transmission over the bridged tap). The quarter wavelength bridged tap length, in feet, is about

$$\frac{\lambda}{4} = \frac{v}{4f} = \frac{1.5 \times 10^8}{f}$$

where f is the frequency (in hertz) at which the bridged tap causes a null in the spectral response. Nulls are also created at odd multiples of this frequency.

The illustration of Figure 5.12 is unlikely to resemble the real world measurements. Experts analysis or advanced algorithms of the output waveforms are generally required to analyze the returned signal owing to the fact that impedance mismatches have a variety of possible causes: cable gauge changes, splices, faults, and bridged taps. Such techniques are being developed for accurate loop identification [Galli 2002].

FDR gives accurate return loss measurements, and is useful in DSL as a pre-qualification tool for finding discontinuities that may exist before the final termination. It is somewhat less useful for in-service lines as the end reflection is "missing" if the line is terminated well and may be intrusive on the service provided.

5.6.2   TDR

Another single-ended reflectometer technique is TDR, which is different from FDR, in that it injects a pulse or stepped signal and then records time-samples of the returned echo instead of injecting a broad spectrum of frequencies at similar amplitudes. The pulse travels the length of the pair, and on encountering a mismatch of impedance the portion of the incident power not absorbed is reflected back to source. The time between the pulse being applied to the line and receiving a reflection is directly proportional to the distance between the source and the discontinuity. The magnitude of the reflected pulse is dependent upon the loss of the line, the impedance mismatch, and of course the size of the source pulse. The transmitted pulse is generally also recorded with the received echo and should be compensated for.

Conventional TDRs can detect some faults and estimate the length of some short loops. New TDR measurement and analysis techniques have recently been developed that can identify the entire loop make-up of the majority of all non-loaded loops [Galli 2001]. For the cleanest measurements, differential probing TDR is used by sending balanced positive and negative pulses on tip and ring. By ensuring only one mode of propagation and rejecting common mode, differential probing drastically enhances the quality of measurements compared to conventional TDRs which use unbalanced probing.

Various pulse shapes can be transmitted by the TDR. However, it has been found that a simple square pulse affords performance as good as most other shapes. Square pulses have a lot of low-frequency energy that experience low attenuation and can find a distant loop end. They also have sufficient high-frequency energy for closely resolving smaller and closer discontinuities. A short pulse (about 1 μs) is best on loop lengths less than a few kilofeet, with a longer pulse (about 5 μs) better on longer loops.

5.6.2.1 Effects of Impedance Discontinuities

The impedance discontinuities that must be estimated to identify properly the loop make-up are end of loop, gauge changes, and bridged taps. Using the well-known ABCD coefficients notation, the transfer function of a loop can be expressed as follows (see chapter 2 of [Golden 2006]):

$$H_f(f) = \frac{Z_L}{A_1 Z_L + B_1 + C_1 Z_S Z_L + D_1 Z_S} \tag{5.1}$$

where Z_S and Z_L are the source and load impedances, respectively.

Whenever the traveling signal encounters a change of impedance, part of this signal (echo) is reflected back. An index of the "amount" of signal that is reflected back is given by the reflection coefficient $\rho(f)$ [Galli 2001]:

$$\rho(f) = \frac{Z_a - Z_b}{Z_a + Z_b} \tag{5.2}$$

where Z_a and Z_b are the characteristic impedances after and before the discontinuity, respectively. The reflection coefficient $\rho(f)$ is defined as the ratio of the reflected wave to the incident wave and, usually, it is a complex function of frequency.

When a signal encounters a change of characteristic impedance, part of the incident wave is reflected back along the line and filtered by $\rho(f)$ and part will travel on to the next loop section. The refracted part of the signal that travels on to the next loop section will be modified by the so-called voltage transmission coefficient $\tau(f)$:

$$\tau(f) = 1 + \rho(f) = \frac{2Z_a}{Z_a + Z_b} \tag{5.3}$$

The transmission coefficient is defined as the ratio of the voltage that is refracted by the junction to the incident voltage.

The shape of an echo depends on the kind of discontinuity (through the reflection coefficient $\rho(f)$) and on the loop sections preceding the discontinuity that generated the echo (through the transfer function $H(f)$ and the transmission coefficient $\tau(f)$). The reflection coefficient $\rho(f)$ is different for each discontinuity and puts a sort of "discontinuity signature" on the echo. Different discontinuities yield echoes of different shape, so the type of a discontinuity can be estimated from the shape of the echo that was generated. The echoes created by different discontinuities (end of loop, gauge change, and bridged tap) are shown in the following subsections.

5.6.2.2 End of the Loop

The input impedance of an on-hook telephone is generally very high compared to the impedance of the cable [Galli 2001], so the loop end is modeled by an open circuit, and then the reflection coefficient is the following expression (Z_a is an infinite impedance):

$$\rho(f) = \lim_{Z_a \to \infty} \frac{Z_a - Z_b}{Z_a + Z_b} = 1$$

Therefore, an infinite impedance will simply send back the signal with no attenuation and no sign change. In this case, the only distortion will be caused by the transfer function and the transmission coefficients of the loop sections on which the signal has traveled. Figure 5.13 shows a measured reflection from the end of a long 18.5 kft loop. Figure 5.14 shows a calculated echo response from the unterminated end of a 9 kft 24 AWG loop.

5.6.2.3 Change of Gauge-Type

A connection between different gauge wires results in a relatively small echo. The magnitude of the reflection coefficient $\rho(f)$ versus frequency

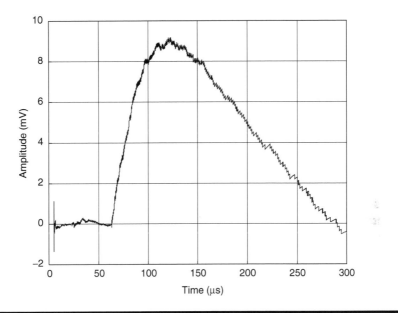

Figure 5.13   Measured echo response of the end of an unterminated 24 AWG gauge loop of 18.498 kft. As predicted by theory, a positive echo indicating the end of the loop is present. Probing signal: 5 V and 5 μs square pulse. The calculated slowly decaying signal has been subtracted from the measured TDR trace to enhance it.

for a gauge change becomes highly attenuated within a few hundred kilohertz. In general, when a signal travels on a loop and passes from a section with higher characteristic impedance to a section with a lower one, the echo will always be negative. And vice versa, when a signal travels on a loop and passes from a section with lower characteristic impedance to a section with a higher one, the echo will always be positive. There is a negligible change in the refracted signal that travels on to the next loop section. Figure 5.14 shows calculated echo responses of gauge changes: Gauge change (positive) is 9 kft of 24 AWG followed by 6 kft of 26 AWG, Gauge Change (negative) is 9 kft of 24 AWG followed by 6 kft of 22 AWG.

5.6.2.4   Bridged Taps

In the case of a bridged tap, the traveling wave sees the parallel of the impedance of the bridged-tap and of the following loop-section. This means that the echo generated at the junction will be negative, because the signal passes from a medium with higher characteristic impedance to a medium with lower characteristic impedance.

216 ■ *Implementation and Applications of DSL Technology*

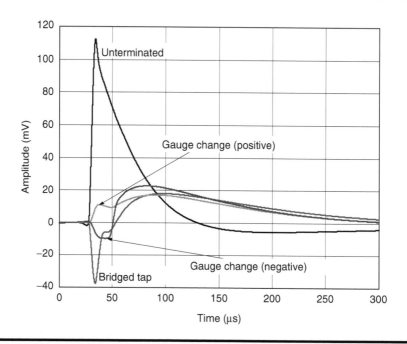

Figure 5.14 Simulated echo responses representing the shape of the echoes generated by four different discontinuities located at a distance of 9 kft from the CO. Probing signal: 5 V and 5 µs square pulse.

The junction will also generate two refracted waves: one wave will travel along the bridged tap and the other one will travel in the next loop section. The end of a bridged tap is normally an open circuit, so in this case, the traveling wave will be totally reflected because $\rho(f) = 1$ for an open circuit, and the polarity of the echo will be positive. This means that a bridged tap creates a negative echo followed by a positive echo.

If the loop section preceding the junction, the bridged tap, and the section following the junction are all of the same kind of gauge and if these sections are sufficiently long, the reflection coefficient boils down to the following expression (Z_a is the parallel of two equal characteristic impedances $Z_b = Z$):

$$\rho(f) = \frac{Z_a - Z_b}{Z_a + Z_b} = \frac{Z/2 - Z}{Z/2 + Z} = -\frac{1}{3}$$

So, in this simple case, the echo will not be distorted by the discontinuity but will only be attenuated and reversed in sign. Also, $\rho(f)$ approaches $-1/3$ with asymptotically increasing frequency.

In contrast to the gauge change case, the value of the transmission coefficient $\tau(f)$ is no longer negligible. In fact, when $\rho(f) = -1/3$, it

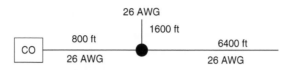

Figure 5.15   An example loop make-up.

follows that $\tau(f) = 2/3$. The absolute value of $\tau(f)$ may be considered flat over the whole bandwidth with an attenuation of $10\log_{10}(2/3)$. Figure 5.14, "bridged tap," shows the calculated echo response of a bridged tap on a loop of 9 kft of 24 AWG, followed by a 2 kft bridged tap of 24 AWG, followed by another section of 6 kft of 24 AWG.

5.6.2.5   Entire Echo Response

All the constituent echo responses displayed above can be combined with loop response simulations to accurately model the TDR echo response of any loop. An example loop make-up is shown in Figure 5.15, and its calculated overall echo response is shown in Figure 5.16. This calculated response was practically identical to the measured response.

It is important to compensate for the slowly decaying signal, which drops from the trailing edge of the transmitted pulse, as caused by the

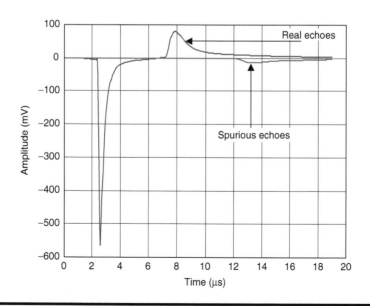

Figure 5.16   The calculated overall echo response, including all echoes and spurious echoes, of the loop in Figure 5.15.

distributed RLC nature of the loop [Galli 2001]. This slowly decaying signal may be neglected when dealing with strong echoes, i.e., those generated by close discontinuities, but becomes manifest when weak echoes, i.e., those caused by gauge changes or far discontinuities, are considered.

As discussed in Section 5.6.2, TDR analyses can be harnessed by overlying algorithms to estimate loop make-ups. These algorithms must account for the slowly decaying signal and spurious echoes which are caused by successive reflections. Spurious echoes are generated because each discontinuity generates both a reflected and a refracted signal, so that a part of the signal travels back and forth on the line, bouncing between discontinuities, before it arrives back at the CO.

5.7 DSL-Specific Line Test

Unlike POTS, DSL extends up into megahertz frequencies and so the spectra of channel properties and impairments may need to be measured or otherwise estimated across all DSL frequencies. Section 5.6 focuses on broadband test and measurement of copper loops for DSL-specific impairments. As mentioned in Section 5.3.1, POTS test equipment can be used to identify certain resistive fault conditions, excessive capacitance, imbalances, and also make insertion loss estimations.

It is quite common in present day deployments that only a single frequency or parameter such as loop length or insertion loss is tested to qualify a line as a Yes or No for DSL service. As DSL grows to offer more broadband services, simple Yes or No provisioning may not suffice, and individual service levels may need to be identified. Using more advanced DSL tests and measurements across broadband, frequencies can provision the maximum achievable bit rates on individual lines and isolate the impact of each impairment. In the absence of broadband test and analysis, DSLs that fail because of the environment at high frequencies can sometimes be repaired by knowledgeable technicians with expensive manual tests, or the DSL service may simply be abandoned.

There are many impairments specific to DSL transmission as discussed in Part A of this handbook, with loop loss and crosstalk first and foremost. As explained in chapter 5 of [Golden 2006], DSL signals are attenuated and distorted differently by transmission through each loop, particularly at high frequencies and on loops with bridged taps. Crosstalk between different DSLs on loops in the same cable is regularly the most severe noise at high frequencies. There can also be radio ingress and impulse noise, which are on occasion even more degrading than crosstalk. Electromagnetic interference (EMI) owing to radio ingress appears as narrowband noise spikes in the frequency domain, and impulse noise occurs as brief spikes in the time domain. All these impairments vary considerably at high frequencies, with noise and crosstalk levels typically differing by 20 dB or more on different loops, and are difficult to predict without electronic measurement.

The attenuation and distortion of a loop is readily calculated at all frequencies if the loop make-up (including gauge types, bridged taps, and cable section lengths) is known, as explained in chapter 2 of [Golden 2006]. This then allows precise calculation of the received DSL signal. Then, if the received noise is known or measured as a function of frequency, the DSL bit rate and performance level can be precisely and unambiguously calculated [Kerpez 2003]. All this data can be measured, or some can be gleaned from databases and by querying DSL modems. Analyses can vary the loop make-up and noise components to determine their individual impact and debug the DSL line.

This section is concerned with layer 1 testing issues. It has been estimated that on a properly screened physical loop, a significant amount of all installation problems involve provisioning of higher layer services. Hence, some automated test equipment provides the facility to connect a "Golden Modem" at the DSLAM, which can synchronize with the customer modem and do ATM loopback testing to verify VCI/VPI (virtual circuit identifier/virtual path identifier) (see Chapter 13). In addition, this modem can look into the DSLAM to do ATM loopback tests, PPP (Point-to-Point Protocol) authentication, IP pings, IP throughput tests, and HTTP (Hypertext Transfer Protocol) queries to verify circuit provisioning up to layer 7.

5.7.1 Diagnosing DSL-Specific Loop Impairments

Many of the conceptually simple bugs that plagued early DSL service offerings have been worked out, but fundamental impairments to high-speed loop transmission have not disappeared. Accurate loop test of specific impairments at DSL frequencies can quantify the impact of many transmission impairments that cannot be determined at POTS test frequencies.

Figure 5.17 illustrates DSL-specific loop impairments. These are mainly loop and bridged tap loss, crosstalk, EMI radio ingress, impulse noise, and background noise. Although often overshadowed by crosstalk, measurements have found many locations with high enough levels of radio ingress or impulse noise to halt DSL service if not handled properly. Impulse noise can be measured by long term (approximately an hour or more) monitoring of raw bit errors. Background noise is typically present as low-level additive Gaussian noise. Other measurements are briefly discussed in the following subsections.

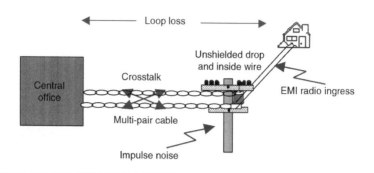

Figure 5.17   DSL impairments. Crosstalk occurs in multi-pair shielded cables, radio ingress couples into unshielded drop and inside wire.

5.7.1.1   Loop Identification

The spectral response of a loop is easily determined with a double-ended measurement with equipment at both CO and customer ends. For example, a signal sweep generator can be connected to one loop end with a spectrum analyzer on max hold connected at the other end. Alternatively, the loop response may be inferred from a single-ended measurement of one port parameters in the frequency domain [Galli 2002]. Single-ended loop measurements using enhanced TDR techniques [Galli 2001] can even determine the loop make-up "stick diagram" showing the lengths and gauges of all sections.

As explained in "TDR," TDR involves the transmission of probing signals onto the loop and the analysis of echoes reflected by impedance changes in order to infer the unknown loop topology.

Analyzing returned TDR echoes to determine loop response can be very involved [Galli 2001]. A slowly decaying signal echo is returned by the pair, various echoes are returned by bridged tap, gauge changes, and loop ends, and spurious echoes bounce between mid-loop impedance discontinuities before returning.

As an illustration, loop identification using broadband differential probing TDR was performed using measurements of 19 loops at a wire center, with each loop picked to have working length such that 5 percent, 10 percent, ..., 95 percent of all loops at the wire center were shorter. The difference between the bit rates of an ADSL on a loop with the actual, and the estimated, loop make-ups were calculated to quantify the error from a DSL qualification perspective. ADSL bit rates were calculated with 12 spectrum management Class 1 disturbers plus 12 self-crosstalk disturbers as defined in the American spectrum management standard T1.417 (see Chapter 7). The difference in downstream ADSL bit rates with the actual loops, and the TDR estimated loops (including bridged taps, gauges, etc.), is shown in Figure 5.18.

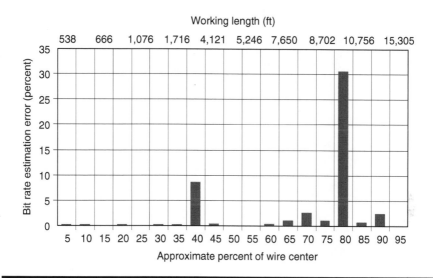

Figure 5.18 **Error in downstream ADSL bit rates from using loop make-ups estimated by single-ended loop identification on 19 loops at a CO. The 15,305 ft loop has zero bit rate.**

5.7.1.2 Crosstalk Identification

Different types of crosstalk sources (i.e., HDSL, ADSL, T1 lines, etc.) have different transmit spectra, so they may be identified from the crosstalk spectra received on a loop. If one denotes a crosstalker's transmit PSD as $D(f)$, the pair-to-pair crosstalk power coupling as $|H(f)|^2$, and the received crosstalk PSD as $Y(f)$, then the received single-disturber crosstalk PSD is $Y(f) = |H(f)|^2 D(f) + N(f)$, where $N(f)$ is noise from other sources. Transmit PSDs of a given type of system are often all the same, but pair-to-pair crosstalk coupling and received crosstalk PSDs vary, as shown in Figure 5.19 for T1 crosstalk.

It can be advantageous to identify the type of crosstalkers, for example, to spot which line is causing a crosstalk problem. One approach to estimating crosstalker types [Zeng 2001] is to create a set of predefined PSDs that may be viewed as a crosstalk "basis set," which are chosen to cover the subspace of received crosstalk PSDs of a given type as much as possible while having minimal cross-correlation with basis sets of other crosstalker types. Identification of a single disturber is achieved by identifying the basis set that maximizes the correlation between it and the measured crosstalk PSD. This technique can be extended to the case of multiple disturbers by using spectral subtraction methods, multiple regression, or other techniques. It has been found that a single high-power crosstalker can almost always be identified, but identifying multiple low-power crosstalkers is sometimes difficult.

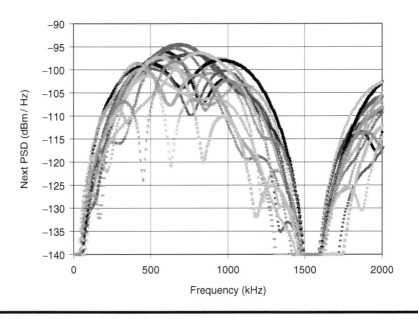

Figure 5.19   Received NEXT from a T1 line over various pair-to-pair near-end crosstalk (NEXT) coupling paths.

Knowledge of the pair-to-pair crosstalk couplings can be used to avoid crosstalk incompatibilities and even to jointly optimize DSL spectra. These can be measured directly by test equipment accessing two pairs simultaneously. One issue can be that test systems are limited to accessing one pair at a time. The pair-to-pair crosstalk couplings can be reconstructed by first estimating the type of crosstalker as discussed above. The crosstalker's transmit PSD is then estimated by quantizing it as $D_{est}(f)$, one of a set of original candidate transmit PSDs. Finally, the pair-to-pair crosstalk coupling can be estimated as $|H(f)|^2 = Y(f)/D_{est}(f)$ assuming that the background noise $N(f)$ is small compared to the crosstalk.

Crosstalk couplings can also be identified directly in the time domain by accessing transmitter and receiver sequences simultaneously, or at a single receiver receiving known sequences such as synch symbols [TR-069].

5.7.1.3   EMI Radio Ingress Identification

EMI, also called radio ingress, is radio signals coupling into unshielded cable, drop, and inside wiring (see chapter 13 of [Golden 2006]). EMI can also couple into multi-pair cables with improperly grounded shields. AM radio ingress is common from 535 to 1605 kHz. In addition, there are short-wave broadcasts, HAM, and other signals at higher frequencies.

Radio ingress is generally narrowband spikes in the frequency domain which can be separated from the broader and more continuous crosstalk and background noise spectra (see chapter 13 of [Golden 2006]). Thus, the power and impact of the radio ingress spectra can be calculated.

5.7.2 DSL Analysis Engine

Low-level test parameters such as loop response and PSD levels can be converted into high-level descriptions of possible service offerings and service assurance levels by an analysis engine for DSL. Models and routines for analyzing DSL transmission have been finely honed over the last couple of decades to accurately determine margins, bit rates, and other performance measures of any type of loop transmission system. This accuracy is greatly aided by the fact that copper loops are largely time invariant (temperature variations can change loop attenuation by a few decibels, but can be modeled). Standards based models of DSL performance [Kerpez 2003] can be tweaked to closely match the performance of actual DSL equipment. Measured noise and loop responses can be input for the most accurate analysis, or certain elements can assume typical model parameters.

The received DSL signal is determined by the loop, and the received noise can be broken down into crosstalk and background noise, EMI radio ingress, and impulse noise. Further, algorithms can identify the individual sources of crosstalk [Zeng 2001]. The routines can input individual noise components to determine their individual impact, as shown in Figure 5.20.

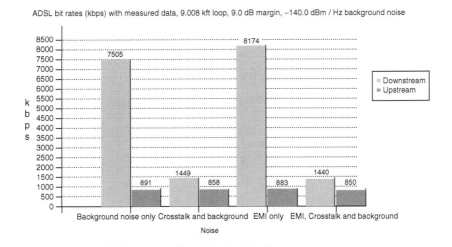

Figure 5.20 **ADSL bit rates calculated with data measured on a particular 9.008 kft loop. The dominant noise on this line is seen to be crosstalk.**

The figure shows ADSL performance calculated with various components of measured noise, including −140 dBm/Hz background noise for comparison as a best case. The calculations are as described in chapter 2 of [Golden 2006], assuming 6 dB margin and 3 dB implementation loss. Many parameters such as transmit power, bandwidth, and ability to reduce certain noise sources, etc., can be varied and the resulting impact on DSL service determined analytically.

Today, sophisticated DSL analysis engines are generally used only very early in the DSL planning process by scientists and engineers. They typically input statistical models of plant characteristics to determine simplified rules that are then used for DSL provisioning, such as a limit on loop length, or on loop attenuation at a single frequency. However, DSL provisioning is becoming more sophisticated and incorporating more variables. Also, detailed DSL analysis routines are creeping into software for DSLAM management and test head analysis. Such routines will eventually be used more for per-line provisioning and automated maintenance.

5.7.3 Implications for Remediation

Using DSL-specific measurements and analysis, the impact of each constituent noise component can be determined and the major trouble can be identified. The type of remediation is then narrowed to a short list as briefly outlined in Table 5.5. The potential improvement offered by each type of remediation can be calculated to see what makes sense on a given loop. For example, the effectiveness of removing bridged tap from the loop can be determined analytically by temporarily changing the loop model to have no bridged tap.

"Electronic remediation" could be administered from a central maintenance station or even implemented automatically. Determining the proper remediation through analysis is more cost effective than actually performing multiple fixes until the right one is found. If it is determined that signals transmitted over the measured loop and received with only background noise can at best achieve poor performance, then this can be noted rather than wasting effort trying to fix an irreparable situation.

5.7.4 Data Collection

An indispensable component for managing DSL is a database of loop information for DSL provisioning and maintenance. As it is gathered, loop measurement, test, and repair information should be stored in a database so that repeating troubles can be easily identified and fixed the next time they occur.

Table 5.5 Outline of Some DSL Impairments, Their Identification, and Possible Remediation

Impairment	Identification	Plant Remediation	Electronic Remediation
Bridged tap	Identify loop make-up, calculate performance with and without bridged tap	Remove bridged tap	Re-allocate spectral power away from bridged tap nulls
Crosstalk	Calculate performance with and without crosstalk; Identify crosstalker types and powers	Swap pairs	Lower crosstalker's power; Implement DSM, perform joint DSL spectral optimization
Electromagnetic interference (EMI) or radio ingress	Identify EMI power; Calculate performance with and without EMI	Upgrade drop or inside wire, ground shields	Window DMT signals; Implement EMI cancelation
Impulse noise	Long-term (hours) error monitoring	Upgrade inside wire	Increase interleaver depth
Distortion	Identify harmonics from upstream coupling into received downstream	Install microfilters or splitter	Decrease upstream transmit power

This database can grow into having a wealth of information on loops, noise, and the histories of deployed DSLs extending far beyond existing loop databases, all of it invaluable for maintaining or deploying new DSL services. It can store loop make-ups or loop responses, data on deployed DSLs, binder information, measured noise, information on crosstalk between lines, etc. This database can be used by operators to run "what-if" scenarios, such as determining how many subscribers could get video-rate service by installing ADSL2+.

To populate such a database, data must of course be gathered. There are three sources for this information: measurements from dedicated DSL test equipment, data from DSL modems or DSLAMs, and existing loop and DSL databases.

5.7.4.1 Dedicated Automated Broadband Test

Automated loop testing systems for POTS testing were described in the first part of this chapter, and are implemented on voice switches of all types throughout most local exchanges. The metallic access through the switch does pass high frequency test signals, and so automated broadband test through the switch is possible. Switches generally use metallic relays with fairly flat responses up to a few megahertz. However, it has been found that the bandwidth of the metallic test bus in some switches is limited, probably because of their switch matrix. In particular, it has been found that the metallic test bus in the 5ESS generally has high attenuation above about 30 kHz. This enables enhancements such as that described in Section 5.3.4, however narrowband test heads give no information about noise at DSL frequencies. For this, an upgrade to a broadband test head is needed [Bostoen 2002], enabling single-ended measurements such as wideband noise spectra and loop make-ups. Also, by accessing two pairs simultaneously, crosstalk couplings could be directly measured.

Dedicated automated broadband test equipment can be well calibrated to accurately perform single-ended tests, and may access any working loop from the CO. This may not be the least expensive solution however, because it requires a physical test device in the CO. Also, sometimes attenuation and distortion are introduced by the metallic test bus, and by the CO wiring that is attached to it. In addition, some switches introduce a considerable amount of noise. A direct connection at a DSLAM, somewhere near the MDF, or at an intermediate dedicated POTS splitter or other DSL frame, avoids the metallic test bus and could allow more accurate broadband measurement but with increased complexity (see Section 5.4).

Systems are available to backhaul metallic test signals between a digital loop carrier (DLC) and the CO. Broadband test from an NGDLC (next generation OLC) or DSLRT (remote terminal) is not widely employed today but is steadily growing as service providers discover the value of this capability. Manual test or backhauling data from the remote DSL modems themselves over their data pipes may be preferable. However, new remote DSLAMs may be equipped with internal switching matricies, and low-cost plug-in cards may enable dedicated wideband test.

5.7.4.2 Extracting Data from DSLAMs and DSL Modems

Because telephone loops are highly variable at high frequencies, DSL modems are adaptive and inherently "learn" the channel response and noise within their bandwidth. Data from DSL modems is double-ended, with upstream and downstream noise data, at both the CO and the customer end. Data on multiple DSLs may be retrieved by querying a DSLAM at a CO.

ADSL modems use discrete multitone (DMT) modulation to subdivide the 1.1 MHz channel into 255 narrowband (4.3125 kHz) channels or tones (see chapter 7 of [Golden 2006]). Receivers must know the received signal power and signal-to-noise ratio (SNR) of each tone, and because the transmit signal is known, the loop magnitude response and noise power spectrum are known with a fine granularity. These spectra can similarly be deduced, to some extent, from the gain control and equalizer coefficients of single-carrier transceivers.

DSL Forum TR-069, "CPE WAN Management Protocol," defines a management interface for broadband modems [TR-069], and standardizes XML commands for remote configuration and line diagnostics. DSL diagnostics can be reported such as signal and noise levels, settings, noise PSD, signal PSD, vendor ID, etc., as well as higher layer diagnostics such as cell error counts, retransmission counts, and loopback. Commands can also report on and configure the in-home LAN. TR-069 enabled modems were just beginning to become popular at the time of this writing, and this should become a very valuable tool.

New ADSL2, ADSL2+, and VDSL2 (ITU standards G.992.3, G.992.5, and G.993.2) modems must be capable of reporting line diagnostics and enabling control as specified in ITU-T G.997, "Physical layer management for Digital Subscriber Line (DSL) transceivers." These include upstream and downstream power spectra: transfer function $H(f)$, quiet line noise $QLN(f)$, and $SNR(f)$. Aggregate parameters are also reported: loop attenuation, signal attenuation, SNR margin, attainable net data rate, and aggregate transmit power (far-end). Loop data can be garnered in "Initialization mode" using standard ADSL training, or "Diagnostics mode" can be invoked on-demand and can run single-ended. Additionally, G.997 specifies control of the following parameters: the power transmitted by each tone, the bit loading (number of bits and gain on each tone), the total transmit power, and the minimum or target or maximum bit rate and SNR margin.

Single-ended loop test (SELT) enabled DSL transmitters can report single-ended measurements from a single DSL DSLAM port, before DSL service is activated or to analyze DSLs that are not working. SELT may measure the following: frequency dependent impedance, TDR signals, noise spectrum at the CO, impulse noise counts, etc. This may be done to determine the following: loop length, loop make-up, crosstalker types, crosstalk couplings, radio ingress, impulse noise, linearity, SNR, bit-rate capacities, load coils, etc. SELT data from DSLAMs can be provided to a separate analysis engine, which interfaces with a DSL OSS.

A SELT DSL transceiver could be used similar to dedicated automated broadband test equipment, switching between lines not yet in DSL service to analyze them. Calibrating the analog front-end of a DSL transceiver is difficult however, and not likely to be nearly as accurate as dedicated test equipment.

5.7.4.3 Database Mining

Existing loop plant databases contain information on loop make-ups and deployed services [Bostoen 2002]. These databases are traditionally used to provision POTS. An example of such a database is the Loop Facilities Assignment and Control System (LFACS) which stores a view of the loop plant for the Regional Bell Operating Companies. In general, there are multiple segments of cable between the MDF and the subscriber, and each segment's length and gauge should be in the loop database. There may be additional complexity that could be recorded, such as aluminum conductors in certain cable types.

The service and physical loop-related information in these databases may be usefully mined for DSL loop qualification. For example, loops need to be disqualified for DSL if they are loaded, or if they are served by only narrowband DLC. This service-related information is available in LFACS, and is uniquely useful for determining a significant percentage of the causes that disqualify loops for DSL.

For many loops, the complete loop make-up is available in the loop plant database. Often, however, the loop make-up is incomplete or not entered in the electronic database. Rearrangement of a loop by repair craft are often not entered into the database. Repairs that need to be made may be carried out using copper wire gauges and lengths that do not match that which is recorded in the records database. A few examples are as follows:

- More economical to carry only one gauge type for repairs, 0.4 mm copper could be replaced with 0.6 mm without necessarily updating the database.
- In an urgent need to repair a cable segment, it may be that the repair crew will find the first cable from the nearest stores that give them the ability to complete the task. Again, the records database may not be updated to record the fact that the length and gauge of the repair does not match its original designation.
- In the case where a duct is full and there is a need to increase the number of subscribers carried by cables in the duct, then it may be that 0.6 mm cables are changed to 0.4 mm to fit duct size.

Paper loop records are frequently available and are typically fairly accurate, they can be read into the database as needed. An ideal DSL qualification engine would combine service-related parameters and whatever loop make-up data is available from the existing loop plant database with automated test and DSL modem data.

Ambiguities between loop records and measurement results can be resolved via intelligent correlation and decision, yielding a new, more accurate loop make-up. Loop records can be groomed with test results off-line, then stored, and are then available during provisioning flow-through.

5.7.5 Broadband Testing and Dynamic Spectrum Management

Dynamic spectrum management (DSM), discussed in Chapter 8, incorporates parameters of the loop plant environment and loop transmission systems that are time or situation dependent, particularly individual crosstalk sources and couplings. Crosstalk varies because there may be different types of crosstalk sources (ADSL, T1 lines, G.shdsl at different bit rates, etc.), different numbers of crosstalkers, and different crosstalk couplings between loops. Measurements of pair-to-pair near-end crosstalk (NEXT) loss show substantial variation, with an 11 dB standard deviation. Actual crosstalk couplings vary substantially with frequency and are often 20–30 dB below the worst-case model assumed in static spectrum management discussed in Chapter 8. Using DSL-specific broadband loop test and measurement to identify the PSDs of individual crosstalk sources, and crosstalk couplings at all DSL frequencies, can greatly assist DSM, allowing bit rates to increase significantly while simultaneously lowering service failures.

5.8 Summary

DSL deployments were initially plagued by myriad of relatively simple practical provisioning errors. Many of these have been solved and are now easy to handle, what remains are largely fundamental transmission and configuration problems. Mechanized loop test systems are widely employed at local exchanges that efficiently test lines at low frequencies. These systems can identify and help repair many impairments that affect both POTS and DSL, and can provide an estimate of loop insertion loss. Simple enhancements to existing POTS test systems can be advantageously applied to DSL testing.

A combination of broadband loop measurements, a DSL database, and DSL-specific analysis routines are capable of identifying and helping to eliminate many of the remaining DSL faults [Bostoen 2002]. Precise loop qualification and service activation may be administered from a central station, lowering costs of DSL provisioning. Truck rolls can be greatly reduced. DSL-specific broadband loop test can help ensure that DSL deployment does not get bogged down with provisioning inefficiencies from crosstalk conflicts, or other noise sources and loop impairments.

Significant ongoing maintenance savings are also expected by DSL test systems that automatically identify the most costly and difficult to diagnose problems. The element that is causing a problem can be isolated: loop, noise, modems, etc. The correct remediation (i.e., remove bridged tap) can be determined analytically before expending effort in the field. Automated DSL test is far less expensive than manual testing, storing measurement data can avoid rework, and potential problems can be corrected before a customer ever sees them. Monitoring the transmission environment of a particular DSL line can ensure carrier-grade service levels that otherwise could only be guessed at within some tens of decibels.

Bibliography

[Bostoen 2002] T. Bostoen, P. Boets, M. Zekri, L. Van Biesen, T. Pollet, and D. Rabijns, Estimation of the transfer function of a subscriber loop by means of a one-port scattering parameter measurement at the central office. *IEEE Journal on Selected Areas in Communications*, pp. 936–948, Vol. 20, No. 5, June 2002.

[Galli 2001] S. Galli, C. Valenti, and K. Kerpez, A frequency-domain approach to crosstalk identification in xDSL systems. *IEEE Journal on Selected Area in Communication* (JSAC), pp. 1497–1506, Vol. 19, No. 8, August 2001.

[Galli 2002] S. Galliand and D.L. Warin, Loop makeup identification via single ended testing: Beyond mere loop qualification. *IEEE Journal on Selected Areas in Communications*, pp. 923–935, Vol. 20, No. 5, June 2002.

[Golden 2006] P. Golden, H. Dedieu, and K.S. Jacobsen, Eds., *Fundamentals of DSL Technology.* Auerbach Publications, Boca Raton, Florida, 2006, p. 457.

[Kerpez 2003] K. Kerpez, D.L. Waring, S. Galli, J. Dixon, and P. Madon, *Advanced DSL Management. IEEE Communications Magazine*, pp. 116–123, Vol. 41, September 2003. See also http://net3.argreenhouse.com.

[TR-069] DSL Forum TR-069, *CPE WAN Management Protocol*, May 2004.

[Zeng 2001] C. Zeng, C. Aldana, A.A. Salvekar, and J.M. Cioffi, Crosstalk Identification in xDSL Systems. *IEEE Journal on Selected Area in Communication* (JSAC), pp. 1488–1496, Vol. 19, No. 8, August 2001.

Chapter 6

DSL Planning Rules, Line Qualification, and Deployment Issues

Nigel Evans and Mark Fletcher

CONTENTS

Abstract   This chapter describes various methods for determining whether individual metallic pairs in the local loop (or "last mile") are capable of supporting Digital Subscriber Line (DSL) services. The chapter describes the following:

- Typical approaches to the derivation of planning rules for different DSL systems.
- Importance of network survey data to enable an incumbent operator to determine the characteristics of their network.
- Use of different line loss estimation data sources (e.g., cable records, line test information or earth capacitance, geographical location or postcode information).
- General methods for assessing the accuracy of different line qualification data and the important factors to consider when setting GO and NOGO thresholds (or red, amber, green thresholds) for deployment. Such factors include the risk of deploying DSL systems above their system limit and also the risk of saying no to customers whose circuit loss is, in reality, within the DSL system limit.

In addition to the reach requirements, this chapter also covers the "quality" aspects of the metallic loop required to give error free transmission of DSL systems; it is known that DSL systems are generally very tolerant to hard or stable network faults (earth contacts, loops, etc.), which can cause analog telephony to fail well before the DSL system. However, DSL systems do tend to be more susceptible to dynamic, often intermittent faults in the local loop such as high resistance joints, dynamic rectified loops, bridge taps, split pairs, etc. Faults of this nature can yield deployment and early life failures for DSL systems, so it is important to understand the mechanics and effects of these faults on the performance of different DSL systems and to ultimately have the ability to detect these DSL-affecting faults in the network via exchange-based and field-based test equipment.

6.1 Line Qualification for DSL Services

In this section the concept of qualifying lines for DSL services is discussed. As the network to which this will be applied will vary from country to country, or even region to region, the discussion is necessarily general. In addition to the network size, average length, and detailed topology varying from telco to telco, any additional data about the network will vary: the accuracy, the type, the format (paper or electronic), and the collection method (automatically, manually, third party purchase)—these and more can affect the way a line qualification programme may progress. "DSL services" cover Asymmetric DSL (ADSL), High Bit-Rate DSL (HDSL), Very High Bit-Rate DSL (VDSL) etc., and each service individually needs a line qualification process.

It is true to say that nearly all lines from the exchange to customers were provided to serve the humble telephone. Deployment of DSL services is a retrofit onto the access network. DSL services can be deployed in one of three ways: (1) a wholly rate adaptive version with the upstream and downstream rate adjusting to local conditions, (2) with one direction fixed and the other rate adapting, or (3) both directions fixed.

In (1), the rate the customer receives depends on the conditions on their line and this rate will vary with time. Different customers will receive different rates because of the varying conditions found across the network. However, there is always a minimum performance, below which service is not viable, either because of an inability to train-up successfully or because of contractual bounds on minimum data speeds. It may be that failures here (generally at long distances) are so infrequent that the cost of running the planning rule to avoid them is more expensive than the failure costs of visits to resolve those failures. Elements of what is outlined in the following sections are needed to establish if this is the case. If this is confirmed, the planning rule becomes trivial. This scenario is quite possible in many parts of Europe but in regions of North America this is not so likely because of some very long lines.

In (2) and (3), there will be situations where the line has too much loss to support a particular DSL service. Two issues need to be addressed: (1) defining a system limit of the DSL service, and (2) determining whether a customer is within that range. These are explored further.

6.1.1 Purpose and Benefit

Why is this necessary? Why not deploy it and see if the service works? With such an approach the customer base divides into three broad groups, of differing sizes depending on the DSL in question. The first group are those customers who receive a good service without any problems; these are predominately near to the exchange. The second group are those for whom the service never works. The customer has expectations of a new service, only for the service provider to spend time and effort to confirm that it cannot work; the customer feels let down and gets a negative perception of the service and the service provider. The last group suffer a worse fate as their lines are close to the limits of the DSL service; the service is intermittent, possibly available in some parts of the day more often than in others. Maybe the service was initially very good but has deteriorated over some months to the point where customer thinks it is faulty. Engineers spend significant time and effort but are unable to repair the service properly and eventually the frustrated customer cancels the service. These latter two scenarios cost the service provider time and money for very little gain. As the prime function of a service provider is to make money, it would be better to deploy only to customers in the first group, so offering a good quality of service to those that are within the range of the DSL service. A threshold value, a system limit, needs to be set that includes the customers in the first group and excludes those in the other two groups.

6.1.2 Setting the System Limit and Noise Margins

The term "system limit" is often used to denote the maximum insertion loss of a local wire pair that a DSL system should be deployed on. The system limit is generally derived following detailed transmission and performance testing using predetermined noise models and a specified noise margin. The decision as to how much noise margin to allow for a given DSL system is very much one for the individual telecoms operator or service provider to decide. Consideration needs to be given to the characteristics and topology of the network, the current and predicted future growth and mix of DSL technologies in the network (NEXT or FEXT limited), and the deployment risk that the telecoms operator or service provider is prepared to take. Specifying a small noise margin may lead to deployment problems and early life failures. In some cases, initial DSL cable fill may be low and the first few DSL systems in a particular cable route may operate without any problems at all. However, for every new DSL system added to the cable, the remaining noise margin is reduced and a critical point could be reached which would result in the ultimate failure of some DSL systems in the cable route. Contrarily, specifying a large noise margin will provide greater safety from deployment and in service failures, but would result in a lower system limit being specified, hence diminishing the number of customers who can ultimately be reached by the DSL service (i.e., an increase in lost business).

The noise models include the effect on the DSL from other systems in the same cable and are covered in chapter 3 of [Golden 2006]. The effects of incompatible products can be seen by the adverse effect they have on the resulting system limit.

Recently some countries have opened up the local network to other licensed operators (OLOs) to the extent that the incumbent telecom operator is no longer the sole user, i.e., the network has become unbundled.

In principle, this means that any operator (with a license from the government's regulatory body) can supply service over the network. In some countries, this has led to the development of an "access network frequency plan," which limits the spectral power distribution that any of the operators can inject into the network. This is designed to prevent all operators from using too much power to provide service to their remote customers and consequently emitting too much noise into their competitors' lines. All DSL products in the noise models should therefore conform to the plan.

Setting the system limit is a prediction of the maximum insertion loss of a line that should support the DSL service at some future time when DSL "saturates" the network (while operating at an acceptable bit error rate). For the purposes of the discussion that follows, it is assumed that the fraction of lines that fail to operate in this environment is negligible, such that we define the suitability of a line for DSL by the system limit. It should, however, be noted that some lines that are predicted to be within the system limit may not work in a satisfactory manner owing to adverse line conditions and transients that are not included in the models.

6.1.3 Is a Line within Range?

The most accurate solution would be to have a large list containing a precision measurement of every line taken when the line was installed. This could be at a single key frequency, say 100 kHz, or over a range of frequencies if the network contains features like bridge taps as the case in North America. The deployment decision would then be merely a comparison between the recorded value and the system limit. However, as most lines were already installed before DSL became a reality, this list is unlikely to exist. Retrospectively sending engineers to measure each and every line is a possibility, but the cost is prohibitive. Sending an engineer to qualify a line after an order has been placed is an option, but this is still expensive and there is also a delay in replying to the customer, who would prefer to know straightaway whether or not they will receive the service.

For many customers of DSL, the precise knowledge of the line loss is unnecessary, so sending an engineer to measure it would be an expensive overkill. If the system limit is, say 30 dB, then knowing that a line has a loss of 10 or 50 dB is unnecessarily precise—a method that could estimate the loss would be sufficient in these two cases.

What is needed is an indirect measure of loss, a property whose numerical value generally increases as the line loss increases but the correlation will not be exact. An example would be the capacitance of the line which is available from exchange-based line test systems—in general, when the capacitance is small the line loss is small, and when the capacitance is large the line loss is high. Another possibility is to use the length of the line if this is readily available from line records or can be estimated from line test. The important feature of this indirect measure is that it is relatively cheap to get this information for most of, if not all, the potential customer base.

The capacitance value needs to be "converted" into loss, or rather how likely the loss would exceed the planning limit. How is this achieved? Generally, a one-off network survey is required to build a knowledge base. A representative sample of plain old telephone service (POTS) customers, both urban and rural, are visited by a field engineer to have their line losses precisely measured, an expensive process but it is only required on a tiny fraction of the network. Additional indirect measurements are also captured for these lines. For the purposes of illustration, we will use capacitance as an example of the possible indirect measures that can be used (see Figure 6.1). The distribution of capacitance and measured line loss is used to characterize the whole network. In particular, the expensive measure of line loss is now replaced by a much cheaper measure of capacitance and a single threshold value of capacitance is used to distinguish between lines marked suitable for deployment and lines marked unsuitable for deployment. These

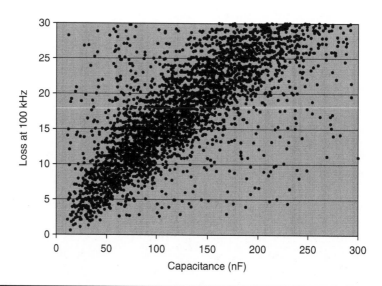

Figure 6.1   Distribution of capacitance against insertion loss at 100 kHz. This type of data enables a capacitance value, measured before Digital Subscriber Line (DSL) installation, to be used to estimate the likelihood of the insertion loss exceeding the planning limit, without the need to send an engineer to actually measure it. There are data points that are obviously incorrect and it is tempting to remove them. This should not be done as the distribution mimics the real network—the errors introduced by this defective data are part of the line qualification process.

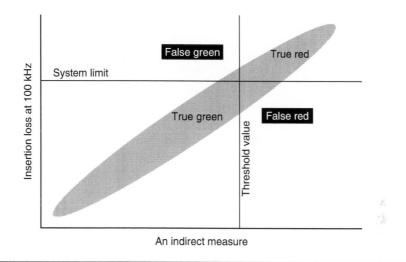

Figure 6.2   A simple distribution of an indirect measure and the insertion loss. For a value of the indirect measure, there are a range of possible insertion loss values. The intersection of the system limit and the threshold value divides the area into four zones.

lines are often color-coded green and red respectively, representing GO and NOGO scenarios. The end result is prequalification of all lines before orders are received, enabling an immediate response as to whether the order can be fulfilled.

How is the threshold value of capacitance determined? Using a single threshold value and the single loss threshold will divide the customer base into four groups (see Figure 6.2). They are as follows:

- True green lines—judged to be in range and actually are in range
- True red lines—judged to be out-of-range and are actually out-of-range
- False green lines—judged to be in range but are actually out-of-range
- False red lines—judged to be out-of-range but are actually in range

Some of these line types have financial implications. True green lines and true red lines do not have any penalty costs as the correct business decision was made—service was provided on true green lines and was not provided on true red lines. False green lines are different because they incur a cost to the business. There is a maintenance engineering visit sometime in the future to determine that it cannot work, as well as waste of the background costs of supplying the order. False red lines also incur a cost, because revenue has not been received from a customer who would have taken the service. In addition, both false costs include intangible costs, such as damage to the brand reputation. Of course, the difficulty is that false green lines cannot be distinguished from true green lines until some indeterminate time after the order is deployed – so there is a risk of failure associated with deploying lines marked green, the degree of risk being dependent on the position of the threshold. False red lines will never be distinguished from true red lines as there would be no further check.

Of course, the precision measurement defines the system limit, so can only give true red lines and true green lines. There are no false reds or false greens (from a deployment decision point of view).

The position of the capacitance threshold is a balance between the false green and the false red costs: if the threshold position is low, then there will be a small false green cost but a large false red cost, whereas if the threshold position is high, there will be few false red costs but large false green costs. There will be an optimum position for the capacitance threshold that depends on the financial values associated with the false green and false red lines. The network survey assists in showing how many false green and false red lines occur as the single threshold moves. The lowest total failure costs mark this optimum position (see Figure 6.3).

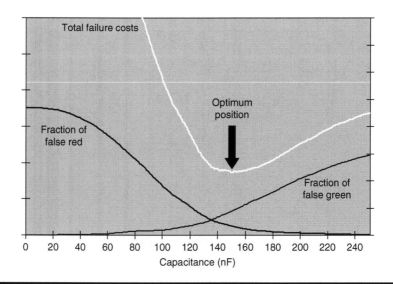

Figure 6.3   Determining the optimum cost. The fraction of false green and false red lines varies as the capacitance threshold sweeps across. By attaching a cost penalty to the two groups, a position of optimum cost may be found. The position is dependent on the ratio of the costs—here a false green to false red cost of 0.5 has been used.

The fraction of DSL orders that fall in the false green and false red sectors will depend on several factors, such as, the topology of the network, the indirect measure being used, and the expected penetration of the DSL service. A decision will need to be made as to whether the percentage of false red and false green given by this method is acceptable.

6.1.4 Improving the Process

In the previous section, the expensive engineering measurement was exchanged totally for a less expensive capacitance-based estimate. Unfortunately, some penalty costs are incurred in the region of the threshold. A potential improvement can be made to the process by combining the best parts of the capacitance-based estimates with the precision measurement, as illustrated in Figure 6.4.

In this scenario, the lines in the region of high false red and false green costs are now labelled with a different colour (amber). In this region, the capacitance based measurement is ignored and the line is passed to the precision measurement team—the line qualification process becomes a two-stage process. With this

1. Number (and cost) of capacitance-based false greens is reduced.
2. Number (and cost) of capacitance-based false reds is reduced.

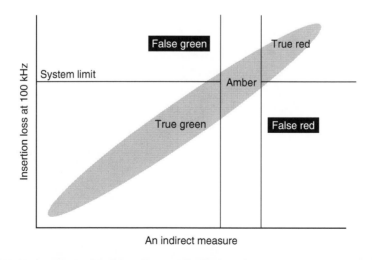

Figure 6.4 **Improving the line qualification process. An improvement to Figure 6.2 is that the region of high false green and false red costs is now called amber. These lines are assessed using a more accurate (and more expensive) method.**

3. Expensive precision measurements are made but the number and cost of false green and false red are nominally zero (because lines below the system limit are assumed to work and those above are not).

So the majority of the false red and false green costs are exchanged for expensive precision measurements, which unfortunately will incur a time penalty as well as a cost penalty.

How are the two capacitance thresholds to be set? Previously, the single threshold was set by balancing two opposing cost-based forces, false green costs and false red costs. Now it will be a false green (or false red) cost against a precision measurement cost. The network survey provides the relative quantities of lines used to weigh these costs. Consider just the green–amber boundary in Figure 6.4 without an amber–red boundary. As the green–amber boundary moves to the right, the false green costs increase, but total precision measurement cost will decrease. There will be a point at which the costs are minimized. There will be a similar optimum point for the amber–red boundary. By combining together the capacitance and the precision measurement, the easy decisions are made using the cheap capacitance measure and the more difficult ones are made with a more expensive process—targeted to where it can add most value.

A consequence of this is that not all lines are prequalified; there are now three flags, green, amber, and red. Orders on green and red lines are handled immediately, orders on amber lines have a delay until a precision measurement turns them true green or true red.

It may be possible that the green–amber boundary is higher than the amber–red boundary. What does this mean? It means that the additional effort of using a precision measurement on targeted lines is not cost effective; a capacitance-based solution on its own is more cost effective.

6.1.5 Indirect Measures for Qualifying DSLs

In the previous section, a capacitance-based measure was used as an illustration of an indirect estimate of insertion loss. It is not the only one available.

For an existing customer, there are several sets of information that can be collected that can prove useful in qualifying their line. These need to be cheaper, but unfortunately less accurate, than the two-ended measurement that an engineer could make. They must also be available for a large number of lines (ideally all), without the need for expert interpretation. These measures relate to (1) measurements on their line conducted from the exchange (single-ended measurements), (2) records made during the construction and deployment of their line (a plant record), and (3) measurements derived from the location of the customer and the exchange.

So what kinds of line loss estimation data are available? The capacitance of the cable for a fixed gauge wire increases in proportion to the length and the line loss. Capacitance is measured between the A leg (or B leg) and earth, but usually the average value would be used. If a single type of cable has been deployed throughout the network then this could potentially be a very useful measure. Most real networks use more than one gauge size and the loss is no longer uniquely proportional to capacitance. As longer lines could use cables of differing capacitance characteristics to those on short lines, it may be that the general trend of loss is nonlinear with capacitance.

Another source is the loop resistance which, in a similar way to capacitance, increases as the cable gets longer and also varies on the differing gauges that make up to whole line.

A recent addition is an enhancement to the traditional line test for POTS. Here the loss is inferred by performing many measurements in the 300–4000 Hz window to deduce the loss at DSL frequencies. These enhancements are now available from the line test suppliers.

Plant records detail the route and the build of the cables from the exchange to the customer. The loss can be reconstructed from knowing the cable gauges and length and, with an associated loss constant for each gauge, calculating the loss. This is potentially more accurate than the capacitance-based measure as the differing losses of different wire sizes can be taken into account. Even if the differing gauges are not taken into account, a simple cable length measure can still prove an effective measure.

The radial distance is the simplest form of postcode-derived distance, deriving a distance based on the geographic coordinates of the customer and the exchange. It is also probably the least accurate as it includes no electrical characteristic of the circuit to the customer—merely the fact that the circuit connects the customer to the exchange. Such a measure can be improved by incorporating features of the network. For example, the network may have large feeder cables to a cross-connection cabinet before being connected to smaller distribution cables to a distribution point. A more accurate estimate of loss may therefore be derived by adding the radial distance from the exchange to the cabinet to the radial distance from the cabinet to the distribution point. No allowance will be made of local features, such as rivers, or the fact that the cables usually follow the roads. In grid-based cities, the obvious improvement would be to follow the grid pattern.

With each of these potential sources there are limitations, sources of error that may or may not be screened out. Figure 6.1 shows the presence of errors, but it is not obvious which value for a data point is incorrect. For example, for capacitance, there is an inability to detect the aerial cables. The two legs are not always identical so a certain degree of mismatch is permitted before a reading is rejected as suspect. Some types of equipment interfere with the capacitance measurement of the line, sometimes to such an extent that the returned reading is meaningless. Some network features, such as bridge taps, also confuse the measurements. In both these cases, there may be other records that identify these types of lines so that the erroneous data can be ignored. If not, the errors, such as those shown in Figure 6.1, will have to be accepted and tolerated, otherwise the data type will have to be ignored. For plant records, the line constants might have to be a single value for each gauge, even though there are several designs of cable of varying ages from several manufacturers. The records need to be up-to-date, particularly if there are potentially multiple routes back to the exchange. Even with postcode data errors occur, putting the customer in completely the wrong location. If one or more of these data sources are to be used, then the errors incorporated in them have to be accepted. In Figure 6.1, it is tempting to remove the data that is so obviously incorrect, but if the network survey is to be an accurate representation of the network, this data should remain as it is part of the characteristics of the network.

Which of these indirect measures should be used, assuming some of these and others are available? The aforementioned network survey can, for each of these indirect measures, be used to compare the numerical value of the measure against the system limit, then formulating a pair of thresholds to mark the green–amber and amber–red boundary. The accuracy of each method would then be dictated by the fraction of lines that are flagged as amber, assuming the same financial penalties were used throughout.

Does just one of these indirect measures need to be used for prequalification? Should it be the most accurate, or the one with the widest customer coverage, which is unlikely to be the same? The proposal is to use all of the available measures and use the most accurate indirect measure for which a definite answer is available. A pecking order of indirect line qualification methods is used to make the prequalification table, as shown in Figure 6.5.

In this way, each individual in the customer base has the best prequalification answer. There will be a fraction where amber is the most accurate answer. These are passed to the on-demand test. In the previous examples, only a precision measurement has been used, but this is not the only option. For example, during the early days of building the line qualification process not all the indirect measures may be conveniently available. It may therefore be expedient to have one (or more) indirect measures applied on-demand, with the precision measurement last (as it is both the most accurate and the most expensive) (see Figure 6.6).

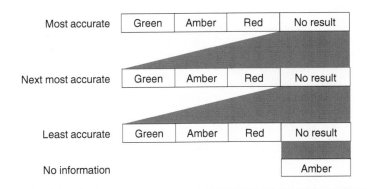

Figure 6.5 A sorting sieve for building the prequalification table. For each customer check, an assessment is available from the most accurate method used. If not, the customer drops to the next level and the process repeated. In particular, if the answer is amber then that is the result—it is most likely to be amber by any of the lower methods.

6.2 Factors Affecting the Successful Deployment of DSL Systems

6.2.1 *Impact of Local Network Faults on DSL Performance*

Planning rules are used to maximize the number of correct deployment decisions based on the agreed maximum line loss (the system limit) that the DSL system would be expected to work over. In addition to this reach requirement, the quality of the metallic pair is an important factor to

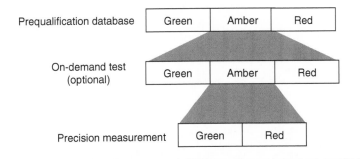

Figure 6.6 The line qualification process. The majority of customer orders are resolved by the prequalification database. The ambers are assessed by whatever on-demand process that has been set up, with some passing for precision measurement which can only return true green (GO) or true red (NOGO).

consider, because deployment failures may be attributable to network faults or impairments rather than excessive line length. This section gives a brief overview of some of the other factors associated with the condition of the actual metallic pairs that can degrade or prevent DSL operation.

6.2.1.1 Split Pairs

Split pairs occur where one leg of a twisted pair is crossed with one leg of an adjacent twisted pair. The metallic connection of the circuit is made but each of the two wires constituting the access network circuit belong to different individual twisted pairs. For switched analog telephony lines, split pairs can cause crosstalk (overhearing), but this will only occur when both customers on the same split pair are making a call. The crosstalk increases for increasing length of split. For digital systems, split pairs are generally more of a problem because (i) excessive numbers of errors are generated when calls are made on associated split circuits, and (ii) digital systems tend to be used for longer periods of time thus increasing the probability of being "hit" by activity on adjacent split lines. Two DSL systems trying to use such lines will experience abnormally high levels of crosstalk, and could easily disable each other.

Many split pairs arise owing to cable installation practices pre DSL, where the individual wires of consecutive cable segments are simply joined together with little care taken to preserve the correct pair twist along the cable route. At the end of the cable route, the wire pairs are then "corrected" by pair identification tones sent from the exchange or customer premise.

Split pairs can exist at any joint in an access network. One of the most difficult network impairments to identify and locate are split pairs that exist between two joints part way down a cable route, i.e., the circuit is split at a joint and then corrected (back onto a genuine twisted pair) for the remainder of the route. This means that at the exchange end and the customer end, the circuit looks like it begins and ends on a normal (un-split) twisted pair. The fault is, however, located somewhere between joints in the cable route.

6.2.1.2 Bunched Pairs

Bunched pairs were sometimes used in an attempt to reduce the DC (direct current) loop resistance of a POTS circuit. The term "bunched pairs" means that two pairs are wired in parallel, joined at the customer end and the exchange end to form one circuit. Although this does reduce the loop resistance and low frequency loss, it does not reduce loss at the higher frequencies which DSL systems operate at. In fact, the practice can substantially impair the impedance match between the line and the DSL equipment.

6.2.1.3 Cable Balance (Longitudinal Conversion Loss)

Transmission paths can never be entirely isolated from the external world. Crosstalk mechanisms not only couple unwanted signals from other communication channels but can also couple in noise from power lines, radio stations, switching offices, and many other sources of electromagnetic radiation. The major advantage of a well-balanced pair is that interference induced equally in both wires of the copper pair balance out. Currents flowing in the same direction along the pair are longitudinal, whereas, currents flowing in opposite directions are differential and this latter mode is usually used to transport DSL signals using balanced driver circuits in the transceiver output stage. However, induced interference may also reach the receiver. This can occur, e.g., if the coupling between the source and one of the wires of the pair differs from the coupling between the source and the other wire or if there is an impedance unbalance with respect to ground within the pair owing to component aging or insulation or conductor irregularities. If the voltages induced in the pair are not equal then the interference will appear as both a longitudinal and a differential current. The balance of the cable, sometimes referred to as the longitudinal conversion loss (LCL), is the degree of rejection the cable has to this longitudinal interference and is given by

$$\text{Balance} = 20 \log_{10} \left(\frac{\text{open circuit longitudinal voltage}}{\text{terminated transverse voltage}} \right)$$

Ideally, the network cable should have a minimum balance of 60 dB in the voiceband to be of good quality. Note that a well-balanced line is possible even in the presence of a split pair.

6.2.1.4 Rectified Loop Faults

Corrosion (or verdigris) across PCB tracks or between adjacent pins on telephone sockets are prime examples of this type of fault condition. Whilst stable linear loop faults do not give much cause for concern to DSL systems, rectified loop faults are more of a problem because they are nonlinear and introduce cross modulation (a sort of self interference which cannot be equalized away). They are also often very dynamic in nature and can stay in an "inactive" state until they become activated, e.g., by a DC line polarity reversal which may be associated with the application of ringing current to the associated analog interface. Essentially, these types of faults are "turned-on" when an incoming call is received at the ADSL customer's analog interface – the customer has no problem making outgoing POTS calls but may experience problems with premature ring trip (false answer) and high ADSL bit errors on incoming calls.

6.2.1.5   Battery or Earth Contact and Insulation Resistance

Earth and battery contact faults from either leg of the network pair and low insulation resistance faults (between A and B wires) can cause failure of a DSL system, if they are too severe. In the case of ADSL, for instance, battery-B and earth-A contacts (see Section 6.2.3) are more likely to cause transmission errors during incoming ringing to the associated analog interface, because the DC reversal typically associated with call arrival causes a sudden change in current flow in the contacting leg of the ADSL wire pair. If sufficient current is drawn then premature ring trip will occur. It should be noted that these types of fault need to be tested for using equipment with a high internal impedance so as not to mask the fault by effectively short-circuiting the line condition causing the problem.

6.2.1.6   High-Resistance Joints

One of the biggest network impairments for digital systems is poor quality or high resistance (HR) joints in the local network, on the MDF, or exchange or customer wiring. HR joints generally occur in a single leg of a pair of wires and are caused by either mechanical breakdown (possibly because of poor workmanship when the joint was made), water ingress into the joint, or corrosion of the joint over time. The typical symptoms of HR joints include the following:

- Wire pair becomes unbalanced thereby allowing external noise onto the local network pair.
- Lines carrying analog telephony can suffer from noise or crackling whilst digital services often suffer intermittent and high bit errors.
- Impulsive noise may be created by very short breaks in the transmission path (micro-interruptions).
- Measured insertion loss of local network pair can increase because of the presence of the HR joint—this is obviously particularly relevant when discussing planning rules and the capability of a given wire pair for supporting DSL.

HR joints can typically be identified by measuring the single-leg resistance of each leg (using a third wire as a common return path) and comparing the difference in measurements. An AC balance test can also identify an unbalanced pair which could be attributable to a HR joint. A moderate HR joint will be visible on a time domain reflectometry (TDR) as a pronounced cable-gauge change, and some modern test systems attempt to exploit this to determine the presence of HR joints.

6.2.1.7 Excessive Impulse Noise or RFI

Crosstalk is generally a steady-state noise source that limits the information capacity of the cable, and hence the maximum useful DSL operating range. Thus, it is crosstalk limits that are usually used when specifying the maximum operating range of a DSL system (see chapter 3 of [Golden 2006]). Within the operating range limit, modern DSL systems equalize to accommodate such steady noise, and typically retrain only if the noise level increases substantially. Intermittent noise sources such as impulse noise and some forms of radio frequency interference (RFI) are less deterministic and quantifiable. However, they do affect the actual performance achievable on a given copper pair connection. It is for this reason that an additional safety margin (on top of the crosstalk noise margin) is often allowed for when setting the maximum operating range or system limit of the DSL system.

6.2.1.7.1 Impulse Noise

Impulse noise is a key impairment for advanced access transmission on metallic twisted pairs. It is caused by a variety of sources producing short electrical transients. Such transients typically come from power switching events. Known examples include the following: mechanical switching in the household (e.g., thermostats, fluorescent light starters), semiconductor switching (especially, light dimmers and faulty switch-mode power supplies), telephony itself (e.g., on-hook or off-hook and ringing), nearby arc welders, electric cattle fences, and lightning and electric trains. These disturbances are electromagnetically coupled into the access network and may cause error bursts in digital transmission. It is well known that the levels of impulse noise on the network tend to correlate very well with human activity (i.e., worse on weekdays and during working hours). Faults investigations have also determined that a significant number of impulsive noise related problems on DSL are attributable to poor wiring practices within the customer environment—such examples include running DSL wiring in same trunking as mains power wiring and running DSL wiring over fluorescent tubes as a shortcut in ceiling spaces.

Some impulsive noise events have been found to be repetitive in nature, related to the local AC power supply frequency. This type of impulsive noise is known as repetitive electrical impulse noise (REIN). Although each event is typically less severe than isolated impulse noise events, sources of REIN are very widespread, e.g., dimmer switches, switched mode power supplies, Christmas tree lights, etc. In more extreme cases of REIN, however, field investigations have identified that a single source of REIN has severely affected DSL performance for many customers within 100 m of the source. REIN often causes a high number of errored seconds to be reported alongside a high noise margin. Diagnosis of REIN can be confirmed using a medium-wave radio at the customers premises where a loud buzz will be heard through the radio when tuned between stations at the low frequency end of the medium wave band (instead of the more usual quiet hiss or crackle).

Forward error correction (FEC) and data interleaving are used in many DSL systems to mitigate the effects of impulse noise and REIN (at the cost of increased transmission delay or latency). DSL systems which do not include such signal processing are more vulnerable to transmission problems attributable to impulse noise and REIN, even at low-moderate levels. Additionally, REIN poses a particular challenge to DSL systems operating in rate adaptive mode because it impairs the CPE's ability to estimate the usable channel capacity.

6.2.1.7.2 Radio Frequency Interference

RFI is a source of noise which affects almost every access wire pair. It varies considerably from pair-to-pair and has characteristics which show temporal variations. Digital transmission systems must tolerate certain levels of RFI as prescribed by legislation on electromagnetic compatibility (EMC).

Varying amounts of RFI will undoubtedly exist in the same frequency range as the transmitted signal. This interference is often in-band and therefore cannot be filtered out. Metallic access transmission systems should not interfere with radio transmissions. This places a limit on the transmitted power spectral density. It is fortunate that access network pairs are well balanced and do not pick up or radiate RFI easily, particularly at DSL operating frequencies. The longitudinal balance (common mode to differential coupling loss) of most wire pairs is extremely large at low frequencies (>60–70 dB) decreasing to around 30 dB at 30 MHz. This means that RF signals will suffer quite large attenuation before coupling into the receiver of a transmission system.

6.2.2 Local Network Components That Affect DSL Deployment

DSL systems are designed to operate over pure metallic cable connections between the DSL modems. In the local "access" network, there may be active and passive components that can prevent or degrade digital transmission systems from working satisfactorily. Some of these are now briefly discussed.

6.2.2.1 Bridged Taps

Bridged taps are open circuit pairs connected in parallel with a service pair, either intentionally placed along a main cable route in anticipation of possible service demand at another location or resulting from past service disconnections or network re-arrangements. In the United States, e.g., many telephone exchanges have more than one switching system, and for flexibility, connections to the MDF may be made to allow easy connection from the loop plant to more than one switch. This parallel wiring at the central office (CO) appears as bridged taps that are not recorded on any line records. Bridged taps cause impedance mismatches and discontinuities which result in reflections and increased duration of echoes. If these are too severe (in terms of delay), they will fall outside the span of the DSL transceivers echo cancellation circuitry and will therefore degrade transmission performance. Ripples in the channel response are caused by the frequency-dependent constructive and destructive interference of the echoes from the bridged taps. These also vary with the length and position of the bridged taps. Many DSL systems are able to operate in the presence of bridged taps as long as the bridged taps obey certain criteria and constraints. Bridge taps also contribute to the measured capacitance of the line, so the line measures longer than it really is.

6.2.2.2   Loading Coils

Loading coils are inductive devices (typically 88 mH) introduced along the length of a cable at regular intervals to improve the frequency response of the line in the 4 kHz voiceband (flattening the response by reducing relative attenuation at the upper band-edge). Unfortunately, in doing so, they also seriously impair the transmission characteristics at the higher DSL frequencies by increasing attenuation at these frequencies. Hence, loading coils are incompatible with DSL deployment on the cable and must be removed prior to DSL deployment. Loading coils are predominantly used on long loops in excess of 18 kft, about 5.5 km, which were once used for trunk transmission and that are now assigned for local loop use. Approximately 20 percent of US lines are estimated as having loading coils. Few non-loaded loops exceed 18 kft.

6.2.2.3   Digital Loop Carrier Systems

Digital loop carrier (DLC) systems use nodes located in the distribution network between the CO and the customer, fed by digital transmission circuits, to provide a concentrating function for local access lines. DLCs were introduced to reduce the length distribution of local access loops and to make the loop plant more amenable to new digital services. Some DLC vendors are introducing DSL cards that can connect to the customer from DLC remote electronics. In cases of severe induced noise problems, DLC systems are the only solution for customers to be able to make and receive calls successfully.

6.2.2.4 Electrical Protection Devices

In some modern networks, modular protectors for pole-top connector blocks or for installation in cable joints can use (partially) semiconductor protectors instead of gas discharge tubes (GDT). This is done to give more precise clamping voltages to protect customer premise equipment (CPE) better. Unfortunately, these semiconductor devices have very much higher capacitance than GDTs (50–100 pF versus 0.5–1.0 pF). This capacitance is sufficient to lower signal levels for VDSL signals by several decibels although the impact on HDSL would be less. Even more unfortunately the capacitance is bias voltage dependent (unlike GDT), so that the A-leg (normally biased at ~ 0 V, where $C \sim 100$ pF) will have very much higher capacitance to ground than the B-leg (normally biased to ~ -50 V, where $C \sim 50$ pF). This will cause unbalance, severely affecting the RFI immunity performance of the cable. Also because the protectors' common ground will be all commoned but only earthed probably through a high HF impedance (e.g., long ground cable and earth spike), the protectors will also result in worse crosstalk. Where they are used, the in-cable protectors may only normally be installed on long distribution side runs.

6.2.2.5 RFI Filters

RFI filters can be found connecting the customer drop wire to the internal customer wiring, especially for customers near AM radio transmission masts. On short noise-free loops these filters can effectively disable DSL systems (system can consistently fail to train-up or yield excessive downstream errors). Obviously, this gives scope for increased number of failures at provision for DSL systems particularly as local records may not accurately reflect the presence of these filters (many may have been installed years ago and also may be hidden away in roof spaces, under floorboards, etc.). It is therefore important to establish methods and field procedures for detecting such devices using line test or field test equipment.

6.2.2.6 Maintenance Terminating Units

There are some devices which can be used to improve fault location and line test assessments. These devices, typically called maintenance terminating units (MTUs), are designed to allow line test systems to test the line without the results being affected by different types of CPE. MTUs are solid-state devices which have high-inline resistance when no current is flowing (on-hook) and low-inline resistance when in the off-hook state. Bypass capacitors allow for the transmission of calling line identity (CLI) in the on-hook state. The line characteristic change between on-hook and off-hook are severe enough to typically cause an ADSL to retrain whenever the customer goes from on-hook to off-hook (and vice versa).

6.2.3   *Pathological Loop Testing*

Pathological loop tests can also be used to evaluate the performance of DSL equipment prior to deployment in the local network and in particular, to ensure new DSL systems are resilient to potential service-affecting network fault conditions. The primary objectives of these tests are as follows:

- To investigate the resilience of DSL systems to simulated line fault conditions and to highlight any major deficiencies or areas of concern. The types of fault conditions which can be simulated include split pairs (where one leg of a pair is crossed with one leg of an adjacent pair), contact faults with an adjacent pair, and rectified or hard resistive loop faults. Typically, the effects of incoming and outgoing call activity (ringing, ring trip, loop disconnect dialling, call clear, etc.) on the DSL systems downstream and upstream bit error performance can be investigated for each fault condition.
- To ensure that DSL systems which also support analog telephony on the same wire pair do not affect the charging integrity and exchange performance with regard to premature ring trip faults (i.e., fault conditions in the local network or customer environment which can cause any incoming analog calls to trip ringing and charge the calling customer).
- To assess the types of fault conditions which could cause, for instance, an ADSL system under test to fail (produce unacceptable error rate downstream or upstream) whilst the customer's analog telephony service remains unaffected and vice-versa.
- To investigate whether certain access network components such as RFI filters, electric fence filters, and various protection devices affect the performance of the DSL system.
- To ensure that the portfolio of test equipment carried by the incumbent operators field force is able to identify and locate the types of fault conditions and access network components which are found to cause service-affecting problems on the DSL system or on the analog telephony service. Also, to verify that the use of cable pair identification (CPI) and TDR test equipment does not affect DSL performance when faulting or testing on adjacent pairs within the same cable route.
- To determine if any additional test equipment is required by field engineers to diagnose and locate DSL-affecting line fault conditions.
- To suggest best working practices to support the provision and repair of DSL services.

It should be noted that some of the laboratory simulated network faults (such as contact faults to adjacent pairs) are not truly representative of the wide range and characteristics of real network faults of this type. However, Pathological loop or provocative testing, as it is sometimes known, is an effective way of stressing a given DSL system in a controlled manner to highlight the types of network fault conditions which are likely to cause transmission problems for the DSL system or result in the DSL system failing whilst the customer's POTS remains unaffected (and vice versa). In addition, the results from the tests can be used to develop efficient and practical field and diagnostic processes for both the provision and repair of digital systems.

6.2.3.1 Typical Pathological Loop Test-Bed Configuration

Figure 6.7 shows the typical test setup for performing pathological loop tests (in this case on ADSL equipment). Simulated crosstalk noise (N) is injected at the line interface of the exchange and remote ADSL units at a 0 dB reference power level. Bit error rate monitoring equipment is connected to the exchange end network interface and to the remote end user interface.

Simulated fault conditions such as contact faults to earth or battery, split pairs, and loop faults are easily connected (Figure 6.7) onto the test pair at either the exchange (EX), an intermediate network joint (NJ), or network termination (NT) points. For each fault type and fault location (EX, NJ, NT), the effect on the ADSL downstream and upstream bit error performance is recorded when incoming (i.e., ringing, ring trip, call clear) and outgoing (loop seizure, loop disconnect dialling, call clear) calls are made or received on the associated ADSL POTS circuit.

To facilitate easy connection of the simulated network faults at the EX, NJ, and NT points, the cable test-bed should be arranged so that these points are all located in the same test room. This is achieved by looping back the cable runs as shown in Figure 6.8.

In addition to the ADSL circuit under test, two other POTS circuits are required—one to be used as an adjacent POTS line for setting up contact faults and split pairs with the ADSL, and the other for making (and receiving) test calls to (and from) either the ADSL or the adjacent POTS circuit.

6.2.3.2 Benefits of Pathological Loop Testing

Pathological loop testing enables the telecommunications operator or service provider or equipment vendor to evaluate the behavior and performance of their DSL systems, in a controlled manner, when subjected to simulated network fault conditions on the local network pair. This leads to a much better understanding of the types of network fault conditions which could cause transmission problems for the DSL system when deployed in a real access network. It also yields valuable information on the likely

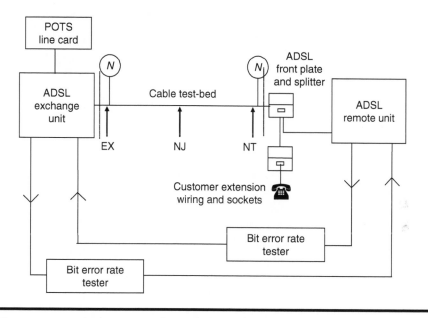

Figure 6.7   Typical test setup for pathological loop testing.

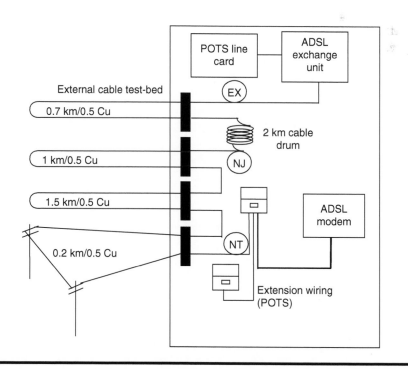

Figure 6.8   Connection of simulated fault conditions at EX, NJ, and NT points.

symptoms and failure modes which may be exhibited by the DSL equipment for different types of fault conditions (e.g., DSL producing errors but POTS working OK). These findings can ultimately be built into practical field and diagnostic processes to aid the provision and repair of the DSL system.

By way of an example, the general findings from a number of ADSL systems evaluations are summarized below:

- For the majority of line fault conditions, analog POTSs will fail before the customer's ADSL service, although significant ADSL bit errors may occur during active ringing cycles on incoming calls to the ADSL (or to an adjacent POTS line in the case of contact faults).
- Only simulated fault conditions found to cause excessive downstream ADSL errors (before the customer's POTS is affected) were forward and reverse rectified faults at the customer end (typically caused by corrosive action at the NTE or DP). Errors were found to occur primarily during incoming ring bursts to the ADSL customer's phone for rectified faults more severe than typically $10\,k\Omega$. Similarly, rectified faults (more severe than approximately $7\,k\Omega$) at the exchange end were found to be the main cause of upstream errors on ADSL when receiving an incoming POTS call.
- Single-leg battery and earth contact faults to adjacent POTS circuits were found to cause minor downstream errors if the fault condition was located at the customer end and minor upstream errors if the fault was located at the exchange end. Again, errors were only found to occur for very severe faults (less than $3\,k\Omega$) and during incoming ring bursts (to either the ADSL or the adjacent POTS line in contact with the ADSL), on- or off-hook activity or loop disconnect (LD) dialling.
- Fact that fault conditions near the exchange tend to cause upstream errors and fault conditions near the customer tend to produce downstream errors is predictable because in both cases the received signal is attenuated by the access network pair.
- Split pairs were found to cause excessive numbers of ADSL upstream and downstream errors particularly for incoming ringing, on- or off-hook activation and LD dialling. The error rate and the direction of these errors were found to be dependent on the position and length of the split.
- ADSL POTS circuit was found to cause premature ring trip (false answer) on incoming calls for severe A-wire to B-wire loop faults between approximately 4 and $5\,k\Omega$. Although the window in which this occurs with ADSL is wider than for a standard exchange line,

it is not significantly greater. In addition, the ADSL POTS circuit was resilient to false answer for rectified faults and earth-A or battery-B single leg contact faults (which have, in the past, been found to cause premature ring trip on other exchange systems).

- RFI filter, which may be found in customer premises connecting the drop wire to internal customer wiring, was found to cause excessive downstream ADSL errors on lines longer than 1.2 km (0.5 mm Copper). Above 2 km, the ADSL system fails to synchronize. Whilst this means that ADSL is likely to fail on a high percentage of lines which have RFI filters installed, a method of detecting the presence of these filters using standard field test equipment has been devised. ADSL transmission problems may also occur on access network lines beyond approximately 3.3 km which have electric fence filters fitted at the customer premises.

6.3 Conclusion

Planning rules are used to maximize the number of correct deployment decisions based on the agreed maximum loss line (the system limit) that a DSL system would be expected to work over. The use of a network survey on a small random selection of the network enables the series of indirect (estimation) measures of loss to be ranked in terms of accuracy for a particular DSL service. The width of the amber zone is determined by balancing failure costs (false red and false green) against the precision measurement costs—the most accurate method has the smallest amber fraction. Indirect measures are used to pre-qualify the majority of the customer base. For lines marked amber, an on-demand estimation or measurement sequence is only used when an order is raised. The whole process does make some incorrect deployment decisions, however, the overall business cost is significantly reduced.

In addition to the reach requirement, the quality of the metallic pair is an important factor to consider, because deployment failures may be attributable to network faults or impairments rather than excessive line loss. Network faults and impairments are typically attributable to poor working practices (e.g., split pair, bunched pairs, etc.), physical faults on the wire pair caused by mechanical failure, or water ingress at a joint, flexibility point or network termination (e.g., high resistance joints, rectified loop faults, etc.), or degradation of the network cable making it susceptible to external factors such as impulse noise.

Some legacy network components connected to the wire pair, such as RFI filters, can also cause transmission problems on DSL.

Pathological loop testing is an effective way of stressing a given DSL system in a controlled manner to highlight the types of network fault conditions which are likely to cause transmission problems for the DSL system or result in the DSL system failing whilst the customer's POTS remains unaffected (and vice versa). In addition, the results from the tests can be used to develop efficient and practical field and diagnostic processes for both the provision and repair of digital systems.

Reference

[Golden 2006] P. Golden, H. Dedieu, and K.S. Jacobsen, Eds., *Fundamentals of DSL Technology*, Auerbach Publications, 2006, p. 457.

Chapter 7

Spectrum Management

Rob Kirkby and Ken Kerpez

CONTENTS

Abstract This chapter will discuss how spectrum management is achieved. Approaches to spectrum management differ among countries and networks, this chapter focuses on spectrum management in the United States, in the United Kingdom, and in Europe.

7.1 Introduction

For a telephone access network, "spectrum management"[*] is concerned with coordinating the behavior of Digital Subscriber Line (DSL) systems sharing the cable environment, to realize the intrinsic data transmission capacity of the cables. For DSL, crosstalk is the dominant noise source (see chapter 3 of [Golden 2006]), so a system's data rate is limited[†] by coupling of signals from neighboring lines. Crosstalk is of two types: near-end crosstalk (NEXT) and far-end crosstalk (FEXT). Spectrum management is about control of NEXT and FEXT. In practice, "spectrum management" means limiting the levels and bandwidths of signals that may be transmitted on lines in shared cables so that different DSL types can coexist in the same cable.

This chapter will discuss how spectrum management is achieved. Approaches to spectrum management differ among countries and networks, this chapter focuses on spectrum management in the United States, in the United Kingdom, and in Europe. First, however, the motives of spectrum management and the common features to be expected in all regimes are examined.

---

[*] The term "spectrum management" is inherited from radio spectrum allocation, in both cases the problem is control of interference and is addressed by control of transmissions.

[†] The other aspects that limit performance are the transmission properties of the line in use, which is often addressed by deployment limits. However, performance cannot be predicted until the noise environment can be estimated, so logically spectrum management comes first.

7.2 Common Features

All telephone networks have in common that they were originally laid down to support voice telephony, and that the later arrival of DSL offers the chance of greatly enhancing the value of this resource at low cost.* However, DSL brings the new problem of interaction between the signals on the separate lines, and this interference must be managed to extract the most benefit from the network. Therefore, compatible behavior must be forced on the disparate users.

Lest it not be obvious why such restrictions to freedom are needed, consider an evolution scenario in their absence: each user individually can enhance his own service by increasing the output power of his modem's transmitter. Each user will, of course, need to increase power frequently, because his neighbors will be doing the same thing, and so between upgrades each user will experience service degradations. The process of progressively increasing the transmitter power (sometimes referred to as the "cocktail party effect") continues until a user's pair fails. Fortunately, this progression is unattractive[†] to most users,[‡] so spectrum management is easily acceptable. It does, however, require the involvement of an administrative body with authority over all the disparate users.

The objective of the management of the spectrum is invariably "best" use of the network, and strategic objectives are usually low-cost service and high data rates to customers. It turns out that some decisions on use are necessary: the optimal ways of exploiting a network for some systems are not compatible. Different administrations typically come to different solutions for their networks.

7.2.1 General Methods

The practice of spectrum management invariably reduces to prescribing what may be connected to the network and what may not. Limits have been proposed based on one or more of the following schemes.

List of approved systems: In the past, specifying approved systems has been a common method, either by requiring equipment to have a certificate of conformance to specified tests from a test house[§] or by specifying a

---

* It also has an attractive investment profile: most costs are "incremental," i.e., not paid until needed,
[†] "Unattractive" means expensive, offering no net gain, and eventually a total loss of service.
[‡] Those users on the shortest pairs will survive the process, and eventually they will have stable connections.
[§] In the United Kingdom, this was called "type approval" and was necessary for safety reasons.

procurement standard. Modern procurement standards usually have many operating modes, so for spectrum management it may also be necessary to prescribe which modes are permitted.

Total power limit: In this case, a limit is applied to the total transmit power. Such a limit is the natural choice for protecting DSL technology. Proponents of power limits argue that the limits are fair in the sense that they allow the user a choice of data rates (for example, based on line length), while giving a universal level of protection because the worst case that occurs is when all neighbors use the maximum power in the same way as the victim line.

Spectral masks: Spectral masks limit the transmit power spectral density (PSD), specifying power limits at each frequency independently. The limit is a continuous function of frequency, often shown as a graph. Formally, it can be specified either as a list of corners of a piecewise plot, or as an algebraic function. The specification of spectral masks is a modern approach common in standards.

Cable fill: In this case, the administration limits the number of pairs that may carry active systems. The cable fill approach attempts to reduce noise by lowering the worst-case crosstalk by decreasing the factor of $(^N/_{49})^{0.6}$, where N is the number of crosstalk disturbers. The successful limitation of cable fill requires the administration to have good cable records, and generally results in gains of only a few decibels.

Adjacency control: With this approach, if one pair is already in DSL use, the administration prevents any other pair in the same cable binder from being used. If successful, this reduces 1 percent worst-case FEXT and NEXT by about 10 dB. The technique is only successful if the cables maintain adjacency at all the joints between cable sections, which requires the administration to have very good cable records.

Signal direction: In this case, different rules are prescribed for the different directions of transmission. Signal direction requirements are commonplace with frequency-division duplexed (FDD) systems such as Assymmetric DSL (ADSL) and are always needed to suppress NEXT.

Line length: With this approach, the rules vary as a function of line length or of line attenuation. For example, it has become clear that for Very High Bit-Rate DSL (VDSL) to achieve a useful reach, all VDSL systems sharing a cable are required to control their upstream power based on the line length [Kirkby 1995].

System location: It is already common to consider different network designs based on fiber-to-the-exchange (FTTEx), fiber-to-the-cabinet (FTTCab), and fiber-to-the-curb (FTTCurb). As a result, different spectrum management rules would seem appropriate, especially because the extension of fiber into the network is desired specifically to enable higher bandwidths in the telephone cables. The United Kingdom's access network frequency plan (ANFP) uses a form of this methodology, where each end of a line has its limits set based on its location (irrespective of where any other modems might be).

Time: Instead of separating downstream and upstream transmissions in frequency, they may be separated in time. In such a case, transmissions are synchronized, and different limits may apply at different times of the day. This has been proposed both on a diurnal cycle—different network uses during and outside of office hours—and very short cycles, such as the "ping pong" style time-compression multiplexed (TCM) integrated services digital network (ISDN) used in Japan. Ping-pong was the basis for an early* proposal for VDSL (see Chapter 17 on Standardization).

Dynamic spectrum management (DSM): In a DSM approach, the time or situation dependent parameters of individual lines and cables can be actively managed and controlled to jointly optimize performance for each individual situation (see Chapter 8 on DSM).

Pollution licensing: In this approach, the administration sells licenses to inject a given amount of power into a given line. The licensee can resell his license. This approach is only a theoretical suggestion at this time; it is an analog of certain methods of chemical pollution control.

7.2.2 Power Spectral Density

Spectrum management is about management of signal levels to reduce crosstalk. The coupling from line to line typically exhibits random phase, so PSD limits are an appropriate management method for spectrum management. Spectrum management limits should be in terms of harm done, and it is therefore wise to cast spectral limits in terms commensurate with the susceptibility of DSL.

The PSD of a random process that is wide-sense stationary is well-defined and easily computed as the long-term time average of the power in each infinitesimal frequency band. Communications signals are typically

---

* The proposal was not accepted, despite its many virtues, because a fault in any one system's synchronization could disable the whole cable.

cyclo-stationary and not strictly wide-sense stationary. However, by assuming that the phase is random, the process becomes stationary and has a PSD. PSDs for spectrum management should be measured by equipment that is not synchronous with the transmitted symbols. The measurements should contain samples with a sufficiently large number of different phases to provide an accurate average.

Mathematical definitions of continuous spectra imply integration over all time. Practical measurements are necessarily finite, but long averaging times are often necessary for two reasons: detailed resolution of frequency, and accuracy. The appropriate resolution bandwidth is of the same order as that of the system under study. ADSL has a subchannel spacing of 4.3125 kHz, which suggests a resolution bandwidth (RBW) near 4.3125 kHz. A resolution bandwidth of 10 kHz is usually specified, because it is available on standard spectrum analyzers. Averaging many such observations is needed for accuracy because any one observation is typically exponential distributed.[*] To measure a PSD within 0.1 dB accuracy, at least 9300 individual observations need to be measured in each sub-band.[†]

Currently, it is common to specify a much wider bandwidth when measuring the out-of-band noise floor: the wider bandwidth is desirable to allow low-level clock leakage from a modem to be acceptable.

For intermittent signals, measurements should only be made during their "on" phase.

7.2.3   Crosstalk Impact

Estimating the practical impact of crosstalk is a key element of spectrum management. (The "spectral compatibility" calculations of a method used in North America are discussed in some detail in Section A.7.2.) It is a well-established subject, having been investigated for plain old telephone service (POTS), analog carrier, and other loop technologies deployed in the past. Particular concern has always been on characterizing telephone subscriber loop signaling and crosstalk in the frequency domain, hence the designation "spectral compatibility."

---

[*] The signal is random, both as transmitted and after coupling by crosstalk, so the observation is random. It is still random after Fourier analysis. Indeed, the transmitted signal is approximately Gaussian, and after coupling into another pair, it is typically even nearer to Gaussian. After Fourier analysis, it is very nearly Gaussian and independent in each of the real and imaginary components at every frequency individually (except near regular features such as pilot tones).

[†] A superhet spectrum analyzer only observes one sub-band at a time, an FFT based instrument observes them all in parallel, which is why they are faster. The idea of an "individual observation" is literally true for FFT based instruments, for a superhet instrument the equivalent is the rate at which the detector produces independent values.

Early copper transmission systems were all symmetrical, and their analysis generally only considered NEXT and ignored FEXT. FEXT is less severe than NEXT because FEXT is attenuated by the cable. However, FDD systems such as ADSL and VDSL experience no NEXT at some frequencies, and therefore FEXT becomes significant. Also, on short loops (for example, about 1 kft), FEXT can have nearly the same power as NEXT. Modern analyses now include all sources of NEXT and FEXT.

It is a standard industry practice to engineer crosstalk using the 1 percent worst-case crosstalk coupling factors. This is to ensure a low probability of service failure. Then if $S(f)$ is the transmit PSD of the crosstalk source in milliWatts per hertz, and $X(f)$ is the dimensionless 1 percent worst-case crosstalk power coupling, then $S(f) \cdot X(f)$ is the received 1 percent worst-case crosstalk PSD in milliWatts per hertz. With the Unger model assumed (see chapter 3 of [Golden 2006] as well as [ANSI T1.417] and [Valenti 2002]), the power couplings are as shown in Figure 7.1 along with some individual measurements [Valenti 2002].

The 1 percent worst-case crosstalk power coupling $X(f)$ varies as $6 \cdot \log_{10}(n)$ dB, where n is the number of crosstalk disturbers in the cable binder. It is difficult to count or control n, so it is typically assumed that the binder is filled with crosstalkers and $n = 24$ or $n = 49$, which is at most 10 dB pessimistic compared to $n = 1$ disturber. The received noise is the sum of all crosstalk, NEXT and FEXT, plus a low-power background noise of -140 dBm/Hz. NEXT, and separately FEXT, from different crosstalker

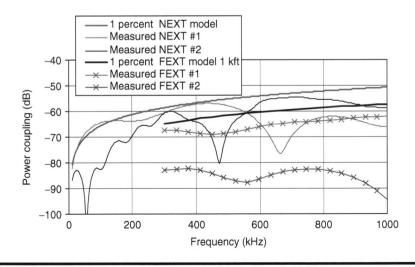

Figure 7.1 **1 percent worst-case single crosstalk disturber power coupling models and measurements of pair-to-pair hear-end crosstalk (NEXT) and far-end crosstalk (FEXT). FEXT is on a 1 kft 24 gauge loop.**

types is summed using the Full Services Access Network (FSAN) crosstalk summation method [Galli 2002]. (See chapter 3 of [Golden 2006].)

7.3 American Spectrum Management

7.3.1 Need for Spectrum Management in America

The bit rates and reaches of early DSL technologies, such as basic-rate ISDN and High Bit-Rate DSL (HDSL), were considered to be limited by NEXT from the same types of systems as the victim system at the same end of the cable as the victim receiver. The impact of this so-called self-NEXT was mitigated by using low-frequency baseband transmission. Later, ADSL avoided most self-NEXT by transmitting upstream and downstream signals in different frequency bands, i.e., by using FDD. The fact that ADSL and other emerging systems were not limited by self-NEXT, coupled with an increasing number of different DSL types, increasingly resulted in alien crosstalk becoming a limiting impairment, where alien crosstalk is defined as crosstalk originating from a different type of DSL than the victim system. Spectral compatibility occurs when alien crosstalk from one type of loop transmission system does not overly degrade the performance of another type of loop transmission system. DSL spectrum management involves controlling alien crosstalk to ensure spectral compatibility.

A common misunderstanding of many individuals who have recently become involved with DSL is that there is a "crosstalk problem" for only a few DSLs on only a few loops. However, every DSL has been designed and built based on, and runs with, crosstalk limitations. For example, crosstalk limits HDSL to a maximum bit rate of 1.5 Mbit/s on two pairs of 26 AWG cabling up to 9 kft in length. With no crosstalk, approximately 10 Mbit/s is achievable with almost the same technology on one pair of this length. Other DSL bit rates, loop reaches, and reliability would also increase dramatically without crosstalk. As time progresses, more DSLs will be deployed and crosstalk will increase, which will cause failures if the plant is not managed so that it functions with crosstalk.

The received crosstalk PSD, which characterizes the power as a function of frequency, equals the PSD transmitted by a crosstalk disturber plus the crosstalk power coupling in decibels. Spectral compatibility is often enforced by limiting the transmitted PSD to be below some defined PSD mask at all frequencies, which in turn limits the received crosstalk PSD. Different PSD masks are defined for different classes of technologies and different loop reaches.

DSL spectrum management requires knowledge of all DSL types and how their crosstalk affects other systems. Spectrum management also requires defining the level of degradation from crosstalk that is considered spectrally compatible. This can be hard to define when some DSL types are favored by some companies (for example, service providers or equipment vendors) and not by others, such as was the case during the development of Symmetric DSL (SDSL) and ADSL. (See Chapter 17 on standardization.)

Competition in the local loop caused a need to standardize DSL spectrum management. In the past, a service provider could choose to deploy only a set of loop transmission technologies that were spectrally compatible and ignore all others. This is still true in some locales. Now, however, many loop plants are unbundled, and a competitor may lease any loop and deploy a number of different systems. In the United States, many competitive local exchange carriers started providing broadband Internet access in the late 1990s by deploying DSL types that were potentially incompatible with existing services. A national standard emerged to provide a technical definition of spectral compatibility and allow competition to progress in an orderly fashion.

In the United States, ATIS standards committee T1E1.4 (now called ATIS NIPP NAI) has issued the spectrum management standard T1.417-2003 [ANSI T1.417], which contains a detailed definition of spectral compatibility that has broad industry consensus. This standard was forged at a time of much competition in the local loop, and it represents a compromise between incumbent and competitive carriers. It provides an unambiguous yardstick for determining whether and under what circumstances (i.e., loop reach) new DSL technologies may be deployed in the loop plant.

There are other impairments to DSL transmission besides crosstalk: impulse noise, radio frequency interference (RFI), etc. However, these impairments are almost totally independent of the generally dominant crosstalk impediment, and are not usually explicit in spectrum management. This section describes managing the spectra of different DSL types and the crosstalk between them. A brief history of American spectrum management, spectrum management standard compliance, and some future directions are presented.

Beyond the current rules, DSM holds the possibility of greatly increased bit rates and reliability, by treating crosstalk as man-made interference that can be measured, understood, and mitigated with multi-user transceiver techniques. DSM is addressed in Chapter 8.

7.3.2 Creation of a Technical Definition of DSL Spectral Compatibility in America

This subsection discusses the processes, people, organizations, and history of the American spectrum management standard, T1.417.

7.3.2.1   DSL Spectral Regulation

The Federal Communications Commission (FCC) sets nationwide telecommunications regulations and allocates radio spectra in the United States. Specific tariffs and regulations are determined individually in each state by the State Public Utility Commission (PUC). Because the Internet spans state boundaries, the FCC has some authority over DSL regulations. At this time, adherence to the American spectrum management standard, T1.417, is not strictly required by the FCC to deploy DSL in the public network. The FCC does, however, mandate that T1.417 will be used to resolve any disputes over DSL spectrum management.

Much of the impetus behind the spectrum management standard was the explosive birth of the competitive local exchange carriers (CLECs) who began providing DSL service on unbundled loops leased from the incumbent local exchange carriers (ILECs) in the late 1990s. More recently, many CLECs have disappeared. ILECs may soon be relieved of U.S. requirements to unbundle "line-shared" loops, where the CLEC provides DSL on the high frequencies of a loop while the ILEC continues to provide POTS on the low frequencies. However, even with these changes the regulations still mandate that ILECs allow unbundling in their loop plant. Also, many new vendors continue to develop new DSL types. In this changing environment, the ILECs have generally taken the safest and simplest method of administering their loop plant by always requiring conformance to T1.417. So, although T1.417 is not specifically required by law, conformance to it is nearly always required. On the other hand, actual testing and independent certification of T1.417 conformance is scant. Conformance is instead declared analytically.

7.3.2.2   Writing the American Spectrum Management Standard

Spectral compatibility was a study project in standards committee T1E1.4 (now NIPP-NAI, see Chapter 17 on standards) in the mid-1990s with a draft Technical Report, but it received little interest because the committee was busy with other standards such as ADSL and HDSL2. An early definition, in 1997, of DSL spectral compatibility was conformance to one of three "composite PSD masks," one for all downstream transmissions, one for upstream within CSA range, and one for upstream beyond CSA range [Kerpez 1997]. The composite PSD masks encompassed all DSL technologies. This approach was very coarse-grained, spectral compatibility calculations assumed that a system would fill the entire composite PSD mask, which was allowed, but which led to pessimistic projections of crosstalk impact.

In the United States, the 1996 Telecommunications act gave the FCC the task of creating an environment to foster competition in the local loop.

Competition was to be driven by unbundling the local loop, allowing CLECs to provide broadband DSL services. ILECs could no longer control the technologies that were deployed in their loop plants. It was in no one's interest to deploy incompatible technologies that were likely to cause service outages. In late 1998, spectral compatibility became a hot topic. Standards committee T1E1.4 was informally given the task of creating a technical definition of spectral compatibility by the FCC, and FCC members began to attend T1E1.4 meetings. The goal was to generate an industrywide standard to allow successful DSL deployments by all parties. Spectral compatibility became an official T1E1 standards project in 1998 and the draft Technical Report became a draft standard.

The concept of spectrum management classes first appeared in late 1998 and was documented in T1E1.4 contributions in early 1999. This approach is still in use, it defines a number of separate spectrum management classes. A class represents a set of technologies whose spectra and crosstalk impact are approximately the same, and membership within a class is proven through conformance with the PSD mask specific to that class. For example, spectrum management class 3 (SM3) represents HDSL systems from all different vendors. There were originally five spectrum management classes. Many individuals and companies contributed to the spectrum management standard, and a version incorporating agreed PSD templates for all defined SM classes, with complete test procedures, conformance criteria, and other electrical specifications was sent out for letter ballot in June 1999.

In mid- to late 1990s, the Internet bubble was fully inflated, and numerous new DSL service providers began to attend T1E1.4 spectrum management meetings. Some parties viewed many of the assumptions in calculating spectral compatibility as overly pessimistic. Others countered that performance in the field was often worse than the calculations predicted. Meetings became highly contentious. Resolution of the spectrum management standard letter ballot was deferred pending agreement on some basic principles. Most issues eventually were resolved through compromise agreements.

One issue was the set of existing DSL systems with which a new technology must demonstrate spectral compatibility. Those interested in protecting existing DSL systems wanted very strict spectral compatibility requirements imposed on future systems, whereas those who were more interested in developing new systems proposed looser requirements. This issue was so contentious that the name of the set of existing systems to be protected changed several times, from "protected" services to "guarded" systems and finally to the neutral language of "basis systems." At one point, T1 lines, which use a widely deployed but older technology, were taken off this list.

A major conflict was defining spectral compatibility between ADSL and the original version of SDSL, which was a proprietary system using the

2B1Q line code. (See Chapter 17 on standardization.) Most DSLs are limited to some performance level by self-NEXT and this performance should be maintained with alien NEXT. But because most deployed ADSL systems use FDD, they do not suffer from self-NEXT, and thus ADSL has no such naturally defined performance level. Some operators were largely deploying ADSL to residential customers, and were concerned that crosstalk from Symmetric High Bit-Rate DSL (SHDSL) would greatly lower ADSL bit rates. Other operators were generally deploying SDSL, often to businesses, and thought that overly pessimistic assumptions and unrealistic guarantees of 6.7 Mbit/s downstream ADSL bit rates had led to over-protection of ADSL and overly restrictive spectrum management limits on SDSL. This impasse was broken by the January 2000 "Fort Lauderdale Agreement," which specified that SDSL is allowed to be deployed at the maximum reach possible in the presence of 49 self-NEXT, and then the protected performance levels of ADSL are defined as the ADSL performance with that resulting level of SDSL crosstalk. The ADSL bit rates thus determined are now in Annex A.9 of T1.417-2003 [ANSI T1.417]. Moreover, SDSL spectral compatibility was specified at every loop length in increments of 500 ft, and the SDSL PSD template was tightly defined.

This definition of spectrally compatible SDSL was ensconced as a new method of spectrum management standard conformance, known as the "technology specific" guidelines. SDSL was categorized with 20 different bit rates and spectrally compatible loop deployment guidelines. This fine-grain structure allows, for example, 320 kbit/s SDSL at 15.5 kft, whereas the nearest SM class is SM2 and is allowed only to a reach of 11.5 kft. Technology specific guidelines for SHDSL and HDSL4 were subsequently added to the standard.

After settling these major issues, the spectrum management standard became standard T1.417-2001 in January 2001. In order for this timely completion, this issue of the standard, known as issue 1, only addressed systems that were deployed from the central office (CO), and deferred defining spectral compatibility of deployments from remote terminals and repeatered lines.

Issue 2 of T1.417, T1.417-2003 defined spectral compatibility of repeatered lines, SHDSL, and VDSL, it became an ANSI standard in 2003. Models for crosstalk from repeatered lines, crosstalk over short exposure lengths, and crosstalk from adjacent binders are the new normative text in T1.417-2003. However, T1.417-2003 does not have a complete definition of the difficult problem of spectral compatibility of remote terminal deployments with CO-based systems.

T1.417-2003 defines methods of calculating spectral compatibility of repeatered lines. Evaluations with crosstalk from repeatered lines relax margin requirements by 1.8 dB, equal to halving the number of disturbers, because repeatered lines are sparsely deployed. Every possible repeater location, in 500 ft increments, needs to be evaluated. There is a test specifying that repeatered lines must also demonstrate compatibility with basis systems in adjacent binders.

VDSL is declared a basis system and a detailed method of calculating compatibility with VDSL is defined in T1.417-2003. This methodology essentially mandates the use of spectra very close to that of the VDSL band plan known as "998" (see Chapter 17 on standardization) at high frequencies, although there is some flexibility. The spectrum management class for VDSL is SM6. G.lite is also a basis system in T1.417-2003. In addition, SHDSL is a basis system, analytical criteria for spectral compatibility with SHDSL are defined, and SDSL is no longer a basis system in T1.417-2003.

It is difficult to ensure spectral compatibility of remotely deployed, cabinet, or remote terminal (RT)-based DSL with CO-based DSL in all cases. ADSL or VDSL deployed from an RT can generate FEXT into CO-based DSL that is almost as powerful as NEXT if the distance from the RT to the CO-based subscriber loop end is a few kilofeet or less. The FEXT is usually severely debilitating to any CO-based ADSL in the same binder. However, loop lengths of RT-based DSL and CO-based DSL are such that the two systems will infrequently share the same binder, especially, if service provider engineering does not allow it.

Spectral compatibility with ADSL2plus (G.992.5) is not defined in T1.417 issue 2. However, it is possible to force RT-based ADSL, or ADSL2plus, not to use some frequency bands, leaving these bands free of crosstalk so the CO-based ADSL can use them. However, disabling portions of the spectrum of systems on shorter loops to improve bit rates on longer loops is a difficult balancing act that can lower the performance of both systems if not implemented properly.

The determination of spectral compatibility of remotely deployed DSL is not defined in T1.417-2003, nor in any other American standard, at the time of this writing. On one hand, T1.417-2003 defines the deployment reference site as the network end of all new technologies and all basis systems. However, there is no requirement to calculate the spectral compatibility of mixed RT and CO deployments. T1.417-2003 also specifically declares that RT-based spectrum management classes 4 (HDSL2), 5 (ADSL), 6 (VDSL), and 9 (overlapped ADSL) are all spectrally compatible. Overall, T1.417-2003 is ambiguous on the spectral compatibility of RT-based DSL.

ADSL2plus and some VDSL2 profiles are not included in T1.417, and at this time of writing there is little ongoing work on ANSI T1.417. The United States appears to be drifting away from the formal "static" spectrum management guidelines in T1.417, at least in part because local loop unbundling has been largely a failed experiment in the United States. The standardization effort in the U.S. standards now focuses on DSM.

7.3.3 T1.417 Spectrum Management Standard Conformance

This section gives a simple explanation of the requirements to comply with T1.417 [ANSI T1.417], but only the standard itself can be used to demonstrate the compliance of a system. Compliance with the standard only determines the spectral compatibility of a system, it does not determine whether a system can or cannot be legally deployed in any particular jurisdiction. T1.417-2001 inherently assumes that crosstalk is reasonably well behaved, DSL receivers treat crosstalk as an independent Gaussian noise, and other generally true assumptions. Issue 2 of T1.417 (T1.417-2003) was completed at the time of this writing, so this section includes details of issue 2.

7.3.3.1 Basis Systems

The general concept of the spectrum management standard is to require spectral compatibility with all members of the set of "basis systems." Basis systems are typically standardized systems that are expected to be widely deployed. The basis systems are as follows:

- Voicegrade services
- Enhanced Business Services (P-Phone^{TM})
- Digital data service (DDS)
- ISDN basic-rate (BRI)
- HDSL
- HDSL2
- ADSL, nonoverlapped
- Rate-adaptive DSL (RADSL)
- Splitterless ADSL (G.lite)
- SHDSL
- VDSL

A new DSL technology must not generate crosstalk that significantly degrades the performance of any basis system. Ideally, crosstalk from all the basis systems should not significantly degrade the new DSL technology's performance, but this condition is not required. The definition of a fixed set of basis systems allows a new technology to demonstrate spectral compatibility without requiring knowledge of all other new technologies, which is potentially unobtainable. Being a basis system in no way assures conformance with T1.417, and basis systems themselves may not be spectrally compatible on many loops. However, by virtue of having been developed and deployed prior to the development of formal spectrum management requirements, the basis systems have a form of "squatter's rights" in the cable plant.

7.3.3.2   PSD Masks, Templates, and Conformance

A system cannot transmit more power at any frequency than the PSD mask that it conforms to. To include as many systems as possible, the PSD mask of a given class should be loose, allowing relatively high power levels. But, to ensure the best spectral compatibility, the PSD mask should be tight. These conflicting requirements led to the definition of a PSD template [Kerpez 1997] [Russell 1999] [Schneider 1999], one case of which is illustrated in Figure 7.2. The PSD template approximates the actual transmitted PSD of the system it represents, and crosstalk is modeled as arising from the PSD template. The PSD mask is usually defined as PSD template plus 3.5 dB (to account for rounding of the number of bits per subchannel in a discrete multitone (DMT) system and to account for transmitter filter ripple; see chapter 7 of [Golden 2006]). The PSD mask sets a hard upper limit on transmitter PSDs but is relatively loose. To offset this looseness, a sliding-window, normalized power constraint must also be satisfied. The sliding-window requirement is that $10 \log_{10}$ of the sum of the ratio of the measured power (in megawatt) divided by the PSD template (in megawatt) cannot exceed 1 dB, in any 100 kHz sliding window. Some classes and technologies have slightly different PSD conformance criteria. For example, HDSL2 and SHDSL need only comply with a hard PSD mask that is about 1 dB above the PSD template. (HDSL and SHDSL are based on single-carrier modulation and do not need the 2.5 dB of headroom for rounding. See Chapter 17 on standardization and chapter 6 of [Golden 2006].)

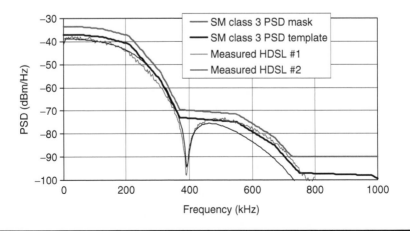

Figure 7.2   Spectrum management class 3 (SM3) power spectral density (PSD) mask, PSD template, and two measured High Bit-Rate Digital Subscriber Line (HDSL) transmit PSDs. Different PSD masks and templates are defined for each SM class and are illustrated in the T1.417 standard [ANSI T1.417] and in the help section of [Kerpez 2001].

Section 6 of T1.417 defines PSD template conformance criteria, and Annex M explains why the specified sliding-window constraint is used. It is based on equivalent noise, which was derived using the fact that channel capacity is directly proportional to received signal power in decibels, and inversely proportional to received noise power in decibels [Schneider 1999].

The total average transmit power across the entire bandwidth of a system is also limited. This limit is often a decibel or so below the total average transmit power of the PSD template, which means the template cannot be entirely filled. Compliance with the T1.417 has generally been demonstrated analytically so far, but complete lab measurement procedures are defined in its section 6 [ANSI T1.417].

7.3.3.3 Deployment Guidelines

Some technologies and classes are spectrally compatible only within a certain radius from a CO. For example, HDSL, HDSL2, and SHDSL have wideband upstream spectra that create NEXT which can debilitate downstream ADSL on long loops. Telephone loops often have unterminated sections attached to them between the normal endpoints, called bridged taps. The working loop length (the length of all loop sections excluding bridged tap) on which a crosstalker transmits is sometimes limited so that it may not disturb the highly attenuated signals of basis systems on very long loops. This length, rounded to the nearest 500 ft, is a deployment guideline. The bridged taps are ignored, because what matters is the loop of a basis system that receives crosstalk, which usually has about the same working length as the loop of the crosstalker, but not the same bridged tap. There are other deployment guidelines, such as not allowing reverse ADSL (i.e., an ADSL system that transmits upstream in the "downstream" spectrum and downstream in the "upstream" spectrum).

Loop length deployment guidelines are expressed in terms of the equivalent working length (EWL). In T1.417-2003,

$$\text{EWL} = L_{26} + 0.75 \cdot L_{24} + 0.60 \cdot L_{22} + 0.40 \cdot L_{19}$$

where L_{26}, L_{24}, L_{22}, and L_{19} are the working lengths of 26-, 24-, 22-, and 19-gauge cable in the loop. The attenuation of a loop's working length approximately equals that of a pure 26 gauge loop of length EWL.

7.3.3.4 Time Domain Requirements

By definition, a PSD is a long-term time average. Start-up signals and instantaneous transmit voltages are not explicitly limited by T1.417. Crosstalk samples are usually well-approximated by a Gaussian probability distribution [Kerpez 1993], and this is assumed but not mandated in T1.417.

There are some systems that transmit bursts of data and are quiet in between. These are known as short-term stationary (STS) systems, and their PSDs are measured while the transmitters are continuously transmitting. There are some requirements in T1.417 for STS systems: intentional synchronization is not allowed, the minimum duration of each burst is 246 μs, and systems must transmit at least 1 percent of the time in any 4 s. At start-up, ADSL modems typically measure noise for 4 s, and the 1 percent requirement helps enable adaptation to STS crosstalk.

A DSL system may initialize while STS crosstalkers are off, and then the DSL may be mis-adapted while the STS crosstalkers are on, potentially causing many decibels of degradation compared to stationary crosstalk. However, STS vendors have presented extensive results showing that STS crosstalk only causes minor degradations. STS crosstalk may be modeled with many different assumptions about traffic and victim systems' adaptation mechanisms, and the impact of STS crosstalk is still under study. Recent findings [Kerpez 2001] show that typical STS crosstalk may cause a few decibels more degradation than stationary crosstalk, and a single STS disturber appears to have worse effect than many STS disturbers, which is the opposite of stationary crosstalk. Several STS disturbers tend to have less impact because of the assumption that they are not synchronized, which means that the crosstalk they cause collectively is distributed in time and with less amplitude variation than would result if they were synchronized.

7.3.3.5 Other Requirements

In addition to PSD conformance, T1.417-compliant systems must meet defined limits of transverse balance and longitudinal output voltage. The frequencies for which these requirements apply are defined for each class or technology in section 5, with testing methodology in section 6 of T1.417. Transverse balance and longitudinal output voltage requirements ensure that the signal transmitted on a pair is balanced, so that the metallic voltage between the two conductors of the pair only weakly couples into longitudinal voltages from a conductor to ground. Transverse balance is a ratio relative to the transmitted metallic signal, and longitudinal output voltage is an absolute measure.

7.3.3.6 Three Methods of SM Standard Compliance

There are three methods of complying with T1.417:

- Belong to one of nine spectrum management (SM) classes.
- Satisfy technology-specific guidelines.
- Pass all analytical evaluations defined in Annex A.

Table 7.1 T1.417 Spectrum Management (SM) Classes, and Spectrally Compatible Technology Specific Guidelines. Non-Loaded Loops Have No Load Coils

SM Class or Technology	Deployment Guideline EWL 26-Gauge (kft)	SM Class Members
SM class 1 (SM1)	Any non-loaded loop	ISDN, SDSL ≤ 300 kbit/s two- and four-line pair gain
SM class 2 (SM2)	11.5	SDSL ≤ 520 kbit/s
SM class 3 (SM3)	9	HDSL, SDSL ≤ 784 kbit/s
SM class 4 (SM4)	10.5	HDSL2
SM class 5 (SM5)	Any non-loaded loop	Nonoverlapped ADSL
SM class 6 (SM6)	Not defined	VDSL
SM class 7 (SM7)	6.5	SDSL ≤ 1568 kbit/s
SM class 8 (SM8)	7.5	SDSL ≤ 1168 kbit/s
SM class 9 (SM9)	13.5	Overlapped ADSL
Technology:		
2B1Q SDSL	20 different, vary with bit rate	2B1Q SDSL ≤ 2320 kbit/s
SHDSL	19 different, vary with bit rate	SHDSL ≤ 2320 kbit/s
HDSL4	Any non-loaded loop	TC-PAM 776/784 kbit/s asymmetric PSD

The first two methods are very similar, involving conformance to predefined PSD templates, and are summarized in Table 7.1. The requirements for SDSL and SHDSL have PSD templates and deployment guidelines that vary with transmitted bit rate. It is important to reiterate that the deployment guidelines in Table 7.1 only ensure that crosstalk impact will be acceptable, and the SM class or technology will function within loop lengths only in rough correspondence.

Annex A compliance, also known as "method B," involves running a well-defined set of computations to analytically demonstrate spectral compatibility with each basis system (details of the computations are discussed in Section A.7). Annex A conformance can be difficult to verify. Members of the standards committee worked to ensure that all calculation parameters were fully specified and generated repeatable results, but people who have not gone through this process could get different results. There is a publicly available tool, at http://net3.agreenhouse.com [Telcordia], that can perform all Annex A computations, but this or other software must be properly used to demonstrate compliance, and other requirements must also be satisfied.

Annex A assumes full binders (i.e., all loops have active DSLs), and 1 percent worst-case crosstalk. Typically a reference crosstalker type is defined (for example, SM3 is the reference for compatibility with HDSL) and two crosstalk scenarios must be simulated: (1) 49 or 24 "new technology" crosstalkers, and (2) 24 or 12 "new technology" crosstalkers plus 24 or 12 reference crosstalkers. Signal-to-noise ratio (SNR) margins with 49 or 24 "new technology" crosstalkers must be within "delta" (decibel) (typically 0 to 1 dB) of the margins with 49 or 24 reference crosstalkers. If the evaluation initially fails, then loop lengths of both the crosstalkers and the basis system may be decreased until it passes. Evaluations are performed separately for each basis system, and the deployment guideline is the minimum loop length for which all evaluations pass.

Using a DSL's exact PSD in Annex A calculations will often allow a longer deployment guideline (maybe 1 kft) than using its PSD template, because the exact PSD is often below the PSD template that it conforms with.

7.4 Spectrum Management in Europe

This section describes European DSL spectrum management in general terms.*

Historically, each country in Europe has had a monopoly operator, which was originally part of the country's post office and usually a part of the government. Over the last 20 years, governments have tended to sell off their operators, but until recently the operators were the only operators using the loops in their networks. Then, in about 1998, the European Commission mandated local loop unbundling (LLU) in Europe. The crucial difference between LLU and loop unbundling in the United States is that LLU happened after the monopoly operators had begun installing their own DSL. As a result, it was typical that only a small set of DSL types had been deployed in each network. Furthermore, the operators had gained some experience in managing interference between the different DSL types. Of course, all the populations of DSL were different, and the management regimes differed widely, too, but the basic concepts were in place.

7.4.1 Regulators

In response to the LLU directive, each European government formed its own telecommunication regulator. Each regulator is mostly concerned with commercial issues (such as that competition shall be fair), but all are aware that LLU can only encourage trade insofar as their national network is actually useful. The regulators have the authority to act, and they will stop objectionable or illegal behavior, such as cartel formation and network

---

* Europe is diverse, and there will be individual exceptions to the generalities here.

pollution. Typically, the regulators delegate technical issues to a committee of their industry players.

The committee considers their (several) future aspirations and the network inherited from the former monopoly operator, and seeks a consensus of how the network is to be used in future. Where consensus is possible, a regulator will usually accept it.

7.4.2 Rules

Because of the historical development, each country in Europe has its own spectrum management rules, seeking to exploit the inherited network and its inherited systems. Often the rules aim for conservation, accepting those DSLs present in the network and seeking to exclude interferers not already in the network.

Because a country gets value from the absence of those systems it did not inherit, trying to impose any single regime over the whole of Europe would be harmful to the technical performance of every country's network. It is expected that the diversity of regimes will continue, and be defended rigorously, despite any paper benefits of a "harmonized" regime.

It should be noted that the LLU directive was motivated by commercial considerations, and in the interests of free trade almost all limitations on connections to the network are forbidden. Rules forbidding connection are only acceptable if aimed at protecting the network.

7.4.3 Standards

Formally, there are no European standards on spectrum management. A "technical report" has been developed in ETSI (the European Telecommunications Standards Institute), and at present is aimed at assisting the technical construction of spectrum management plans without coercing any particular construction method. The standard [ETSI TR 101 830] is developed in three parts that were produced separately.

Part 1, entitled "Definitions and signal library," defines a common set of terms and the relevant properties of known systems. It includes a discussion of formal reference models and the features that may be found in a network (although real networks need not have all the features). This document was published in 2002 and was revised to include ADSL2plus in 2005.

Part 2, entitled "Technical methods for performance evaluations," is concerned with the estimation of the impact of crosstalk from arbitrary populations of interferers. Its principal use is in estimating harm simulations, these are the technical basis for comparing options when constructing a particular frequency plan. It was published in 2003. Publicly available software exists that covers much of the same subject and implements the same algorithms [FTW].

Part 3 has not yet been published. It is entitled "Construction methods for spectral management rules," and was originally envisaged by some ETSI participants as a means of justifying to a regulator why a particular approach is respectable. Perhaps its time has passed, given that the regulators tend to accept the committee consensuses. If Part 3 is ever published, it is almost certain that every country will want it to endorse its own methods.

One significant absence in the ETSI technical report is any definition of "spectrally compatible" systems. The definition is left as a value judgment for the industry committees as they debate in the process of reaching consensus.

7.4.4   British Spectrum Management

In the United Kingdom, spectrum management is effected by specifying spectral masks that apply to both ends of the lines. These are prescribed in the* ANFP [NICC ND1602].

The following sections describe the ANFP rationale, construction method, and provide an outline of the masks. The extension for subloop use (ANFPS) will also be described.

The ANFP is presently undergoing development, and the anticipated changes are discussed at the end.

7.4.5   Need for Spectrum Management in the United Kingdom

British Telecom (BT) had recognized its need for spectrum management while it was a monopoly service provider: like any big company, its divisions had sometimes differing agendas, and when that difference led to incompatible deployment desires, there was a need to resolve a way forward. For some years, this resolution was based on analysis of mutual harm amongst all existing system types and a proposed newcomer, similar to the methods in the American standard, T1.417.[†] In about 1994, a less laborious method was developed, which was a forerunner of the

---

* Actually, there are two versions of the specification. [NICC ND1602] is properly entitled "Specification of the Access Network Frequency Plan applicable to transmission systems used on the BT Access Network," but British Telecom (BT) is one of two incumbent operators. For historical reasons the city of Kingston upon Hull is served by its own private telephone company, and it has an analogous document that uses the same construction but is different in detail because the Hull network is different in detail.

[†] Although the approach was similar to the approach in T1.417, it was wider ranging, it involved a search over all mixes of neighbors to find the most harmful one.

ANFP. In about 1998, the European Commission mandated LLU in Europe, and suddenly what had been a monopoly's need to manage its resources became a matter for public policy.

United Kingdom's need for spectrum management is the same as any other country's: efficient use of the telephone access network as a shared resource. Unchanged is the need to resolve conflicts of interest, and explain to those who are displeased with the resolution why the decided way forward is reasonable. LLU increases the stress on the process, in that an error would be harder to rectify. (Within one company, the process benefited significantly from the inherent goodwill of the participants.)

7.4.6 Access Network Frequency Plan

The ANFP is the public policy document that defines the spectrum usage rules in the United Kingdom.

7.4.6.1 Authority

The document is owned and authorized by the UK regulator, Ofcom,* and is considered stable. It was produced by a cross-industry committee, acting to advise Oftel. When the committee agreed, Oftel accepted its recommendations; when it could not agree, the committee proposed choices from which Oftel chose.

7.4.6.2 Scope

The ANFP applies to all connections to BT's telephone access network, whether or not the connection is made by BT. All connections are subject to the same ANFP limits.

The present ANFP only permits use of the spectrum up to 1.1 MHz, which is the spectrum of interest to all technologies except ADSL2plus and VDSL. The ANFPS [NICC ND1605] is an extension to enable VDSL trials on subloops (i.e., using FTTCab).

7.4.6.3 Controlled Interfaces

The ANFP identifies those interfaces where connection to the shared cables is permitted, and for each such connection point it specifies a single PSD mask. The ANFP limit is that the spectral power flowing through the interface toward the cables shall not exceed the mask in any frequency band.

---

* Ofcom was formerly known as Oftel.

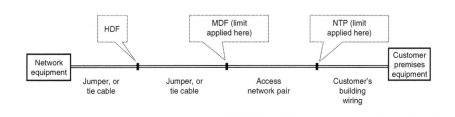

Figure 7.3   Network Interfaces to which the access network frequency plan (ANFP) applies.

The same limit applies irrespective of who is using the interface, and for what purpose.

These interfaces are the main distribution frame (MDF)* at the exchange[†] and the network termination point (NTP)[‡] at the customer's premises. The MDF is not a legal demarcation point, so nonBT connections also have a third interface, the handover distribution frame (HDF).[§] The ANFP does not directly address the HDF, and managing crosstalk interference in the cabling between the HDF and the MDF is the responsibility of the network operator(s) using that cabling. The ANFP is applied at the MDF and at the NTP.

The ANFP treats each end of the loop separately, based on where that end is, any other terminations do not change the approach. As it happens, the lines in BT's access network have two ends, faults excepted[¶], typically, one end is at the exchange and the other is at the customer's premises, as shown in Figure 7.3.

The ANFP is only concerned with possible impact on other lines in a cable, it does not explicitly consider what sort of system may be connected to the interface. The "spectral power flowing" could be equally well from, for example, a single modem connected directly to the interface, several instruments sharing the line via splitter filters, several instruments sharing the line via house wiring, or noise picked up via house wiring. The ANFP

---

* The main distribution frame (MDF) is the connection equipment with which BT terminates each of its access network cables at the switched network end.

[†] "Exchange" is an obsolete term, here meaning the premises containing the MDF. Once upon a time these were all exchanges in the sense of containing switching equipment, but now often they only contain a remote multiplexer. (However, the sign on the door still often says "telephone exchange.")

[‡] The network termination point (NTP) is the legal demarcation between the network provider's cabling and the customer's in-house wiring. On a telephone line this point often has a master socket or NTE (network termination equipment).

[§] The "handover distribution frame" is the equipment that terminates the tie cables.

[¶] A bridged tap would be considered a fault. Also, formally, one could claim that a broken line is two one-ended lines. The ANFP still prescribes each end its same mask and the same limits apply even under fault conditions.

Table 7.2 Access Network Frequency Plant (ANFP) Classes

Mask Name	Defines Transmitted PSD Permitted at
Down exchange	MDF of the exchange
Up extra short	NTP of near customers
Up short	NTP of near customers
Up medium	NTP of mid-distance customers
Up long	NTP of far customers

applies equally to signals intentionally injected into a line with intent to communicate over it and to noise leaked into it.

7.4.6.4 Masks

Under the ANFP, an NTP falls into one of five classes, one downstream and four upstream, that are broadly based on the electrical length of the line. This classification was initially because of BT's historical deployment of HDSL systems. The classes are "extra short," "short," "medium," and "long." BT has declared the classification for each NTP in its network [IBTE 2001].*
Together with the MDF interface, there are five distinct classes of interface, and the ANFP prescribes one mask for each of them. The classes are detailed in Table 7.2.

Each mask is defined numerically by a table of values[†] that specify corners of a polygon (on axes of log frequency versus the PSD in decibels). Figure 7.4 illustrates the ANFP masks.

7.4.6.5 Deployment Rules

One view of spectrum management is that it exists to allow performance to be predicted. However, the ANFP only permits connection to the loop plant, it does not provide performance guarantees. The risk is left to the system operator, who must define his own deployment rules, for example, by specifying a longest line on which a particular service can be deployed. The particular deployment rules an operator chooses will depend on many factors, including, which modems they are using, which market they are pursuing, how they handle their customers, and their own attitude to risk. The deployment rules are private property, and they are usually kept confidential.

---

* This classification was based on the best information at hand at the time of the declaration, there is a process to allow changes, perhaps owing to new cable installation, or perhaps owing to discovering errors in the original information.

[†] The values are not reproduced here. The reader is referred to the latest published ANFP, which is definitive (is currently [NICC ND1602]).

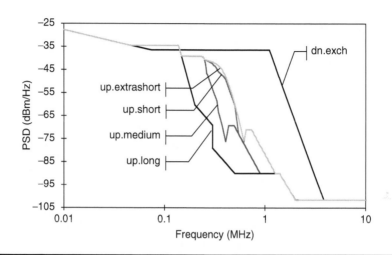

Figure 7.4   The ANFP masks presented graphically.

7.4.7   ANFP Construction

A theoretical treatment of DSL spectral compatibility might start by considering the space of all possible lines and all possible DSL systems using them, and seek to characterize the "attainable region,"[*] i.e., the points in the space where all the DSL systems work. This approach would be ambitious: even in the less ambitious space of line models and idealized DSL systems, the attainable region has complicated boundaries. In practice, simplification has been sought to identify a useful portion of the attainable region.

In the ANFP construction, the simplifying approximation is that if any of a given system type may be present at a particular place then any number of that system may be present. In terms of the Unger models (see chapter 3 of [Golden 2006]), this is equivalent to approximating $\left(\frac{N}{49}\right)^{0.6} \approx 1$.

When multiple system types are allowed, crosstalk is assumed to be from a 100 percent fill of a hypothetical system type with a PSD that is the envelope of all the PSDs potentially present. This approximation is conservative, no real mix can be worse. It is, however, surprisingly close to reality, in a full cable, the approximation typically overestimates harm by less than 2 dB [France 2004]. Furthermore, the approximation buys great simplicity.

This underlies the superficially simple procedure by which the ANFP masks were constructed:

---

[*] The attainable region was first discussed by [Shannon 1948].

1. Network interfaces were grouped into clusters of mutual neighbors. Ideally, the cluster would be the line ends that share a common entry point to a shared cable.
2. System types to be admitted a priori were identified.
3. For each given system type, the historical deployment rule was identified.
4. Taking the deployment rules together, the network interface clusters were categorized. The category indicates which ends of given systems might be installed on at least one line end in the cluster.
5. For each end of each given system type a spectral mask was obtained.
6. For each category, a composite mask was produced. The composite mask is the envelope of the spectral masks of all system ends that might be found installed at the cluster.

Note that this construction allows each network interface to be assigned a single particular mask. The simplicity and transparency of this process enabled an informed cross-industry debate, and regulator buy-in to it [OFTEL 1999], [OFTEL 2000].

The construction process is discussed in more detail in [France 2004] and in the public documentation with [NICC ND1602].

7.4.7.1 Post-Construction Check

The systems used to construct the ANFP included the bulk of the system types BT had already installed—it was considered* too expensive (and too time-consuming) to replace any of them. The same economic argument that induces acceptance of the existing deployment of systems should also, logically, motivate protection of the existing systems, as disabling them is also too expensive. It happens that BT's existing deployment[†] remains viable under the United Kingdom ANFP, a fortunate inheritance from the conservative nature of BT's deployment history.

This was not an automatic outcome: the European directive on LLU strives to increase deployment of more systems, and if it is successful then networks all over Europe will become more heavily populated. In turn, existing systems will get more live neighbors, and a louder noise environment. It is not automatic that an existing system can survive this noise increase. This reasoning is independent of how any particular spectrum management rules are written, so finding this problem while attempting to construct the rules does not necessarily invalidate the method.

---

* However, replacing systems was considered, particularly, for the spectrally incontinent HDSL systems.

[†] This includes those systems properly deployed under BT's deployment rules. Any system deployed outside these rules may well fail in the future.

Early in the development of the ANFP, it appeared that BT's deployment of HDSL systems would be compromised: the longest lines of HDSL as BT had deployed it would not survive in a cable full of systems that fully populated the spectral mask as the HDSL standard defines it. This particular problem was found to be in the generosity of the masks in the standard: the masks were deliberately set high so that real systems would pass that test unless a genuine fault occurred. However, real systems did not fully exploit the mask. The ANFP subsequently used a tighter mask (which still admits HDSL as actually implemented).

There may generally be a design decision in the step "identify which systems are deployed in significant numbers," because there is a potential trade-off between the systems admitted and the deployments that will work. For example, in some other network where there are old-style PCM systems* (perhaps G.703 carrying T1 or E1), there would be a very poor noise environment for more modern DSL systems. To develop an analog of the ANFP for such a network, one possible way forward would be "grandfathering," i.e., to consider the system types obsolete and exclude them from the plan, tolerate the existing systems unless they cause actual harm, remove or modernize them if they do, but do not permit them to be replaced, repaired, or installed anew.

7.4.7.2   Why Spectral Masks Only

Section 7.2.2 discussed why PSD limits are appropriate in spectral management. In principle, one could introduce other limits, too, such as a limit on the total power sent to the line. This was not done in the ANFP principally because the worst case noise environment would not improve significantly with such limits, and supernumerary limits might exclude future uses of a line for no good reason. Such future uses include sharing a line between two systems via splitter filters (for example, with POTS and ADSL on one line).

7.4.8   Current Developments

In the United Kingdom, any signal injected into the access network needs a frequency plan document to permit it. To support VDSL experiments on the subloop, a trial use plan called the ANFPS [NICC ND1605] was produced. At that time, the principal use anticipated for the subloop was VDSL, and it was expected that "from the exchange" uses and "from the cabinet" uses would

---

* BT has none in its access network. At one time it had a lot of these systems, mostly deployed in the "junction" network that links telephone exchanges. Those few systems deployed from exchange to customer always had their own separate cables, and they are not considered a part of the access network. Nowadays, such needs are met by optical fiber systems.

use frequency bands that did not overlap. The eventual incorporation of VDSL into the main ANFP was expected to be simple.

Since then, it has been desired to use ADSL2plus from the exchange, and the ADSL2plus does overlap the VDSL spectrum. Furthermore, the desired variant of VDSL has also changed. In the revised ANFP, it seems likely that the categorization will use three lengths: the distance from the exchange to the drop wire terminal, affecting ADSL band uses as at present; the distance from the cabinet to the drop wire terminal, affecting VDSL upstream power back-off (UPBO, see Chapter 17 on standardization); and the distance from the exchange to the cabinet ("cabinet assigned loss"), controlling VDSL downstream signals to protect ADSL2plus. Furthermore, some of BT's cables go directly from the exchange to the customer without passing a cabinet, it is being considered whether VDSL can safely be allowed on these cables.

7.5 Summary

DSL spectrum management requires knowledge of all DSL types and how the crosstalk they cause affects other systems. Spectrum management also requires defining the degree of degradation from crosstalk that is allowed by a spectrally compatible system. The scientific knowledge couples with the agreed rules in a process that should allow new DSL rollouts, competition, and new services, while minimizing conflicts and maintaining vital legacy services.

Spectrum management is often mentally associated with LLU, but it is more correctly associated with DSL. Even monopoly telephone companies need spectrum management to ensure successful deployments of DSL, particularly, if they deploy multiple DSL types from different manufacturers.

In the United States, committee T1E1.4 has issued the spectrum management standard T1.417-2003 [ANSI T1.417], which contains a detailed definition of spectral compatibility that had broad industry consensus. This standard was forged at a time of much competition in the local loop, and it represents a compromise between incumbent and competitive carriers that was forged over several years by committee. T1.417 evolved from an original specification of a small set of "composite PSD masks" to its current, somewhat complicated, form. T1.417 does not specify which systems may be admitted into the loop plant, but rather only states levels of allowable impact on each of the defined "basis systems." There are different methods of complying with T1.417, including the calculations-based "method B," which can allow great flexibility but is difficult to verify. T1.417 is not strictly mandated by the FCC, and compliance is not generally strictly measured, but is instead more conceptual and theoretical. Nonetheless, operators in the United States often mandate conformance with T1.417, because the alternative is an uncontrolled and error-prone access network.

In Europe, each country has its own regulator and its own spectrum management rules. Typically, the rules are developed by considering the current needs and future aspirations of the industry and the network inherited from the former monopoly operator. Often an industry consensus is reached, and a recommendation is made to the regulator. Because of the disparate needs of European countries, there is no European standard equivalent to T1.417, and no expectation there will ever be one. A technical report has been developed, aimed at assisting the construction of spectrum management plans, but the report is not binding on any particular regime.

In the United Kingdom, the spectrum management regime is a relatively straightforward definition of hard PSD mask limits, and compliance is more strictly defined than in T1.417. The strictness is believed to give improved operation of those systems admitted, at the cost of denying those systems not admitted. The clarity of the relationship between the regime limits and identifiable harm proved to be an advantage in the committee discussions that developed the ANFP.

In the future, DSM holds the possibility of increased bit rates and reliability, by treating crosstalk as man-made interference that can be measured, understood, and mitigated with advanced DSL management systems and multi-user transceiver techniques.

References

[France 2004]   P. France, Ed., *Local Access Network Technologies*. IEE Telecommunications Series 47, book published in July 2004 by the Institution of Engineering and Technology. ISBN 0852961766. Chapter 4 of [France 2004], *DSL Spectrum Management—the UK approach*, by J.W. Cook, R.H. Kirkby, and K.T. Foster, pp. 37–82, is particularly relevant here.

[Galli 2002]   *Methods of Summing Crosstalk from Mixed Sources—Part I: Theoretical Analysis. IEEE Transactions on Communications*, pp. 453–461, Vol. 50, No. 3, March 2002.

[Golden 2006]   P. Golden, H. Dedieu, and K. S. Jacobsen, Eds., *Fundamentals of DSL Technology*. Auerbach Publications, 2006, p. 457.

[Kerpez 1993]   *Near End Crosstalk is Almost Gaussian. IEEE Transactions on Communications*, pp. 670–672, Vol. 41, May 1993.

[Kerpez 1997]   *Generic Approach and Common Specifications of Transmitter Power Spectral Density Templates for Twisted-Pair Loop Transmission Systems*. T1E1.4 Standards Contribution T1E1.4/97-294, September 1997.

[Kerpez 2001] *Simulations of Short-Term Stationary (STS) Traffic Creating Crosstalk into ADSL.* T1E1.4/2001-279, November 8, 2001.

[Kirkby 1995] R. Kirkby, *FEXT Is Not Reciprocal.* ANSI T1E1.4/95-141, November 1995 [Orlando meeting]. (Idea owing to John Cook, of BT Laboratories).

[Russell 1999] *Proposed Text Clarifying the Use of Templates and Masks.* T1E1.4/99-220, April 20–23, 1999.

[Schneider 1999] *Sufficient Conditions for a PSD Template to Represent PSDs of a Spectrum Management Class.* T1E1.4/99-141, March 8–11, 1999.

[Shannon 1948] C.E. Shannon, *A Mathematical Theory of Communication. Bell System Technical Journal*, pp. 379–423, 623–656, Vol. 27, July, October, 1948.

[Valenti 2002] *NEXT and FEXT Models for Twisted-Pair North American Loop Plant. IEEE Journal on Selected Areas in Communication* (JSAC), pp. 893–900, Vol. 20, No. 5, June 2002.

[ANSI T1.417] ANSI NIPP Standard Document T1.417-2003, *Spectrum Management For Loop Transmission Systems.* September 2003.

[ETSI TR 101 830] ETSI TR 101 830, Rob F. F. van den Brink, Ed., *Transmission and Multiplexing (TM); Spectral Management on Metallic Access Networks.* Published in three parts. TR 101 830-1 *Part 1: Definitions and Signal Library.* TR 101 830-2 *Part 2: Technical Methods for Performance Evaluations.* TR 101 830-3 *Part 3: Construction Methods for Spectral Management Rules* under preparation.

[FTW] FTW, Telia, FSAN et al., *FTW xDSL Simulation Tool.* Available at http://xdsl.ftw.at/xdslsimu/

[IBTE 2001] *Journal of the IBTE*, Vol. 2, No. 1, January–March 2001.

[NICC ND1602] NICC ND1602:2002/11 Issue 2; November 2002 *Specification of the Access Network Frequency Plan Applicable to Transmission Systems Used on the BT Access Network.* Oftel Technical Requirement OTR004:2002 Issue 2 available at http://www.nicc.org.uk/nicc-public/publication.htm

[NICC ND1605] NICC ND1605:2001/10 *Specification of the Trial Access Network Frequency Plan Applicable to Transmission Systems Used on Subloops in the BT Access Network*, Issue 1, October 2001, available at http://www.nicc.org.uk/nicc-public/publication.htm

[OFTEL 1999] *Access to Bandwidth: Delivering Competition for the Information Age.* November 1999. May be downloaded from http://www.ofcom.org.uk/static/archive/oftel/publications/1999/consumer/a2b1199.htm

[OFTEL 2000] *Access to Bandwidth: Determination on the Access Network Frequency Plan (ANFP) for BT's Metallic Access Network.* October 2000. May be downloaded from www.ofcom.org.uk/static/archive/oftel/publications/broadband/llu/anfp1000.htm

[Telcordia] *The Telcordia DSL Spectral Compatibility Computer.* Available at http://net3.agreenhouse.com

A.7 Appendix on Spectral Compatibility Calculation

The U.S. spectrum management standard [ANSI T1.417] defines "spectral compatibility" in terms of the impact of one DSL system type on each of a set of others. This appendix discusses the details of one such impact calculation.

A.7.1 DSL Types

Spectrum management requires detailed knowledge of the PSDs of the different DSL types that may share the same cable. Sometimes it also requires descriptions of time-domain behavior. Table A.7.3 gives a brief listing of these different DSLs. There are also a number of proprietary DSL technologies which are not included in Table A.7.3.

A.7.2 Spectral Compatibility Calculations

The amount of degradation of a DSL's performance caused by crosstalk from another DSL can be accurately forecast by a computer simulation calculation. A statistically worst-case environment is simulated by using the 1 percent worst-case crosstalk coupling, highest probable number of crosstalk disturbers, and long loops. Then, the transmission performance (i.e., bit rate) is accurately calculated with computer programs. If the performance meets some target then nearly all deployed DSLs (at least 99 percent) will also meet that target. This is more efficient than lab testing many combinations of two DSL types in a cable until 99 percent have passed. Furthermore, there are many different DSL types which need to be crosschecked. Moreover, computer simulation is repeatable, and simulation parameters have been standardized so that results from multiple parties are in agreement. The simulation parameters have been calibrated with lab measurements.

Generally, there are three types of DSL performance calculations: total crosstalk power, single-carrier equalizer equations, and multi-carrier "water filling." Specific details of these calculations are in Annex A of the spectrum management standard [ANSI T1.417]. Spectral compatibility with POTS and other narrowband services is calculated simply by determining if the total received crosstalk power is above or below a certain threshold.

Wideband single-carrier DSLs almost universally employ a receiver with a decision-feedback equalizer (DFE), and if trellis coded, then the feedback portion is implemented in the transmitter with a precoder. Spectral compatibility with single-carrier baseband PAM and passband QAM DSLs

Table A.7.3   Common American Broadband Copper Loop Transmission Systems

Acronym	Description	Standard(s)	Modulation	Number of Pairs	Line Bit Rate	Approximate Passband Frequencies
ADSL	Asymmetric DSL	ANSI T1.413, ITU-T G.992.1, ITU-T G.992.3 (ADSL2)	Discrete multi-tone (DMT)	one	Up to ~1 Mbit/s up-stream, up to ~8 Mbit/s down-stream	25–138 kHz up-stream, 25–1104 kHz down-stream
ADSL2+	Extended band-width Asym-metric DSL	ANSI T1.413, ITU G.992.5	Discrete multi-tone (DMT)	one	Up to ~1 Mbit/s up-stream, up to ~20 Mbit/s down-stream	25–138 kHz up-stream, 25–2208 kHz down-stream
ISDN, BRI, or BA	Integrated Services Digital Network (ISDN) basic-rate (BRI), or basis access (BA)	ANSI T1.601, ITU G.961	2B1Q	one	160 kbit/s symmet-ric	0–80 kHz
HDSL	High Bit-Rate DSL	ITU-T G.991.1, ETSI TS 101 135, ANSI T1.TR.28,	2B1Q	two	1.544 Mbit/s symmet-ric	0–370 kHz
HDSL2	High Bit-Rate DSL, 2nd genera-tion	ANSI T1.418, ITU-T G.991.2	16-level Trellis coded (TC) PAM	one	1.544 Mbit/s symmet-ric	0–300 kHz up-stream, 0–440 kHz down-stream

is determined by calculating the SNR at the output of the DFE with the modified Wiener–Hopf equations derived by Salz. The SNR margin is the computed SNR minus the SNR required for 7–10 bit-error rate (BER). A positive SNR margin, usually 6 dB or more, is needed to ensure reliable service with unknown impairments and temperature variations [ANSI T1.417].

ADSL, G.lite, and ANSI-standard VDSL are modulated with DMT modulation, which transmits up to 256 orthogonal QAM sub-channels at tones with 4.3125 kHz spacing up to 1104 kHz. Receivers generally incorporate a front-end pre-equalizer, and it is safe to assume that each sub-channel is flat across its narrow 4.3125 kHz bandwidth. The Shannon capacity of each tone is computed assuming 6 dB margin, 3 dB coding gain, and 9.75 dB SNR gap, then these are all summed to compute the system bit rate.

A.7.3   Example: HDSL Crosstalk into Downstream ADSL

Figure 7.5 shows an example of downstream ADSL transmitted on a 15-kft 26-gauge loop with two types of crosstalk: self-FEXT crosstalk from other ADSLs, and alien-NEXT crosstalk from HDSLs. The HDSLs would be repeatered on such a long loop. The upper left plot shows the transmit PSDs of HDSL and downstream ADSL. The lower left plot shows the power couplings (power transfer functions) of the 15-kft loop, and the 1 percent worst-case 24-disturber NEXT and FEXT for the 15-kft loop. Adding the ADSL transmit PSD to the loop coupling (in decibels) gives, the received ADSL PSD in the upper right plot. Adding the ADSL transmit PSD to the FEXT coupling gives the received FEXT PSD from ADSL, and adding the HDSL transmit PSD to the NEXT coupling gives, the received NEXT PSD from HDSL, in decibels. Subtracting the appropriate crosstalk PSD from the received ADSL signal PSD gives the lower right hand plot, the received SNR.

Downstream ADSL has a passband extending from about 138–1104 kHz. At a frequency in the ADSL passband, as the SNR increases more bits can be transmitted by using a constellation with more signal points at that ADSL frequency. The minimum constellation is quadrature phase-shift keying (QPSK), requiring at least about 17.5 dB to be useful [ANSI T1.417]. The bottom right plot in Figure 7.5 shows that few frequencies are at all useful with HDSL NEXT, whereas with self-FEXT all ADSL frequencies are useful. Adding −140 dBm/Hz background noise to the crosstalk, and performing the calculations in the spectrum management standard [ANSI T1.417], ADSL can transmit 1102 kbit/s downstream on the 15-kft 26-gauge loop with self-FEXT, but it can only transmit 96 kbit/s downstream with HDSL NEXT.

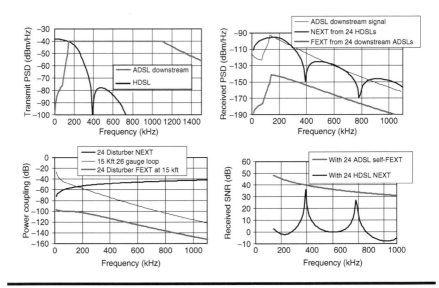

Figure 7.5   Downstream ADSL performance with self-FEXT from 24 other ADSL crosstalk disturbers, and with 24 HDSL NEXT disturbers on a 15 kft 26 gauge loop.

HDSL generates strong levels of crosstalk relative to the weakly by received ADSL signal on this loop. This results in poor ADSL performance compared to ADSL self-crosstalk, and so HDSL can be said to be incompatible with ADSL on such a long loop. HDSL is compatible with ADSL on loops no longer than about 9–12 kft [ANSI T1.417].

Chapter 8

Dynamic Spectrum Management

Seong Taek Chung

CONTENTS

Abstract   Dynamic spectrum management (DSM) dramatically improves the speed of Digital Subscriber Line (DSL) services. When the highest level of coordination is available, up to 500 Mbps can be realized for a single twister pair. With the medium or low level of coordination, the data rate can still be significantly improved over current data rate. In practice, the level of attainable coordination varies depending on each service provider's situation: deployment scenario, government regulation, etc. This chapter provides a guideline for each service provider so that 500 Mbps per user becomes viable in the near future.

8.1 Introduction

Dynamic spectrum management (DSM) provides an evolutionary path toward the goal of ubiquitous single-line Digital Subscriber Line (DSL) service that delivers 500 Mbit/s to each customer. This chapter provides an overview of various DSM technologies that offer a promising lead to this ultimate speed.

The performance* of any DSL system [Starr 1999] depends on the channel and the noise level at the receiver. The noise at the receiver is composed of additive white Gaussian noise (AWGN), impulse noise, any AM radio signals that couple into the twisted pair line, and crosstalk from other lines in the same physical cable. The capacity, which is the maximum data rate achievable in a given channel at a particular noise level (see chapter 4 of [Golden 2006]) is expressed as

$$C = \frac{1}{2} \log_2 (1 + \text{SNR}) \quad \text{bits/dimension.}$$

Here, SNR is the signal-to-noise ratio at the receiver. Therefore, to increase the data rate, either the transmit signal power must be increased or the noise from other sources must be decreased. DSM is an approach to improve the performance of all transceivers in the entire DSL system by dynamically managing the spectra of all DSLs such that the crosstalk from other lines is mitigated.

At present, the spectrum of any standardized DSL is controlled by an upper limit ([ANSI T1.417], method B). All the DSL standards clearly dictate the maximum power spectral density (PSD) that the DSL system is allowed to transmit. This limit, which is a function of frequency, is known as the PSD mask. As long as the PSD mask is not exceeded, DSL modems are allowed by standards to transmit any spectrum that they choose. There are two problems with this approach.

First, this approach does not always provide reasonable data rates for all the lines. To determine the static PSD mask for each standard, various loop topologies were considered. However, it is impossible to consider all possible deployment topologies in advance, and therefore the standardized PSD masks are suboptimal for at least some unforeseen deployment topologies. For example, when asymmetric DSL (ADSL) was first deployed, all the DSL access multiplexers (DSLAMs) were located in the central office (CO). However, as the number of ADSLs increased, service providers

---

* The performance of DSL systems can be characterized by any of the following metrics: the bit rate versus reach curve per line, the bit rate distribution among lines in the same physical cable, etc.

began installing remote terminals (RTs); hence, now quite a few DSLAMs are located in the RT as well as in the CO. The static spectrum mask approach significantly reduces the bit rates of lines from the CO because of the strong interference from the transmissions on lines emanating from the RT.

Second, the data rate achievable by static spectrum management is much lower than the optimal data rate* that the network can provide with a dynamic approach. This is due to the fact that any static spectrum must be designed assuming the worst-case deployment scenario; in other words, after considering various channel and noise conditions, the PSD that can guarantee the minimum data rate with the worst combination of channel and noise conditions should be selected. Therefore, the current design approach necessarily leads to DSL deployment that is overly restrictive in terms of both coverage and speed. Moreover, static spectrum management hurts other lines unintentionally; some lines (short line or low-noise lines) can achieve reasonably good data rates even though they do not transmit all the power. By transmitting at the full power level (as in static spectrum management based on the PSD mask), modems on these lines unnecessarily limit the data rate achievable on other lines. This phenomenon happens because the static spectrum management does not take into account the potential to optimize the data rate for all the users.

Suppose that 25 copper lines exist in a cable bundle. Depending on the time (day versus night, weekday versus weekend) and the location (commercial versus residential area), the demand for DSL services varies. For example, the data traffic load at an office is heavy during the daytime and may be close to zero at night. Furthermore, depending on specific loop topologies, the interference caused by one line to another varies. Some pairs might be physically close in the binder over a very short distance; hence, the interference between these pairs might be very small. On the other hand, other pairs might be proximate along most of the loop length, which leads to a very strong interference between the two pairs.

In static spectrum management, ignoring all these specifics, it is assumed that all 25 copper lines are always used at their full capacity, and that the interference path gains are always at the maximum level in all cases. This assumption is inevitable because only one (a "static") solution has been specified. In reality, it is very unlikely that all the lines are always used at their maximum bit rates, and it is impossible simply because of cable geometry that the interference path gains are all at the maximum level. It is possible that only one subscriber is using his or her DSL service while no other users are using the DSL service at the moment. Likewise, perhaps

---

* For multiple lines, no one scalar value can fully describe the trade-off among different lines; hence, instead of data rate, the data rate region (the region where achievable rate pairs are marked) is preferable for multiple lines.

the interference level between two active pairs is minimal because of their locations in the cable binder. Even in these cases, with a static spectrum management method, the bit rates on all loops are limited.

The limitation caused by static spectrum management is an especially serious problem for expanding DSL service range. The transmitted signal is attenuated as it propagates along the copper loop. Hence, to support customers in a wider area, either a stronger signal must be transmitted from the CO, or the loop noise must be reduced. Under a static spectrum management scenario, the transmitted signals are typically already at the upper limit because the standards presume the worst-case scenario. Reducing the interference is not feasible under static spectrum management because crosstalk does not decrease unless modems transmit at lower power levels, which is contrary to the objective of transmitting at the maximum allowed power level. Hence, using static spectrum management, the DSL service area is reduced, as is the revenue of service providers.

DSM [ANSI T1E1.4-R4] is an evolutionary step in DSL spectrum management strategies. DSM adaptively allocates the available resources among multiple lines, depending upon the channel characteristics and the crosstalk activity on the lines. By guaranteeing spectral compatibility among different services, DSM promises faster and more reliable information delivery.

Revisiting the example where 25 copper lines exist in a telephone bundle, suppose that in the middle of the night at school, one PhD student is trying to download a large paper from a website using DSL. If static spectrum management is used, the speed is as slow at night as during the daytime when many other students are also using their DSLs. However, if the DSL network uses DSM, the speed at night can be much faster than during the daytime because the bit rate on the line in use can be optimized to take advantage of the low-noise conditions.

Two principles guide the improvements resulting from DSM. First, DSL modems should not transmit more power than necessary to achieve their target data rates with good quality of service. Second, DSL modems should not use more bandwidth than necessary. When applied, these two simple principles allow for enormous rate and reach improvements. Because the spectrum is dynamically controlled by these two principles, DSM can provide a satisfactory solution for channel and noise environments that static spectrum management could not handle properly. At the time of this writing, DSM is an ongoing project of NIPP NAI [ASSIA Inc. 2006].

8.2 History of DSM

To understand the role of DSM at each development stage of DSL systems, it is useful to present the context in which DSM evolved.

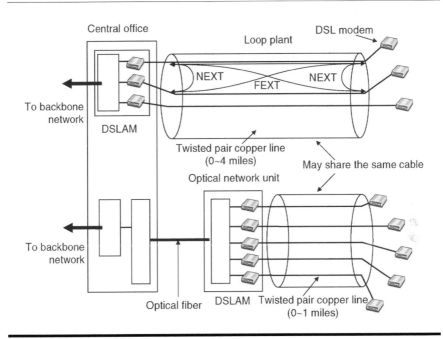

Figure 8.1   Digital subsriber line (DSL) environment.

As described in detail in [Golden 2006], DSL transmission occurs between the modems at the customer side (the customer premises equipment (CPE)) and the modems at the CO (or RT or equivalent), which are connected through twisted-pair copper lines. Several twisted pairs are physically bundled together in the same cable, resulting in electromagnetic coupling, which induces interference known as crosstalk in the neighboring lines. Near-end crosstalk (NEXT) refers to the interfering signals at the receiver originating from transmitters at the same side of the line as the affected receiver. Far-end crosstalk (FEXT) signals originate from transmitters at the other side of the line from the affected receiver. Both types of crosstalk are illustrated in Figure 8.1.

Because the spectral incompatibility among different DSL types (for example, ADSL and SHDSL, see Chapter 17) and among different service providers can aggravate the crosstalk problem, existing static spectrum management enforces overly conservative PSD constraints for all DSL services, thereby severely limiting the achievable performance. Clearly, more optimal spectral allocations can be determined. DSM, even in a non-coordinated situation (to be described), immediately achieves a remarkable improvement in both the DSL service coverage area and the transmission speed by enabling the DSL modems to "autonomously" adapt their own spectra without causing harm to the existing DSL systems.

The autonomous operation of modems on copper loops became a prominent issue in the United States in 1996, when the Telecommunications Act of 1996 required incumbent local exchange carriers (ILECs) to lease unbundled network elements (UNEs) to certified competitive local exchange carriers (CLECs). The UNEs include lines from the CO to the customer. Therefore, multiple service providers may share the copper lines in a telephone binder. For example, in Figure 8.1, lines operated by different service providers could reside in the same binder. In this "unbundled" scenario, the data rate can be poor, especially for the longer loops, when static spectrum is used.

On the other hand, the increasing investment in fiber-to-the-curb (FTTC) installations is gradually resulting in a topology where the optical fiber is deployed between the CO and an optical network unit (ONU), which is located closer to the customer premises. Then, twisted pairs emanate from the ONU to provide the "last mile" connection to the customer. The ONU equipment includes a DSLAM performing the appropriate multiplexing or demultiplexing operations between the fiber link and the DSLs. The corresponding inverse operations are done at the CO. Very High Bit-Rate DSL (VDSL) will mainly be deployed in such "fiber-assisted" environments (see Chapter 17).

The extension of optical fiber toward the customer premises implies higher DSL data rates because the shorter lines attenuate signal much less than the longer lines do. A potentially more important observation, however, is that with an ONU, a single DSLAM may be attached to all of the DSL modems. This change of paradigm allows the possibility of coordinating the DSL modems at the DSLAM, enabling additional significant data rate gains beyond those achievable with autonomous DSM [Ginis 2002]. The loop configuration suggested in the lower part of Figure 8.1 illustrates this scenario.

Another advantage of deploying DSL from an ONU or the like is that a completely coordinated DSL service is possible if all the lines from the ONU are owned by a single entity, which is the case in most countries.

8.3 Level of Coordination in DSM

There are four levels of coordination defined in DSM. All the levels inherently have one common theme in the sense that the power and rate allocations for each line are dynamically controlled. However, each method assumes different level of coordination among lines. Some schemes assume an autonomous balancing of line spectra to improve DSL service data rates, ranges, reliability, or symmetries. Other schemes assume coordination of line signals maintained by a single service provider. DSM is

Table 8.1   Dynamic Spectrum Management (DSM) Levels

DSM Level	Description
	No DSM
0	Completely autonomous spectrum management
	Single-line
1	Spectrum balancing
	Multiple-line
2	Spectrum balancing (spectra controls)
	Multiple-line vectored
3	Coordinated LT-side downstream transmission and upstream reception

Source: From J. Cioffi et al., *Vectored DSLs with DSM*. To appear IEEE WCNC, 2006.

categorized depending upon the level of coordination among DSLs. DSM is classified into four levels of coordination: level 0, level 1, level 2, and level 3. Table 8.1 summarizes the four levels.

Before explaining each coordination level in detail, it is useful to understand the DSM reference diagram. The basic reference diagram appears in Figure 8.2. The DSM-D interface carries data from vendor equipment through an element management system (EMS) for LTs (line terminals or DSLAMs) or an auto-configuration server (ACS) for NTs (network terminals or CPE modems). The DSM control interface, denoted DSM-C, carries commands usually known as "profile settings" from the management system (either the EMS or ACS) to the DSLs.

8.3.1   Level 0 or 1

Level 0 coordination means that fully autonomous methods are used for spectrum selection, and level 1 coordination means that autonomous spectrum determination is performed with data rate or power imposition. Most

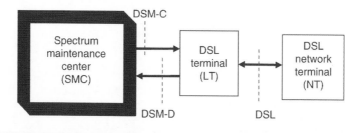

Figure 8.2   Dynamic spectrum management (DSM) reference model.

of the currently deployed DSL modems use level 0 coordination, although level 1 systems have been deployed in some networks. Because these two levels are closely related, they are explained together here.

Autonomous spectrum management is achieved using the technique known as "water filling" [Yu 2000] [Yu 2002]. Water filling is a well-known optimal power distribution algorithm for the single-user communication channel [Starr 1999] that provides the basis for the power and bit allocation schemes in most discrete multitone (DMT)-based modems.* The single-user water filling concept is illustrated in Figure 8.3a. Given the channel SNR in the frequency domain[†] (denoted as SNR(f)), the optimal spectrum, $S(f)$, that maximizes the data rate is obtained by allocating more power to frequency bands with higher channel SNR. The strategy is pictorially the same as "pouring" the total power in the bowl of the inverse SNR(f) curve, hence the name water filling.

Level 0 DSM is the case when each DSL modem does water filling independently. A power-minimizing (or margin-fixing) version of water filling is commonly deployed to practice DSM, in which case the target data rate is fixed and the objective is to minimize the total power used. This power minimization version is especially attractive because the modem only uses the amount of power necessary to support its own performance requirements and thus does not cause excessive interference to other lines.[‡] In addition to the margin-fixing schemes, both margin maximization and rate maximization schemes have been suggested [Starr 1999].

In each of these cases, the optimal power allocation is still as achieved by "pouring the power" into the inverse SNR(f) curve. However, the water level is dictated by the target data rate rather than by the total power constraint.

Currently, most modems perform bit swapping, which is a simplified version of water filling; bit swapping is a practical technique to implement water filling in a time-varying noise environment. When the SNR versus frequency changes, all the bits and powers allocated in the affected frequency

---

* DMT is a multi-carrier modulation method adopted in a number of existing DSL standards. In DMT, the information bits are mapped into a set of quadrature amplitude modulated (QAM) signals, where these QAM signals are transmitted through independent subchannels in the frequency domain. A detailed description of this method can be found in [Starr 1999] or in chapter 7 of [Golden 2006]. It is possible to perform DSM with a single-carrier modulation scheme, but the implementation of DSM is much easier for DMT schemes because the power per subchannel is controlled directly. (See chapter 6 of [Golden 2006] for details of single-carrier modulation.)

† In general, SNR is defined as the signal power divided by the noise power at a specific frequency or within a particular frequency band. Channel SNR refers to the SNR with unit signal power across the entire frequency band, which concisely characterizes the communication channel. For notational simplicity, SNR(f) denotes the channel SNR at frequency f.

‡ Note that in the power minimization version, the water level of Figure 8.1 might not always be constant over frequency.

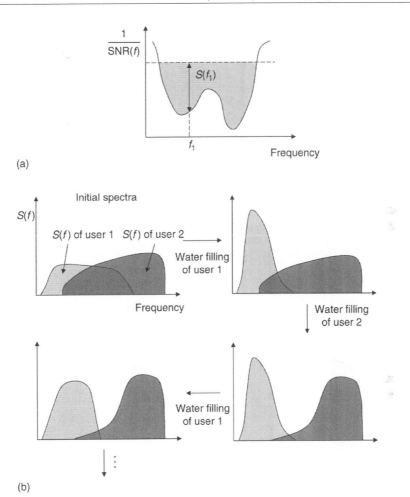

Figure 8.3 **(a) Water filling for single-user; (b) iterative water filling for two-users.**

band should be recalculated to achieve the optimal data rate. In most cases, the resulting bits and powers over frequency after the water filling are similar to the bits and powers over frequency before the noise environment changed; by swapping bits from the subchannels where SNR decreased to the subchannels where the SNR increased, the same results as water filling may be obtained. Practically, bit swapping is preferred to "real" water filling because the modems do not need to recalculate the bits and powers over all the subchannels. Instead, after identifying subchannels in which the SNR significantly changed, the bit swapping is performed over only the affected subchannels. (See chapter 7 of [Golden 2006] for details of bit swapping.)

Level 1 DSM is the case where all DSL modems use an algorithm called "iterative water filling" to allocate power. Iterative water filling is an extension of the water filling process for a multi-user communication environment [Yu 2000] [Yu 2002]. This algorithm is based on formulating power allocation in the multi-user interference network as a competitive game, in which each user tries to maximize its own data rate in the presence of crosstalk interference from other users as noise. Starting from any initial spectrum, the water filling procedure on each line is independently performed across all the DSLs within a binder. It can be experimentally shown that this distributed iteration process converges to a competitively optimal equilibrium point, sometimes called a Nash equilibrium in game theory. Although multiple Nash equilibrium points may exist in the multiuser interference channel, such a case has not yet been found over ADSLs tested, and indeed can be mathematically proven not to exist in frequency bands up to 2 MHz on copper lines.* Furthermore, the convergence and uniqueness of the iterative water filling solution from any initial spectra are guaranteed for ADSLs [Chung 2003a]. Power-minimizing iterative water filling is also possible, in which case the target data rate on each line must be provided to the modems in advance. Maximum data rates are established today even in static spectrum management by provisioning rules known as profiles; thus, the need for DSM to impose a target data rate maintains the existing autonomous nature of DSL deployment today.

For most ADSLs, the data rates achieved by iterative water filling are very close to those achieved by the optimal rate and power allocation [Chung 2002]. In [Chung 2002], various channel path gain scenarios (not just ADSL channel path gains) have been investigated to see under what channel conditions the power allocation from iterative water filling performs close to the optimal power allocation. It turns out that in the worst case, for certain channels, the data rate for one user from iterative water filling can be just one half of the data rate from the optimal scheme. However, in typical ADSL cases, the performance of iterative water filling is in fact very close to the optimal case. This result is highly encouraging; specifically, it implies that iterative water filling, in which the bit rate and power allocation for each line are adapted autonomously with no central control, performs essentially the same as when the spectra of all the lines are controlled by a centralized network entity. It should be noted that the total power (or rate) still needs to be determined and communicated by the centralized network entity.

Figure 8.3b illustrates the iterative water filling process for two lines. At each iteration step, each line's spectrum "moves away" from the frequency region characterized by strong interference owing to the water filling

* The detailed mathematical conditions can be found in [Chung 2003a].

process explained above, thereby achieving better performance step by step. For DMT modems, the iterative water filling algorithm can be summarized for "off-line" analysis as follows:

1. Initialize IterNum $= 0$
2. Repeat
   For $i = 1 \ldots L$
   (a) Compute water filling spectrum, S_i, for line i.
   (b) Compute crosstalk spectra into all other lines as

$$j \neq i, \quad \sigma_{j,n}^2 = \sum_{i \neq j} |H_{ij,n}|^2 \cdot S_{i,n} + \text{Thermal Noise}$$

   until IterNum $=$ max

Here, L is the number of DSLs, max is the maximum number of iterations allowed, $S_{i,n}$ is the i-th line's spectrum at the n-th subchannel, $\sigma_{j,n}^2$ is the crosstalk variance of the j-th line at the n-th subchannel, and $H_{ij,n}$ is the crosstalk transfer function from line i into line j at the n-th subchannel.

This procedure assumes that the water fillings occur in a specific order. However, it should be emphasized that for practical DSL channels, the same convergence result is obtained whether or not the water fillings occur in a specific order or simply successively. In other words, each modem only needs to autonomously execute its own water filling process.

Because each modem performs water filling via bit swapping, as long as the total rate (or power) is communicated to the modem at the CO, iterative water filling could be implemented in the network without requiring any modems to be replaced. However, the conservative rules often imposed by spectrum management standards would not allow the benefits of iterative water filling to be fully realized. If these rules were lifted, the benefits of iterative water filling would be much more pronounced. Furthermore, such lifting combined with iterative water filling would not lead to performance degradation on any existing line that continues to use static spectrum management. The iterative water filling would result in spectra that are no more "harmful" to existing lines than the ones they would use under the spectrum management standards. However, the more the lines use iterative water filling, the greater the advantage becomes. A "single" static line, for instance, a 998 VDSL system with a fixed PSD mask [ANSI T1.417], can unnecessarily limit the performance of all future lines. (See Chapter 17 for details of VDSL systems using band plan 998.)

DSM is permitted in the existing North American spectrum management standard [ANSI T1.417] under the definition of "new technology." This

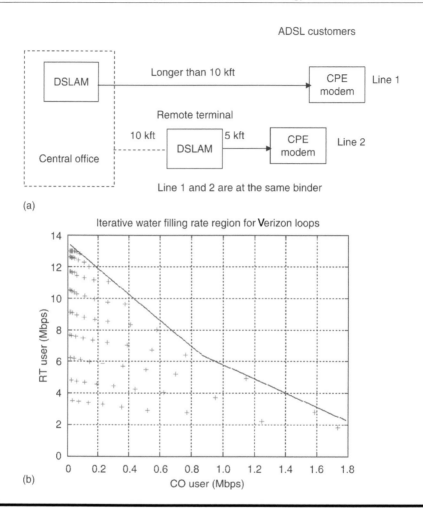

Figure 8.4 **(a) Asymmetric DSL (ADSL) transmission scenario; (b) achievable rate region by iterative water filling.**

accommodation essentially paves the way for future DSM use by allowing new systems to be deployed as long as they do not harm any existing systems more than those systems are allowed to harm themselves. As more DSM-enabled DSL are deployed, DSM will gradually "clean up" the loop plant in a graceful and consistently performance-improving manner. Service providers will be motivated to use DSM-enabled modems instead of static modems as they deploy more remote terminals and more DSLs because line maintenance will decrease and data rates will dramatically increase.

Figure 8.4 illustrates a classic field problem in current ADSL deployments, for which iterative water filling can achieve dramatic data rate increases. In Figure 8.4a, the ADSL receiver on line 1 experiences large FEXT from line 2 because of the downstream transmitter of an ADSL modem installed in the RT, which is only 5 kft away from the ADSL receiver on line 1. Under the existing static spectrum management rules, line 1 simply does not work; the achievable data rate is only 300 kbit/s in the downstream direction, and field personnel must be dispatched to investigate the reason for the poor data rate.

Crosstalk measurements for a typical configuration as in Figure 8.4a were reported by the Verizon Broadband Integration Lab [Cioffi 2002]. The worst-case crosstalk coupling pair is considered here to investigate the benefits of iterative water filling. With static rate-adaptive training, the Verizon Broadband Integration Lab measured downstream bit rates of 9 Mbit/s on the short line but only 100 kbit/s on the long line. Figure 8.4b shows the achievable rates when modems on both lines use power-minimizing (minimum power for a given rate) iterative water filling. For example, denoting the bit rate on the short line as A and the bit rate on the long line as B, rate pairs $(A, B) = (2, 1.8)$, $(4, 1.4)$, $(6, 0.9)$, and $(8, 0.6)$ Mbit/s are achievable. At the expense of reduced rates on the short line, enormous improvements are possible in the long line. The spectrum on the short line "migrates" to higher frequencies, yielding to the need for the long line to use the lower frequencies.

The iterative water filling algorithm can "optionally" be augmented by an outer control loop when centralized resource allocation is used, thus providing more flexible services to customers in the form of a broader range of data rate choices. Clearly, this would require a higher degree of coordination, and each modem would need to report the relevant DSM data, such as the achieved data rate, the noise margin, and the associated transmit power level, to the spectrum maintenance center (SMC). Based on this information, the SMC would determine the achievable data rate pairs, or, more generally, the achievable data region, based on which it can command each modem to perform iterative water filling with a more optimal transmit power level and target data rate.

Even this small amount of cooperation among the modems offers a great deal of flexibility. For example, upon the request of a customer or a higher level of network hierarchy, the SMC can command a modem operating with an excessively large noise margin to increase the data rate, thereby providing a higher-speed service, or to decrease its transmit power, which results in less crosstalk interference to other modems. This concept of flexible service provisioning is not feasible in the static spectrum management environment, under which each modem is forced to ensure its transmissions remain below the levels allowed by a fixed PSD mask, regardless of where the modem is located, and to "accept" the resulting data rate whether or not it is satisfactory.

Although the well-known water filling algorithm results in a bit distribution with continuous values, all practical DMT systems require that the number of bits assigned to each subchannel is an integer. (See chapter 7 of [Golden 2006].) Methods of discrete bit allocation have been developed in the past for single-user DMT systems [Starr 1999]. After a connection has been established, a bit swapping procedure is used to adapt the bit allocation to changes in the channel or noise that do not significantly degrade the channel capacity. However, the number of bits that can be swapped during each bit swap iteration is limited in ADSL1, the mechanism does not allow hundreds of bit changes at one bit swap. On the other hand, both ADSL2 and VDSL2 allow many more bits to be swapped during a single iteration, thus enabling a faster response to a time-varying noise environment.

For interested readers, several theoretical works in this field on various aspects of iterative water filling are briefly introduced. In [Chung 2003a], the power allocation strategy is modeled as a rate maximization game, and iterative water filling is a distributed power allocation scheme in the game. In the paper, it is shown that the Nash equilibrium always exists.* At the Nash equilibrium point(s), no user needs to change its power allocations to maximize its own rate. If a Nash equilibrium does not exist, then it means that at least some users are changing their strategies during each time instant, even though the channel and total power budget are fixed for a long time. This behavior is not desirable in practice because it implies that even though the channel and power budgets are fixed, some users might be changing their power and rate allocations. Therefore, it is a significant benefit that in iterative water filling, the Nash equilibrium always exists.

Theoretically, it is also important to enumerate all the Nash equilibrium points, because each equilibrium point can be an operating point for the non-cooperative rate maximization game. Hence, by obtaining all the Nash equilibriums, the number of operating options is known. Enumeration of the Nash equilibrium is not straightforward in general. For example, in bi-matrix games, the most elementary type of nonzero-sum games, there have been many discussions regarding the enumeration of Nash equilibriums. For this rate maximization game, [Chung 2003b] found that Nash equilibriums can be easily enumerated.

8.3.2 Level 2

8.3.2.1 History of Level 2

Level 2 DSM [Cioffi 2006] assumes that the binder topology is known to the SMC. Binder topology can be identified through a number of mechanisms

---

* Regardless of the channel conditions, the existence of the Nash equilibrium is guaranteed. The uniqueness of the Nash equilibrium, however, can be guaranteed only under certain conditions [Chung 2003a].

within the SMC. Interference path gains between lines can be inferred via sophisticated correlation methods based upon reported data from the lines, telephone-number and address correlation, and other published information about topology. Inference of suspected crosstalkers and types can be calculated for the same DSL service provider's network as well as for any competitive service provider's network. Binder identification is beyond the scope of this section, but the discussion that follows presumes it can be estimated with some reasonable tolerance.

The optimal condition in level 2 coordination is called the optimum spectrum management (or optimum spectrum balancing) situation, in which all the modems know all the other modems' practices and exact crosstalk coupling. Optimum spectrum management (OSM) bounds were investigated by Cendrillon of Alcatel in [Cendrillon 2006]. OSM is called OSB (optimum spectrum balancing) in Annex A of [ANSI NIPP-NAI]. The OSB technique essentially describes a procedure to maximize a weighted sum of data rates for all DSL subscribers within a binder. The OSB procedure requires central knowledge of the allowable spectral masks for all the lines and the crosstalk power coupling transfer functions between all pairs of lines. The optimization procedure assumes that crosstalk occurs only on the same subchannel between lines, and the SMC proceeds to allocate all the subchannel energies (equivalently, power spectra) and the number of bits per subchannel for standardized ADSLs and VDSLs. By varying the relative weights of all the lines (so that the non-negative weights all sum to one in each case), the rate region is obtained. The rate region is a multidimensional graph describing the achievable rate combinations for all the users in the same binder. The number of dimensions equals the number of lines. For example, when there are two users in the same binder, a two dimensional graph will show the pair of rates (R_1, R_2) where R_1 is for line 1 and R_2 is for line 2 that is achievable in the given loop condition.

OSB methods provide a large gain in situations where crosstalk is strong, particularly when there is a mixture of long and short lines in the binder as shown in [ANSI NIPP-NAI]. The algorithm of OSB is computationally complex; the complexity increases exponentially with the number of lines and number of subchannels. A few approximations have been derived to simplify the OSB calculations and essentially achieve the same performance. All these OSB methods presume a central implementation of bit swapping by the SMC. As such, unfortunately, all these methods then require a bundled implementation that is almost always contrary to regulatory practice. Furthermore, to respond to time-varying noise conditions on the lines, OSB requires a rapid SMC response, high-bandwidth control, and data flows from both the modem at the operator's end of the line and the modem at the customer premises to the SMC. Hence, even if bundling is allowed in certain areas of the world, the high-speed central reaction to noise changes is not desirable and may not be feasible in most situations. Additionally, there is no facility in current standards for such centrally controlled bit swapping.

To overcome this issue, band preference spectrum management (BPSM) was proposed [Cioffi 2004]. In this scheme, the spectrum is managed per band (which consists of a number of adjacent subchannels) instead of per subchannel. In its original form, BPSM proposed that the SMC distributes the new spectral mask to each CPE modem and then indicates that band preference should be used, then CPE modems would autonomously control each subchannel by water filling according to the spectral mask specifically assigned to each modem. This approach is very similar to level 0 or 1 DSM, the only difference is that in level 0 or 1, the spectral mask is defined by the standard. In contrast, in the original form of BPSM, the spectral mask is calculated by the SMC depending on the binder topology. By adjusting the spectral mask, the interference to other users can be properly controlled.

Compared to OSB, BPSM has significantly lower overhead and a more distributed resource control mechanism. This scheme, however, is not proper when the number of active lines or the noise environment is time-varying. Suppose that the spectral mask for a certain line is calculated when it causes large crosstalk to another line, to protect the other line, a very strict spectral mask must be imposed. However, this strictness is unnecessary when no other users are operating in the same binder.

Thus, in the current BPSM, a power scaling factor is introduced instead of a spectral mask. The power scaling factor conceptually is a penalty given to each band. When the spectral mask is controlled, depending on the binder condition, the level of the spectral mask is adjusted. Depending on the binder condition, the level of power scaling is adjusted. In this chapter, the focus is on BPSM using power scaling, which is the most advanced transmission scheme in level 2.

8.3.2.2 BPSM Using Power Scaling

Band preference (BP) as standardized in [ANSI NIPP-NAI] can be selected by one-bit indicator from an SMC to all DSLs controlled by that SMC, in other words, the BP is either "on" or "off." Different service providers' SMCs can make independent BP decisions if the binder or cable is unbundled. When BP is off, the DSLs operate as they would otherwise (presumably using level 0 or level 1 DSM). When BP is on, DSL transceivers independently and locally run an algorithm that is a modified version of water filling. In standardized DSLs, there is a quantity known as the PSDMASK that is distributed by an SMC (or set somehow by an operator or standard) to DSL modems. The PSDMASK, which is a function of subchannel index n, represents an upper limit on the power that can be allocated to each subchannel. When BP is on, the SMC uses PSDMASK to provide a BP weighting factor to the loading algorithm. This BP-on weighting factor is typically between the values of two and eight in different bands, for example, if the weighting factor is two, then half of the optimal power allocation is used for that particular band. These weighting factors are the novel idea in the BP loading algorithm compared to the iterative water filling algorithms. The modems continue to load, and if noise changes exceed a threshold, then the algorithm reverts to normal (BP-off) water fill loading, this change in procedure is the result of the observation that if the line's noise changed too much, the weighting factor previously calculated is not optimal anymore. The modems then wait for any future subsequent control from the SMC to use a new set of BP weighting factors with BP reset to "on." This adaptation is devised so that the CPE modem can autonomously move to a new optimal point when the binder conditions change.

To understand the simplicity of the BP loading algorithm, it helps to restate the basic water filling algorithm that assigns an energy E_n to each subchannel as a function of a measured channel to noise gain ratio g_n on each subchannel and a constant gap Γ such that the total energy on the line does not violate the total power, maximum margin, or PSD constraints. The basic equation is

$$E_n + \frac{\Gamma}{g_n} = \text{constant on all subchannels} \qquad (8.1)$$

subject to the energy being non-negative and satisfying the constraints. Note that Equation 8.1 represents level 0 or 1 DSM water filling. This simple concept leads either to the maximum data rate when the margin and total transmit power are fixed (rate-adaptive DSLs), or to the minimum transmit power when the margin and data rate are fixed (fixed-rate polite DSLs).

With band preference and the supplied set of weighting factors $\alpha_n \geq 0$, the water filling in Equation 8.1 is modified to

$$\alpha_n \cdot E_n + \frac{\Gamma}{g_n} = \text{constant on all subchannels} \qquad (8.2)$$

Equation 8.2, which represents BP (level 2 DSM) water filling, is also subject to the energy being non-negative and satisfying the PSD and total power constraints. Rate-adaptive or fixed-rate and fixed-margin water filling can also be implemented within the BP water filling just as in the normal case in Equation 8.1. The only difference is the use of the energy scaling factor α_n that is supplied by the SMC. If the noise changes substantially on a line so that

$$\frac{g_{n,\text{new}}}{g_{n,\text{old}}} < \text{threshold},$$

then the modem resets $\alpha_n = 1$, in other words, BP is turned off for this modem and the scale factor is reset so normal water filling is again used. The SMC can tell whether the CPE modem has reset the BP by examining the reported PSD(n) and SNR(n), because the reporting mechanisms of these parameters are standardized. The quantity α_n can be inferred from the difference in supplied PSDMASK(n) quantities when BP is on and off.

Two important issues in the BPSM are how to divide the band and how to select the weighting factors. The most obvious optimal solution is to perform an exhaustive search over all possible combinations of bands and weighting factors. Close-to-optimum solutions are currently being investigated, details are provided in [Lee 2006].

The performance of the various DSM methods is illustrated in Figure 8.5. For the loop configuration suggested in Figure 8.5a, the rate regions for OSB, interworking function (IWF), and BP are provided in Figure 8.5b. It is interesting to note that BP provides a rate region as large as that of OSB; hence, BP is much more desirable than OSB, considering all the benefits of BP.

As previously mentioned, most DSL loading algorithms use discrete (integer bits per subchannel) methods like the Levin–Campello procedure described in [Starr 1999]. Such practical loading algorithms typically require the flexibility to add an incremental amount of energy to a subchannel to round the number of bits to the next higher integer. The incremental energies ΔE_n required to add each additional bit to a constellation might be stored in a table. (There may be many efficient ways to implement such a system without storing a full table of incremental energies for each subchannel.) For BP water filling, the table of incremental energies is scaled by α_n, and then loading proceeds as in the normal discrete-loading situation. The presence of unequal α_n will tend to favor loading in some bands over others.

BP enables the use of level 2 DSM in unbundled environments where binders of DSLs may be excited by the equipment of different service providers. Each service provider's SMC can independently make a decision on whether to set BP $= 1$ and supply a set of α_n to any or all of its controlled lines. If there is no SMC for any particular service provider, then α_n is set to one for all that service provider's DSLs. If multiple service providers are using BP, then they all may experience yet larger average DSL data rates, but none may exceed existing applicable standardized (or regulator imposed) PSD masks. Thus, BP represents a significant and practical improvement from a regulatory and implementation viewpoint.

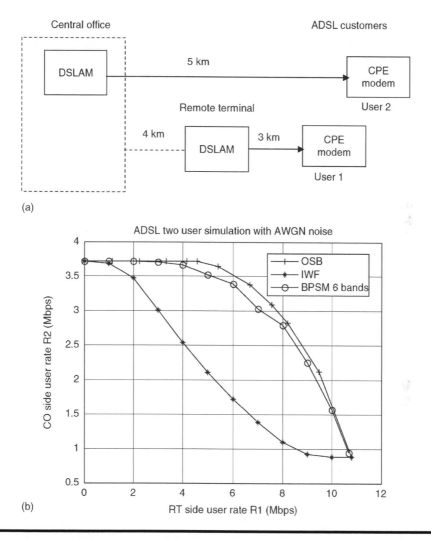

Figure 8.5 (a) ADSL transmission scenario; (b) achievable rate region by optimum spectrum balancing (OSB), interworking function (IWF), and band preference (BP).

8.3.3 Level 3

In this level of coordination, the SMC may observe, process, or supply corresponding packets of channel input and output samples for some or all of the DSLs connected to it. The most advanced spectrum utilization is achieved by the vectored mode, wherein synchronized signals can be co-transmitted and co-received at one side (typically the ONU side). In this mode, the SMC is located behind the DSLAM and controls the transmission and reception of signals. Coordinated advanced DSM systems essentially execute the following steps, perhaps periodically, depending on the specific transmission scenario:

- Network information acquisition phase, where loop characteristics, transmission parameters, and traffic information as well as external user commands are collected by a single service provider.
- Negotiation and optimization phase, in which all the information acquired in the previous step is used to distribute the network resources among the service provider's customers, and the communication parameters are determined to best serve those customers.
- Coordinated operation phase, in which the optimized communication and network parameters are provided to the modems within a single service provider's DSLAM to achieve the "cooperative" data transmission and reception.

Vectored DMT can achieve FEXT-free transmission in a coordinated environment in which multiple DSLs "share" their physical signal information [Ginis 2002]. In level 3 coordination, the modems on both sides of the loop (transmitter and receiver) can be coordinated, or only one side (i.e., transmitters) can be coordinated. When both sides are coordinated in level 3, then by performing a singular value decomposition (SVD), the interference from others can be completely removed. In most DSL deployment environments, however, it is necessary to assume that only one side can be coordinated in level 3, and, therefore, this is the model on which this chapter concentrates.

One-sided coordination can occur when a fiber-assisted DSL system (shown in Figure 8.1) is deployed. In this case, the DSLAM is capable of generating and receiving synchronized DSL signals from multiple lines. By employing frequency-division duplexing (FDD), which allocates upstream and downstream transmissions to disjoint frequency bands, NEXT is easily mitigated; the major interference, then, is due to FEXT. The performance loss caused by FEXT becomes much more severe when the lengths of the DSLs differ significantly. For instance, if all the modems at the customer side transmit the same power level in the same frequency band, FEXT caused by short lines at the upstream receiver of a longer line may severely degrade the upstream bit rate on the longer line.

Vectored DMT exploits the optimum multiple-input multiple-output (MIMO) signal processing structure, which is the generalized decision feedback equalizer (GDFE). The GDFE offers a unified view of signal detection in the interference channel. In the DSL environment, major interference sources include intersymbol interference (ISI), a signal distortion caused by the frequency selectivity of the line, and crosstalk interference,

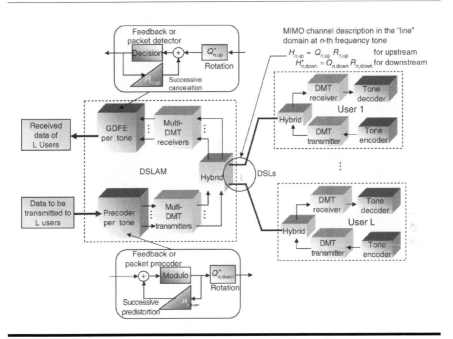

Figure 8.6   Vectored DMT diagram.

more generally referred to as inter-user inference (IUI). In its most general form, the GDFE thus views ISI and IUI in the same domain and tries to eliminate both simultaneously. DMT is a special case of GDFE wherein only ISI is eliminated.

When DMT is employed as a modulation technique, the signals from multiple lines can be processed separately at each sub-carrier frequency. The GDFE structure then can be derived to eliminate the crosstalk. Figure 8.6 illustrates the block diagram of a vectored DMT system. One should note that all DSL modems at the customer side are essentially the same as before, except that the transmission and reception now occurs synchronously. Synchronous transmission is feasible in a one-sided coordinated situation with a common DSLAM and is accommodated in the VDSL1 and VDSL2 standards as an option called "synchronous mode" [ETSI TS101 270-2]. The DSLAM processes the signals to be transmitted on each line or to be received from each line. It is assumed that all the channel information, such as signal attenuation and crosstalk characteristics on each line, is reported to the SMC and is known to the DSLAM in advance.

In the upstream direction, the DSLAM becomes the cooperative receiver. The received signal on each line is first processed via the usual DMT techniques. After the DMT processing stage, the signal at each line contains crosstalk interference from other lines. For each subchannel, all the signals from multiple lines are then collected, and interference cancelation is performed: a "rotation" operator ($Q^*_{n,up}$ in Figure 8.6) is first applied such that the signal to be decoded in the current line contains interference only from the previously decoded symbols in other lines. Finally, decision-assisted successive cancelation ($R_{n,up}$ in Figure 8.6) is performed.

In the downstream direction, the DSLAM is the cooperative transmitter, in which case the corresponding GDFE has a precoder structure. The multi-line signals to be transmitted are first collected for each subchannel, and appropriate pre-distortions or precodings (characterized by $Q_{n,down}$ and $R_{n,down}$ in Figure 8.6) are successively introduced across all the lines. These distortions are such that, after passing through the channel, the received signal at the modem in the customer premises is free of crosstalk.

Although the concept of the GDFE is applicable to any system, the use of DMT as a modulation technique greatly reduces the computational complexity of transmitter and receiver processing because each subchannel can be separately processed in the frequency domain. Moreover, the major crosstalk sources for each line are usually because of, at most, two to four neighboring lines in the cable. Therefore, the complexity increase at the DSLAM is the additional GDFE (or precoder) processing block associated with, at most, 4×4 matrix operations per subchannel for each line. There is essentially no increase in complexity in the CPE.

In most DSL deployment scenarios, two-sided coordination is not possible because modems at the customer premises are geographically separate; however, in some private network environments utilizing multiple lines, modems at both the transmitter and the receiver sides could cooperate. In such cases, instead of deploying FDD, NEXT cancelation methods can be used, thus allowing the use of the entire frequency band for both the upstream and downstream directions.

8.3.3.1 Simulation of Vectored DMT

The need to simulate vectored DMT to investigate the potential benefits in performance has revealed a deficiency in the channel modeling approach that has traditionally been used in the DSL industry. For a single isolated line, the channel insertion gain can be accurately modeled using the primary parameters of the associated cable (i.e., 24 AWG, 0.5 mm, etc.). However, in most existing channel models, the geometry of lines in the same binder is not considered explicitly. When the interference from other lines is treated just as an interference as in level 0, 1, and 2, the traditional channel models provide a reasonable approximation of a real channel. However, in level 3, the interference is canceled via GDFE operations. In this case, exact channel models are necessary to provide accurate results, because small phase and gain changes make a big difference in the performance.

It should be emphasized, however, that more accurate channel models are necessary only to predict the performance of vectored DMT using simulations. In a real DSL deployment, of course, the channel path gains are fixed by the loop plant geometry. The gains are estimated by the modems, and a channel model is not necessary.

8.4 Information Necessary for DSM

DSM specifications address informative methods and requisite accuracies for identification of crosstalk and line information. Such specifications also address the format of line and crosstalk information for use by a service provider's DSL maintenance entity, as well as the format for optional specification and control of DSL transmit spectra where permitted by the DSL modems. It is difficult to predict the final format of all the necessary information. In this section, the possible formats of the information for the ADSL and VDSL standards are provided. More recent results can be found in [ASSIA Inc. 2006].

Here the ADSL2 list of channel-identifying parameters is enhanced. The ADSL2 recommendation [G.992.3] specifies that parameters be computed in two modes:

1. Diagnostics mode can be invoked on-demand to compute diagnostics parameters accurately and exchange them reliably between the modems, even in conditions of poor SNR that would not normally allow the line to go into the state known as "Showtime." In fact, diagnostics mode is a stand-alone mode of operation that is not followed by Showtime.

2. Initialization is the procedure by which the two ADSL2 modems establish a steady-state connection. Successful completion of initialization is followed by Showtime. The computation of test parameters during the initialization procedure may be less accurate than in diagnostics mode, because such computation is constrained by a shorter period of training than is used in diagnostics mode. Moreover, during initialization the computed parameters are not exchanged between modems by default as they are in diagnostics mode. Instead, these parameters are stored in modem buffers and are transmitted through higher layers on demand. Some of these parameters are updated during Showtime, and again on demand: when an external request for that parameter is issued, the modem recomputes the parameter, updates its buffer, and transmits the new value, fulfilling the request.

The reason for this on-demand computation requirement is that DSM or the DSLAM itself may use this information for management of the loops. In the case of DSM, the information may be used to coordinate the transmitted spectra. To realize its greatest potential, DSM may recognize and respond to changing conditions in the binder, such as modems being turned on and off, and nonstationary CPE ingress noise sources, without having to wait for the next retrain or the next diagnostic test. On-demand reporting of operational data has been standardized by the ITU-T in [G.997.1] and is implemented in the SNMP MIBs (Simple Network Management Protocol Management Information Base). Tables 8.2 and 8.3

Table 8.2   Per-Subchannel Parameters Specified in [G.997.1], Which Enable Level 2 DSM with Discrete Multitone (DMT) Modems

	Diagnostics Mode	**Initialization**
$H(f)$		
Range	(Scale)$\times(2^{-15}-1)$, linear complex, scale is $2^{-15}-2$	$-102.2-0$ dB
Granularity	Scale	0.1 dB
Size	Single 16 bits scale + 2 × 16 bits per bin	10 bits per bin
Filter removal	Best effort	Best effort
Measurement period	1 s	0.5 s
Showtime availability	N/A	Yes
Showtime update	N/A	No
$QLN(f)$		
Range	$-140-0$ dBm/Hz	$-140-0$ dBm/Hz
Granularity	0.1 dB	0.1 dB
Size	11 bits per bin	11 bits per bin
Filter removal	Best effort	Best effort
Measurement period	1 s, rms value	0.5 s, rms value
Showtime availability	N/A	Yes
Showtime update	N/A	No
$SNR(f)$		
Range	$-32-96$ dB	$-32-96$ dB
Granularity	0.1 dB	0.1 dB
Size	11 bits per bin	11 bits per bin
Measurement time	1 s	0.5 s
Showtime availability	N/A	Yes
Showtime update	N/A	Yes

Table 8.3 Aggregate Parameters Specified in [G.997.1], Which Enable Level 2 DSM with DMT Modems

	Diagnostics Mode	Initialization
Line Attenuation		
Range	0–63 dB	0–63 dB
Granularity	0.1 dB	0.1 dB
Size	10 bits	10 bits
Showtime availability	N/A	Yes
Showtime update	N/A	No
Signal Attenuation		
Range	0–63 dB	0–63 dB
Granularity	0.1 dB	0.1 dB
Size	10 bits	10 bits
Showtime availability	N/A	Yes
Showtime update	N/A	Yes
SNR Margin		
Range	−63–64 dB	−63–64 dB
Granularity	0.1	0.1 dB
Size	11 bits	11 bits
Showtime availability	N/A	Yes
Showtime update	N/A	Yes
Attainable Bit rate		
Range	$0–2^{32} − 1$	$0–2^{32} − 1$
Granularity	1 bit/s	1 bit/s
Size	32 bits	32 bits
Showtime availability	N/A	Yes
Showtime update	N/A	Yes

list the test parameter specifications that are also mentioned in the [G.992.3] recommendation.

8.5 Application of DSM to Other Systems

Although DSM was developed for DSL, it may have potential applications to other types of systems, such as wireless systems. In particular, DSM methods may be of use in wireless multi-user MIMO systems. When the channel is changing very fast, then the channel path gain cannot be updated quickly, so level 0 or 1 DSM schemes could be suitable. When the channel is changing slowly, then level 2 and 3 DSM schemes might be applied.

References

[ASSIA Inc. 2006]   Assia Inc., Stanford University, *Proposed Text for Annex C in DSM Report*. ANSI NIPP-NAI Contribution 2005-217R1, Irvine, California, January 2006.

[Cendrillon 2006]   R. Cendrillon, W. Yu, M. Moonen, J. Verlinden, and T. Bostoen, *Optimal Multi-User Spectrum Balancing for Digital Subscriber Lines*. Accepted for publication in IEEE Transactions on Communications.

[Cioffi 2002]   J.M. Cioffi, S.T. Chung, and J. Lee, *To 2001-273R1 Using Measured Verizon DSL SNRs*. ANSI Contribution T1E1.4/2002-069, Vancouver, British Columbia, February 18, 2002.

[Cioffi 2003]   J. Cioffi and M. Mohseni, *Preference in Water-Filling with DSM*. ANSI contribution T1E1.4-2003-321R1, San Diego, California, December 8, 2003.

[Cioffi 2004]   J.M. Cioffi, W. Rhee, M. Mohseni, and M.H. Brady, *Band Preference in Dynamic Spectrum Management*. Eurasip Conference on Signal Processing, Vienna, Austria, September 2004.

[Cioffi 2006]   J. Cioffi et al., *Vectored DSLs with DSM*. To appear IEEE WCNC, 2006.

[Chung 2002]   S.T. Chung and J.M. Cioffi, *Rate and Power Control in a Two-User Multicarreir Channel with No Coordination: The Optimal Scheme vs. Suboptimal Methods*. IEEE VTC 2002 Fall, Vancouver, pp. 1744–1748.

[Chung 2003a]   S.T. Chung, S.J. Kim, J. Lee, and J.M. Cioffi, *A Game-Theoretic Approach to Power Allocation in Frequency-Selective Gaussian Interference Channels*. IEEE ISIT 2003, p. 316, 2003.

[Chung 2003b]   S.T. Chung, *Practical Transmission Techniques for Frequency-Selective Gaussian Interference Channels*. PhD Thesis, Stanford University, 2003.

[Ginis 2002]   G. Ginis and J.M. Cioffi, *Vectored Transmission for Digital Subscriber Line Systems*. IEEE JSAC, Vol. 20, No. 5, June 2002.

[Golden 2006]   P. Golden, H. Dedieu, and K.S. Jacobsen, Ed., *Fundamentals of DSL Technology*. Auerbach Publications, p. 457, 2006.

[Lee 2006]   W. Lee, Y. Kim, M.H. Brady, and J.M. Cioffi, *Band-Preference Dynamic Spectrum Management in a DSL Environment*. Submitted to GLOBECOM, 2006.

[Starr 1999]   T. Starr, J.M. Cioffi, and P.J. Silverman, *Understanding Digital Subscriber Line Technology*. Prentice Hall, 1999.

[Yu 2000]   W. Yu and J.M. Cioffi, *Competitive Equilibrium in the Gaussian Interference Channel*. IEEE ISIT 2000, p. 431, 2000.

[Yu 2002]   W. Yu, G. Ginis, and J.M. Cioffi, *An Adaptive Multiuser Power Control Algorithm for VDSL*. IEEE JSAC, Vol. 20, No. 5, June 2002.

[ANSI NIPP-NAI]   DSM Draft Report, ANSI NIPP-NAI Contribution 2005-031R5, October 2005, Las Vegas, Nevada (J. Cioffi, Editor).

[ANSI T1.417]   American National Standard, *Spectrum Management for Loop Transmission Systems*. ANSI Standard T1.417-2001.

[ANSI T1E1.4-R4]   DSM Mission Statement, *Proposed Scope and Mission for Dynamic Spectrum Management*. ANSI Contribution T1E1.4/2001-188R4, Greensboro, North Carolina, November 2001.

[ETSI TS101 270-2]   ETSI, *Transmission and Multiplexing (TM), Access Transmission Systems on Metallic Access Cables, Very High Speed Digital Subscriber Line (VDSL), Part 2: Transceiver Specification*. ETSI Standard TS 101 270-2, 2001.

[G.992.3]   Asymmetric Digital Subscriber Line (ADSL) Transceivers 2, ITU-T Standards G.992.3, 2002.

[G.997.1]   Physical Layer Management for Digital Subscriber Line (DSL) Transceivers.

Chapter 9

DSL Architecture: Opportunities for Differentiation and Evolution

Gavin Young and Michael Brusca

CONTENTS

Abstract   Broadband Digital Subscriber Line (DSL) access networks can be designed in a variety of ways to adjust to specific end user demands in particular market areas. The customers of DSL access networks are Internet service providers (ISPs), application service providers (ASPs), and other network operators together with the end users. This wide-ranging customer base can have diverse requirements. Fortunately, there are a variety of protocols that ride above the transport layer to facilitate flexibility of their DSL product choice. Such protocols can employ capabilities such as authentication, quality of service (QoS), configurable latency, as well as allowable bandwidth at the lower layers. These capabilities can enable support of applications such as video conferencing, gaming, streaming, Voice-over-IP (VoIP), and Video-on-demand (VoD). This chapter describes how the various layers of protocols and interconnecting element interfaces provide opportunities for service providers to differentiate DSL products beyond the simple "best-effort Internet access" paradigm that has been predominant in the first phase of DSL deployments. The chapter also indicates ways in which this Internet focused architecture may evolve in the future to support enhanced services beyond commodity Internet access.

9.1 Introduction

Many analysts are fond of saying that network transport is a commodity business and that the only value (and hence, better Average Revenue Per User [ARPU] and margin) lies in offering differentiated services above the network layer. This may be true to a degree in some sectors of the network transport market. For example, the core or backbone fiber networks often only differentiate themselves on features related to network speed and resilience. Hence, the real differentiator to the customer then rapidly becomes one of price. As a consequence, this section of the market has already degenerated to commodity products with commodity pricing and over-capacity in the market with multiple networks existing on many intercity and intercountry routes. Consequently, the market cannot support all of the operators and many have hit hard times.

Some analysts may be inclined to extrapolate this scenario to the broadband access market. However, the broadband access market is very different. For a start, there is much less competition. Although there are a variety of technologies that can provide broadband access (DSL, cable modem, satellite, broadband wireless), many consumers do not have much choice. DSL is the most globally prolific broadband access technology, yet it is only available to around 60–75 percent of the population in many developed countries. Cable franchise areas cover an even smaller footprint. The broadband fixed-wireless access spectrum auctions and deployments have not been a resounding success in many countries, consequently, this technology has very limited availability. Satellite is available everywhere including rural areas, but because of its higher costs and high latency (which can reduce throughput on some applications including Web surfing), it is often regarded as the technology of last resort.

There are of course a number of dimensions to the differentiation of broadband DSL services. There is the fundamental business model philosophy that separates the approach of various players. For example, wholesale versus retail, distribution channels and associated support (tiering), residential versus business focus, branding versus "white-labeling" of products, etc. Another key dimension to service differentiation is in the "service surround" which generally encompasses factors that are dictated by processes and systems as opposed to network connectivity. Such service features can include service level agreement (SLA) parameters, availability, provisioning time, fault response time, ordering interface (fax, Web, electronic data interchange [EDI], eXtensible Markup Language [XML] for batch orders, etc.), online access to systems to track order progress and trouble tickets, as well as enabling end users to perform their own line prequalification processes. However, such factors, whilst important, are beyond the scope of this chapter. To constrain the focus, the discussion of service differentiation is limited to attributes associated with the network architecture and connectivity.

Many consumers usually only have a choice of one DSL network provider within each technology class where the technology is available to them. However, the end users are not the only customers for broadband access networks. Service providers such as Internet service providers (ISPs) and application service providers (ASPs) together with corporate IT divisions also want to procure such connectivity on a wholesale basis. Again, choice can be limited. For example, the vertically integrated nature of the cable industry (and lack of current regulation to force open interfaces here) means that independent ISPs often cannot deliver their services over a third-party cable modem infrastructure. DSL fares better in this respect and owing to regulation is much more open in certain markets e.g., the EU, to a competitive broadband services market. However, to truly maximize customer (consumer or service provider) choice, there needs to be competition based on product features at the network infrastructure level* not just across access technology types.

No single operator is likely to provide all of the broadband access product functionality that the market demands. This can be owing to a variety of factors ranging from the DSL network providers market segment focus, geographic target, and consideration of their own cross-portfolio cannibalization. This chapter explains why broadband DSL access is not just a simple "fast pipe," but is highly customizable to the needs of different market segments because of the huge range of differentiating features that can be provided.

In this chapter, the various opportunities for service differentiation are highlighted with reference to the alternative configurations, including the employment of alternative protocol stacks and network element interfaces. Many services can be delivered over a single platform, consisting of common elements such as the DSL access multiplexer (DSLAM), layer 2 switch, and broadband remote access server (BRAS), which are configured

---

* A point emphasized by the work of the United Kingdom broadband stakeholders group, a government initiated "think tank."

accordingly to support various levels of service requirement. The chapter points out how the various protocol layers and interconnecting interfaces provide opportunities for service providers to differentiate DSL products beyond the simple best-effort Internet access paradigm that has been predominant in the first phase of DSL deployments. The chapter also indicates ways in which this Internet focused architecture may evolve in future to further support enhanced services beyond commodity Internet access.

9.2   Early DSL Access Architectures

This part first gives a high level description of the typical DSL network architecture, which was first deployed primarily to provide best-effort Internet access. It describes how the architecture evolved from the dial-up Internet access model and the end-user options for connecting their customer premises equipment (CPE) and local area networks (LANs) to the DSL network. It then goes on to describe each layer of the architecture in greater detail together with options for service providers, such as ISPs to interconnect with the DSL network platform.

9.2.1   Overview of Best-Effort Internet Access Architecture

The desire for "always on," mass market, broadband Internet access services has had a significant impact on how DSL network operators design their broadband network platforms. Many requirements have had to be taken into account, the most important of which are as follows:

- An always on service in the strictest sense of the word implies that the Internet Protocol (IP) addressing scheme has to be static. This is virtually impossible to achieve for mass market broadband IP services given that public IPv4 addresses are in very short supply. Hence, a dynamic addressing scheme as employed by most of the dial IP service providers is necessary.
- Bandwidth contention or end-user "overbooking" of some description needs to be employed within the network to cut down on network costs, thereby making a viable business case for offering broadband IP services.
- For some sectors of the market, it may be desirable to be able to offer a service provider selection capability, e.g., for an end user to switch directly between ASPs or between their ISP and corporate network, without traversing the public Internet.

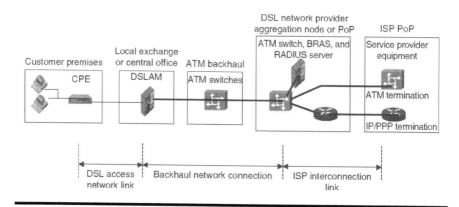

Figure 9.1   Overview of originally deployed Digital Subscriber Line (DSL) architecture.

Broadly speaking, as shown in Figure 9.1, three main network components are required to be able to offer DSL service to an end user. The first network component is the local DSL link connecting the end user to a DSLAM in their local central office (CO) by use of DSL transmission over copper telephone pairs to a dedicated end-user port on a DSLAM. The second network component is the backhaul link, which connects the aggregated traffic from the end-user ports on the local DSLAM to a broadband IP point of presence (PoP) or aggregation node. The aggregation node usually contains an ATM switch or a BRAS. This DSLAM to PoP link is usually over a high speed fiber core network connection. The aggregated end-user traffic from a number of DSLAM end-user ports is usually grouped (e.g., on the basis of ATM traffic class, contention ratio or ISP) and transported within an ATM virtual path (VP). This second network component usually has a built-in bandwidth contention (on the fiber backhaul) to take advantage of the bursty nature of IP traffic and the fact that not all end users will be active at the same time. This fiber bandwidth sharing allows a significant reduction in network infrastructure costs. These first two network components in combination usually contain a number of "product" options with regard to access speeds and can also contain options for different contention ratios. The first two network components also combine to convey ATM cells across the copper and fiber bearer transmission systems. They transport an individual end user's access "pipe" (based on an ATM permanent virtual circuit [PVC]) from the end user's premises to the broadband PoP or aggregation node.* The BRAS in this PoP may be used to terminate

---

the ATM PVC. Hence, ATM traffic engineering and quality of service (QoS) can be used across this part of the broadband access connection.

The third major network component in the overall end-to-end connection between end user and their chosen ISP is an aggregate "fat pipe" link to provide interconnection between the DSL network provider's PoP (or aggregation node) and an ISP's premises. These network components are more like leased lines, because they are symmetric and can have various speeds and perhaps resilience options. They may also have different interface and protocol options such as Layer 2 Tunneling Protocol (L2TP), IP, or direct ATM (discussed later). The number and speeds of such interconnect components that an ISP chooses to take will depend on factors such as: (a) how many concurrent end users they wish to support, (b) whether or not they wish to impose further bandwidth contention on their services (over and above those already imposed by the DSL network provider's end user access component), and (c) whether or not they wish to offer their end users different levels of service (say three offerings labeled, for argument sake, as "bronze," "silver," and "gold" representing three differently contended services).

Within the generic architecture shown above there may be different business entities operating. For example, typically an end user will contract with an ISP for high-speed Internet access. The ISP may contract with a DSL network provider for the broadband access network connections. The DSL network provider may in turn contract with regional fiber network operators for backhaul fiber ("middle mile") between their DSLAMs and their metropolitan PoP (where the DSL network provider locates their ATM switches, BRAS, etc.). If the DSL network provider is not the incumbent operator, then it may also contract with the incumbent for CO space to locate their DSL equipment together with rental of the copper pairs. For an incumbent telephone company, the DSL network provider and ISP may be separate divisions within the same company. However, for brevity in the following network architecture discussions, these distinctions are avoided. Hence, where the generic term "service provider" is used, this could denote an ISP or ASP, or a teleworker's (end user's) corporation (for "Intranet" access). Where the term DSL network provider is used, this implies both local access and regional backhaul network service provision.

9.2.2 Evolution from Dial-Up Internet Access: A Session-Based Architecture

To resolve issues related to IP addressing and service selection, a session-oriented approach has been taken to the provision of early DSL Internet access services [Enrico 2000]. This is the same approach that was used by ISPs, when they deployed dial-up Internet services using voiceband-modem or basic rate Integrated Services Digital Network (ISDN) access. A benefit of using this approach in initial DSL deployments was that ISPs were already familiar with the basic architectural principles, hence they had the necessary in-house skills and equipment (such as RADIUS servers) to operate such a service.

The Point-to-Point Protocol (PPP) is used to establish end-to-end sessions between an end user and their chosen ISP over which IP traffic can then be transported. PPP is the protocol that facilitates the familiar "logon" via the use of a username and password. It is also used for authentication of the end user before an IP address is allocated. Similar to dial IP, end user's IP addresses can be dynamically allocated during the establishment of a PPP session from an address pool administered by the ISP. The DSL network provider does not need to take part in the address allocation process, thereby allowing it to avoid the significant overhead that would otherwise be associated with obtaining and administering vast allocations of IP addresses. In addition, the ISP can reclaim IP addresses on termination of a PPP session (which can be end-user initiated or as the result of an idle timeout), thereby allowing it to make efficient use of public IP address allocations.

The use of PPP sessions denotes that the resulting broadband IP services are not always on. However, the time taken to establish a PPP session over a broadband network is typically very short (usually less than one second). In addition, provided that the underlying network has been dimensioned such that there are always a usable minimum set of (albeit contended) resources available to every end user, then they should never, under normal conditions, encounter the broadband equivalent of a "busy tone." In this case, the service is said to be "always available."

A key feature of the architecture is the broadband PoP or aggregation node, the key component of which is one or more broadband BRASs. These act as a brokering point with regard to the establishment of PPP sessions allowing end users to be able to select their ISP on a per session basis. From the PoP, IP tunnels* are used to provide point-to-point connectivity across a high speed IP interconnection link to the ISP. This interconnection is usually either an aggregated stream of IP packets from the ISP's end users (PPP Terminating Architecture [PTA]) or PPP sessions from the ISP's end users (L2TP architecture). In the PTA approach, the DSL network provider's BRAS is terminating the PPP sessions. In the L2TP approach, the DSL network provider's BRAS acts as an L2TP access concentrator (LAC) and forwards an aggregated stream of PPP sessions over an L2TP tunnel to

---

* The IP tunnels are implemented using the IETF recommendation for the Layer 2 Tunneling Protocol or L2TP (RFC2661).

an L2TP Network Server (LNS) which can be implemented on another BRAS or a router in the ISPs domain. The PPP sessions originating from end users are extended via the appropriate tunnel to the ISP and terminated (following successful authentication) on the ISP's LNS. It is these tunnels along with appropriately dimensioned and provisioned transmission pipes that constitute the aggregate fat pipe interconnections described above. There are two main approaches to implement PPP functionality on broadband networks. These are termed "PPPoA" and "PPPoE" and are described in more detail later.

9.2.3 End-User IP Connection to the DSL Service

DSL service presentation to business end users is predominantly via a 10/100 BaseT Ethernet interface on a DSL router (CPE). This allows a customer LAN to be used to share the DSL connection amongst a number of host machines (PCs, etc.). In a PPP over ATM (PPPoA) environment, the PPP session is terminated in the DSL router. A simple Web server built into the DSL router can facilitate easy configuration (of the username, password parameter, and network settings) and control of the PPP session. The IP address obtained from the ISP address pool during session negotiation is assigned to the network-side port.

There are then two options with regard to the IP addressing used on the LAN side of the DSL router. One option is to use private addressing (as described in RFC1918) and use network address translation (NAT) to hide the entire private IP subnet behind the single network-side IP address. The advantages with this option are that it makes efficient use of an ISP's address allocation and offers the end user a level of security against hacking. However, NAT is a double-edged sword—it does not allow unsolicited IP access from the network-side of the DSL router. This is problematic for an end user subscribing to an Internet access service that wants to host Internet services (Web servers, etc.) on their own premises.

There is an expedient workaround that can be used to address this issue. An approach variously known as "static PAT (Port Address Translation)" or "punctured NAT" can be used to define specific UDP/TCP (User Datagram Protocol/Transmission port numbers that will be Control Protocol) "listened for" on the network-side interface. These are mapped to specific IP addresses on the LAN-side subnet that correspond to hosts upon which the required Internet services are running. The problem here is that as each new Internet service is added by an end user, it is necessary to make a change to the configuration of the end user's DSL router.

Another aspect to NAT, which makes it undesirable for many end users, is that it is not transparent to all higher layer applications. In particular, H.323-based video conferencing applications and many network games may not work.

The other option for addressing used on the LAN, is to avoid running NAT on the DSL router—thereby eliminating the problem of application non-transparency described above. In order for this to work, an IP subnet that is part of the ISP's address allocation has to be statically assigned to the end-user LAN and a suitable IP route established on the BRAS. Of course, this approach means that the IP addresses spanned by that subnet are unavailable for use with other end users' PPP sessions. This option is therefore expensive with regard to IP address usage. In addition, the DSL router no longer has the security that comes with NAT (unless firewall functionality is put in its place). Many ISPs selling DSL access services offer both static IP address (no NAT) and dynamic IP address (NAT) options and usually charge more for static IP addresses.

The preceding arguments are not applicable to all consumer end users because many consumers will use universal serial bus (USB) modems instead of Ethernet routers to connect to their DSL service. USB DSL modems extend the ATM PVC from the DSL modem into the host PC or MAC itself (in essence, the USB DSL modem is acting as a layer 2 bridge). This allows the PPP session to run all the way from the ISP's BRAS to the end user's PC. The IP address allocated by the ISP during session negotiation is assigned directly to the host as in dial IP. (In fact, the method of managing the PPP session is very much like the legacy "dial-up networking" client.)

9.2.3.1 Device Types

The CPE used at the end users premises to connect them to their DSL service is available in a number of forms. The simplest (and hence among the cheapest) is a peripheral component interconnect (PCI) card ADSL modem which can be installed directly on a card slot within a PC. Many service providers have avoided offering this approach directly, because of liability concerns when an end user is required to "crack open" their PC to get service. However, one way they have mitigated such issues is to bundle a new PC (including a pre-installed ADSL modem card) as part of a DSL service package.

The next simplest product is a USB modem. Like the PCI DSL card, this provides DSL transmission and can support PPP and ATM protocols via the use of software drivers installed on the end users PC. However, it is an external modem and so does not require opening up the PC case for installation, it simply plugs into a USB port. Some USB modems also have drivers for Apple Mac machines. It is possible to share a DSL connection with other devices using the simple modems described above if one PC is acting as a master gateway. However, this has security issues if Internet connection sharing is enabled on a gateway PC, which also directly uses the publicly accessible IP address allocated by the ISP to the DSL end user. For more than one end user, it requires the gateway PC to be always switched on (because it powers the DSL PCI card or USB modem) in order for other devices to access the DSL connection.

A DSL router is the best way to share a DSL connection among a number of end users. In addition to the modem functionality, these devices provide Ethernet ports to enable multiple devices to connect to the router either directly or via a hub. The router can contain a range of functions including a Dynamic Host Configuration Protocol (DHCP) server to allocate private IP addresses to devices on the LAN, such as multiple PCs and other peripherals like printers and scanners ensuring that they are addressable and reachable to and from each other. The DSL router will also include NAT functionality to translate between private LAN IP addresses and the publicly accessible wide area network (WAN) IP address allocated to the router by the end user's ISP (during PPP log-on session). This gives better security than the simple DSL modems described above, because it helps to hide the IP address of any connected PCs from potential intruders. Some manufacturers refer to DSL routers as residential gateways or premises routers. Some manufacturers include advanced functions such as firewalls, virtual private network (VPN) termination, and print servers in their CPE products. DSL routers that also include voice ports are generally termed integrated access devices (IAD). All this additional functionality beyond the basic DSL modem has the potential to make the DSL device a portal in the home or business, through which a diverse range of services are provided. This DSL gateway market is being targeted by manufacturers and vendors from a range of industry segments including modem, router, consumer electronics, games console, and PC industries.

Home networking technologies have been developed to support the transparent transmission to and from the DSL network, without having to use Ethernet cabling to wire every device to the router. Hence, some DSL routers include home phone networking alliance (HPNA) interfaces to connect to home devices over existing internal telephony extension wiring. Another even more popular approach is to use wireless LAN interfaces to facilitate radio links between the DSL router and LAN devices, e.g., via use of the IEEE 802.11g standard (Wi-Fi). For such wireless networks, security can be provided by appropriate configuration of the encryption and access control and even via use of additional overlay VPNs, to prevent interception of local signals and unauthorized access to the LAN or DSL connection.

As these customer premises devices are developed to complement and interwork with the evolving DSL network architecture, additional functions will become included. These may include the shaping of IP flows per downstream devices such as IP phones. Sessions need to be prioritized for each of these IP-based devices. Some devices may communicate with proxies that then can relay or translate state or configuration information for the end device. Data, voice, and video can be delivered from a single DSL router or gateway device. However, these differing services can place different and perhaps conflicting requirements on the DSL access delivery mechanism. For example, streamed video is characterized by being tolerant of delay but very intolerant of transmission errors. Voice requires low latency and is relatively tolerant of errors. Data is generally tolerant of both delay and transmission errors, however, in the context of TCP/IP (Transmission Control Protocol/Internet Protocol), the delay must be minimized to prevent a detrimental effect on data throughput (see Section 9.2.5.1.2).

The majority of initial applications delivered over the DSL network to the customer premises were focused on the PC. Thereafter, new applications have increasingly targeted a diverse range of devices such as televisions, set top boxes, games consoles, telephones, cameras, Web appliances, home automation equipment, and security devices—as the "data origination and consumption points." As an overall trend, the DSL router is becoming more sophisticated and will include increasing capabilities to allow remote management as part of a DSL service. This could be used, e.g., to enable the VPN and firewall functionality of a teleworkers DSL router to be centrally configured, in accordance with corporate IT security policies. It could also be used to enable additional voice ports to be temporarily enabled on a residential IAD for a weekend visit by a relative. The signaling and management capabilities of DSL CPE will become increasingly sophisticated as the architecture evolves from best-effort Internet access to delivery of multiple, simultaneous, and real-time multimedia services.

9.2.3.2 CPE Interconnection Approaches for DSL Services

In the early days of DSL deployment, interoperability of CPE with DSLAMs was not as mature as it is today; therefore, the only viable approach was for the DSL service provider to provide the end user with CPE that they knew would interoperate with their installed DSLAMs. With this approach, the CPE was often owned by the DSL network provider as an NTE (Network Terminating Equipment). Some operators increased DSL CPE choice for end users by performing their own rigorous testing on vendor equipment, and then selling or renting CPE from a limited range of equipment whose interoperability and range was proven.

The development of unified DSL standards by the ITU-T together with interoperability test plans and "plug fests" programmes developed by the DSL Forum have resulted in interoperability of DSL CPE and DSLAMs from disparate vendors. This has enabled some DSL service providers to move to a "wires-only" model where the end user is free to supply their own DSL CPE as long as it meets the appropriate standards. This has consequently allowed a retail DSL CPE model to develop in those countries where the service providers have confidence in the vendor community's ability to produce standards compliant products (thus meeting a minimum transmission range performance standard).

To take advantage of interoperable DSL CPE and facilitate a mass consumer market for DSL services, the other key requirement is self-installation of equipment by the end user. Without this, installation req-uires a DSL network provider engineer to install the equipment, which is clearly inappropriate for large scale deployment. Developments in DSL CPE interoperability and ease of configuration have avoided the need for an engineering visit to install and configure the DSL modem or router. How-ever, it was also necessary to be able to avoid an engineering installation of the ADSL splitter that separates and isolates the ADSL signal from the voice-band POTS signals. The development of microfilters facilitated this so that self-installation of both microfilter and ADSL CPE has become the de-facto approach. It also gives the end user the flexibility to plug their ADSL modem into any telephony extension socket in the premises. The user simply needs to place a microfilter into every socket into which a telephony device is present. The microfilter is essentially a low pass filter that allows the 4 kHz POTS signal to reach the telephony CPE but blocks the ADSL upstream and downstream signals as shown in Figure 9.2. The microfilters also isolate the ADSL modem from the impedance variations of the telephony CPE which can present an open circuit in the ADSL band (especially, when off-hook) that could result in error bursts or even complete loss of synchronization (of the modem to the DSLAM). The ADSL

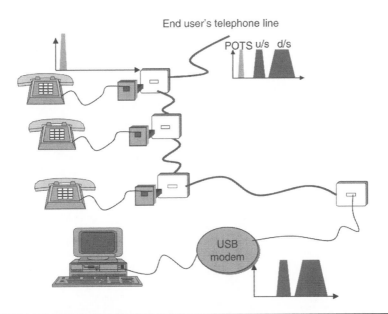

Figure 9.2   Asymmetric DSL (ADSL) customer premises equipment (CPE) instal-lation using distributed microfilters.

CPE contains a high pass filter to block the POTS signal to only allow the high frequency ADSL upstream and downstream signals to pass.

For consumer focused ADSL, the key issues associated with self-installation of microfilters, variation in DSL equipment performance, and ease of configuration of higher layer protocols have been largely solved. For business focused DSL services, especially those using Symmetric High Bit-Rate DSL (SHDSL) routers, SHDSL leased line replacement services (e.g., using circuit emulation to present T1, E1, X.21, or V.35 to the end-user or VoDSL IADs) will still be offered by service providers who provide installation and perhaps ongoing CPE management options as part of the DSL service.

9.2.4 Overview of Protocol Layering in DSL Architectures

Network architectures are often described in terms of the various protocol layers. DSL transmission systems work at the physical layer of the Open Systems Interconnect (OSI) model* and only span the access link between end-user premises and the DSLAM associated with a local CO or remote terminal (RT). However, additional network links and protocol layers are required to deliver an end-to-end service between an end user and an ISP. Examples of the links and protocol layers are described in more detail in the remaining parts of this section. An example of the protocol layers used to transport e-mail (using Simple Mail Transfer Protocol [SMTP][†]) across a DSL network is illustrated in Figure 9.3 showing the mapping to the various layers of the OSI model.

There are actually seven layers in the OSI model. In the above e-mail example, the presentation and application layers will be determined largely by the e-mail package that the end user is running on their PC, to communicate with their mail server. The five layers above are those required to be provided by the DSL network provider and ISP to convey the e-mail application across the network.

The applications and higher layer protocols are successively encapsulated by the lower layer protocols. Depiction of this way of visualizing the protocol "stack" is often shown in a "horizontal" protocol encapsulation diagram with lower layer protocols on the left. For an over-simplified example, see Figure 9.4.

---

* http://www.webopedia.com/quick_ref/OSI_Layers.asp

† SMTP is often used for an outgoing mail servers, while POP3 servers are used for incoming mail servers.

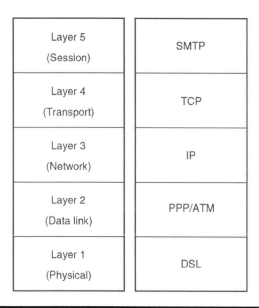

Figure 9.3   Example of DSL architecture protocol stack.

An end user's application data may be encapsulated within a session layer "transfer protocol" such as SMTP for e-mail, File Transfer Protocol (FTP) for file transfer or Hypertext Transfer Protocol (HTTP) for
Web surfing. This is then encapsulated within a transport layer protocol
such as TCP which can provide flow control (if required by the application) ordering guarantees, and reliability (in the form of handshake and
retransmission). The transport protocol is then encapsulated within IP
packets where IP is essentially acting as the unifying "network glue" of
the Internet.

IP network layer encapsulation may involve further partitioning of the
data stream (fragmentation) and adding "overhead" headers and trailers,
such as source and destination addresses to each fragment of the data.
These packets can then be conveyed between the end user's PC or DSL
router and an ISP using PPP sessions. The PPP sessions enable authentication (by username and password) and also enable the ISP to provide the
end user with a publicly accessible IP address. In this way, data can be

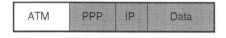

Figure 9.4   Example of simplified protocol encapsulation.

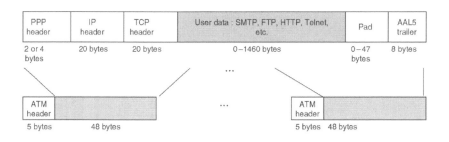

Figure 9.5   User data transported via Transmission Control Protocol/Internet Protocol (TCP/IP) in Point-to-Point Protocol (PPP) sessions over ATM.

routed between the ISP/Internet back to the end user. The PPP sessions can then be "adapted" to ATM cells (e.g., via PPPoA protocol and ATM adaptation layers such as AAL5) for transport over an ATM network with the appropriate ATM traffic class and quality of service (QoS). This involves further segmentation of the packetized data stream into 53 byte ATM cells with their own headers for switching across the ATM infrastructure (as illustrated in Figure 9.5). The ATM cells will then be transmitted over some form of physical layer transmission system for each link in the connection, such as via DSL over copper and via SDH or SONET over fiber.

The overheads associated with each successive layer of encapsulation reduce the ratio of end-user data to actual number of "bits" transported across the physical transmission system, hence reducing effective throughput as perceived by the end-user's application. There is often a functionality versus efficiency trade-off in DSL system architecture design with respect to the protocol stacks employed.

In practice, there may also be different protocol layers present on different links within the end-to-end connection. For example, as will be described later, the link between a DSL network provider and an ISP may group all the PPP sessions of end users associated with that ISP into an L2TP tunnel, which ultimately may be conveyed to the ISP over either ATM or Ethernet infrastructure as illustrated in Figure 9.6.

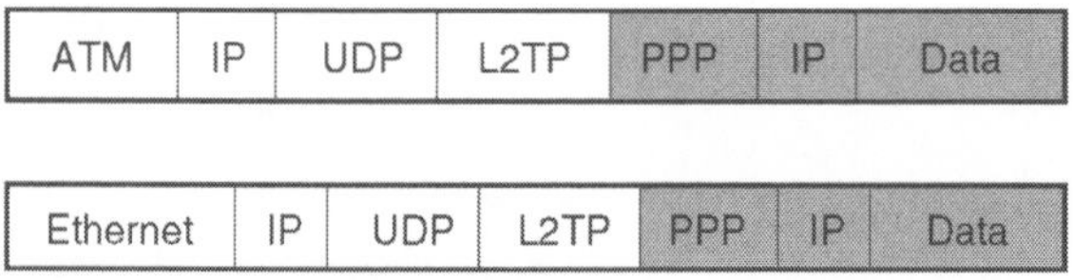

ATM	IP	UDP	L2TP	PPP	IP	Data

Ethernet	IP	UDP	L2TP	PPP	IP	Data

Figure 9.6   Example of protocol encapsulations on the link between DSL network provider and Internet service provider (ISP).

The various protocol layers involved in end-to-end DSL architectures are described in greater detail in the following sections.

9.2.5 Physical Layer

9.2.5.1 DSL over Local Copper Access Network

The physical layer in the local access network consists of twisted copper pairs of wire bound in binder groups of cables, which was originally used just to connect end users telephony equipment to the local CO. An end user's DSL CPE is connected to a DSLAM via these same copper wires. DSLAMs can be both CO-based and RT* based. The various DSL transmission types and modulation techniques are described in detail in other chapters of this book. However, one characteristic of the transmission technique does have a notable impact on the performance of the higher layer protocols and its impact that is covered here.

9.2.5.1.1 Impact of DSL Latency on Applications

ADSL was originally conceived as a delivery mechanism for streamed, compressed video over the copper access network (e.g., for Video-on-Demand [VoD]). The high tolerance of video delivery to delay meant that for this application ADSL was able to employ physical layer forward error correction and interleaving techniques to mitigate the effects of impulse noise on the integrity of the bit stream. Interleaving in particular is critical in the defence against impulse noise events, commonly encountered in the copper access network, as interleave processing can translate a single significant noise event into many recoverable insignificant events. However, its downside is that it introduces significant delay. Hence for ADSL delivery, low delay is synonymous with burst error events, whilst if higher delay can be tolerated (as for VoD), improvements toward more error-free transmission can be achieved. Typical round trip delay figures for ITU-T compliant ADSL are up to 45 ms if interleaving is employed and 5 ms if it is disabled.

However, as the driving "killer" application for ADSL shifted away from VoD and toward high speed Internet access and in particular gaming, the deployed architectures generally only used the fast latency path. This is because of the use of TCP above the end-user IP layer, which can affect data throughput (see next section).

When considering transmission of multiple, mixed services over an ADSL access system, the link between interleaving depth (i.e., delay) and

---

* Remote DSLAMs placed closer to the end user effectively reduce the length of copper over which the DSL needs to operate. They are often used in parts of the network where local communities are a long distance from the CO or can be located in the basement of multi-tenant units or on campus networks.

error rate leads to the logical conclusion that different interleaving depths may be required for different services. This was considered during the development of ADSL, hence the ITU-T ADSL standards allow for two different interleave "paths" through the ADSL link whereby the data stream can be split between a "fast path" and an "interleaved path," a concept referred to as "dual latency." However, dual latency is optional and was not used in initial DSL deployments, which were all single service (i.e., either VoD or Internet access but not simultaneously). Furthermore, it is not possible to dynamically vary the share of the available bandwidth between the two paths, hence, the bandwidth allocation is usually fixed under management control. This leads to limitations for multiple service delivery, where it may be desirable to dynamically share ADSL transmission capacity between multiple services each with differing latency requirements. One possible way around this dual latency problem is to carry all traffic over the same low latency (non-interleaved) ADSL path and rely on appropriate transport layer protocols (e.g., TCP) to ensure error-free delivery for error sensitive applications. However, TCP is not appropriate for real-time services that cannot tolerate any delay owing to retransmission and buffering. Hence, there may then be a requirement for an application layer forward error correction (AL-FEC) code. In summary, the ADSL "slow" interleaved transmission path is expected to be only really necessary for a video service, where low bit error rate (BER) (resulting in less Moving Picture Experts Group [MPEG] artefacts and corrupted video frames) is more important than delay. This will be especially true for transport of high-definition TV over DSL. For data, additional delay can affect throughput where TCP is used. For voice, any additional delay can affect subjective quality, hence latency should be reduced. Thus, the vast majority of ADSL deployments today and in the future will probably either be ADSL fast path based or they will use a "compromise" interleaving depth on a single slow path (e.g., between 10 and 16 ms) to ensure good impulse noise immunity, line stability, and video quality without unduly compromising data throughput or voice quality.

Note that for SHDSL transmission, there is no choice of latency at the DSL transmission layer, and the DSL transmission latency is low. The ITU-T standard requires a one-way delay of $<500\,\mu s$ for bit-rates $>1.5\,Mbit/s$ and a one-way delay of $<1.25\,ms$ for bit-rates $<1.5\,Mbit/s$. Hence with SHDSL, voice and TCP protocols do not experience any performance impact.

9.2.5.1.2   Impact of Physical Layer Latency on TCP/IP Data Throughput

The TCP receive window size is the amount of receive data (in bytes) that can be buffered at one time on a connection. The sending PC or server can send only that amount of data before waiting for an acknowledgement (ack) and window update from the receiving PC or server. Larger receive windows improve performance over high delay, high bandwidth networks. For greatest efficiency, the receive window size should be an even multiple of the TCP maximum segment size (MSS). The maximum TCP window size is determined by a parameter in the PC's registry, not something the average end-user PC will be capable of configuring.

The TCP/IP stack is designed to "self-tune" in most environments. Instead of using a hard-coded default receive window size, TCP adjusts the window size to even increments of the MSS negotiated during connection set up. Matching the receive window to even increments of the MSS increases the percentage of full-sized TCP segments used during bulk data transmission. By default, on many PCs the TCP window size is set to 8 KB, rounded up to the nearest MSS increment for the connection. If that increment is not at least four times the MSS, it is adjusted to four times the MSS, with a maximum of 64 KB. The maximum window size is 64 KB because the field in the TCP header is 16 bits long.

The maximum sustained TCP/IP throughput is the maximum throughput that can be achieved by an end user sending a very large data file. In assessing maximum sustainable throughput, it can be assumed that the TCP window on the end user's PC is open to its largest possible size and that there are no lost or errored packets causing retransmission and associated decrease in TCP window size. This is reasonable if there are low error rates on the physical layer transmission systems.

The maximum sustained DSL throughput is given by the minimum of the rate allowed by the upstream bandwidth, the rate allowed by the downstream bandwidth, and the rate allowed by the maximum window size.

TCP throughput can be constrained by the rate at which acknowledgement packets can be returned:

$$\text{Throughput} = \text{UpStreamBandwidth} \times \text{SegmentSize} \times \frac{\text{SegmentsperACK}}{53 \times \text{CellsperACK}}$$

Note that in the early days of ADSL trials, where it was used only for VoD, the ADSL was highly asymmetric (e.g., 2 Mbit/s downstream and 16 kbit/s upstream). This did not work well for Internet access, because the acknowledgement packets on the upstream path could not be returned fast enough to allow TCP to make most effective use of the downstream bandwidth (low upstream bandwidth also increases round-trip time).

TCP throughput cannot be larger than the downstream bandwidth with allowance for TCP, IP, AAL5, and ATM headers:

$$\text{Throughput} = \text{DownStreamBandwidth} \times \frac{\text{SegmentSize}}{\text{SegmentSize} + \text{Overhead}}$$

TCP throughput may be constrained by the TCP window size (in bits) and round trip time (in seconds):

$$\text{Throughput} = \frac{\text{Window Size}}{\text{RoundTripTime}}$$

(Window Size is the minimum of the sender's buffer size, the receiver's buffer size, and the congestion window. With no congestion, this is restricted by sender and receiver buffer sizes, typically 8 KB for a PC.)

The resulting maximum sustained TCP throughput is given by the minimum result from these equations. Much work on this topic was undertaken in the early days of the DSL Forum. Typical results are that for a 2 Mbit/s link (net of ATM and DSL overhead), TCP and IP overheads cause maximum downstream data rates of around 1.7 Mbit/s (maximum segment size 536 bytes) or 1.9 Mbit/s (maximum segment size 1400 bytes). Segment sizes of 1400 bytes are typically more common than 536 bytes.

It can be shown that a TCP window size of 8 KB means that end-to-end round trip delays must be kept low (e.g., <20 ms for a 2 Mbit/s connection, but can be greater for lower bit-rates) to achieve the maximum DSL throughput, because throughput decreases exponentially with increasing round trip delay. It should be noted that because the TCP throughput decreases exponentially, the drop in throughput as delay increases from 50 to 100 ms is much greater than the drop in throughput as delay increases from 200 to 250 ms.

A TCP window size of 64 KB can give the maximum throughput as long as round trip delays are less than 250 ms.* Some versions of Microsoft Windows may have a TCP window re-scaling option, which allows these large window sizes but this would not be straightforward for end users to configure to maximize their DSL throughput. If there are segments that are errored or lost, the TCP congestion window shrinks dynamically anyway.

A consequence of the above is that if a larger ADSL interleaving depth is used to improve video performance (BER), then any simultaneous data transmission will only avoid TCP throttling, if it is at low bandwidth. Video transmission together with simultaneous delivery of high-speed data (e.g., >1 Mbit/s) or voice will need to use a reduced interleaving depth to avoid impacting the TCP data (or voice) quality.

This analysis has concentrated on maximum sustained TCP throughput. This is the measure often used for testing connections (as per the ftp download tests used by many Web sites) and is a measure of interest to users

---

* This is why some broadband satellite modems include proprietary TCP ack spoofing, and why they are better for ftp transfer than Web surfing. It is also why transcontinental landlines could impact perceived DSL throughput.

transferring large files, such as PowerPoint presentations or MP3 files. However, for interactive Web surfing, the time taken to download a small file is more appropriate. Differences in maximum TCP window size become less important if file transfer is often completed before the TCP window attains its maximum size. Because Web pages are made of many small files of size 1–3 kbytes, which are downloaded with parallel TCP sessions, it may be concluded that the download of these Web pages would not be affected by the maximum TCP window size allowed. However, it must be remembered that because the transfer takes place in the ramp up phase of the TCP window, the download rates will be far from the DSL downstream maximum, because in practice there is typically a ramp up from a small TCP window at the beginning of the data transfer. Therefore, data transfer is probably completed before the TCP window attains its maximum size.

In conclusion, it is important for the TCP/IP throughput over a DSL network that the underlying physical layer performance is adequate. The key parameters are latency and underlying error rates to reduce TCP re-transmission.

9.2.5.2   SONET and SDH over Regional Fiber Backhaul

DSLAMs are usually connected to the broadband IP PoP or aggregation node via optical fiber links. In the early days of some deployments, the backhaul from the DSLAM may have operated at 34 Mbit/s (E3) or 45 Mbit/s (DS3). However, as DSL service became popular, more DSLAMs now operate at a minimum of 155 Mbit/s (STM-1 or OC-3) backhaul. The exception to this is in lower-density areas such as rural deployments or in the basements of multi-tenant units, where very small DSLAMs may be deployed with 2–8 Mbit/s backhaul using inverse multiplexing over ATM (IMA) with backhaul via 1–4 × 2 Mbit/s (E1) circuits. To accommodate higher service take-up and demand for higher bandwidth video DSLAM vendors then developed 622 Mbit/s (STM-4 or OC-12) backhaul transmission cards. The state-of-the-art has now moved even beyond 1 Gbit/s to examine 10 Gbit/s technology.

DSLAMs can either directly "drive" the fiber backhaul connection, or they can have a local connection to a colocated SONET/SDH Add-Drop Multiplexer (ADM) to exploit an existing SONET/SDH infrastructure. The direct drive approach involves less equipment in the CO. It would need to make use of 1 + 1 protected fiber transmission directly from the DSLAM to improve resilience. The use of SONET/SDH ADM for backhaul requires that additional network elements have to be managed. Nevertheless, SONET/SDH ring's resilience can be employed for fast fail-over protection. It is also useful when a different operator is being sub-contracted to provide the fiber backhaul and needs to specifically manage this link to meet SLA targets set by the primary DSL network provider.

Both CO and RT-based DSLAMs can connect directly to both the local narrowband Class 5 voice switch (for interconnecting the POTS in the base-band of an ADSL link) as well as a SONET/SDH ADM. SONET/SDH fiber-based backhaul transport can be used to connect DSLAMs to an ATM switch. The ATM switch connects to a BRAS, which can also be via SONET/SDH if the two are not colocated in the same PoP.

The above is the dominant approach with DSL architectures, but a new generation of DSLAMs with integrated ATM switching/or IP routing capabilities may lead to greater use of "sub-tended" architectures. Here, a "hub" DSLAM may act as an aggregator of traffic from remote subtended DSLAMs, as well as providing DSL cards for DSL connectivity to local copper lines.

9.2.6 ATM Layer

9.2.6.1 Role of ATM within the DSL Architecture

The ATM layer is used to connect the end user's DSL CPE to the DSLAM, the DSLAM to the ATM switch, and the ATM switch to the BRAS. This is accomplished in the form of ATM PVCs. The number of PVCs that are used to connect from the DSL CPE through these network elements on to the BRAS can range from 1 to a maximum of 8 PVCs (4 is also common), depending on the configuration of the architecture and product options offered by the DSL network provider. A single PVC is most common (especially on ADSL), because it eases the ATM network provisioning task, and multiple PVCs are not required for simple best-effort Internet access services. For a brief while (pre-Ethernet architecture dominance), multiple PVCs looked set to become more common as voice was bundled with ADSL services and SHDSL was deployed to business customers. This selection of the number of PVCs is a determining factor for the DSL network providers approach to managing QoS (i.e., whether to use ATM QoS or IP QoS), as well as the attributes of the services to be supported.

ATM can support several different ATM service classes and associated QoS levels. The unspecified bit rate (UBR) service class has predominantly been implemented in networks to support best-effort service. Other ATM service classes can provide "beyond best-effort" service. They include CBR (good for video streaming or leased line substitutes), variable bit rate–real time (VBR-rt) (good for voice), VBR–non real time (VBR-nrt) and UBR with a minimum cell rate (MCR) guarantee (both good for Internet access with minimum throughput guarantees). The ATM standards specify the appropriate ATM adaptation layer that best suites these service classes. Owing to its lower costs and the large demand for data services, AAL5 has been implemented throughout most networks. AAL1 has been implemented in network elements for applications such as leased line or circuit emulation.

More recently, AAL2 has been employed for its inherent capability to support voice. Hence, the primary role of ATM within DSL architectures is to manage the traffic and QoS on the bandwidth-limited DSL over copper link and the expensive backhaul link from the DSLAM to an aggregation node.

9.2.6.2 Rationale for Use of ATM within the DSL Architecture

Throughout the last several years, silicon and systems development of today's broadband access technologies have contributed to the IP versus ATM debate of recent years. In some parts of an end-to-end network, bandwidth can be used to solve problems associated with QoS. For example, dense wavelength division multiplexer (DWDM) can be deployed in the core of the network and gigabit Ethernet can be deployed in both the core and the customer's building. However, apart from using fiber or coax bearers, most access delivery media do not have the luxury of excess bandwidth. When today's broadband access systems and silicon for ADSL, Local Multipoint Distribution Service (LMDS), and satellite were first being developed several years ago, ATM was the only way of managing and policing traffic and offering absolute QoS mechanisms that could underpin service level guarantees. Hence, use of ATM was subsequently embedded in ADSL silicon and IP was seen more like "just another application" to be transported rather than a comprehensive networking technology. It was only after the initial broadband access standards and silicon development that IP developments such as Resource Reservation Protocol (RSVP), Differential Services (Diffserv), and Multiprotocol Label Switching (MPLS) came along to move IP-centric networks toward equivalent functionality.

The use of ATM in much of today's DSL silicon is, of course, not to the exclusion of IP. In the DSL Forum's system guidelines for ATM-centric architectures (as opposed to its packet-based recommendations), IP is carried over ATM. Several network operators took advantage of this IP over ATM approach by using an end-to-end architecture, enabling product offerings for either layer 3 IP or layer 2 ATM services over a common platform. Hence, their ISP customers then have the choice of which network product best suits their services. For example, some companies may procure the IP product to construct IP VPNs, and others may procure the ATM product, e.g., for VoD. In fact, in some parts of the world, the regulators give the operators little choice but to offer a wholesale layer 2 service, to which ISPs add their own IP layer.

There is logic (technical and marketing) in the history behind why ATM functionality has ended up in broadband access silicon and systems, and there appears to be very real markets for both ATM and IP delivery over such systems today. However, the more important issue is what will the markets require tomorrow (e.g., multicast) and how should the technology evolve to best serve those markets? The communications world is moving toward an increasingly IP-centric future. It seems less likely that in the near term the ATM layer will be completely removed from say ADSL, becasue the impact on silicon and interoperability of mass market ADSL products would slow down broadband access deployment to the masses. What is possible is that as the new developments for IP QoS and connection oriented capability are developed in the Internet Engineering Task Force (IETF), the ATM layer within broadband access systems may become "dumbed down," so as not to employ addressing and signaling mechanisms at both layers.

Whilst many European and North American operators have deployed the ATM-centric "standard" DSL architecture described above, several operators in the Asia Pacific region have avoided the use of ATM on the backhaul from the DSLAM, instead, they used Ethernet metropolitan area networks for the backbone. This works well in geographies where metropolitan fiber Ethernet connectivity is deployed close to the DSLAM locations (COs, basements of multi-tenant buildings, or on campus networks). With the advent of gigabit Ethernet metropolitan area networks and gigabit Ethernet cards in DSLAMs, such approaches are now gaining greater interest. Ethernet backhaul foregoes the traffic management (and operations and maintenance) capabilities of ATM. For example, when using virtual LAN (VLAN) tagging to segregate traffic flows, the ability to enforce QoS guarantees is reduced. However, if the price of over-engineering Ethernet capacity justifies it, then more operators may investigate this approach—particularly, because it can avoid the need to configure ATM PVCs across the network when an end user is provisioned with service.

9.2.6.3   ATM Traffic Management

In the backhaul network from the CO to the ATM switch, there could be contention among end users to gain access to the services that are transported. The first aggregation point of all of this traffic is the DSLAM, however, many DSLAMs are incapable of upstream traffic shaping. At the egress of the DSLAM (upstream toward the ATM switch), there may be multiple ATM VPs carrying different services and each containing multiple users (PVCs). The DSL network provider can manage end-user contention by controlling how many end users are put into each VP. VP usage can be monitored. Once predetermined VP capacity thresholds are reached, additional VPs can be created (as long as they are still within the bounds of the backhaul bearer circuit).

Note also that the VPs (connecting the DSLAMs to the ATM switch and BRAS) can contain end users from different ISPs. The ISPs may be extremely sensitive to end users of other (competing) ISPs bursting and affecting the service performance of their own end users, hence a diligent traffic management approach is vital.

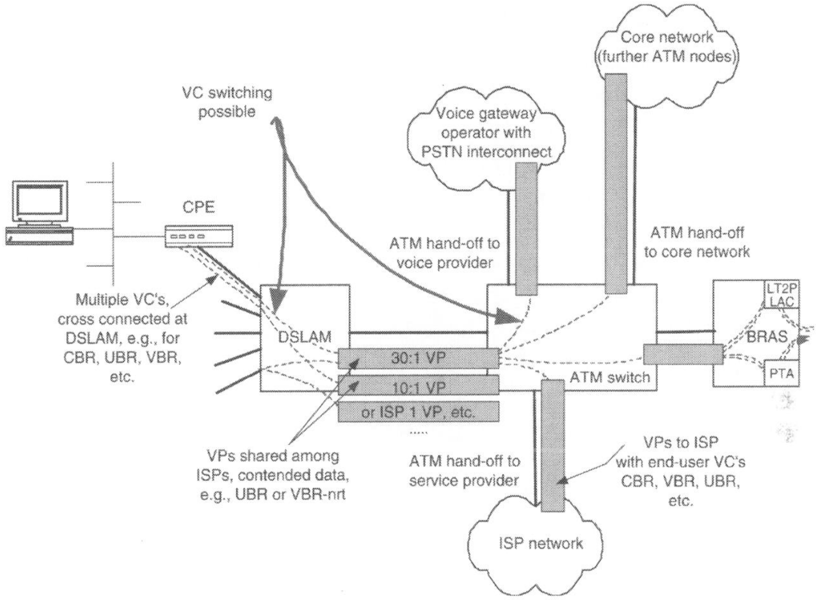

Figure 9.7   ATM virtual circuit (VC) switching possibilities within the DSL network architecture.

It would be desirable for the BRAS to shape both VPs and VCs in the downstream direction toward the ATM switch. The ATM switch can be set to police the aggregate VPs upon ingress (for both upstream from DSLAMs and downstream from BRAS). However, traffic shaping may be more desirable, because some traffic from ISPs may come directly into the ATM switch and not via the BRAS (if they have an ATM interconnect to the DSL network provider), and the ISP may have insufficiently abided by their traffic contract. Also, DSLAM upstream traffic shaping toward the ATM switch may be limited on some vendors DSLAMs. It is also desirable to shape traffic at the egress of the ATM switch toward the DSLAM to control contention within the VPs. The ATM VC switching possibilities are illustrated in Figure 9.7.

Traffic shaping and policing can have a profound impact on end-to-end performance of the network. For example, ADSL upstream cell delay variation (CDV) from access to backhaul could cause cell loss at a backhaul network switch policing point. There are solutions to such issues including design options of increasing core network CDV target, introducing shaping from access VCs into backhaul VPs, or introducing shaping rather than policing at the VPs ingress to the ATM switch.

Traffic shaping allows an effective data transmission rate to be limited with traffic beyond this rate being buffered, until such time as the buffer reaches a predetermined upper limit—at which time it is discarded. Traffic shaping can be done at the ATM VP or VC levels. ATM QoS can be used to prioritize servicing of traffic on a port, therefore, it affects which traffic will be discarded. ATM QoS is applied to VPs or VCs as they connect to a port. The purpose of traffic shaping and the use of QoS is to manage contention of limited bandwidth links in the DSL network architecture. Contention only occurs between traffic types when multiple traffic types are used. Hence, this is not an issue in the initially deployed DSL architectures that transport only data for best-effort Internet access.*

In today's prevailing architectures, the mechanisms used for controlling end-user traffic are the DSL profiles for the DSLAM and ATM profiles in the ATM switch. In most DSL networks, profiles allow a preset bandwidth and ATM traffic characteristics for upstream and downstream traffic.

The ATM switch within the DSL architecture is capable of traffic shaping and policing on the end-user VCs, and switching them "downstream" into the appropriate VP on the ATM switch port for onward connectivity toward the DSLAM and associated end user. In the upstream direction, the ATM switch can extract the end-user VCs from the many different DSLAM VP connections and merge them (via cross-connection) into a single VP for "upstream" connection toward the ISP hand-off interconnection point. This ATM VP can be traffic class and contention ratio specific, thus enabling the ISP to easily extract say their "gold" premium end users from the rest. For an ATM-interconnected ISP, the VP can emanate from a switch port that is dedicated to the ISP interconnect, so these VPs inherently only contain traffic for this single ISP. The DSL CPE will provide some upstream shaping, and the line rate also imposes limits on peak cell rate. The ATM switch can use VC shaping on its ingress ports to smooth excessive bursts, etc.

The network management system may perform call admission control (CAC) on the VP links between the DSLAM and the ATM switch. This can enable it to reject and flag any VC provision requests that would "overload" the VP causing violation of product traffic contracts.

Some DSL network providers support both best-effort and guaranteed services offerings using ATM service classes providing service class differentiation with associated QoS; ATM shapes traffic to minimize jitter. Buffering of traffic takes place when traffic arrives too quickly. The shapers delay some or all of the cells in a stream to bring the ATM cell stream into compliance with a traffic profile. The DSL loop inherently shapes the upstream traffic from the end user with the non-guaranteed service, via line profiles.

---

* It would however become an issue for an evolution to a combined data and voice product. To transport voice at the same time as data over the access connection, the DSL network Provider could add a second ATM VC with a VBR-rt traffic class. The DSLAM and IAD/CPE would need to incorporate an extended form of "per VC" queuing prioritization method to ensure that the voice cells have priority that the data queue is the first to go into discard mode.

The BRAS is configured to shape traffic downstream. With an end-to-end ATM architecture (direct PVC ATM interconnect between DSL network provider and ISP), the downstream traffic is shaped from the ISP's ATM switch over the end-to-end PVC to the end user's DSL CPE.

For non-best-effort guaranteed services using the ATM layer, the ATM traffic parameters are provisioned at each element to support the traffic of the end user's service. More specifically, using VBR-nrt as an example, the ATM traffic parameters for VBR-nrt PVCs are a peak cell rate (PCR), a sustained cell rate (SCR), and a maximum burst size (MBS).

When multiple PVCs to a single end user are implemented, the DSL bandwidth must be partitioned among the PVCs. The DSL network provider provisions the ATM PVCs' traffic parameters at each element accordingly.

9.2.7 Higher Layer Protocols

9.2.7.1 PPP Layer

Within the PPP family of protocols are the Link Control Protocol (LCP), Password Authentication Protocol (PAP), the Challenge Handshake Authentication Protocol (CHAP), and the IP Control Protocol (IPCP).

Inherent to PPP, LCP is typically used to configure a link layer. Either PAP or CHAP are used to authenticate the end user by sending a username and a password to his ISP, once the link layer is in place. IPCP is used to configure the network layer, upon successful authentication by PAP or CHAP. Once IPCP is configured, packets can be exchanged between the end user's PC and the ISP (or corporation). IP packets carrying application layer packets are encapsulated in PPP frames and communicated between the PC and the ISP. These IP packets are recovered at the opposite end from which they were originally encapsulated, i.e., PC or ISP. Once they are recovered, the IP packets can be routed to their ultimate destination such as a server on the Internet or on a corporate LAN. The DSL CPE* passes the username and password over the PPP session toward the network. The structured username appendix or network access identifier (NAI) (e.g., @FastISP.com), also called the realm or domain name, is used by the BRAS to determine where to forward the PPP session.

Integral to this configuration procedure is the dynamic assignment of an IP address from an address pool to the end user's PC. With PPP, this assignment is considered dynamic, because the IP address is assigned and maintained only for the duration of the PPP session with the ISP. Once the procedure for the configuration of the PPP session and address assignment

---

* For PCI cards, Ethernet, and USB DSL modems, the PPP protocol stack resides on the PC. For a DSL router, it resides within the router.

is complete, packets at the application layer (e.g., e-mail or Web pages) can be exchanged between the ISP and the end user.

With DSL, PPP is typically implemented for mass market services. A DSL network provider aggregates PPP sessions as well as routes them to ISPs at the BRAS. This allows the end user's PPP session to be carried from the end user's premises to the BRAS using PPPoA or PPPoE (PPP over Ethernet) over ATM.

9.2.7.1.1 PPPoA versus PPPoE

PPP can be carried directly over the ATM layer (PPPoA), which is typical in most parts of Europe, or it can be carried over Ethernet and then over the ATM layer (PPPoE), which is typical in North America. With PPPoA, the PPP session is connected between the ISP and the end user using PPPoA carried over an ATM PVC. Hence an ATM PVC is provisioned between the DSL modem/router and the BRAS equipment, to transport a single PPP session. With PPPoE, the PPP session is connected between the ISP and the end user using PPPoE encapsulated in Ethernet prior to being carried over the ATM PVC. Multiple PPP sessions can be thus transported over a single ATM virtual channel connection (VCC).

The PPPoA approach is based on the PPP over ATM recommendation (RFC2364) drawn up by the IETF. Although PPPoA is well established and many ISPs have deployed it, it nevertheless has a disadvantage, because it does not support the multiplexing of a number of concurrent PPP sessions into a single ATM VC. This means that in an ATM environment which only supports "permanent" connections (PVCs), it may not be possible to rapidly configure a new concurrent instance of an IP service given that the lead time to configure a new end user PVC over the ATM network (via management systems) may be in the order of days. One way to overcome this limitation is to operate PPPoA in a "switched" ATM environment that supports SVCs. The full PPPoA recommendation supports the signaling protocols necessary to dynamically set up an ATM SVC during the establishment of a PPP session. However, even though SVC functionality has been built into some DSLAMs and other operators' ATM infrastructures, it is not widely deployed in DSL architectures. SVCs may now never gain widespread deployment in DSL architectures because of the trend toward addressing, network routing, and "signaling" at the IP layer.

The other approach to PPP is to make use of the IETF recommendation for PPPoE (RFC2516), which as its name suggests facilitates PPP sessions over an (connectionless, broadcast) Ethernet LAN. This then enables the establishment of multiple concurrent PPP sessions and the auto-discovery of a BRAS that supports PPPoE. Whilst offering some advantages, PPPoE

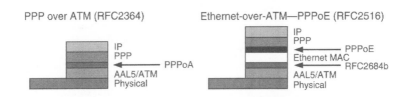

Figure 9.8   Comparison of PPP over ATM (PPPoA) and PPP over Ethernet (PPPoE) protocol stacks.

has greater protocol overhead, so it is a less efficient protocol stack than PPPoA (i.e., smaller ratio of end-user data to overall bits transported on the physical layer). This is illustrated in Figure 9.8.

9.2.7.2   RADIUS

The Remote Authentication for Dial-In User Service (RADIUS) protocol is the prevailing mechanism for the end user to be authenticated by the ISP, to check that he or she is a valid paying end user (and hence one who should be allowed to access the Internet or local services such as e-mail servers and news feeds). RADIUS may also be employed to obtain IP address information from the ISP. Finally, RADIUS can be used to capture an end user's session duration to generate accounting information or usage statistics.

RADIUS works in conjunction with PPP. The end-to-end DSL access connection is established by an end user initiating a PPP session. This will generate a RADIUS request that will be used by the ISP to fully authenticate the end user and establish the end-to-end PPP session. The two protocol options within RADIUS for authentication are known as PAP or CHAP. The authentication information is exchanged between the end-user PC or DSL router and the ISP's RADIUS server. The DSL network provider's BRAS and RADIUS server passes the username and password information to the selected ISP's RADIUS server. The DSL network provider's BRAS selects the ISP based on the structured username appendix (e.g., @fastISP1.com, also referred to as a qualified domain name or "realm"). During this session establishment, upon successful authentication, an IP address (or subnet) is sent to the PC (via PCI card, Ethernet, or USB ADSL modem) or DSL router from an allocated address pool via either the ISP's RADIUS server or the DSL network provider's RADIUS server (if it is managing the IP addresses on behalf of the ISP). The end-user CPE is also provided with primary and secondary Domain Name System (DNS) addresses by the ISP's RADIUS. Hence, the role of the DSL network provider's BRAS is to terminate the ATM part of the end user's broadband DSL or backhaul connection. The BRAS then supports authentication and routing based on the

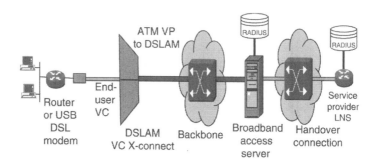

Figure 9.9 **Example of Remote Authentication Dial-In User Service (RADIUS) functionality within Layer 2 Tunneling Protocol (L2TP) interconnect architecture.**

domain name or realm. When an end user with the ISP qualified domain name logs in, the DSL network providers BRAS parses the domain name. The domain name is checked against the domain names listed in the DSL network provider's BRAS domain database. To explain these various steps during the PPP session establishment using RADIUS, here is an example using an L2TP interconnect between the DSL network provider and ISP.

For an L2TP interconnect to the ISP, when the ISP domain is identified, the BRAS effectively routes the end user PPP session to their ISP by forwarding the session down the appropriate L2TP tunnel thereby forming a L2TP interconnection between the DSL network provider and ISP. In this model, the ISP is responsible for maintaining their database of usernames and passwords. The DSL network provider determines the correct target ISP L2TP tunnel (and hence, RADIUS servers responsible for ultimate authentication of each user). An illustration of where the network and ISP RADIUS servers fit within this architecture is given in Figure 9.9.

The username and domain name are captured from the end user when they initiate the connection, and the DSL network provider's BRAS (acting as a LAC to originate the L2TP tunnel) replays them to the ISP's LNS (which terminates the L2TP tunnel). An example of such a session setup procedure is illustrated in Figure 9.10.

9.2.7.3 End-User IP Layer

The IP layer that carries the end users' applications and data (e.g., Web traffic between the end user's PC and a Web server on the Internet) needs to have certain associated settings established to function properly. The ISP can assign the appropriate settings using RADIUS as described above, e.g., a RADIUS access-accept packet may be used to download the end user's IP address together with additional settings such as one or more DNS address (e.g., primary and secondary). This information is then forwarded (via IPCP) to the end user's DSL router or PC (for Ethernet and USB DSL modems).

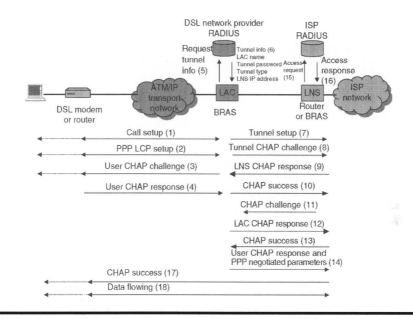

Figure 9.10   Example of PPP session setup (L2TP interconnect shown). (The numbers in parenthesis indicate the sequence of information flows.)

If the ISP allocates IP addresses to end users directly from its own RADIUS server, the ISP then has direct control and can then size this address pool according to a contention ratio of end users to addresses. Addresses can be allocated and de-allocated to end users dynamically, session by session. The ISP can then decide whether they wish to assign static addresses to certain end users, i.e., the end user gets the same address each time they initiate a broadband session. Alternative approaches are to give a pool of IP addresses to the DSL network provider to allocate to end users on the ISP's behalf, typically at the BRAS. Another option is to use a proxy RADIUS where the DSL network provider RADIUS server interacts with an ISP RADIUS to perform authentication and address allocation.

Normally, where the end-user CPE connects to a LAN (i.e., Ethernet interface, not USB) many-to-one NAT is used to translate the private end-user LAN addresses to the dynamically assigned DSL CPE WAN address associated with the active broadband session. The IP subnet on the LAN port of the router will need to be globally registered if unrestricted access to the Internet is required. However, the LAN can use private ISP-supplied addresses (or a default subnet) on the LAN if restricted access is required (e.g., to a corporate Intranet or an ISP using NAT for Internet access). In both cases, the LAN addresses could be DHCP served, static, or a combination. Also, for these fully routable solutions, the ISP RADIUS will still need to download an IP address and give a route to the DSL router pointing to the static LAN-side subnet at session setup. For the no-NAT option of a static address, the ISP must set aside a valid IP subnet from their own IP address allocation for assignment to the LAN side of the DSL CPE. These addresses on the subnet are not then available for assignment to any other end users. The route to this subnet needs to be downloaded for the DSL router via the ISP during PPP session establishment in the RADIUS access-accept packet. This route, which enables the end user's LAN to become addressable, will be automatically removed when the end user's PPP session terminates.

9.2.8 Service Provider Interconnection Options

The interconnection between DSL network provider and ISP can vary in terms of bit-rate and resilience. However, the major differentiation is in the nature of the protocols used across the interconnect. Some of the most commonly deployed options are now described.

9.2.8.1 L2TP Interconnection

The L2TP interconnection between a DSL network provider and an ISP is described as the example in the above descriptions of the protocol layers in the DSL architecture. The ISP's equipment (such as a router or BRAS acting as an LNS) terminates L2TP tunnels from the DSL network provider. The DSL network provider and ISP must agree upon tunnel setup parameters (name, password, etc.). With an L2TP interconnect, an ATM end user data path provides transport between the end user's premises and the BRAS of the DSL network provider. The BRAS terminates the ATM connection and then places the end users PPP session into an L2TP tunnel (based upon the log-on realm) for hand-off to the ISP. The BRAS actually provides the L2TP interconnectivity between the DSL network provider and the ISP. This form of interconnection allows the ISP to have complete end-to-end control over their end user's PPP sessions. The DSL network provider's BRAS performs the function of an LAC which originates the L2TP tunnel and concentrates PPP sessions from end users into this tunnel. The ISP provides an LNS that terminates the tunnel. An example of L2TP interconnect using PPPoA is illustrated in Figure 9.11.

The L2TP connection to the ISP customer can typically either be a 100 Mbit/s or 1 Gbit/s Ethernet connection or a 155 Mbit/s STM-1 or 622 Mbit/s STM-4 connection (using ATM). At the highest layer, the end user IP data path is connected from their DSL CPE across the network to the ISP's LNS, where it can be forwarded to the Internet via a transit or peering connection or used to access local services such as e-mail. However, more importantly for ISP interconnection, is the fact that the end user's IP data path (encapsulated within a PPP session) is interconnected between the DSL network provider and the ISP via an L2TP tunnel. This L2TP tunnel contains the PPP sessions from the various DSL end users.

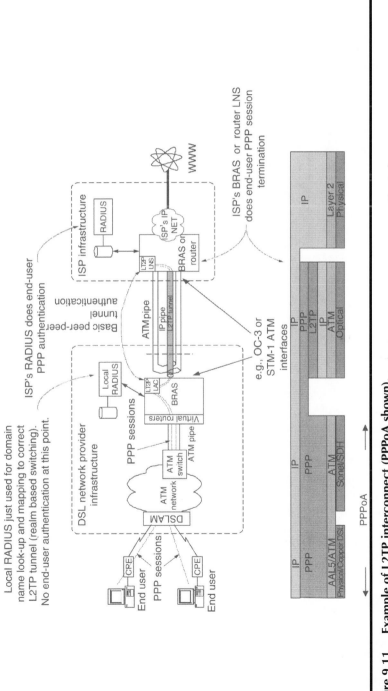

Figure 9.11 Example of L2TP interconnect (PPPoA shown).

For Ethernet interconnection, the L2TP tunnel is carried over an IP transport layer between Ethernet ports on the DSL network provider and ISP equipment.

There can be two IP layers in such an end-to-end DSL connection. The highest one carries the end users' data. A lower layer can be used to carry the end users' PPP sessions within an L2TP tunnel. This transport IP layer can be carried over Ethernet or ATM. For example, the transport IP layer requires allocation of an IP address range for each ATM PVC in the interconnection to provide an address for each end (DSL network provider LAC and ISP LNS) of the ATM PVCs transporting the L2TP tunnel(s). Static routes can be used together with simple IP ping techniques to debug interconnection issues.

9.2.8.2   Routed IP Interconnection—PPP Terminating Architecture

An alternative to the L2TP interconnect described above is for the DSL network provider's BRAS and RADIUS to terminate the PPP session and simply forward the recovered IP packets to the ISP. This is known as the PTA. However, the resulting IP interconnect model foregoes the benefits to the ISP of having explicit end-to-end visibility and control over each end-user PPP session with its associated authentication and control of IP address allocation. These are "subcontracted" to the DSL network provider with a degree of control possible via proxy RADIUS techniques.

With PTA interconnection, both PPPoA or PPPoE over ATM can be supported from the end user. In this architecture, both the PPP session and the ATM PVC are terminated at the BRAS. The BRAS is configured with logically separate routing processes or virtual routers for each ISP hand-off.

After the end users' PPP sessions are terminated at the BRAS, IP packets are routed based on the routing table associated with each virtual router. The virtual router to which a PPP session is assigned can be based on the structured username appendix (or domain or realm) associated with a preselected ISP. However, an end user can optionally reach multiple ISPs using the appropriate domain name, to indicate to which ISP the traffic is to be forwarded (achieved by routing the IP packets to a BRAS port connected [via Ethernet or ATM] to a particular ISP). The RADIUS protocol then provides a mechanism for the DSL network provider's BRAS and RADIUS (acting as a proxy RADIUS server) to communicate with an ISP's RADIUS server for authentication and optionally, IP address management. With this architecture and interconnect approach, the BRAS can aggregate end-user traffic at the IP layer (thus facilitating multicast, etc.). The virtual routers within the BRAS ensure separation of ISP traffic. Again, simple IP ping techniques can be used to debug interconnection issues.

9.2.8.3 ATM Interconnect

An ATM interconnection to an ISP is often an STM-1 (SDH) or OC-3 (SONET) link operating at 155 Mbit/s or STM-4 (STM) or OC-12 (SONET) link operating at 622 Mbit/s and presenting an ATM user network interface (UNI). This link can carry an ATM VP for each ATM traffic class, perhaps with each contention ratio combination for the ISP procured DSL access connections from the DSL network provider. The VPs contain separate VCs for each end user. The main benefit of this approach is that it gives full control of the IP and PPP layers to the ISP. They only rely on the DSL network provider for ATM layer connectivity. The ISP can decide how many end user VCs to pack into a given VP, thus determining the contention ratio of the backhaul from the DSLAM to BRAS. However, the ISP then has full responsibility for provision of BRAS, RADIUS, and IP address management functionality. An example of ATM interconnect using PPPoA is depicted in Figure 9.12.

An ATM interconnect offers the prospects of exploiting ATM techniques across the interconnect for debugging purposes, etc. For example, an ATM connectivity check can be performed across the DSL network provider to ISP interface to ensure ATM connectivity as part of the interconnect commissioning. This can also be used as part of any necessary repair process. An ATM loopback check can be undertaken using ATM OAM (operations, administration, and maintenance) flows. The end-to-end ATM layer

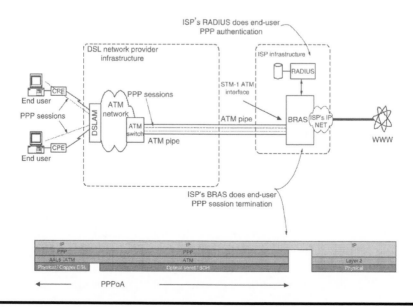

Figure 9.12   Example of ATM Interconnect (PPPoA shown).

F5 OAM cells as defined in ITU-T recommendation I.610* can potentially be carried transparently by the ATM network between the end user (if the CPE supports it which is less likely with USB modems than DSL routers) and the DSL network provider interface. F5 OAM flows for connectivity checks between the DSL network provider ATM switch and the ISP ATM interface can be undertaken for both provision of new service and also repair. It may also be possible to use additional continuity check functionality using F5 continuity check (CC) cells for monitoring of circuits.

The downstream ATM traffic (from the ISP's network) being offered to the network via the interconnection point should be shaped by the ISP, so that it does not exceed the VP's ATM traffic contract. Ideally, the ISP should also shape the traffic on each VC such that the line-rate and VC bandwidth on the DSL are not exceeded, thereby resulting in dropped cells.

9.2.8.4 Bridged Ethernet Interconnection

With the bridged Ethernet architecture, each end user is provisioned to a single VLAN of their selected ISP. Ethernet frames generated by the end user's PC are carried over a PVC that connects the end user to the ISP. The DSL network provider does not process any protocols above the Ethernet layer. In this architecture, the BRAS bridges Ethernet frames received from the end users to a PVC that has been preprovisioned to the preselected ISP. The BRAS aggregates PVCs from the various end users, thereby reducing the number of PVCs to be terminated for each end user as compared to the ATM interconnection architecture.

The ISP can implement either IP or PPP above the Ethernet protocol layer, depending on how the ISP specifies configuration of the client software on the PC. For both protocol options, end user's Ethernet frames are encapsulated (using RFC 2684 [which "obsoletes" RFC 1483]) and sent over an ATM PVC by the DSL CPE to the BRAS.

Secure VLANs are based on an end user's subscription to an ISP. The BRAS prevents the bridging of one end user's Ethernet frame to another end user within a given VLAN. Because inter-ISP communication is also restricted, unintentional connections to another ISP is prevented. Additional security filtering can prevent user-to-user communication through the BRAS. The VLANs facilitate service provider traffic segregation in the BRAS.

Some limitations of bridged Ethernet interconnection include the need for multiple ATM PVCs and the inability for the end user to choose connections to more than one ISP. If an end user has multiple PCs, they would all connect to the ISP supported by the VLAN. Similarly, an ISP's membership in a VLAN restricts to which end user's the ISP's traffic can be bridged.

---

* Also I.361 for payload type field and PTI definition.

Thus, VLANs ensure ISP traffic segregation. Whilst feasible, unicast delivery of streaming media is potentially available to all end users, although in large VLANs performance may be degraded.

Some ATM switches now have additional functionality, whereby a group of end users' PVCs can be assigned to the same VLAN and interconnected to an ISP over a common VLAN trunk (without the need for a separate BRAS for any VLAN functionality). This can then be configured to facilitate intra-VLAN connectivity between end users without the traffic traversing the VLAN trunk. It is typically used to interconnect geographically distant Ethernet LANs over an ATM network, so end users that are connected at remote sites can communicate with each other, as if they were all on the same large LAN. This can be ideal for connecting end user branch office sites and teleworkers with the corporate headquarters (housing data and application servers). The headquarters could be interconnected to the DSL network provider over the VLAN trunk (high speed symmetric circuit carrying Ethernet, possibly transported over fiber [or bonded SHDSL]), and the branch offices and teleworkers could be served with DSL access lines. This type of equipment may also permit point to multi-point connections in conjunction with an Ethernet multicast address to support streaming applications like a corporate presentation "broadcast" (actually multi-cast) to all branch offices.

Ethernet bridging functions are defined in IEEE 802.1d. This has been extended to include the idea of bridging different VLANs, which are identified by a VLAN header inserted into the Ethernet frame, according to the VLAN bridging functions defined in IEEE 802.1q. The use of VLANs over the common trunk or interconnect allows multiple VLANs to co-exist over the same infrastructure, with the end-user site membership of a VLAN being defined by the VLAN identifier used on the appropriate bridge port. Hence, another application for this interconnect could be for an ISP to offer a layer 2 (Ethernet or ATM) VPN capability to different corporations over a single interconnect with the DSL network provider. The different corporations' end-user sites would be assigned different VLAN tags.

9.2.8.5 TDM Interconnect

It is possible to use standardized circuit emulation techniques to deliver time-division multiplexing (TDM) services such as leased line replacements over the ATM-based DSL architecture. The emulated circuits can be handed-off to a service provider in ATM format, or they can be mapped back to TDM and handed-off in an aggregated TDM interconnect. The DSL access products that feed into an aggregated TDM hand-off to an ISP use ATM AAL1 circuit emulation techniques [ATM Forum (a)]. TDM DSL access involves provision of DSL CPE, which maps TDM traffic from an end user's equipment into ATM cells as illustrated in Figures 9.13 and 9.14.

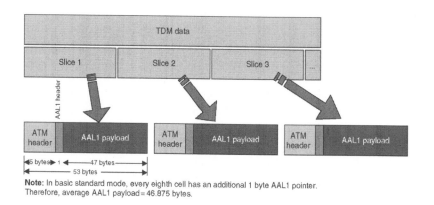

Note: In basic standard mode, every eighth cell has an additional 1 byte AAL1 pointer. Therefore, average AAL1 payload = 46.875 bytes.

Figure 9.13   Mapping of time-division multiplexing (TDM) data to ATM via ATM with Adaptation Layer1 (AAL1).

The encapsulated traffic is transported across the DSL and ATM network to the interconnection point. The ATM switch (with appropriate interface cards and functionality) then removes the ATM encapsulation and presents the traffic (which is aggregated from a number of end users) to the service provider as a TDM interface, e.g., operating at either STM-1, DS3, or E3 rates. Lower speed or non-aggregated hand-off at T1 or E1 rates is possible but will consume a lot of ports on the ATM switch. These DSL access products are designed to transport traffic traditionally conveyed over leased lines and can be used to transport TDM voice channels as illustrated in Figure 9.15.

In addition to providing fully transparent T1 or E1 connectivity over SHDSL (unstructured mode), the network can deliver structured T1 and

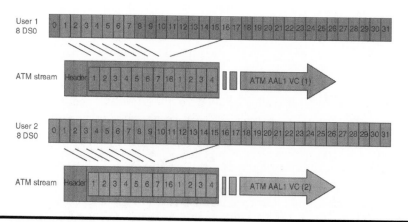

Figure 9.14 $n \times 64$ kbit/s TDM timeslots from E1 frame into ATM AAL1 VCs.

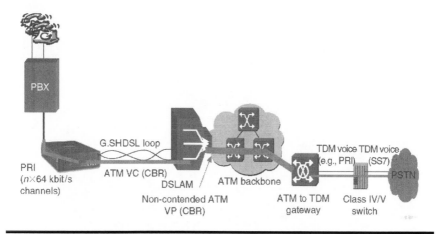

Figure 9.15   Example of ATM circuit emulation over SHDSL to provide private branch exchange (PBX) TDM access.

E1 services (structured data transfer [SDT] mode) giving visibility down to individual 64 kbit/s timeslots, where each (fractional) T1 or E1 can be independently configured. In SDT mode, TS16 can be used for channel associated signaling (CAS) or data.

With a TDM interconnect, the DSL network provider provides the "co-ordinates" of where a particular end user's traffic resides within the aggregated interconnect pipe. An example of an E3 interconnect aggregated to the E1 level is shown in Figure 9.16.

In the above example, it can be seen that more efficient use of the interconnect bandwidth would be made if the E1 frames within the E3 interconnect were "packed" at the 64 kbit/s DS0 level (as shown in Figure 9.17), instead of the (fractional) E1 level.

This greater packing efficiency complicates the process of communicating, where in the aggregated hand-off individual DSL end user's timeslots are located. This end-user traffic identification issue is now further explored below. Here is an example in which SHDSL is used to deliver E1 or fractional E1 services to end users via an STM-1 TDM interconnect to the DSL network provider.

The DSL access connections that are aggregated in the TDM hand-off from DSL network provider to service provider present E1 or fractional E1 services to end users, capable of delivering a range of $n \times 64$ kbit/s channels to each end user. The STM-1 interconnect is capable of carrying a maximum of 63 E1 access connections (and hence supporting 63 virtual E1 interfaces). The mapping of E1 access connections into the STM-1 can be as follows:

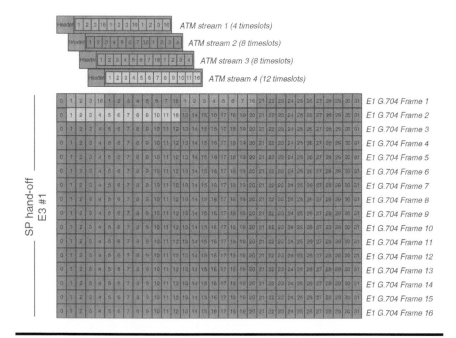

Figure 9.16   E3 hand-off with aggregation at E1 granularity.

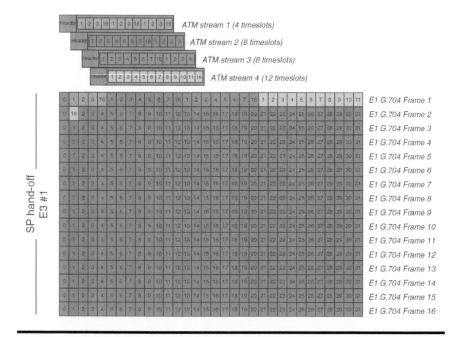

Figure 9.17   E3 hand-off with aggregation at 64 kbit/s DS0 granularity.

STM-1		Synchronous Transport Module
AU-4		Administrative Unit
VC-4		Virtual Container
TUG-3	× 3	Tributary Unit Group
TUG-2	× 7	Tributary Unit Group
TU-12	× 3	Tributary Unit (VC plus pointer)
VC-12		Virtual Container (container + Path OverHead [POH])
E1		
$n \times 64$ kbit/s		

An individual end user's (fractional) E1 DSL connection is referenced within the STM-1 interconnect using the standard SDH notation of (k, l, m) where $k =$ TUG3# $(1 \ldots 3)$, $l =$ TUG2# $(1 \ldots 7)$ and $m =$ TU12# $(1 \ldots 3)$. The (k, l, m) end point would be agreed and communicated to the service provider when the end user connection is provisioned. E1 access connections within the VC-12 can be unchannelized (unstructured E1) or channelized (structured) $n \times 64$ kbit/s.

This approach to transport of TDM over an ATM and DSL network enables such a service to be delivered over the same basic network platform that is used for Internet access. It reuses many of the common network elements (DSLAM and ATM switch), therefore enabling a DSL network provider to better leverage these assets as shown in Figure 9.18.

However, unlike for Internet access services, the DSL or ATM network platform synchronization becomes far more important when used to deliver TDM services. This is vital for some leased line substitution scenarios,

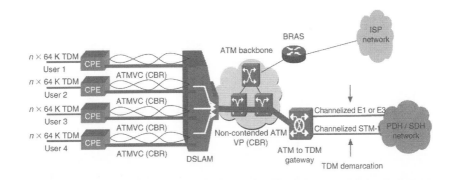

Figure 9.18 **Example of TDM service co-existing on DSL Internet access network platform.**

e.g., for private branch exchanges (PBXs) interconnection or for mobile base-station access. Mobile handover (mobile handset "cutting-over" to a new base station as the end user moves between radio cells) requires synchronized adjacent base stations. Any synchronization problems would cause handover to fail resulting in dropped calls. In mobile networks, radio frequency (RF) stability of base-stations is important in this process. Because the base stations derive synchronization from the fixed network access lines, the DSL network platform synchronization quality is key. Latency is also a very important parameter in TDM over DSL or ATM networks to meet international leased line specifications, especially for transport of voice services.

9.3 Evolution of the Architecture and DSL Delivered Services

This part gives an overview of some of the possibilities for the DSL network architecture evolution. It describes fundamental network planning considerations and explores the more prominent services that are beginning to be delivered over DSL. These services depend on leveraging underlying infrastructure enhancements as well as newer technology platforms. Going beyond best-effort Internet access, Part 2 of this DSL architecture chapter describes how the architecture can continue to evolve from the initially deployed DSL access network platforms and interconnection options. The initial phase of DSL deployment was focused on "build and operate" with the prime objective being geographic coverage. The main service offered was simply fast Internet access. To cross the chasm between early adopters and mass market take up of DSL, a wider range of compelling services and advanced applications needs to be offered and the DSL network platform must be capable of supporting them. Hence, following initial deployment, service providers then embarked on service development and personalization of services with a plethora of bundled packages and options. The prime objective at this stage is customer retention and increasing the average revenue per user (ARPU).

The three key technologies on which DSL depends to deliver advanced services are ATM, IP, and Ethernet. This could potentially lead to some divergence in alternative extensions of the DSL network platform and the technologies that underpin it. In addition to extending the use of ATM, IP, and Ethernet to enhance the capabilities of the DSL network platform, this section also describes other possible directions toward which the DSL network architecture can migrate.

9.3.1   Fundamentals of Network Evolution

Now that substantial DSL infrastructure has been deployed, network operators face the challenge of using it to provide such advanced services as digitally derived voice, video conferencing, gaming, as well as data, audio and video broadcast, streaming, and downloads. However, there are a number of challenges for building upon and interworking with the existing DSL network. Much of this depends on the following:

- Sound engineering economics supporting convergence
- Integrated systems that enable advanced services delivery
- Flexible architectures that can evolve to meet future demands

Some of the specific enhancements to DSL include increased allocated bandwidth, and assigning relative priorities and guarantees for traffic types, at the ATM, IP, and Ethernet layers. The initially allocated or provisioned bandwidth can be increased to support the advanced services requiring additional bandwidth, on an as needed basis. For each PVC between the DSLAM and the DSL modem, the line rate could be provisioned almost at the maximum rate at which the physical layer can achieve synchronization* (usable IP payload can be approximately 15 percent less allowing for protocol overhead of ATM, PPP, etc.). The allowable bandwidth for the end user can then be adjusted by reconfiguring or signaling the provisioned bandwidth or line profile. This can be implemented for higher priority traffic in addition to best-effort traffic. The end user's ability to make use of a larger amount of bandwidth on an as needed basis has been termed "Turbo Button."

Traffic can be aggregated from the ATM through the IP layers. At the BRAS, ATM PVCs can be aggregated. As was described in Part 1 of this chapter, at layer 2 the BRAS serves as an LAC tunneling multiple end user PPP sessions directly to an ISP. For further layer 2 aggregation, a Layer 2 Tunnel Switch (LTS) can be added to reduce the number of interconnections between the ISP's LNS and the regional access provider's LAC, with L2TP tunnels. The traffic engineering of these tunnels provides over-subscription of the network connection to a particular ISP. Going forward, MPLS can be used for aggregation, as well as for IP or ATM integration.

As newer traffic types that go beyond best effort are introduced, relative priorities and guarantees for these traffic types will have to be assigned. This differentiated treatment can be implemented by employing ATM, Ethernet, or IP QoS. In either approach, higher priority and guaranteed traffic can

---

* Factoring in planning rules that allow for a pessimistic crosstalk environment to ensure that the service will continue reliably beyond initial provisioning.

take precedence over best-effort traffic. Future architectures will make use of a single QoS system that will manage multiple traffic types in concert—including IP, PPP, and Ethernet.

9.3.2 ATM-Centric Architecture Evolution Possibilities

Not all of the initially deployed DSL architectures were ATM-centric. Some deployments in geographies with a high number of MTUs (multi-tenant units) used Ethernet-centric backhaul from the DSLAM. However, the majority of large deployments have initially used ATM on both the DSL access network and the backhaul from the DSLAM to an aggregation point of presence. This section overviews some of the evolution options where the DSL network provider had started initial deployment with an ATM architecture and wished to leverage that investment further.

9.3.2.1 QoS at the ATM Layer

If services offered are ATM based, it is clear that ATM QoS is an obvious means of ensuring quality for those types of services. Different ATM traffic classes are specifically designed to support different types of services. For example, an ATM voice service is very well supported using the VBR-rt traffic class. Other traffic classes also have their prime applications such as CBR, which is well suited to transporting leased lines or ATM-based video services using MPEG-2 over ATM for video streaming. For best-effort services that are offered with some level of assurance, VBR-nrt and UBR with an MCR guarantee can be used. Because ATM has been designed to support a mix of high speed data services, real-time, streamed, and downloadable audio and video services, ATM QoS has all the capabilities required for a full service access network. Bundles of services can then be delivered over DSL using multiple ATM VC per end user, whereby each VC traffic class can support a different application (e.g., data, voice, or video). Figure 9.19 illustrates the use of two VCs over ADSL to deliver integrated voice and data service.

Various levels of QoS can be implemented at the ATM layer using the different ATM service classes alone. In the DSL networks, the ATM UBR service class will continue to be deployed as an efficient means of providing best-effort Internet access services. End-to-end DSL networks commonly comprise of a series of links between ATM nodes. Just as a chain is only as strong as its weakest link, an end-to-end connection of an ATM service is only as robust as the "weakest" ATM link. In this way, the guaranteed quality specified by the link with least stringent QoS parameters (e.g., highest contention ratio and hence lowest sustainable cell rate) represents the level of assurance that can be offered to the end users or an interconnecting service provider.

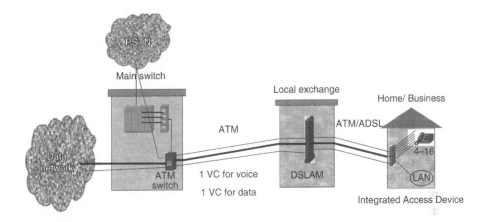

Figure 9.19   Use of two VCs over ADSL to deliver integrated voice and data service.

Note that some ATM switches have the capability to schedule the changing of a VP connection (VPC) to one with a different traffic profile. This can be exploited to offer a "time-of day" based service, where for example, a consumer is connected to a VPC with a high downstream rate during the evenings but a lower rate VPC during the day. This would therefore free-up more of the backhaul bandwidth during the day when businesses get a lower contended service. The DSL port on the DSLAM must be set to the higher of the downstream rates at the physical layer, but traffic "flow" is controlled at the ATM layer. Whilst this approach is technically viable, it does involve a break in the end user's transmission when the ATM switch changes VPC profiles (effectively tearing down one and re-connecting with the other). It is also not very scaleable in that there is a limit to how many end users to which this can be applied at the same time.

For IP-based services that are transported over DSL and backhaul ATM links without leveraging the full QoS and traffic engineering capabilities of ATM, IP QoS (discussed later) may be used in the end-to-end connection to facilitate service bundles with the appropriate prioritization and flow management. Hence, one of the first major decisions in evolving from a best-effort architecture is at which layer should QoS (especially for service bundles) be managed. Often this decision also translates into how many ATM VCs should be provided to the end user. For example, to offer an integrated IP voice and data bundle, some form of QoS needs to be employed to ensure that a large data packet being sent from an end user over the DSL does not hold up a more time critical voice packet. This is because voice is a real-time service whose quality is more dependent on delay and delay-variation than data such as e-mail. If ATM QoS is employed, then use of two ATM VCs (i.e., VBR-rt for voice and UBR for data) naturally fragments* the packets into ATM cells so that the cells carrying voice can be given precedence over cells carrying data traffic. Buffer management and scheduling algorithms in DSL CPE, DSLAMs, and ATM switches ensure that the higher priority VBR-rt voice traffic is naturally expedited through the network to meet the delay and delay variation requirements. Without the use of ATM layer multiple-VC QoS techniques, the IP layer must be used to deal with prioritization and control of delay and delay-variation. This may require fragmentation of IP packets with subsequent increased processing load on the IP routing elements in the architecture such as the BRAS.

Many initial SHDSL deployments were specifically designed with multi-VC bundles to develop services for businesses. For example, a combined VBR-rt VC together with a UBR or VBR-nrt VC facilitates delivery of an integrated voice and data service. A CBR VC combined with VBR-nrt emulates a leased line plus data bundle or video conferencing stream with an associated control channel. Another option is to deliver, e.g., 8 UBR or VBR-nrt VCs to enable service provider's targeting MTUs to offer a VC per office or apartment in the building to segregate and manage the individual traffic streams all carried over the same SHDSL connection.

ADSL, however, has been primarily deployed with a single UBR VC for best-effort Internet access. Some deployments have also deployed multiple VCs for consumer VoD services using, e.g., CBR for the downstream video payload together with a UBR or VBR-nrt VC for control and perhaps simultaneous Internet access. ADSL integrated access devices (IADs) capable of supporting voice (via a VBR-rt VC) and data (via a UBR or VBR-nrt VC) have also been developed by some vendors and are discussed in the next section.

9.3.2.2 Voice over ATM

POTS (existing analog telephony) can be maintained within the ADSL architecture as illustrated in Figure 9.20. Figure 9.21 is a generic diagram of how additional voice channels can be derived from within the ADSL digital payload. There are three key methods for deriving additional voice channels from the DSL access connection (beyond the "lifeline" POTS baseband voice channel available on ADSL). "Channelized Voice over DSL" (CVoDSL) uses a fixed portion of the DSL physical layer bandwidth for voice, therefore, it is relatively inflexible in that this bandwidth is not readily made available for data when the voice channels are not in use. It is however simple and carries no significant protocol overhead on the voice channels. Voice-over-IP (VoIP) multiplexes voice with data at the IP layer. It is inherently flexible but can result in inefficient protocol overhead, which consumes a large amount of the DSL bandwidth unless efficient voice codecs and header compression are used. VoIP is discussed

---

* Via the ATM SAR (Segmentation And Reassembly) function which performs Segmentation And Reassembly (SAR).

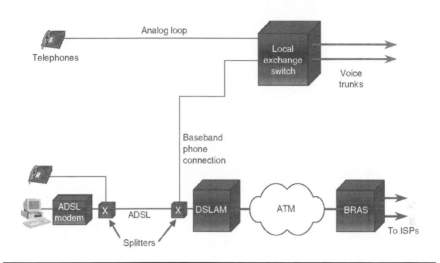

Figure 9.20   Existing analog telephony within an ADSL architecture.

later in this chapter. VoATM can be positioned as a compromise in that it can dynamically reallocate voice channel bandwidth to data without a complex protocol stack. The Voice over ATM (VoATM) protocol stack is illustrated in Figure 9.22 together with CVoDSL and VoIP for comparison.

ATM provides many of the features required for voice support over a limited DSL access bandwidth such as small packet size (cells), prioritization, and call admission control. Part 1 of this chapter illustrates how circuit emulation using ATM with Adaptation Layer 1 (AAL1) and CBR traffic class could be used to emulate leased lines over an ATM-based DSL network. This can be used to transport voice in TDM timeslots, which are transported over ATM (usually in conjunction with SHDSL). Service providers have used this approach to interconnect PBX equipment to establish a virtual private voice network as is attained with traditional leased lines. This approach is

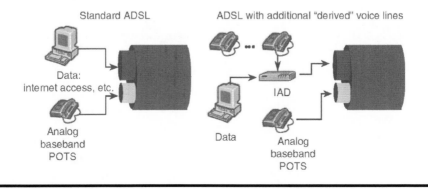

Figure 9.21   ADSL with "derived" voice.

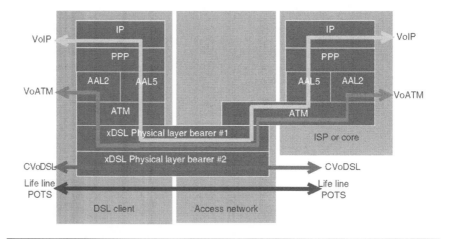

Figure 9.22 **VoDSL protocol stacks (ADSL with "lifeline" POTS shown).**

ideal for providing dedicated voice trunk connections. However, it is not so amenable to integrated voice and data delivery because the CBR traffic class used to emulate the leased line requires that the DSL bandwidth dedicated for voice is allocated during provisioning and so is unavailable to provide extra capacity for data when voice calls are not active. An alternative approach is to use ATM Adaptation Layer 2 (AAL2) with VBR-rt traffic class. This can multiplex a number of voice circuits on a single ATM VC (as shown in Figure 9.23) and the DSL bandwidth occupied by this voice VC can be dynamically "re-allocated" to a second ATM VC that carries data (e.g., UBR or VBR-nrt VC). This makes more effective use of the DSL access capacity for integrated voice and data services.

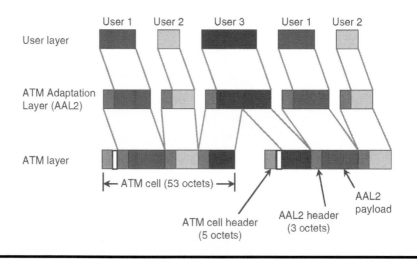

Figure 9.23 **Overview of AAL2 protocol for VoATM.**

With ADSL, in addition to the analog phone that transmits voice in the 4 kHz baseband over the same phone line as ADSL, additional analog phones can be connected via an IAD. Within an IAD, analog phone signals are mapped into ATM for transport over DSL by way of an interworking function (IWF).

Similarly, VoATM is converted to and from audible analog voice at the CO or another centralized location for interconnection to the public switch telephone network (PSTN). A corresponding IWF is also integrated into a VoDSL gateway, which is then connected to the Class 5 switch via either V5.2 (European), GR-303 (North American), or TR-008 (North American) standardized concentrating interface on a PSTN switch. The availability of these interfaces on Class 5 phone network switches is not ubiquitous, so large deployments have been somewhat sparse. DSLAMs are available with integrated VoDSL gateways that enable the voice traffic to be "broken out" at the local CO for direct interconnect to the corresponding local Class 5 voice switch. This sort of distributed gateway architecture is most applicable for incumbent service providers who own the local switches and can afford multiple interconnection points. Most early VoDSL deployments were by alternative service providers that used a more centralized approach to VoDSL gateway location thus increasing the degree of "PSTN bypass" and reducing the number of PSTN interconnection points. An illustration of connections to the PSTN for VoDSL using an integrated access device (IAD) is in Figure 9.24.

VoATM that is supported over DSL has been commonly referred to as Broadband Loop Emulation Service (BLES). The requisite standards and

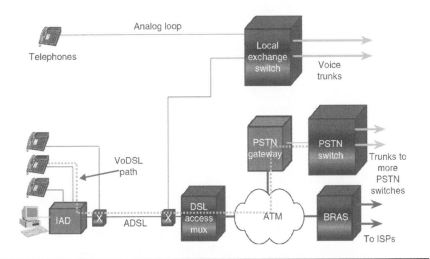

Figure 9.24 Voice over DSL (VoDSL) showing Integrated Access Device (IAD) and Voice over ATM (VoATM) gateway to public switch telephone network (PSTN).

interoperability test specifications have been developed by the ATM Forum and the DSL Forum [ATM Forum (b)] and [DSL Forum TR-039]. As the name implies, by employing ATM technology a DSL emulates a copper loop that carries analog voice. Hence, BLES connects readily to the existing PSTN, without modification of legacy analog phones and Class 5 PSTN network switches. Because BLES emulates analog loop technology, basic POTS and ISDN service as well as many phone features are supported transparently to the end user. BLES provides digitally derived voice that, as the name implies, seeks to emulate existing PSTN voice in terms of both quality and features. It reuses the existing PSTN narrowband switching (beyond the gateway), call control protocols, billing, and CPE (analog phones).

Technology developments may facilitate further evolution of this VoDSL architecture. It is technically feasible to include the PSTN line card function* on the DSLAM's ADSL cards. This avoids the need for an ADSL POTS splitter at the DSLAM and enables the PSTN voice traffic to be backhauled over the same ATM network as the digitally derived voice channels. This could eventually lead to separate local Class 5 PSTN switches or concentrators being redundant, as all voice traffic could be backhauled to a more centralized switch. Service providers may find this attractive for rural or new build areas. Another potential development with respect to ADSL is the "all digital loop." This involves extending the ADSL spectra down to baseband instead of leaving a 4 kHz band free for analog POTS signals. With ADSL2, the additional digital capacity of this approach is up to 256 kbit/s, which is sufficient to transport around three to six toll quality voice channels (depending on the voice codec used).

The challenge of any derived voice technology is building in enough reliability for services such as calls to the emergency services, which is assumed as always accessible over an analog POTS line, even if electrical power is down at the end-user premises. Traditionally, analog phones have been powered remotely from the CO. Today's DSL modems and IADs require local powering at the customer's premises. The requirements for supporting emergency power for telephony on digitally derived voice services in the event of local power failure are usually subject to local and country-specific regulations.

9.3.2.3 Mobile Base Station Feeds

SHDSL using circuit emulation over ATM can be used as an alternative to a leased line, hence it can be used to provide connectivity from the base station controller (BSC) to the TDM-oriented base stations (base wireless terminal [BTS]) of second generation (2G) mobile telephony service

---

* Known as SLIC or end-user line interface circuit.

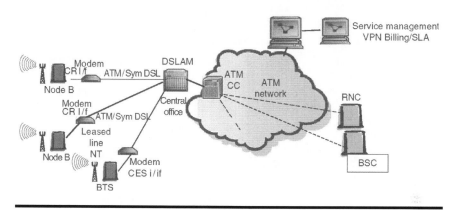

Figure 9.25   SHDSL feeds to 3G cellular base stations.

providers. Figure 9.25 illustrates symmetrical ATM based DSL (i.e. SHDSL) connectivity to 3G cellular base stations. 2G services are characterized by the GSM (Global System for Mobile Communications), CDMA (code division multiple access), and TDMA (time division multiple access) standards. The 2.5G (General Packet Radio Service [GPRS]) data service is also very much a TDM-based technology; hence it can exploit circuit emulation services (CES) over ATM on SHDSL. However, third generation mobile services (3G) use ATM backhaul transport. Relevant 3G standards include Universal Mobile Telecommunications System (UMTS), and CDMA2000 or Wideband-CDMA (W-CDMA). The standards for W-CDMA define that all the elements of the Terrestrial Radio Access Network (known as UTRAN [UMTS Terrestrial Radio Access Network] in UMTS systems) be interconnected over an ATM transport network. Mobile base-station equipment vendors have subsequently incorporated ATM ports in their base-station switches. Consequently, some DSL CPE vendors have developed SHDSL CPE specifically with ATM or "cell relay" (CR) interfaces for direct connection to the 3G base stations (Node Bs). Service providers had to pay a lot of money for the licenses to the 3G spectrum. They therefore need to ensure that the large backhaul networks required to connect the Node B base stations to the Radio Network Controller (RNC) are constructed as cost-effectively as possible. To deliver broadband services over 3G mobile networks, a larger number of small radio coverage cells are required (micro cells and pico cells), as illustrated in Figure 9.26. This makes a compelling case for the use of SHDSL in locations where it can prove more cost-effective than a traditional leased line. SHDSL CPE has been developed specifically for this application to include –48 V power supplies and environmentally hardened equipment design. A number of service providers have explored this application of ATM over DSL, and it has been proven to work well. KPN Belgium conducted trials in 2002 [Cuypers 2002], where the base station connected

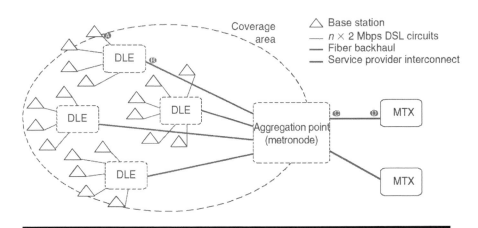

Figure 9.26   DSL backhaul to microcells for 3G wireless.

via SHDSL was monitored for the percentage call blocking and dropped calls. The resulting performance met the requirements. KPN Belgium assumed that a base station typically supports 42 simultaneous mobile voice conversations at 16 kbit/s. Hence, the base station to mobile switching center connection would typically require 672 kbit/s of conversation data plus 128 kbit/s protocol data thus requiring the equivalent of 13 timeslots on a TDM leased line. For larger sites, higher data rates can be achieved using multiple copper pairs bonded using Inverse Multiplexing over ATM (IMA).

9.3.2.4   Video-on-Demand Using ATM

Video-on-demand (VoD) was initially conceived as the driving application for ADSL deployment long before it was used for Internet access. Consequently, there have been numerous trials and deployments of VoD around the world. VoD enables end users to select the content they want, when they want it, without being constrained by broadcast times and content in programming schedules. Hence, numerous end users can all be simultaneously watching different movies, and some may even be watching the same movie but a different part of it. Typically, the downstream video has been delivered at speeds between 1.5 and 2 Mbit/s (achieved via MPEG1 or MPEG2 video compression) over an ATM PVC having a CBR traffic class. A lower rate upstream or bidirectional control channel of around 16–128 kbit/s is used for control purposes, and this PVC can be UBR or VBR-nrt traffic class. The control channel allows the user to select movies or previews as well as start, stop, rewind, and fast-forward the movie as if using a virtual VCR. In some deployments, there has also been a low rate PVC for simultaneous Internet access (e.g., 115 kbit/s emanating from the serial port of a set top box).

The hold time for a VoD session can be a couple of hours (determined by movie length). Unlike Web surfing, the traffic is not bursty. Therefore, data packets for individual user's sessions cannot be as efficiently multiplexed on the same backhaul bandwidth. Hence, where as Internet access may have a contention ratios varying from say 10:1 (business grade) to 50:1 (consumer grade), it is not possible to allocate as many simultaneous VoD users per unit bandwidth. Taking into account how many end users who subscribe to the service may be simultaneously active together with the typical session hold-time, a VoD service may only overbook backhaul bandwidth by a ratio of around 3:1 to 4:1. Therefore, in terms of network resources, it can be more costly to deliver than Internet access.

More recently, service providers have looked to deliver VoD services at the IP layer. This is illustrated in Section 9.3.3.4.

9.3.2.5 Broadcast TV Using ATM

Broadcast television (BTV) services involve delivery of content including a large variety of regularly scheduled programming. This differs from VoD or pay per view, because the video source is not under control by any of the end users.

The delivery of BTV involves a head-end architecture to support multicast capable DSLAMs. The head-end is the part of the network where the video channels are aggregated, encoded, and passed through the network to the DSLAMs and finally to the end users. Encoders stream each of the broadcast channels across the broadband network on a separate PVC to a multicast capable DSLAM. Video content is broadcast to end users using IP packets encapsulated into AAL5 and transported over ATM using multicast connections. The data for each broadcast channel is assigned an IP multicast address. Content channels can be delivered over the allotted number of PVCs to the DSLAMs. Each channel is sent to the DSLAMs within the viewing area. Within the DSLAM, the data is replicated using the multicast capability along with ATM PVCs. The DSLAMs are implemented with capabilities to switch ATM traffic to multiple DSLs for delivery to end users' homes. Each end user who requests a particular broadcast channel has the appropriate content sent to their respective port on the DSLAM. From the DSLAM, the video is transmitted over the loop to a set top box, and then presented on a TV or PC monitor.

At the DSLAM, Internet Group Management Protocol (IGMP) multicast control is terminated on the DSLAM. Several PVCs are required per end user including those for forwarding video multicast traffic as well as for IGMP channel and head-end video on demand traffic. As the PVCs within the DSLAM provide streams to individual end users, the DSLAM only forward a channel to legitimate end users who request it. In the case where multiple set top boxes on the same DSL port request the same channel, only one copy is sent to that port. Also, see the use of IGMP with IP in "IP Multicasting."

9.3.2.6 Full Services Access Network

To effectively address the residential market, many analysts believe that service providers need to offer a "triple play" of video, voice, and data. Given the limited bandwidth on DSL, it is a challenge to enable all three service types to be offered in a manner in which they can be used simultaneously. ADSL is capable of offering such a triple play if operated at higher rates such as 4 Mbit/s. However, to offer multiple simultaneous video channels, higher access speeds are required. This can be resolved by providing access transport using either fiber to the home or via a hybrid fiber or copper access network, whereby Very High Bit-Rate DSL (VDSL) operating at over 10 Mbit/s is used over the final copper pair into the home (see Figure 9.27).

The Full Service Access Network (FSAN) organization, now a group within the ITU-T, has produced a series of specifications to address this triple play delivery over Passive Optical Networks (PONs), Hybrid Fiber (PON), and VDSL architectures. The specifications cover architecture, OAM, CPE, and even aspects such as Digital Rights Management (DRM). Fundamentally, the architectures use ATM QoS approaches but in conjunction with IP-centric approaches for signaling [FS-VDSL].

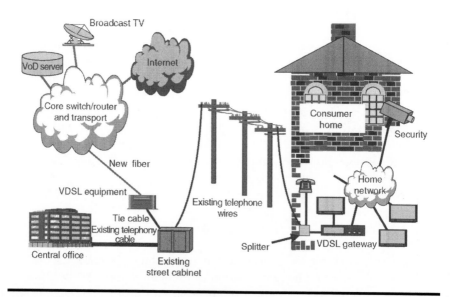

Figure 9.27   Full Service Access Network (FSAN) VDSL access architecture.

9.3.3   IP-Centric Architecture Evolution Possibilities

9.3.3.1   IP Quality of Service

IP QoS can be implemented over an IP routed network architecture to support ASP service offerings, while supporting current best-effort and new ISP service offerings over the same DSL. The PTA described in Part 1 helps facilitate this by extracting the IP packets from PPP sessions, which for example enables "QoS markings" in IP packet headers to be examined. To introduce IP QoS over the existing architecture, a number of prerequisite capabilities are implemented at various parts of the network and customer premises. By employing the concepts defined in Diffserv [Grossman 2002] multiple traffic types can be managed in concert including IP, PPP, and Ethernet.

To extend current network capabilities for support of advanced services using IP QoS and Diffserv queues there is a requirement to preprovision layer 2 ATM (or Ethernet) network capacity between the BRAS and that CPE. Traffic is simply transported over the physical and ATM (or Ethernet) layers to and from the DSLAM, and is throttled and policed at the IP layer at the BRAS. The BRAS therefore manages the IP QoS and polices individual sessions or "microflows." This is a more sophisticated approach than simply a priori "hard" partitioning of bandwidth between applications or relying upon distributed precedence approaches that only ensure "fairness" between traffic classes but not between users within a traffic class. The IP QoS is transparent to any DSLAMs that have already been deployed without IP awareness.

Additional ATM traffic engineering of an existing hierarchy of ATM network elements and their branched connections may need to be re-examined. This hierarchy of ATM network elements can include a DSLAM and possibly subtended DSLAMs, an ATM switch, and a BRAS, as depicted in Figure 9.28. The shaping of traffic across these elements entails buffering the aggregates of traffic at a queue with a specific allowable throughput. The fewer the ATM hops from the DSLAM to the BRAS, the easier it is to manage congestion of the ATM connections and the downstream IP QoS. The BRAS manages potential congestion over these ATM hops to the DSLAM, in part by having Diffserv capabilities. IP packets can be extracted from Diffserv queues in an order according to the relative priority of the packets. For the traffic flowing toward the various end-user premises, the hierarchical shaping of aggregate traffic is based on ATM connection rates, traffic flows of PPP sessions, and IP service classifications.

The IP packets in the Diffserv queues are labeled* using Diffserv code points (DSCPs), which consist of best effort (BE), expedited forwarding

---

* Sometimes referred to as "coloring" the packets.

374 ■ *Implementation and Applications of DSL Technology*

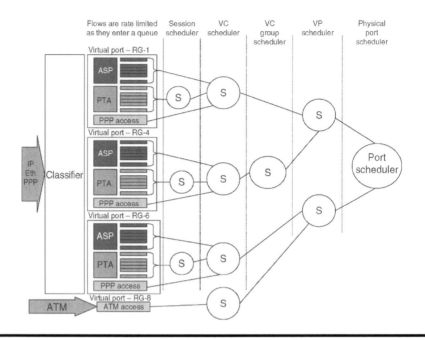

Figure 9.28   Hierarchical scheduling with Internet Protocol Quality of Service (IP QoS) in an ATM-based DSL network. (From DSL Forum TR-059, *DSL Evolution— Arichitecture Requirement for the Support of QoS-Enabled IP Services.***)**

(EF), and assured forwarding (AF). BE is for the default traffic, such as Internet access. EF is typically for high priority, latency-sensitive services such as voice. AF classes provide a minimum guaranteed bandwidth with definable maximum burst, committed burst, and excess burst.

Although QoS is implemented at the IP layer, the architecture may still depend on the use of ATM PVCs over the DSL link and possibly the aggregation network. From an ATM simplification perspective, only one UBR PVC is necessary to support all IP-based services over a DSL. This then dissociates the layer 2 (ATM) connectivity from the QoS design thus simplifying provisioning. Real time media applications such as voice and video require adequate resources along the transport path and hence require specific QoS support. If a second PVC is implemented over the same DSL, a PVC with VBR-rt or CBR QOS may be preferred to support such applications requiring specific delay and jitter, and therefore could be used to transport EF or AF marked packets for improved delivery in comparison to BE marked packets. This is tantamount to fragmenting traffic flows at the ATM layer (e.g., to allow voice to be put in ATM cells ahead of data traffic cells) and can avoid the need to fragment IP packets or Ethernet frames at higher layers on slower DSL links. Figure 9.29 illustrates the IP-routed network architecture with connections to an (ASP).

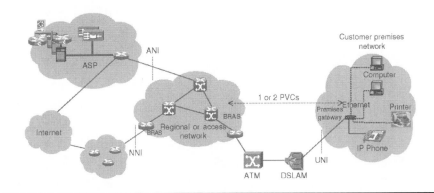

Figure 9.29   IP routed network architecture to an application service provider (ASP).

Alternatively, multiple real-time applications over a single DSL connection can be supported over just one ATM PVC. Voice-over-IP would normally be engineered to utilize EF IP QoS, which has the highest level of priority. When supporting VoIP, the larger frames originating could be fragmented and given higher priority than data packets. This is more important on the lower speed, ADSL upstream link.

Best-effort traffic to ISPs can continue to be transported over L2TP tunnels in this architecture, as well as the bridged Ethernet and other service provider interconnections described earlier.

IP QoS could be provisioned or signaled using session control protocols such as SIP or Resource Reservation Protocol (RSVP [IETF RFC 2205]). A resource reservation protocol gates the admission of the session over the allotted network resources, and provides feedback to the initiating CPE or BRAS. The session control protocols incorporate QoS elements into session signaling messages. There is a trade-off between simply preprovisioning network resources and the higher efficiency (but greater complexity) of dynamically signaled resource requests and allocations. An intermediate approach that is being increasingly adopted is the use of policy servers that dynamically react in a predefined manner, e.g., using hierarchical scheduling and shared shapers in the BRAS to back-off Internet throughput when a video session starts.

As networks grow to support more advanced services over an increasing number of DSLs, IP policies are increasingly being implemented in separate policy servers to enforce QoS relationships. These policies are usually enforced at the BRAS and possibly also the CPE. The policy manager controls network resources and sits above network elements at a common enabling services layer in the DSL Forum generic reference model (see Figure 9.30). In future, the policy enforcement may extend to DSLAMs, application servers, or even deep packet inspection (DPI) devices that are application aware and can be colocated with the BRAS.

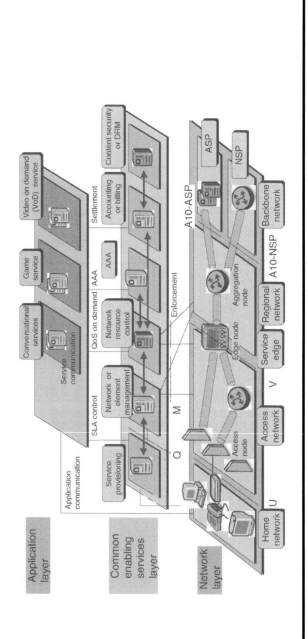

Figure 9.30   **Generic DSL reference model. (From DSL Forum TR-058, Multi-service Architecture and Framework Requirements.)**

When employing IP QoS in the DSL network architecture, the BRAS manages traffic flows in the downstream bandwidth based on the particular end user's service profile, while the CPE manages the upstream traffic shaping. The IP flows are shaped and policed between the BRAS and the end user's CPE, transparently through the DSLAM. The BRAS and CPE (effectively a routing gateway [RG]) are therefore the main IP policy enforcement points. Both support multiple queues per end user with appropriate scheduling mechanisms to effectively implement Diffserv queuing behavior. In addition to supporting BE, EF, and AF behavioral characteristics on a per end-user basis, both the BRAS and RG can have multiple queues per traffic class.

The BRAS manages the admission of sessions toward the customer premises. An external policy server can be used to send both admission control and session policy parameters to the BRAS. Depending on how the policy server executes the user's service policy, a per-flow admission control can be achieved at the network layer or the application layer. The BRAS can prioritize network resources by mapping traffic flows into the Diffserv Per Hop Behaviors (PHB). This is implemented at the BRAS via bandwidth allocation based on IP traffic class (BE, EF, or AF), with appropriate queues per traffic class.

To be an effective IP policy enforcement point, it is preferable for the CPE to have the capability to identify DSCPs on incoming IP packets from end-user applications to make upstream traffic shaping or policing decisions. In addition, the CPE should also be capable of supporting low jitter traffic on a single ATM PVC (e.g., via use of packet or frame fragmentation techniques).

9.3.3.2 IP Multicasting

Multicasting is a network capability for sending information from one source to multiple receiving participants. It involves the replication of channels on demand. Multicast "trees" are built to efficiently route streaming content through the network to avoid duplication of content transmission. Because multicasting involves the replication of channels on demand, it is being considered useful for broadcasting content streams, such as broadcast television (i.e., IPTV). In general, broadcast involves delivery of a single copy to all requesting or participating end users over the entire sub-network. Broadcast television services involve distributing content comprising a large variety of regularly scheduled programming. This differs from VoD because the video source is not under explicit control by any of the end users. With traditional broadcast technology, packets are replicated to all routers and hosts on the network. It is therefore not suited to wide area streaming distribution, because it does not scale well over a wide area. With multicast, routers are prevented from unnecessarily transporting broadcast packets, hence enabling multicast to more efficiently use network bandwidth than a pure broadcast approach—so traffic only flows to those end users that want it.

The multicast capability can also be located in the DSLAM, instead of just the BRAS. When the multicast function is placed in the DSLAM, hence closer to the end user, a single content stream to the DSLAM can then be replicated for distribution to those end users on the DSLAM that have requested the same content. This can therefore provide network transmission efficiencies in locations where there is high take-up of similar services in the same DSLAM coverage area.

In a "single edge" network architecture, IP multicast services (typically video or radio channel broadcasts) are aggregated for delivery toward end users at the BRAS which is therefore the multicast router. However, in more distributed "multi-edge" architectures, multicast services can be injected into the aggregation network via a separate device. Both the BRAS and the DSLAM network elements can be designed to be multicast replication points. The BRAS achieves point to multipoint multicast replication toward multiple DSLAMs at the IP layer and the DSLAM replicates at layer 2 (ATM cells or Ethernet frames) toward multiple users requesting the same channel. A DSL network provider that employs multicast capabilities can potentially sell the ability to inject multicast to an ISP or ASP. An ISP or ASP can also employ multicast capabilities in their own router or BRAS. From a BRAS, multicast capabilities deliver streamed content, as depicted in Figure 9.31.

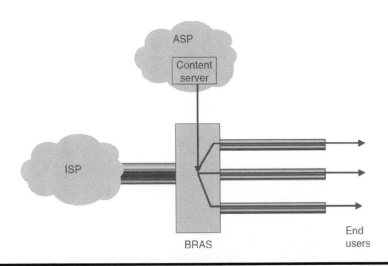

Figure 9.31   Streaming media with multicast at a network access provider's Broadband Remote Access Server (BRAS).

The prevailing protocol used for multicasting is the IGMP. The IGMP runs between hosts and their immediate neighboring multicast router. Based on the end user membership information learned from IGMP, a multicast router (which could be a BRAS) is able to determine if any multicast traffic needs to be forwarded to any of its leaf sub-networks (i.e., connections to DSLAMs). To set up and maintain multicast trees, multicast leafs are created in response to an IGMP join.

Multicast addresses identify particular transmission sessions, rather than a physical destination as in IP routing. The multicast addresses are translated by network routers into host addresses, so the source does not have to know all of the target end-user addresses. The multicast addresses are allocated to prevent collisions.

9.3.3.3 IP Awareness and Routing in DSLAMs and the Aggregation Network

DSLAMs are primarily layer 2 devices that initially provided just ATM aggregations and cross-connection functions. With the evolution toward Ethernet aggregation between the DSLAM and BRAS, DSLAMs have evolved to be able to cross-connect ATM PVCs on the DSL side to Ethernet VLANs on the trunk or aggregation side.

As the QoS and security features of IP are enhanced and routers become ever faster, it is not unreasonable to envisage some broadband IP network platforms evolving toward a fully routed IP network extending out to the DSLAMs (and non-wireline equivalents) and perhaps even as far as end user's CPE. Indeed it makes sense to put IP awareness into DSLAMs (especially multicast capabilities such as understanding IGMP), because their locations are ideal replication points for the distribution of streamed multicast video.

Figure 9.32 illustrates a DSLAM with IP routing. If IP routing is incorporated into DSLAMs, PPP can be terminated in the DSLAM instead of the BRAS. A RADIUS client can then reside in the DSLAM for assisting with the authentication process. A DSLAM may integrate IP capabilities with its existing ATM capabilities, as described in Section 9.3.3.2.

The desire for having IP capabilities distributed to the DSLAM depends on the forecasted community of interest for delivery of IP-based services. This contrasts with the earlier approach to introduce IP QoS, because with that approach more centralized hubbing at the BRAS is appropriate in the near term. A centralized approach may be more appropriate, if other types of broadband access are being offered by the DSL network provider such as VDSL or FTTH. In that case, the DSLAM would be treated like other alternative access technologies as a simple layer 2 aggregation node with broadband transport connectivity. Nevertheless, there is some interest in examining architectures with more capabilities

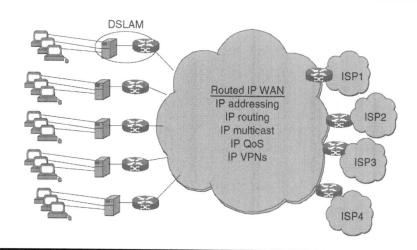

Figure 9.32   Colocation of IP routing functionality with the DSL Access Multiplier (DSLAM).

for handling layer 3 traffic in the DSLAMs for services such as gaming and other peer-to-peer (P2P) end-user services. These services can cause a different concentration of traffic flows on the network between local end users (as opposed to just client to central server communications such as for Web surfing). Without local routing at the DSLAM, local P2P communications between end users within the same CO may result in traffic being backhauled to a centralized BRAS or router before "tromboning" back to a colocated end-user peer. This is illustrated in the Figure 9.33. Hence, network architectures may need to evolve for optimum efficiency if such traffic flows become a significant percentage of overall traffic volumes.

However, moving to a distributed architecture whereby IP routing and perhaps even BRAS functionality resides in the DSLAM brings with it significant manageability issues. A DSL network typically comprises dozens of DSLAMs for every BRAS. Hence, giving the DSLAM IP capabilities beyond their traditional layer 2 (ATM or Ethernet) aggregation role presents an operational challenge.

One of the attributes of moving to an Ethernet-centric architecture is increased flexibility in the location of the BRAS in the network. The BRAS frequently is the layer 2 or layer 3 dividing line defining the "IP edge" and in many ways defines the MAN or WAN boundary. Whereas there may be good reasons to employ increasingly nondeterministic layer 3 behavior in the core, the motivation in the aggregation and backhaul network is more suspect especially as carriers look to converge more and more legacy applications onto a common network. Nondeterministic

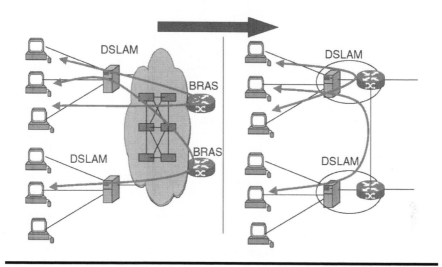

Figure 9.33 Avoidance of "tromboning" for local peer-to-peer (P2P) traffic via colocation of IP routing functionality within the DSLAM.

behavior requires a high degree of aggregation (to obtain a sufficient statistical sample) and a relatively dense mesh. Neither precondition is true in the aggregation nor backhaul network between the RG and the BRAS switches.

This suggests a solution in which the BRASs are pushed out toward the customer, to make the MAN component of the network trivial in networking terms. However,

- Layer 3 is challenged to scale to the metro edge (e.g., impact on efficiency of IP address allocation).
- The closer to the edge that nondeterministic behavior is pushed, the less benefit is accrued because of the reduced aggregation.

The alternative is backhauling customer traffic to a small number of major PoPs at the edge of a highly aggregated WAN core. This then challenges MAN technologies to deliver deterministic behavior over a domain of significant scale, greater than can be addressed by the lowest cost switching technology, Ethernet, in its original form.

This indicates that BRAS placement and determining the relative scales of MAN and WAN are a crucial design decision and understanding the consequences of the various trade-offs is vital to achieve an operable network.

9.3.3.4 Caching and Content Distribution

The earlier description of introducing IP QoS into the network above falls short of providing the capabilities required for a full service access network that enables VoD and broadcast television over DSL (in addition to high speed data services and voice services). To distribute content, capabilities for streaming, and local caching need to be added judiciously to balance the load on the network. Video can be encoded for more efficient distribution via the use of standardized compression techniques such as MPEG-2 and MPEG-4.

Media servers and content switches can work together using content distribution protocols and software, multicast routing, as well as redirection engines. Content servers may be grouped into content server farms. This would enable load balancing with caches to attain high availability, scalability, and performance. Caching involves the temporary storage of content. Its use can reduce IP backbone bandwidth consumption between the content source and the local point of presence (POP). Caching is often used to offload Web traffic from IP backbones. Consequently, faster downloads are discernible because the content does not have to traverse the otherwise larger extent of the network. Caching can be placed local to each BRAS to allow transparent caching for Internet traffic and streaming content. Caching capabilities can be dispersed as far toward the network edge as the customer premises, in set top boxes and PCs (e.g., using local hard disk storage as per Personal Video Recorder [PVR] devices).

In addition to content servers, content switches are employed to ensure optimum load balancing as well as to provide redundancy. Content switches are typically layer 3 and layer 4 switches. Based on server status and application availability, content based routing decisions can be made for distribution over DSL networks. This intelligent routing would be implemented across links between the server farms and the DSL networks. It employs algorithms for routing to the closest content distribution server. Performance is dependent on alternative links and fast fail-over routing algorithms. When an end user enters a standard HTTP request of an origin server, e.g., a uniform resource locator (URL) of the closest content distribution server is returned. A number of available content switches can operate at the IP layer as fast routers and at the TCP layer (layer 4), thereby taking into consideration the type of application as well as the network topology and status. Application-aware routing may show promise for coping with P2P traffic anomalies, as well.

Complementing the caching capabilities, streaming can be implemented to enable the end user to begin to view content, while it is being downloaded into a computer. The content can be injected or hosted locally at servers that are directly connected to the BRAS, and offered as a service to ASPs and other content providers (who may be wholesale customers of the

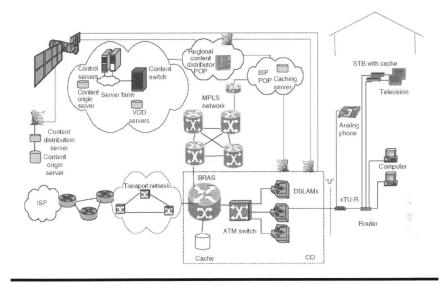

Figure 9.34   Video and content delivery with distributed caching.

DSL network provider). The actual video streaming can be delivered via unicast or multicast. With unicast, a point-to-point connection as used for VoD can be delivered. With multicast, streaming uses a point-to-multipoint distribution of content by a service provider. A service provider can schedule the streaming of live multicasts, or end users can request streaming via VoD.

The DSL network architecture diagram of Figure 9.34 illustrates how content servers and caches and switches are added to deliver video and other media types. In addition, satellite data relay is used to initiate and replicate streaming media for distribution to multiple regional and local POPs, simultaneously. Content providers pay for satellite data relay service, whereby content is sent from a content origin server via a content distribution server to a satellite. It is also possible to add a video server or cache capability on a card within the DSLAM to further enhance the end user's experience of a streaming video service and the efficient use of backhaul network bandwidth.

9.3.3.5   A Comment on Peer-to-Peer File Sharing

One of the early "killer applications" of flat-rate, always-on broadband services has been P2P file sharing. This is where users share files across the Internet between their PCs. Typically, these files are MP3 music files or MPEG encoded movies. New protocol clients facilitate "many-to-many" automated resource sharing.

Hence, P2P subsequently became the dominant component of bandwidth used by residential Internet end users. Many home PCs can be running constantly as data servers "24 × 7" to download files on a global basis. As a result, over 60 percent of the traffic on many broadband networks became P2P traffic, originating from a disproportionately lower percentage of the user base.

P2P clients communicate with other clients in a completely random fashion, disregarding the traffic engineered topology of the network. As a consequence, ISPs are swamped with a flood of traffic that did not originate on their network because of the ad-hoc connectivity with P2P clients, some even from other countries. This impacts the level of service for "innocent" non-P2P users, which can result in churn. Other consequences are the surge in monthly Internet transit bandwidth fees and the need for unplanned expenditures in the network to increase capacity. This financial pressure undermines the business model of flat-rate basic Internet access and managing network costs through over-subscription becomes inadequate.

There are commercial approaches to mitigating the impact of P2P traffic. For example, some broadband ISPs have imposed download restrictions, e.g., targeting people who regularly download more than 1 GB per day. Others have introduced tiered pricing based on bandwidth consumption,* e.g., some ISPs' cheapest DSL services ban use of P2P in the acceptable use policy and to use P2P, one must pay for a higher service tier. However, there are also a number of technical approaches that can be used, predominantly by appropriately "processing" traffic flows at the IP layer. For example, an ISP may choose to apply restrictions by rate-limiting or "de-prioritizing" P2P traffic. An ISP could limit the number of file transfers per P2P protocol. They could also apply least cost routing ("intelligent routing") to P2P traffic to keep it "on-net" (thus avoiding Internet transit costs), if possible. It is also possible to "manage" the P2P protocol "chatter" that consumes bandwidth (up to 5 kbit/s per PC host and 200 kbit/s per P2P supernode or ultrapeer), even when hosts are idle and not actively sharing files. Such IP layer processing can be performed in routers, BRAS equipment, or specialist hardware specifically designed to alleviate P2P traffic problems.

The nature of P2P traffic will increase the prospects of ISPs and DSL network providers employing IP QoS and "high-touch" IP packet and flow handling within their architectures, to optimize network performance and capacity utilization at minimum cost. The use of policy servers to control the BRAS and perhaps adjunct deep packet inspection (DPI) devices may also increase the ways in which P2P traffic is "managed" at the IP and application layers.

---

* Which may be prepaid.

9.3.3.6 Voice over IP

The most promising form of digitally derived voice is VoIP, which can also be transported over the DSL network platform. There are a number of approaches to VoIP user-interfaces for the end user. VoIP telephones convert analog voice to VoIP within the telephone itself. Multi-service access nodes (MSANs) are essentially DSLAMs that can digitize analog voice on the copper line and convert it to IP packets on the DSL card. Alternatively, existing analog telephones can be connected to an IAD at the end-user's premises in which analog voice is converted to VoIP. This is achieved via an analog terminal adaptor (ATA) that plugs into a DSL router, or the adaptor may be contained in a separate device. The final approach is to use a software VoIP client running on the end users PC that can be upgraded for telephony via the use of a headset and a microphone. The major benefit of VoIP over DSL is that it allows further integration of voice with IP-based data and other applications.

Using IP phones, VoIP traffic traverses the DSL loop for connection to an IP-based "Softswitch." The Softswitches are interconnected to traditional switches of the PSTN. With analog phones, analog voice traffic is converted to IP in the IAD or MSAN and sent to the appropriate Softswitch. Unlike BLES (see Section 9.3.2.2), VoIP does not depend on the existing analog PSTN for features. This integration of voice (over IP) with DSL aligns with the telephony evolution toward the use of the Softswitch. Trunk gateways and SS7 gateways are used to interwork the Softswitch with the traditional PSTN (see Figure 9.35).

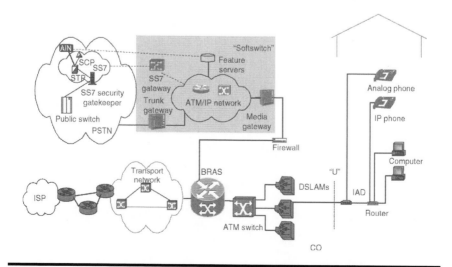

Figure 9.35   VoIP over DSL, connected to a Softswitch, and PSTN.

There are a host of issues that are being addressed in standards bodies including call control signaling and protocols (H.323, MGCP, H.248, SIP [RFC 2543]), CPE, numbering or addressing, NAT traversal, and billing.

Of the protocols being developed, SIP appears to be the preferred strategic choice as the protocol for creating, modifying, and terminating sessions and applications. These sessions are to include conferencing and gaming using Internet multimedia conferences and distribution, in addition to simple VoIP telephone calls. Going beyond traditional analog voice capabilities, members in a SIP VoIP session can communicate via multicast or via a mesh of unicast relations, or a combination of the two.

Some service providers have opted to offer VoIP as a cheaper (and possibly lower grade) voice service. For multisite businesses that have branch offices or employ teleworkers, this option allows for less costly 'on-net' calls, whereby dial tone is drawn from a PBX at a main corporate office (hence, the corporation pays the phone bill instead of the end user). Other service providers have ensured equivalent performance to traditional voice by employing QoS to offer guaranteed telephony quality, similar to that which users would experience on the PSTN. Developments are ongoing to ensure that addressing schemes do not over burden the routing of incoming calls and that voice packets are not delayed by large data packets on DSL links. In addition, DSL CPE such as routing gateways are being designed to overcome delays and jitter on voice calls via the use of fragmentation, prioritization, and scheduling techniques. Nevertheless, VoIP over DSL is in use today, despite contention and lack of QoS guarantees on best-effort DSL networks. For the most part, early VoIP deployments consisted of integrated voice and data enterprise solutions. However, public services began to emerge in 2003. Now that a critical mass of DSL broadband connections has been reached, the access network bottleneck has been removed for many end users consequently enhancing the quality and usability of VoIP.

9.3.4 ATM to Ethernet Architecture Evolution Possibilities

With the trend toward delivering a triple play service bundle comprising video services along with voice and data over the embedded DSL access infrastructure, the DSL industry began to develop architectures that migrate the layer 2 technology from ATM to Ethernet. A number of innovations in the area of Ethernet networking suggest an Ethernet-centric architecture is increasingly becoming a viable option for broadband networks. The specific innovations of interest are as follows:

- IEEE 802.1ag/ITU Y.1731 connectivity fault management which equips Ethernet with carrier class dataplane OAM tools for both performance and fault management

- IEEE 802.1ah Provider Backbone Bridging (PBB also known as MAC-inMAC) which provides Ethernet with the capability to stack customer traffic on provider Ethernets and support up to 2^{24} service instances.
- Provider Backbone Transport (PBT) which allows alternative control planes to be applied to Ethernet adding connection management and traffic engineering capabilities, particularly for private line and backhaul applications.
- Shortest path bridging or IETF trill which replaces spanning tree protocol with link state routing, greatly increasing the scalability, resilience, availability, and mesh utilization of bridged Ethernet.

Ethernet is already becoming the link layer of choice in new deployments and recent innovations in the Ethernet space equip it to take a broader role in network architectures.

Ethernet-based access can provide an improved equipment cost per megabits per second and faster transport mechanisms (gigabits per second instead of the 622 Mbit/s upper limit of many ATM architectures) to support the increase in connection speeds generated from ADSL2plus and VDSL2 DSL connections. Because of the interest in conveying video and primary line voice as part of the triple play service bundle, modern Ethernet architectures also seek to use effective multicast techniques (for broadcast video) and redundancy techniques (especially important for voice).

The Ethernet-centric DSL architectures involve deployment of a new access network topology, whereby the aggregation network connection between the DSLAM (access node) and the first IP layer edge router (referred to in such architectures as the broadband network gateway [BNG] instead of a BRAS) is Ethernet-based rather than ATM-based. The BNG has the edge routing capabilities of a BRAS, as well as the function to terminate the Ethernet layer and corresponding encapsulation protocols. Optionally, a DSL access connection can communicate to two separate BNGs (i.e., multi-homing) for high availability arrangements. A standard approach to Ethernet-centric DSL network architectures is covered in [DSL Forum TR-101]. [DSL Forum TR-101] also covers specific functional requirements for the access nodes, BNGs, RGs, and the Ethernet aggregation network.

The cost and bandwidth incentives to use Ethernet instead of ATM create an opportunity for service providers. Ethernet can be employed to emulate some of the capabilities that were previously inherent in ATM architectures; for example, VLAN tags can be employed instead of ATM's VPI/VCI for use in traffic identification and segregation. Also, the inherent broadcast domain nature of Ethernet can be exploited by moving away from the connection-oriented nature of ATM and effectively putting large groups of users from a single or multiple DSLAMs in a large VLAN.

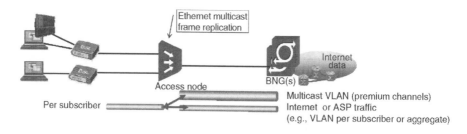

Figure 9.36   Ethernet-based access node connected to broadband network gateway (BNG).

Figure 9.36 depicts this new topology with an access node that is directly connected to a BNG. Alternatively, the access node can be connected to an Ethernet aggregation network consisting of Ethernet switches before connecting to the BNG as is illustrated in the [DSL Forum TR-101] reference model illustrated in Figure 9.37. The approach to QoS that uses hierarchical scheduling on a BRAS (as discussed in previous sections), can also be adapted to the BNG in an Ethernet VLAN environment.

9.3.4.1   Access Node Functionality

The aggregation network that is Ethernet-based consists of access nodes, Ethernet switches, and BNGs equipped with Ethernet interfaces. The fundamental changes from an ATM to an Ethernet-based access architecture

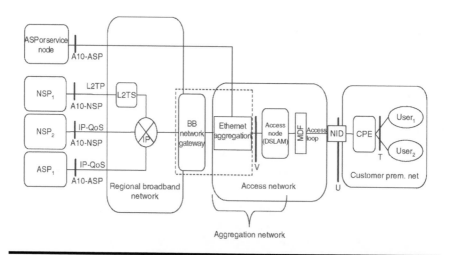

Figure 9.37   [DSL Forum TR-101] Ethernet-based DSL network reference model.

involve direct Ethernet framing at the access node and the support of Ethernet transmission at the trunk interfaces. Instead of ATM cross-connects and switches, the aggregation network now consists of Ethernet switches.

Similar to other access architectures, Ethernet-based access architectures refer to U-interfaces (DSL side) and V-interfaces (trunk side) on the access node (as illustrated in Figure 9.37). At the U-interface, the access loop typically supports IP transport via Ethernet encapsulation as IPoE (IP/Ethernet) and PPPoE (IP/PPP/PPPoE/Ethernet). The Ethernet frames may then be encapsulated within ATM for transmission across the DSL for compatibility with existing interoperable DSL CPE. Alternatively, some of the newer "Ethernet in the First Mile" (EFM) DSL standards may be used for providing an entirely native Ethernet layer 2 connection end-to-end.

In regions where PPPoA is used, an IWF in the access node can be implemented to map PPPoA end-user traffic to PPPoE (see Figure 9.38). The access node is now somewhat like an Ethernet switch with DSL interfaces. Additionally, the access node may have enhanced functions for other protocol interworking scenarios, security, multicast support, along with support of ARP and IGMP processing, end user identification, and end user isolation.

Figure 9.38 depicts the IP-based services as they are encapsulated on the end-user side into PPPoA, transported over the DSL using ATM at layer 2, "interworked" to PPPoE, and then encapsulated into VLANs.

An Ethernet-centric solution that maintains layer 2 end-user separation will minimize the complexity of the DSLAM both functionally and in terms of the amount of required configuration. Solutions are specified in the form of QinQ tag stacking or MACinMAC to provide scalable separation and defer the requirement to implement complex policy and layer 3 packet

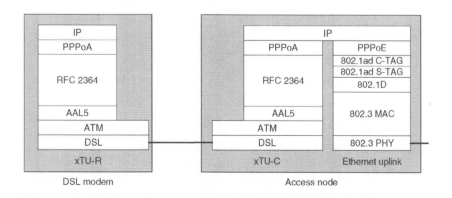

Figure 9.38   Encapsulation of the protocol stack into virtual LANs (VLANs).

processing until higher up in the network (e.g., at the BNG). This becomes an important consideration as the electrical or optical boundary is pushed into outside plant.

9.3.4.2   Ethernet Aggregation Network and VLANs

Ethernet switches (mainly using gigabit Ethernet interfaces) connecting the access nodes to the BNGs form the aggregation network (see Figure 9.39). The transition to an Ethernet-based DSL architecture can entail deployment of Ethernet VLAN switching, Ethernet frame-based prioritization of end-user traffic (QoS), multicast, and Ethernet OAM in the DSL aggregation network.

Access nodes are installed in a distributed manner which depends on local loop lengths as well as CO capacity requirements (number of DSL ports and associated backhaul link speed) which is also determined by local end-user demand. Access nodes may be installed in the CO to serve several thousand loops of longer loop lengths, while more distributed access nodes (remote) may be installed in RTs to serve much lower numbers of loops at shorter loop lengths in a hierarchical topology. This hierarchy distributes the Ethernet aggregation between the elements of the Ethernet switches in the aggregation network and the subtending access nodes (DSLAMs).

The Ethernet frame used in Ethernet-based DSL architectures includes information regarding VLAN identity and priority. The format is defined in IEEE 802.1D/Q standard. This information is transported in a four-byte

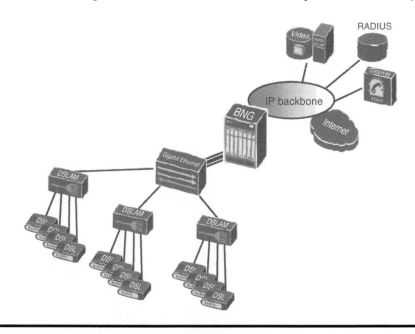

Figure 9.39   Aggregation network and BNG.

VLAN tag. The first 2 bytes have a fixed value $= 0 \times 81 - 00$, followed by 3-bits of priority, 1 bit $= 0$ and 12-bits of VLAN addressing. The 12-bit VLAN addressing space imposes a limitation of 4096 individual VLANs when a single VLAN tag is used. Hence, "stacked" VLANs (per IEEE 802.1ad) are used to increase scalability. This uses an outer VLAN tag (S-tag, sometimes referred to as the "Service" tag) and inner VLAN tag (C-tag, sometimes referred to as the "Customer" tag). A VLAN is referred to as an S-VLAN if there is no inner VLAN.

The Ethernet frame structure is depicted in Figure 9.40. The Ethernet switches within the aggregation network perform S-VLAN bridging, whereby Ethernet frames are forwarded according to the S-tag. Optionally, there may also be a C-VLAN per end user or DSL. VLAN tagging also provides for a marking capability for an Ethernet frame's class of service (CoS). Normally, the Ethernet aggregation network would only switch S-VLAN "trunks" and would not examine any individual end user's C-VLANs encapsulated within the S-VLAN. These would only be affected by functionality at the edges of the Ethernet aggregation network e.g., the access node and the BNG.

Ethernet VLANs can be used to logically group traffic flows. For example, some architectures may use service-based VLANs where voice, data, and video traffic are allocated to specific VLANs. Other approaches involve having a dedicated VLAN for each end user's data (known as the 1:1 VLAN model emulating the ATM VPI/VCI paradigm) or alternatively may limit 1:1

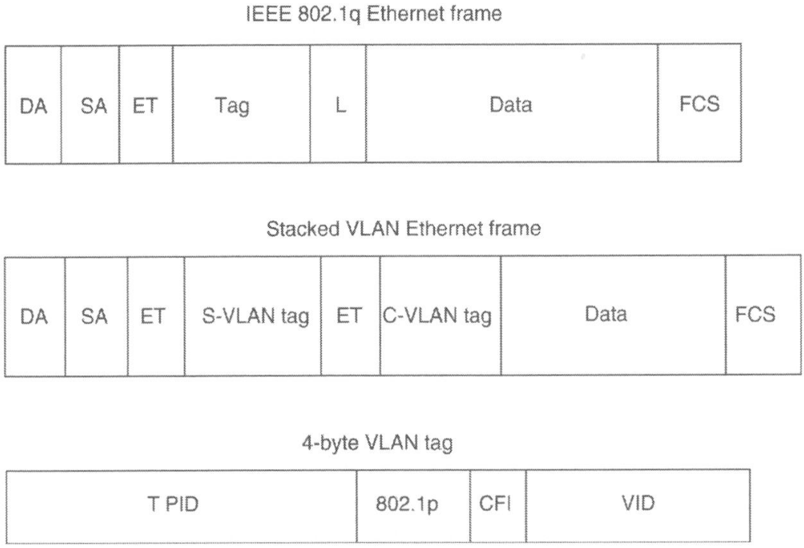

IEEE 802.1q Ethernet frame

DA	SA	ET	Tag	L	Data	FCS

Stacked VLAN Ethernet frame

DA	SA	ET	S-VLAN tag	ET	C-VLAN tag	Data	FCS

4-byte VLAN tag

T PID	802.1p	CFI	VID

Figure 9.40   Ethernet frame structure.

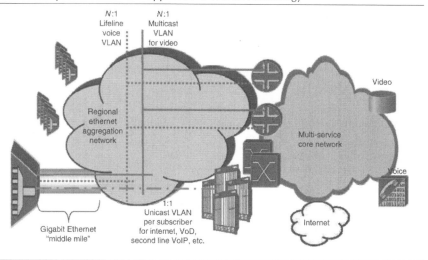

Figure 9.41   Example "triple play" Ethernet architecture incorporating both 1:1 and *N*:1 VLANs.

VLANs just for business users whilst putting all residential end users data in a single VLAN (known as the *N*:1 model). In all Ethernet architecture variants, video multicast traffic has its own *N*:1 VLAN carrying IPoE multicast traffic to a number of access nodes. Hybrid approaches using both *N*:1 and 1:1 VLANs are also possible as illustrated in the Figure 9.41.

Note that when using a VLAN per service, it is not necessarily possible to directly map a QoS level per service such as video. Control and data plane flows within the same service may have different QoS needs. The 1:1 VLAN approach essentially provides a single logical circuit to manage with no requirement for Ethernet MAC address tracking. The end-user traffic isolation provided by this approach also facilitates security, troubleshooting, logging, and accounting.

The end user's device such as an RG sends untagged Ethernet frames, tagged frames, or priority-tagged frames toward the access network. The access node may be preprovisioned to map a C-VLAN to a particular DSL port for every potential user. Alternatively, business users may be able to set their own C-tag within a wider S-tagged VLAN that is exclusively for their use as a layer 2 VPN. The S-Tag or C-Tag pairs have to be unique within the local aggregation network to avoid misdirection of traffic.

9.3.4.3   Customer Premises Equipment: Routing Gateway

At the customer premises in an Ethernet-based access architecture, the RG would interoperate with the existing ATM-based standardized DSL equipment, although new capabilities may be introduced to take advantage of non-ATM options at the U-interface (i.e., DSL technology using EFM). The RG may set priorities on tagged Ethernet frames in the upstream direction. To efficiently support video, RG capabilities are necessary for support of multicast by forwarding IGMP messages from the LAN to the WAN interface. This can entail acting as an IGMP proxy on behalf of a number of set top boxes (IGMP sources) within the home. Other capabilities may include local NAT and firewall features as well as classification according to source IP/MAC address or incoming LAN physical port. These QoS capabilities are used to distinguish between multicast services such as IPTV and best-effort applications.

9.3.4.4 Quality of Service

For QoS in the Ethernet access architecture, priority tags or markings are used to specify the relative priority. By applying VLAN arrangements, as described above, S-VLANs can be used to segregate traffic by service, DSLAM, or CO, whilst C-VLANs can segregate traffic by user. The handling of traffic according to priority can be accomplished by using Ethernet priority bits (IEEE 802.1p). Ethernet normally implements simple priority queuing with the embedded "*P*-bits" in a given packet identifying the per-interface queuing discipline and discard eligibility to be applied. This is sufficient to offer simple classes of service and can be augmented with true call admission control and configured paths using Provider Backbone Transport (PBT) if absolute (instead of relative) QoS is required. Alternatively, some architectures may use MPLS techniques to convey traffic with the appropriate QoS over the Ethernet infrastructure, e.g., MPLS virtual private LAN services (VPLS) for N:1 VLANs and MPLS pseudo wires for 1:1 VLAN trunk groups.

The access node is able to add or modify a priority tag of incoming or outgoing Ethernet frames. Modification may be performed on a per flow basis. Depending on the exact functionality of the access node and BNG, they may be able to classify and then prioritize and schedule traffic according to the following:

- VLAN
- Priority
- L2 (MAC) destination or source addresses
- EtherType
- L3 (IP) destination or source addresses
- TOS
- L4 (TCP/UDP) destination or source ports.

Just as traffic can be sorted where there are multiple applications per traffic class with ATM, the same applies to Ethernet. The flow of applications to and from the end users can be co-ordinated by establishing policies and user service profiles in addition to the use of IP DSCP traffic classification. The BNG and an associated policy manager can be engineered to enforce the policies associated with each end user. User-specific rules can identify associated traffic flows and apply appropriate priorities. Shaping as well as policing of traffic can be implemented for each end user. Similar to the BRAS functions described previously, hierarchical scheduling, policy management, queuing, and shaping may be implemented within the BNG to support QoS.

9.3.4.5 Multicast

In the Ethernet aggregation architecture, multicast traffic (such as video "broadcast" channels) would usually be transported in a dedicated multicast VLAN (see Figure 9.41). The multicast traffic can be encapsulated as IPoE. The delivery of the multicast channels would follow the N:1 architecture with a number of different access nodes being on the same multicast VLAN. Ethernet multicast capabilities are well understood and the mapping of IP multicast to Ethernet offering layer 3 and layer 2 integration and simplification is exploited in [DSL Forum TR-101].

The IGMP protocol (as described in Section 9.3.3.2) could be employed in the access node, the BNG, and also potentially within the aggregation network's Ethernet switches. Snooping and multicast functions can be distributed across the Ethernet aggregation network as shown in Figure 9.42.

An IGMP snooping function involves the examination by multicast devices (access nodes and Ethernet switches) of IGMP reports within IP (or PPPoE) flows toward the IGMP routers. Knowledge of the IGMP

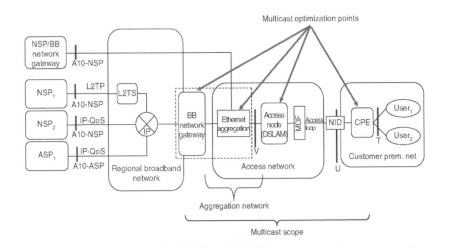

Figure 9.42   Location of multicast functionality within Ethernet DSL architecture.

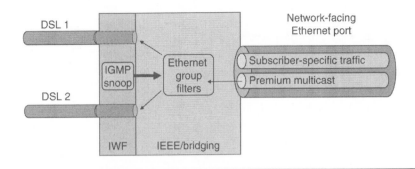

Figure 9.43   Multicast of premium channels within access node.

reports lets the Ethernet switch or access node know whether it needs to start replicating a channel toward a DSLAM or end user that is not currently receiving that channel. The IGMP messages are mapped to VLANs in the access node, which is configured to recognize IP group addresses corresponding to multicast channels (see Figure 9.43).

9.3.5   Management, CPE, and Other Developments and Applications

9.3.5.1   Auto-Configuration of Next Generation CPE

The initial auto-configuration specifications from the DSL Forum [DSL Forum TR-037] addressed auto-configuration of the ATM VPI/VCI, encapsulation (e.g., PPPoA etc.), and allocation of IP address. Subsequently, work focused on how best to automatically configure more sophisticated DSL CPE such as DSL routers that incorporate firewalls, VPN capabilities, and wireless LAN interfaces. The simplification of CPE configuration and management evolved with ongoing developments of communications protocols, particularly XML. This configuration and management for CPE can be addressed from the perspective of both the LAN [DSL Forum TR-064] in the customer premises and the WAN interfacing to the DSL access and broadband core network. Motivated by the desire to avoid any software conflicts on an end user's PC, methods for configuring DSL CPE through software on PCs inside the LAN continue to be improved through newer software developments. Going beyond the LAN, protocols to communicate between the CPE and a network-based Auto-Configuration Server (ACS) enable secure auto-configuration as well as other CPE management functions [DSL Forum TR-069].

CPE with pre-installed software can automatically invoke a sequence consisting of power-up, diagnostic run, interface initialization, and training to the DSL. This avoids the need for an end user to load a service provider supplied CD into a PC. Instead of depending on an application from a CD, auto-runs, CPE discovery, and queries for model and version are communicated automatically. The end user simply has to input the username and password. After initial configuration, the CPE may query servers for additional configuration or to convey updates on the latest CPE status.

The WAN communication between CPE and ACS allows for secure autoconfiguration of a CPE at the time of initial connection and any subsequent reinstall, as well as other CPE management functions. The management functions can span a series of connected customer premises devices such as those in a home network. Some of the other management functions facilitated by an ACS may include dynamic service provisioning, comparing performance against SLA requirements, and software or firmware image management. These developments of more sophisticated communication between CPE and network will lead to different types of service offerings. The first is where the end user has greater capability to order services "on-demand," e.g., via a Web portal interface. The network is then able to communicate with the DSL CPE to ensure that it is appropriately configured for the selected services. This could be used on an as needed basis to enable additional VoIP ports, e.g., when a relative visits a residence or a business has a seasonal peak in telephone orders. This concept is illustrated in Figure 9.44.

The second service offering facilitated by more sophisticated CPE is managed CPE (see Figure 9.45).

This sort of service could, e.g., be used by a service provider to manage all of the DSL CPE on behalf of a corporation. The service provider could then ensure that all DSL CPE at branch offices and teleworker locations is configured with the appropriate corporate firewall policy and VPN access

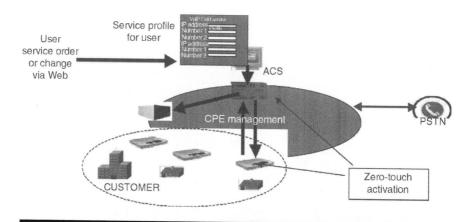

Figure 9.44   User control of DSL service and hence CPE configuration.

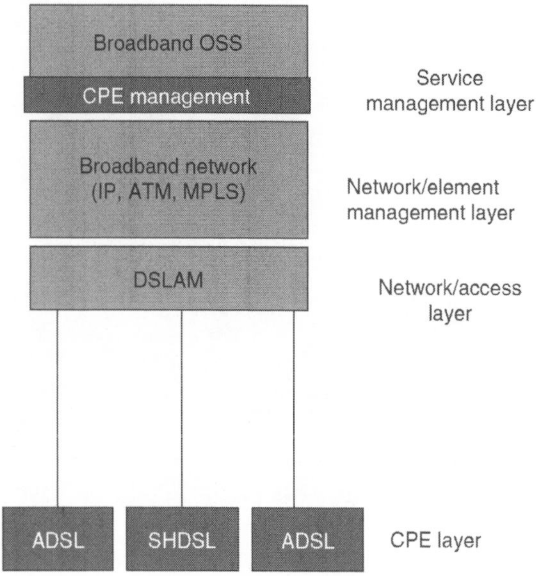

Figure 9.45   Key elements in a managed CPE service.

privileges. It could also be used to configure a VoIP VPN between sites. A more detailed example of such a system is illustrated in Figure 9.46.

The new generation of sophisticated CPE can facilitate bundled service offerings that are remotely configured and managed. This remote management can be used to glean performance information and diagnostics from

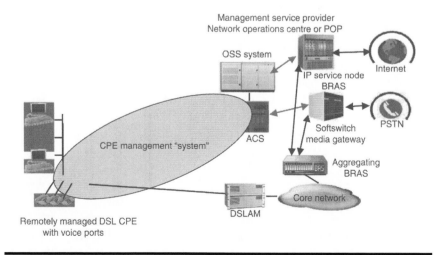

Figure 9.46   Example of a CPE management system within DSL architecture.

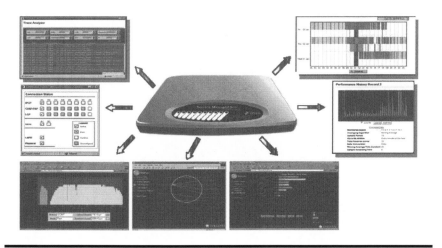

Figure 9.47   Example diagnostics available on "NextGen" managed DSL CPE (Courtesy of Virtual Access.)

the CPE (see Figure 9.47) to monitor usage and performance for audit purposes and SLA monitoring.

As DSL network architectures evolve to become more Ethernet-centric and less ATM-centric, there is scope for DSL CPE to support improved end-to-end Ethernet OAM for service assurance. Further developments may facilitate the specific performance monitoring and fault diagnosis of video traffic.

9.3.5.2   Other Applications and Developments

Some of the new DSL applications that are evolving include newer types of DSL and associated CPE, integration with Wireless LAN (WLAN) "Hot Spots," and extending access to MTU. Other technology developments include bonded DSL, which can be used to increase the speed or range of DSL delivered services.

9.3.5.2.1   New Types of DSL and CPE

End users are connecting multiple computers to DSL modems through premises gateways and routers. Home networks are being designed to complement the capabilities of the core Internet architecture, leveraging a single DSL connection with multiple PCs that share resources, often starting with the sharing of printers and files.

Already, more and more end users are connecting their networks to entertainment equipment. The capabilities that are being developed for home entertainment and automation leverage the evolving DSL architecture. Audio such as MP3 is being stored on PCs and delivered through home networks to be played over stereo receivers. Slide shows of digital photographs are being stored on PCs, later to be shown on TVs and other external monitors. Hence, the PC network serves as an alternative to set top boxes, and as a central point for entertainment systems. The latest wireless interfaces (802.11n) are enabling end users to send video from their computers to their TVs. Also, home security systems are making use of functions centered on computer resources.

To support these services, the various QoS levels are being associated to service levels to support video, voice, music, and gaming applications. These QoS levels are being developed to work over all DSL types and home-networking technologies. Beyond the initial deployment of ADSL and SHDSL, the newer types of DSL include ADSL2, ADSL2plus, extended reach ADSL, and VDSL2.

9.3.5.2.2 Wireless LAN Hot Spots

ISPs are adding WLAN capabilities using Wireless Fidelity (Wi-Fi) to DSL packages as well as optional DSL routers. With a wireless connection to a DSL router, end users can connect to DSLs without Ethernet or pairs of wires. This DSL router capability is available for use in homes or offices or in cafes, parks, and other public areas. Wi-Fi extends the high speed access of DSL through base stations called access points or hot spots, if in public areas (see Figure 9.48). The typical signal range to and from hot spots or Wi-Fi access points is around 100 m, without obstructions such as buildings or other structures made of metal or concrete (walls). Other limiting factors for Wi-Fi transmission include the number of simultaneous end users connecting to the same access point and the distance the PCs WLAN client adaptor is from the access point.

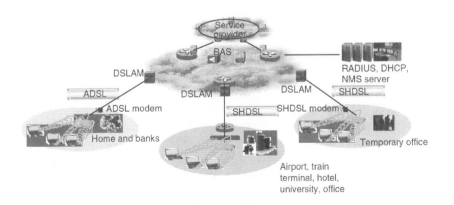

Figure 9.48 ADSL and SHDSL serving wireless "hot spots."

A number of ISPs are offering Wi-Fi hot spot connections to end users that have already subscribed to their DSL service from their homes or offices. In this way, end users can make a high speed network connection, when they are away from their wired DSL connection at their home or office.

The Wi-Fi signals operate in the 2.4 and 5 Ghz radio bands. These frequencies are available without a license in many countries. Most newer portable (laptop) computers as well as some hand held PCs are equipped with built-in Wi-Fi antennas and software.

The standards to which Wi-Fi equipment is being built are IEEE 802.11a, b, or g, with n due for ratification in 2007. Newer products that are compliant with 802.11a operate up to 54 Mbit/s using the 5 GHz frequency and feature the ability to support more end-user PCs per room, while products compliant with IEEE 802.11b operate at up to 22 Mbit/s (typically 11 Mbit/s) using 2.4 GHz and feature the ability to work with existing hot spots. Products which are compliant with the IEEE 802.11g, also feature the 54 Mbit/s higher speeds but operate at the 2.4 GHz frequency and can therefore be backward compatible with IEEE 802.11b systems. IEEE 802.11n products may operate in the 2.4 or 5 GHz band and operate at speeds from 300 Mbps to over 600 Mbps. Many "dual mode" devices are available. The IEEE 802.11e standard adds QoS capabilities to Wi-Fi. This enables certain traffic flows to have priority over others and hence can be used to deliver integrated voice and data (via VoIP) over a combined DSL and WLAN connection. Wi-Fi phones for this kind of application are already available from some vendors. There are significant synergies between DSL and WLAN technology in terms of target markets and data rates. The development of end-to-end IP QoS capabilities in DSL architectures and WLAN technology looks set to perpetuate these synergies and hence the joint use of both technologies in delivering IP applications and services.

9.3.5.2.3 Multi-Tenant Unit (MTU) Applications

Service offerings can be bundled over DSL to provide the communications infrastructure for MTUs such as hotels, apartment houses, condominiums, and commercial office buildings, particularly smaller office buildings. The service providers that target these types of premises are some times referred to as building local exchange carrier (BLECs). Traditionally, BLECs have been forced to use either expensive T1/E1 leased lines or fiber connections to connect to the MTU building. DSL (especially SHDSL) can be used as an alternative to fiber from a CO to provide access to MTUs. The more cost-effective nature of DSL in comparison to fiber has led to smaller buildings being targeted with a lower required "take-rate" by the tenants for the BLEC to break even. DSL can also be used to scout for lucrative businesses by offering it for the initial tenants when a BLEC targets a new building. These

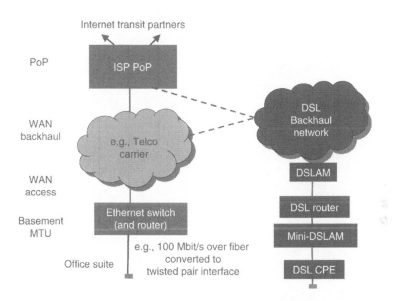

Figure 9.49 **Use of DSL for both MTU access and internal distribution as an alternative to fiber.**

tenants can later be migrated to fiber as demand justifies the increased capacity and cost. Finally, DSL can be used to back up a fiber access bearer once the latter is installed. The use of DSL for both MTU access and internal distribution as an alternative to fiber is depicted in Figure 9.49.

In addition to providing access to the building, a "mini-DSLAM" or Ethernet switch with DSL ports can be used to distribute voice and data over internal copper pair wiring (e.g., Category 5 [Cat 5] cable) to rooms within the MTU. Internal DSL distribution may use ADSL (if baseband POTS is present on the internal wiring), SHDSL, or even VDSL2 (e.g., configured for 10–100 Mbit/s symmetric distribution to match Ethernet LAN speeds). Figure 9.50 shows an example of a mini-DSLAM in MTU.

9.3.5.2.4 Bonded DSL

Further enhancements to ADSL and SHDSL include bonding. The bonding of multiple DSLs enables higher access rates as well as extended serving distances for a required access rate (see Figure 9.51). Bonding can be at the physical, ATM, or PPP layer.

Physical layer bonding usually involves two pairs bonded at the line card on the DSLAM. This results in simpler CPE, than with bonding at the ATM or PPP layers. However, a downside is that if one of the pairs fails the whole access connection is lost.

ATM bonding makes use of IMA [ATM Forum (c)]. IMA also involves bonding the pairs on the line card at the DSLAM, whereby the DSL link

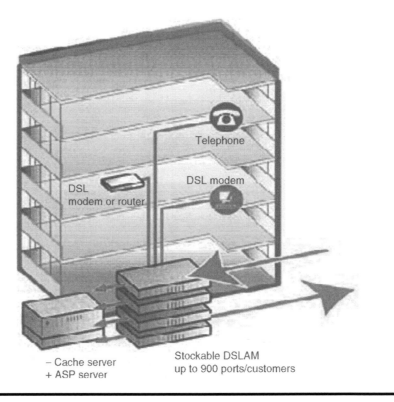

Figure 9.50   Example of a mini-DSLAM in a Multi-Tenant Unit (MTU).

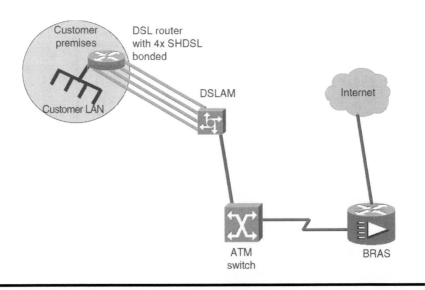

Figure 9.51   Use of bonded DSL to provide higher aggregate DSL access speed.

appears as a single ATM connection to the upstream ATM switch. The cell streams from the CPE router are transmitted across multiple interfaces (DSLs) and then recombined at the DSLAM to appear as a single stream of ATM cells on the backhaul link. SHDSL CPE routers are available that are capable of bonding up to eight copper pairs via IMA. The advantages of the IMA approach to bonding are as follows: (a) ATM QoS can be used over the bonded link, (b) DSLs can be added or deleted from the bonded link without service disruption (useful for up-selling customers to higher bandwidths), (c) and the support for copper pair or DSL failure with automatic recovery without loss of data.

Multi-Link PPP (ML-PPP) can be used to aggregate PPP sessions from a DSL router to the DSL network provider's BRAS. This lacks the QoS capabilities of IMA but can work with DSLs of different rates. The ML-PPP session termination places additional processing load on the BRAS, which could affect scalability if applied to an architecture with a centralized BRAS (instead of distributing the BRAS closer to regional nodes of the edge of the network). Alternatively, an ISP could terminate the ML-PPP sessions on their own router or BRAS, which would make this aggregation approach totally transparent to the DSL network provider.

Finally, an end user could use a load balancing router to connect into two DSL routers (or more) to increase the access bandwidth available to share among their LAN devices. This requires no involvement with the ISP or DSL network provider and also provides resilience if one of the DSLs fails. However, it results in at least a three box solution on the end users premises which is more difficult to configure and manage.

9.3.5.2.5 Service Offerings of New and Bundled Services

With the arrival of new broadband services, DSL access is going far beyond e-mail, Web surfing, file downloads, and on-line transactions. The bundling of these current services, with telephony services and new applications such as interactive gaming and IPTV over DSL, has already become a major motivator for new DSL subscriptions.

Among the additional services expected to be bundled are remote access to corporate networks employing VPN technology, voice telephony using VoIP, audio and video streaming and downloads, public space wireless LAN (hot spots) using Wi-Fi IEEE 802 (a, g, or n), and live high-definition television. Fixed mobile convergence (FMC) is expected to be a big driver for residential DSL and Wi-Fi hot spots, as it enables local mobile calls to be carried over a broadband IP connection and hence offloaded from the cellular network, freeing up capacity for truly mobile, typically vehicular, users.

Some of the newer services, especially FMC will require adding QoS capabilities to existing and new networks. The treatment of the traffic from these services is being handled both on an individual service basis as well as on an aggregate service basis, whereby services are grouped into flows that require similar QoS and other types of treatment. Going forward, services currently offered with best-effort treatment can be given various priorities.

QoS and greater IP awareness in DSL networks facilitates new "high-touch" IP services such as centralized firewalls, integrated voice and data VPNs, and effective distribution of content. It could also lead to different business models from the flat-rate subscription models of initial DSL deployment. This may include usage billing (e.g., on a per gigabyte or per minute basis). DSL network providers have now begun to bundle a range of new services while guaranteeing rates and quality.

9.4   Concluding Comments

Certain products at the lowest end of the DSL broadband access market will inevitably "commoditize." For example, best-effort Internet access for consumers based upon ADSL operating at just a few hundred kilobits per second has already become a commodity. However, as this chapter illustrates, a DSL broadband access network offers a huge range of implementation options that can be grouped and packaged with different product attributes giving a large range of choice for service providers and their end users. As a service provider, you cannot offer differentiated DSL products if you do not control your own broadband access infrastructure or, do not have access to a competitive market of disparate DSL network products.

There is unlikely to be a "one-stop shop" that covers all options and permutations of appropriate feature sets matched to different applications and service requirements. Consequently, broadband access is far from being a commodity business, when the particular needs of the wider market and its various constituent segments are examined. It is clear that a "one size fits all" approach to DSL products will not meet the needs of all service providers and their end users.

DSL network technology and architectures have many degrees of freedom to facilitate both "mass customization" and bespoke design of broadband access products, thus raising the value beyond a pure commodity connectivity play. This chapter has illustrated the capabilities inherent in DSL network architectures and interconnecting interfaces that allow a wide variety of DSL connectivity products to be delivered, hence facilitating opportunities for service differentiation. This chapter has also illustrated how the ATM-based best-effort Internet access architectures deployed in the first phase of global DSL roll-out can evolve. This evolution is predominantly focused around moving toward increasing use of Ethernet and IP technologies and adding QoS capabilities that enable multiple simultaneous services to be delivered.

The relative merits of ATM-, Ethernet-, or IP- centric broadband access systems will continue to be debated with vendors pursuing and further developing their own preferred approaches. Already DSLAMs have been evolved from simple ATM VC cross-connects. Some vendors had previously developed DSLAMs that were SVC-capable ATM edge switches and others have subsequently developed them as Ethernet switches (with DSL cards) that have IP "awareness" or even integrated IP routing and multi-cast capability. MPLS is seen by some vendors as the way to integrate IP and ATM capability to get the best from each. As always, interoperable standardized products are preferred by many operators and progress in this area could dictate the speed of adoption and ultimate success in the market of the ATM, Ethernet, IP, and MPLS approaches to broadband DSL access.

References

[Braden 1997]   Braden, R., Zhang, L., Berson, S., Herzog, S., and S. Jamin, *Resource ReSerVation Protocol (RSVP)—Version 1 Functional Specification.* IETF RFC 2205, September 1997.

[Cuypers 2002]   Cuypers, W., *KPN Belgium Case Study—BTS Backhaul over SHDSL.* The Journal of the Communication Network, Vol. 1, Part 2, July–September 2002, pp. 104–109.

[Enrico 2000]   Enrico M., Billington N., Kelly J., Young G., *Delivery of IP over Broadband Access Technologies.* BTTJ, Vol. 18, No. 3, July 2000.

[Grossman 2002]   Grossman D., *New Terminology and Clarifications for Diffserv.* IETF RFC 3260, April 2002.

[ATM Forum (a)]   ATM Forum af-vtoa-0078, *Circuit Emulation Service Interoperability Specification.* Version 2.0, January 1997.

[ATM Forum (b)]   ATM Forum, af-vmoa-0145.000, *Loop Emulation Service Using AAL2.* July 2000.

[ATM Forum (c)]   ATM Forum, af-phy-0086.000, *Inverse ATM Mux.* Version 1.0, July 1997.

[DSL Forum TR-037] DSL Forum TR-037, *Auto-Configuration for the Connection between the DSL Broadband Network Termination (B-NT) and the Network Using ATM.*

[DSL Forum TR-039] DSL Forum TR-039, *Requirements for Voice over DSL TR-039.* March 2001.

[DSL Forum TR-058] DSL Forum TR-058, *Multi-service Architecture and Framework Requirements.*

[DSL Forum TR-059] DSL Forum TR-059, *DSL Evolution—Architecture Requirements for the Support of QoS-Enabled IP Services.*

[DSL Forum TR-064] DSL Forum TR-064, *LAN-Side DSL CPE Configuration Specification.*

[DSL Forum TR-069] DSL Forum TR-069, *CPE WAN Management Protocol.*

[DSL Forum TR-101] DSL Forum TR-101, *Migration to Ethernet-Based DSL Aggregation.*

[FS-VDSL]   *FS-VDSL System Architecture Specification.* June 5, 2002.

Bibliography

[DSL Forum TR-010] DSL Forum TR-010, *Requirements and Reference Models for ADSL Access Networks.*

[DSL Forum TR-025] DSL Forum TR-025, *Core Network Architecture for Access to Legacy Data Networks over ADSL.*

[DSL Forum TR-032] DSL Forum TR-032, *CPE Architecture Recommendations for Access to Legacy Data Networks.*

[DSL Forum TR-042] DSL Forum TR-042, *ATM Transport over ADSL Recommendation (Update to TR-017).*

[DSL Forum TR-043] DSL Forum TR-043, *Protocols at the U-Interface for Accessing Data Networks Using ATM/DSL.*

[DSL Forum TR-056] DSL Forum TR-056, *Network Migration.* February 2003.

[IEEE 802.1a)] IEEE P802.1AB/D13, *Approved Draft Standard for Local and Metropolitan Networks: Station and Media Access Control Connectivity Discovery.*

[IEEE 802.1b)] IEEE 802.1Q, 2003 Edition, *IEEE Standards for Local and Metropolitan Area Networks—Virtual Bridged Local Area Networks.*

Chapter 10

DSLAM Architecture and Functionality

Michael Clegg

CONTENTS

Abstract   This chapter looks at the genesis and evolution of the Digital Subscriber Line access multiplexer (DSLAM). The DSLAM was a breakthrough device in the success of Digital Subscriber Line (DSL). Its own architecture has changed in accordance with the evolution of the underlying DSL technology and also the change in the services offered over DSL, as applications migrated from pure Video-on-Demand (VOD) to high-speed internet access to integrated triple play services (voice, video, and data). This chapter examines the role the DSLAM plays as a network element, the critical functions it performs, and how it now delivers on the triple play vision.

DSLAMs have proven to be remarkably resilient and adaptable, and much of the architecture and network design is carried over into Fiber-in-the-loop systems and their OLT (Optical Line Terminator) counterpart.

10.1   Introduction

The DSLAM (DSL access multiplexer, with DSL standing for Digital Subscriber Line) is a pivotal element in the end-to-end network system for DSL deployment. In the simplest terms, a DSLAM is exactly what its acronym indicates—a product located at the network side edge of the access network that multiplexes several DSLs together. As described in this chapter its actual functionality can vary from very simple to quite complex.

From the outset, the role and functionality of the DSLAM has been strongly bounded by three external influences:

1. Service mix and deployment cost
2. Regulatory requirements
3. Industry standards

The interplay between these elements is highlighted throughout the chapter. The reader is encouraged to bear them all in mind when focusing on any one particular aspect. As will be seen, an interesting outcome of the regulatory influencer is that the service mix and DSLAM functionality have come full circle.

The DSLAM is just one element of an end-to-end network that delivers a set of services; the Holy Grail combination being the so-called triple play of voice, video, and data. In understanding the role and function of the DSLAM, the whole network and service mix must be considered.

To better appreciate this, it is useful to understand how the DSLAM came about, how the external influencers at the time impacted its architecture, and where DSLAM functionality is evolving to.

10.2 Genesis of the DSLAM

In the early 1990s a confluence of technologies—MPEG (Moving Picture Experts Group) video compression, rising disk drive capacities, higher speed processors, and DSP technology—enabled the delivery of Video-on-Demand (VoD), over existing telephone lines, using newly developed ADSL technology. The use of a splitter to separate the plain old telephone service (POTS) spectrum from the data spectrum was very novel at the time. This enabled the data to be overlaid on the same line as the telephony, allowing both to be delivered simultaneously and independently. The first network solutions were built on the products available at the time. The proprietary Asymmetric DSL (ADSL) modems had a 1.5-Mbps unidirectional downstream video channel and a 16-kbps bidirectional control channel. The downstream rate was chosen to be compatible with T1. The first service trials conducted by Bell Atlantic (now part of Verizon) and Telecom Italia used the architecture shown in Figure 10.1.

As shown in Figure 10.1, the ADSL modems were standalone transmission devices. In the central office (CO), the bidirectional control channels were connected to a X.25 switch that aggregated the control traffic to limit the number of computer ports required. The ADSL modem video channel

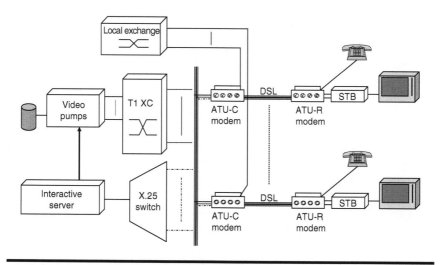

Figure 10.1   Video-on-Demand (VoD) Trial Network.

was connected to an industry standard T1 cross-connect which was in turn connected to a video server. The cross-connect enabled different video streams to be switched between users.

Around the same timeframe BT (British Telecom) also conducted a VoD trial with a slightly different architecture; the VoD control channel used a 9.6-kbps Async protocol and the video was delivered at a 2.048-Mbps E1 rate. The main difference though was that the T1 cross-connect was replaced with an Asynchronous Transfer Mode (ATM)-based line interface module that encapsulated the video and control channel traffic into ATM for transmission back to a separate server center. It also laid the foundation for using ATM to carry the mix of traffic over a single network.

The results from the initial trials were that although the service itself was very attractive to consumers, the system was too expensive for a profitable business case. The two main lessons learned were that

1. Building an entire "broadband" network to deliver a single service would not be profitable.
2. Using standard products was too expensive as they included a lot of unnecessary functionality and interconnects plus the products were not optimized for video traffic characteristics.

It was readily apparent that eliminating intermediate ports and non-optimized equipment would significantly reduce product costs. Specifically, the cross-connect and DSL modems should be combined and ideally integrated with a high capacity transmission interface. Hence, the integrated product that was later known as the DSLAM was born.

The functionality of DSLAM that was first conceived is shown in Figure 10.2.

The DSLAM as initially envisaged integrated a SONET/SDH (Synchronous Optical Networking/Synchronous Digital Hierarchy) terminal multiplexer, a T1/E1 switch, an X.25 concentrator, and the DSL modems. Note that the number of modems exceeds the network port capacity providing the DSL concentration function. The concentration ratio is a reflection of the service mix and usage. For video, which has a long "hold" time and high percentage of simultaneous users correlated with peak viewing time, a 3 : 1 ratio was considered optimum. At the time, backhaul transmission costs were another factor in making the business case unviable. Developments in optical transmission have largely eliminated this as an issue today (although backhaul costs are still an issue).

Another aspect of the VoD network solution shown so far is that it was a closed system, i.e., the network provider is the service provider. Later telecom deregulation required the network operators to allow multiple service providers access to the network, a requirement that had a

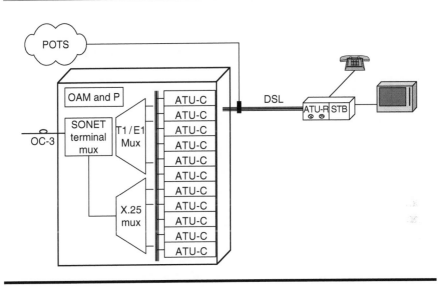

Figure 10.2   **First synchronous Digital Subscriber Line access multiplexer (DSLAM).**

significant impact on how much functionality is integrated into contemporary DSLAMs.

10.2.1 Evolution of the DSLAM

Following on from the initial trials, the ANSI T1E1.4 (now ATIS NIPP NAI) standards group began developing an ADSL standard. Based on the early trial lessons and advances in technology, an enhanced set of service interfaces were embedded directly into the ADSL modems, interfacing directly back to their respective and separate, service delivery networks.

The ADSL ANSI (American National Standards Institute) Issue 1 standard (1995) specified a unidirectional downstream channel running at up to 6 Mbps, plus a separate bidirectional channel running up to 640 kbps. The unidirectional downstream channel was further subdivided into four sub-channels that could be configured at a combination of $N \times 32$ kbps up to the maximum rate supported by the line. An envisaged combination was $4 \times T1$ or $3 \times E1$ channels. The bidirectional channel was also divided into three sub-channels whose rates were as follows: 16 or 64 kbps nominally for use as a video control channel, 160 kbps nominally to be used for ISDN transport, and 384 or 576 kbps channel that could be used for videoconferencing or data. An envisaged network application is shown in Figure 10.3.

It is important to notice that in these early days the services over the DSL (e.g., the video or voice) were still considered synchronous data streams,

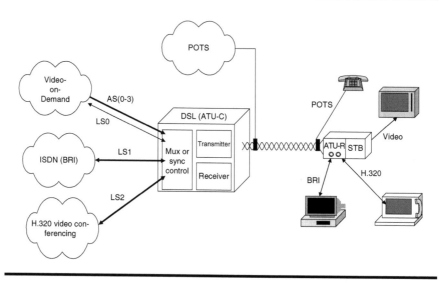

Figure 10.3   ANSI (American National Standards Institute) Issue-1 Asymmetric Digital Subscriber Line (ADSL) network application.

each of the them potentially even with their own data, which could differ from the actual DSL data clock.

Although this standard enabled multiple services, the service set was still too rigid. Interestingly, this standard specified dedicated sub-channels for voice, an option, that would fade away but has made a comeback in the ADSL2+ standard as channelized voice over DSL. The main concern with this standard was that the equipment was now even more complex as multiple network interfaces were specified. The business case for DSL was yet to be made.

Around this time, further DSL deployments hit a lull. Telecom deregulation was taking place around the world and operators were hesitant to deploy networks until the regulatory framework was clear. In many countries, satellite TV and video rentals were taking off, weakening the VoD business case even further.

In the meantime, technology development continued and ATM was gaining acceptance as a method, via adaptation, to transport multiple protocols at variable rates over a single network infrastructure. This meshed perfectly with the multirate, multi-service requirement of ADSL access. As ATM inherently supports statistical multiplexing, it could also be used for concentrating the DSL ports onto a single network port. A further benefit of ATM is that it is independent of the line rate, allowing the DSL modems to train at the maximum rate achievable for that particular loop. This resulted in the "novel" ADSL service offering, which was "best offer" in terms of bit rate, in contrast with High Bit-Rate DSL (HDSL) and Symmetric High Bit-Rate DSL (SHDSL), for which there was a guaranteed standard rate.

The ANSI T1E1.4 ADSL standard was revised and the Issue 2 standard now included an ATM option, i.e., a single downstream and a single upstream channel carrying ATM. The inclusion of ATM was driven by BT, Alcatel in particular, and Westell, following on from the BT VoD trial where Alcatel provided an ATM-based access transport and DSL multiplexer (the Alcatel LIM), that connected into Westell DSL modems. Two sub-channels were also specified: one interleaved (higher latency, lower Bit Error Rate (BER) and one non-interleaved (lower latency, higher BER). All service multiplexing took place at the ATM layer. For video applications, the video and control channels were carried on separate virtual circuits within an ATM virtual path (VP). After the standards issuance, DSL transceiver devices were developed that integrated the ATM cell delineation and idle cell generation function, in an industry standard ATM Utopia bus interface. These DSL devices functioned identically to any other standard ATM PHY (PHYsical layer interface) device.

To revitalize DSL deployments, two significant things happened:

1. Internet access—as the Internet took hold, demand for higher access speeds exploded and DSL finally had a business driver.
2. Telecom deregulation—to encourage competition, new service providers were encouraged and incumbent operators had to open their networks to competing providers. They also had to run their own service divisions at arms-length. To enforce this, the network divisions were confined to offering layer 2 access and all layer 3, specifically Internet Protocol (IP) routing had to be in a separate business unit.

The combination of these two events reinforced ATM as the best technology for transporting services over DSL. It also forced a separation of specific network elements as shown in Figure 10.4.

For Internet applications, to enable users to access multiple service providers the Broadband Access Server (BAS) was developed. A predefined permanent virtual circuit (PVC) was set up between the user customer premises equipment (CPE) and the BAS. Point-to-Point Protocol (PPP) sign-on as proven in dial-up modems was used to steer the user to his or her selected Internet Service Provider (ISP). In video and some other cases, the users were connected directly to their service provider using ATM circuits. As shown in Section 10.5.1 the DSLAM could perform a special form of VPI/VCI (virtual path identifier/virtual channel identifier) address translation to minimize the number of backhaul PVCs that needed to be managed. Essentially, a single VP is established between the service provider point of presence (POP) and the DSLAM.

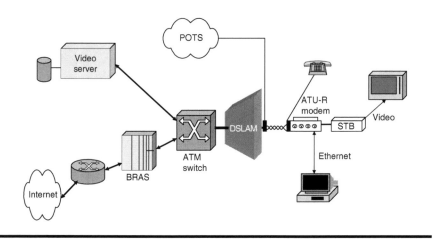

Figure 10.4   Network based on ATM DSLAM ready for deregulated operation over the ATM layer.

Ironically, after the drive for DSL to be multi-service capable, Internet access was initially a single service application. Subsequently, the DSLAM ATM capabilities have been used to deliver derived voice services based on broadband loop emulation services (BLES), which uses AAL2 (ATM Adaptation Layer 2) adaptation. AAL2 is a method for efficiently carrying voice traffic over an ATM virtual circuit (VC). Most recently, network operators are requesting integrated voice or DSL cards that use BLES to migrate the time division multiplexing (TDM) POTS network to packet Softswitch networks. This is described more fully in Section 9.11.2. Coming full circle IP is now becoming the convergent protocol carrying voice, video, and data, and the latest DSLAMs have eliminated ATM and run IP directly over DSL.

In campus type networks and provider networks where layer 2 ATM and layer 3 IP separation was not as rigorously enforced, all IP variant DSLAMs are popular. In other cases, ATM DSLAMs are absorbing the BAS and Multi-Protocol Label Switching (MPLS) router functionality to offer a lower cost, single-managed entity solution. This is described more fully in Section 10.6.

10.3   DSLAM Functionality

Since its incarnation, the DSLAM has and is continuing to evolve; however, the underlying generic functionality is basically the same. A DSLAM performs the following core functions:

1. In the upstream (user to network) direction, it multiplexes a number of DSL modem ports each carrying a single user's traffic into one or more network ports carrying multiple users' traffic.

2. In the downstream direction, it demultiplexes the network traffic onto the correct user port.
3. It concentrates the traffic, i.e., the network bandwidth is less than the sum of the DSL bandwidth. The concentration ratio can be as high as 50 : 1 for residential data users (i.e., Internet users sharing a total bandwidth between DSLAM and network, which is 50 times smaller than the sum of their actual ADSL bit rates), down to 10 : 1 for business data users to as low as 3 : 1 for streaming video users.
4. Inherent in its multiplexer function, it also prevents any direct DSL-to-DSL port switching. This stops one user seeing other user traffic.
5. In the ATM case, it performs address translation so that each user CPE modem sees the same ATM network address allocation, typically $VPI = 0$, $VCI = x$ where x is a limited set that corresponds to different service types.

Most DSLAM also

- Perform some degree of line diagnostics.
- Implement ATM F4/F5 Operation and Management (OAM) functionality (ATM DSLAMs only).

DSLAMs may also optionally

- Integrate the DSL splitter function.
- Integrate analog POTS or Integrated Services Digital Network (ISDN) functionality. Often this is even done on the same card as the DSL modem—this is termed Integrated Voice and Data (IVD); however, a DSLAM can contain dedicated cards for voice or ISDN only.
- Integrate a CO GR.303 or V5.2 voice gateway for IVD and derived voice.
- Integrate the BAS and backbone router functionality.
- Deliver network timing (i.e., the 8-kHz signal, to sample regular voice and to synchronize ISDN signals to the master clock in the POTS/ISDN network).

10.3.1 Statistical Multiplexing

Multiplexing is the process by which a number of separate, lower speed data streams are combined, into a single, higher speed stream. Demultiplexing is just the reverse. Once the streams are combined, some method is required to identify the individual streams. The two basic techniques commonly used are as follows:

1. Time division multiplexing (TDM)
2. Packet multiplexing

10.3.1.1 Time Division Multiplexing

The TDM principle is shown in Figure 10.5.

As the name indicates TDM is time synchronous. The operation is best understood by imagining a switch connected to the multiplexer's output. The switch periodically moves from one input stream (channel) to the other. It connects to each input stream just long enough to allow a predefined amount of data to pass onto the multiplexed stream. On each cycle through the input streams, the switch connects to a framing signal. The framing signal allows the start of multiplexing cycle to be identified in the output stream. The input stream is identified or "addressed" by its location in time, known as a timeslot, relative to the framing signal. Prior knowledge of the frame structure, i.e., how many timeslots there are per frame and how many bit periods there are in each timeslot is needed to correctly demultiplex the frame. In the simplest and most common case, each input stream has the same bit rate and each timeslot has the same number of bits in it.

For statistical multiplexing or concentration, the switch cycles around the input streams as before but skips those input streams that are idle. Each multiplexed frame contains a subset of the input streams. To operate correctly, there must be a mechanism, some form of signaling, to know when a channel wants to send traffic. The maximum number of channels that can be served at any one time is directly related to the number of timeslots per frame.

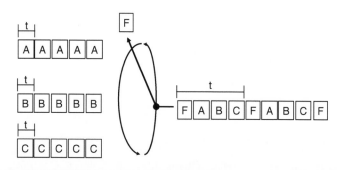

Figure 10.5   Time division multiplexing (TDM).

TDM works best when the bit rate of the data source is constant and is simplest when every stream has the same rate and if the rate of the TDM channels is perfectly locked to the "clock" or symbol timing of the carrying DSL. To minimize the processing requirements for the signaling, the active or busy time for each channel should be greater than several seconds.

When the clock of the TDM channels differs (even slightly) from the bit clock or symbol clock of the DSL, special techniques have to be used to "rate adapt" the data in the TDM channel to the actual channel in the DSL, which will be either slower or faster. Elaborate techniques, including the reconstruction of the actual clock of the TDM channels, have to be used to achieve error free TDM transmission.

The benefit of TDM is that it is simple to implement if the data is the same across all channels and the channels are active for reasonably long time periods.

This is also its biggest disadvantage for applications that send data in bursts. If the gap between the data bursts is small, to minimize signaling overhead, it is most efficient to keep the channel active even though the timeslots are carrying no data between bursts. TDM also requires a separate signaling network and controller (a signaling plane) to assign timeslots on the multiplexed links.

TDM was extensively used for voice traffic, of which all channels are perfectly synchronous over wide areas, and which has a typical hold time of several minutes, notably on T1/E1 trunks, and streamed video, which has a typical hold time greater than one hour. It is also used on SONET/SDH transmission networks where channels are preconfigured and statically assigned.

10.3.1.2 Packet-Based Multiplexing

Packet-based multiplexing differs from TDM, in that each packet carries with it the routing information needed to route the packet from source to destination. Each packet also contains additional information, a header or framing bytes that enable the start and end of each packet to be identified in a stream of bits. The packets can be of variable length or fixed length, in which case they are often termed cells. The routing information can be either the actual destination address as in the case of the IP and Ethernet, or it can be a tag as in the case of Frame Relay (FR), ATM, and MPLS. Tag-based packets require a control plane to configure the intermediate switch modes between the source and the destination to ensure the packet is correctly routed and (if not routed via a constant path) are put in sequence again at the receiving end, before delivery to the end user.

	Fixed Length	Variable Length
End Point Address		IP, Ethernet
Tag	ATM	MPLS, FR

The packet multiplexing principle is shown in Figure 10.6.

The operation is again best understood by imagining a switch connected to the multiplexer's output. As each packet contains its own routing address, there is no requirement to control the order in which packets are switched onto the higher order stream. The switch simply cycles through each input channel in turn, and if data is present, i.e., at least one complete packet or cell, connects just long enough to allow one or more complete packets or cells of data to pass onto the multiplexed stream.

Statistical multiplexing is intrinsic to the operation. Packets are multiplexed whenever available in the ratio in which they are presented to the switch. As packet data normally arrives in bursts, each channel has a queue or buffer to store the data until it can be switched through. More sophisticated statistical algorithms use the queue length to determine how many packets per channel to pass through on each cycle. To avoid packet loss, the only criteria is that the sustained average rate of packets summed across all channels must not exceed the output stream capacity and the burst size must not exceed the queue or buffer capability.

10.3.2 DSLAM Traffic Processing

In any multi-service transmission system, it is important to ensure that each service is prioritized to ensure it has an adequate share of the link bandwidth to avoid traffic loss. In a TDM system, this is simple; each traffic

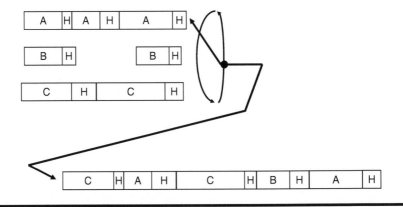

Figure 10.6   Packet multiplexing.

stream is allocated sufficient timeslot capacity for its maximum data requirements. The disadvantage is the loss of any statistical gain from the bursty and variable rate characteristic of data.

In a packet- or cell-based system with its dynamic multiplexing of traffic flows, traffic must be identified, switched between ports, and queued to accommodate data bursts and the contention that occurs when simultaneous bursts on different ingress ports are switched to the same egress ports. Traffic must also be prioritized so that voice or streaming video traffic is not delayed by data traffic. At the same time, it is very important to ensure that traffic sources do not exceed their allocated bandwidth, potentially causing a loss of data on other compliant links. These issues are respectively known as quality of service (QoS) and traffic management (TM).

The importance of QoS stems from the fact that the different traffic types have different transmission requirements in terms of delay, jitter, and bit rate. For example, constant bit rate (CBR) traffic like voice requires packets to be sent with minimal delay and very little variance or jitter in inter-packet spacing. Near real-time streaming video has the same strict jitter requirements but latency is less of an issue as long as it is constant. Data can be best effort as lost packets can be retransmitted and it does not matter if the packet transmission is bursty with high jitter. QoS is implemented through a set of prioritized egress buffers.

Traffic management has three aspects to it:

1. Conditional access control (CAC): CAC is an ingress port function that ensures that only valid traffic enters the network. It can be used to limit the types of services, e.g., it can block a voice call if a user is only signed up to Internet access. It can also be used to block malicious users, e.g., users who send traffic in with invalid address information or attempt to emulate another user.
2. Policing: Policing is an ingress port function that enforces a traffic contract (e.g., in terms of average and peak rate, discussed below) and is used to ensure that users only gain access to the network resources that they have subscribed to.
3. Rate shaping: Rate shaping is an egress port function that ensures that the user traffic is within the bounds of their traffic contract.

The first DSLAMs were very simple and to keep costs down had very little buffering, i.e., TM was also very limited; the DSLAMs simply statistically multiplexed ATM cells together on the upstream and forwarded the ATM cells on the downstream to the correct DSL port. TM and hierarchical shaping were deployed in the ATM edge switches that aggregated traffic from a number of subtended DSLAMs. An added bonus of this was that the user service rates were all controlled in one place.

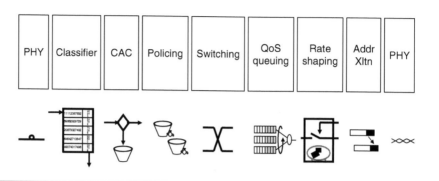

Figure 10.7   DSLAM traffic processing.

For streaming video, which was the initial application, this was not a problem. As data, Internet access, emerged as the major application and with the added advent of Voice over DSL, more and more TM has been incorporated into the DSLAM. The range of ingress to egress traffic processing that takes place in a DSLAM is shown in Figure 10.7. Each of the separate processing functions is discussed in the following sections.

10.3.2.1 Classification

When a packet is received, it first passes through a classifier. For IP packets, the classifier looks at the address and optionally other information in the packet such as the packet type, and may even look at some of the data, to classify the packet. For ATM and frame relay, a simple address lookup is sufficient as the packet characteristics are known when the address is assigned.

The classifier generates a set of attributes and typically a tag that will then be used by other downstream processes. Typically, all packets belonging to the same flow, e.g., a video stream, an Internet session, or a voice call, will have the same tag. Subsequent packet processing functions use the tag to identify the flow, saving the need to do a full address lookup at every processing step.

One of the benefits of layer 2 technologies such as ATM, FR, and Ethernet has been the ability to quickly classify and switch traffic-based on a simple address lookup. The recent advent of multilevel, very long bit field classifiers that can run at gigabit plus data rates has largely eliminated the complexity and cost differential between layers 3–7 and layer 2 switches.

10.3.2.2 Conditional Access Control

The purpose of CAC is to validate the packet. If the packet is a valid packet it is admitted into the DSLAM. If an invalid packet has been received it is dropped. In the ATM instance it is not sufficient just to drop the first cell in a packet. Once any cell of a packet is dropped, all the cells making up the remaining part of the packet must be tracked and dropped.

Dropping invalid packets protects the network from malicious users and prevents network and node resources being expended unnecessarily.

10.3.2.3 Policing

A prime function of the DSLAM is to statistically multiplex several users onto a single link. One way to prevent any single user from getting a disproportionate share of the network link is simply to limit their physical DSL link speed. Another way is to allocate and enforce a traffic contract. This is known as policing. Policing offers the added benefit in that it can be used when multiple services run concurrently over the same physical link and you want to control the bandwidth of each traffic type.

To account for the bursty nature of packet data, the traffic contract has a peak rate, a sustained rate (an average of the packet rate over a time period), and a maximum burst size, which can be several packets or cells. One method for policing is to use a token bucket. Tokens are put into the token bucket, unless the bucket is full, at a regular rate that is proportional to the sustained rate. The size of the bucket is proportional to the peak rate. The operation is as follows:

- During idle periods between bursts, the token bucket will fill up at the sustainable rate.
- When a burst of packets arrives, the burst size is checked, if it is over the limit the packets are dropped, otherwise the packets are put into a FIFO.
- Now if a token is available in the token bucket, the packet at the head of the FIFO is allowed to continue and one token is removed from the bucket for each packet.
- FIFO continues to empty as long as tokens are still available in the bucket.
- If the token bucket empties, then a packet will only leave the FIFO when another token has been added.
- If new packets continuously arrive faster than the tokens are added, the FIFO will eventually overflow and packets will be dropped.
- As long as the average arrival rate is less than the sustained rate and the peak rate is not excessive, no packets will ever be dropped.

In a simple system, policing can be applied per ingress links, in more sophisticated systems policing is applied per traffic flow or virtual circuit.

10.3.2.4   Switching

Switching is the process by which the ingress packets are switched to the egress ports. Switching can be integrated into the DSLAM backplane bus, or dedicated centralized switching devices can be used, as discussed in Section 10.4.1.

To determine the correct egress ports, the switch fabric can look up the original packet address, or if the packet has previously been classified and tagged, the tag can be used to simply do a lookup. A separate control plane is required to configure the switch routing table. DSLAMs typically use static routing tables that are configured through the Element Management System (EMS). An EMS is a software that enables the configuration plus fault and alarm reporting of a physical unit (an element) such as a DSLAM, and interfaces it into the service providers overall Network Management System (NMS). DSLAMs that use Ethernet layer 2 bridging can be self-learning. As each Ethernet packet contains both a source and destination address, the learning algorithm can associate Ethernet addresses with ports and build the routing table dynamically.

For switched digital TV applications, the switch fabric should support multicast so that the same downstream packet can be sent to multiple DSL ports. To handle "channel selection," a tuner function is required, which is most often implemented using the IGMP protocol. The IGMP function and multicast packet replication can be executed on the DSLAM or on the BAS device. The benefit of the latter being that the BAS is already a layer 3 device, the downside being that the capacity of the link between the BAS and the DSLAM must be at least as large as the maximum number of simultaneous streams.

Ideally, the switch fabric should be non-blocking and have low latency. This requires careful selection of the switch fabric and also good design of the internal pathways. Packets should transit the actual fabric as quickly as possible without delay to be queued in the output QoS or rate shaping buffers, where the appropriate network level protocol prioritization and packet handling algorithms are applied.

10.3.2.5   QoS Buffering

Once the packets have transited the switch fabric, they are buffered for prioritization and shaping before exiting the DSLAM. Prioritization ensures that larger non real-time data packets do not delay the real-time critical packets. This process is called QoS buffering.

In general, intelligent prioritized buffering is required whenever several traffic flows are multiplexed together, which is typically done on the egress ports. This is also required whenever several flows of packets make the transition from a high-speed link to a slower-speed link as may happen

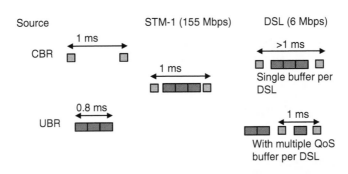

Figure 10.8   Practical example of Quality of Service (QoS) requirement.

within the DSLAM pathways, so QoS buffering is also sometimes used between the ingress ports and the switch fabric.

The importance of QoS buffers is demonstrated in Figure 10.8.

Let us assume a DSLAM, with a user simultaneously receiving data and voice over a DSL. The voice is CBR traffic with a new packet arriving every 1 ms. The data, e.g., Ethernet packets is a burst of uncommitted bit rate (UBR) traffic. On the high-speed network link, there is sufficient bandwidth for the data burst to fit between the two voice packets, causing no delay to the voice. Once the traffic enters the DSLAM, it gets buffered to accommodate the change in speed between the network link and the DSL link. As the DSL link runs at a much slower rate than the network link, the data burst now takes longer than 1 ms to be sent down the DSL link, causing the voice packet to be delayed, potentially for so long that it needs to be discarded.

To ensure the voice QoS requirement is maintained, a separate buffer is needed for each traffic type as shown in Figure 10.9.

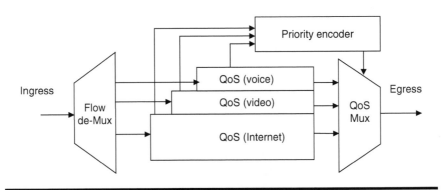

Figure 10.9   Schematic description of a QoS buffer.

In practice, many modern switches maintain a separate queue for every stream—per flow in IP systems or per VC in ATM systems. With separate buffers, the CBR traffic has highest priority, and UBR the lowest.

The method of operation is that the traffic for a particular DSL channel is demultiplexed into separate QoS buffers based on the flow type. The status of each QoS buffer is fed into a priority encoder that controls the output multiplexer.

10.3.2.6 Rate Shaping

In the same way, policing ensures that traffic entering a network node meets its traffic contract, rate shaping ensures that traffic leaving the node is compliant.

One method used is similar to policing and relies on a dual, cascaded leaky bucket approach. The output flow rate of the upper bucket is equal to the sustained rate, although the bucket itself is sized to accommodate the ingress peak rate. The output flow rate of the lower bucket is equal to the peak egress rate allowed, although the size of the bucket itself is equal to the maximum burst size. By opening the taps the output rates are increased. The upper bucket buffers the incoming data and feeds it into the lower bucket at the set sustained rate. The lower bucket fills up waiting to be scheduled. When its turn arrives, it outputs data at a rate equal to the peak rate until either its scheduled time is over or the bucket empties limiting the maximum burst size of the data. Note that if the ingress data is correctly policed, the upper bucket should never overflow. If the output scheduler is correctly configured, the lower bucket should never overflow.

Rate shaping is often combined with QoS, as shown in Figure 10.10.

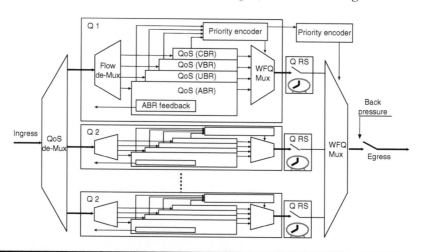

Figure 10.10 **Combined QoS and rate shaping.**

First the traffic flows are separated into separate QoS-based buffers and each buffer is separately rate shaped. All flows destined for the same output port are then grouped together and transmitted at a rate governed by the physical line rate. An egress scheduler governs the allocation of port capacity to the individual buffers. The scheduler ensures that any CBR or traffic is always sent when necessary. Transmits from buffers containing uncommitted or available bit-rate data are apportioned according to a Weighted Fair Queuing (WFQ) algorithm. In WFQ, each buffer is guaranteed a minimum amount of bandwidth, no less than its contracted sustained rate. Any excess bandwidth on the link is allocated to each traffic flow in relative proportion to the amount of data in that flow's buffer with an upper limit equal to the peak rate. On each cycle of the scheduler, multiple packets may exit each buffer up to the maximum burst size limit.

In some instances, it is useful to create logical ports to rate shape a group of flows, and then to have traffic from multiple logical ports egress the same physical port. This is called hierarchical shaping. In hierarchical shaping, a scheduler governs the maximum bandwidth for the logical ports. The scheduler is often implemented using a Calendar algorithm in which the aggregate output capacity is divided into timeslots corresponding to a certain bandwidth increment. Each logical port is assigned a number of timeslots that sum to its desired logical port rate. A sophisticated version of the scheduler will allow overbooking of the physical egress port, i.e., the sum of bandwidth across the logical ports exceeds the physical port capacity. As before, committed bit-rate traffic is guaranteed the necessary bandwidth and UBR traffic contends for the remaining physical port bandwidth using a WFQ algorithm. In this second stage, the weights are apportioned to the amount of data queued in each logical port as a whole.

Hierarchical shaping has been used extensively in DSLAM networks to reduce the complexity and cost of the DSLAM. If a DSLAM is directly connected over a point-to-point link to an upstream ATM switch, by using hierarchical shaping the ATM switch can pace all the downstream DSLAM traffic. For each physical DSL link, a single logical port is created on the ATM switch port connecting to the DSLAM. Each logical port is rate limited according to the subscribed service class, which must be less than or equal to the physical DSL link speed. Now the ATM switch manages all the traffic and the DSLAM only ever receives traffic at a rate less than the destined DSL port. The benefit of this is that the link management takes place on fewer nodes, i.e., in the ATM switch, which generally has more sophisticated management capability than a DSLAM, and the DSLAM itself can be a simple forwarding design with minimal buffering.

10.3.2.7 Address Translation

In ATM, FR, and MPLS systems, where the cell or frame address is localized to the link, the final step before the packet exits the DSLAM is address translation. The flow tag is used to look up the correct address that is then added to the packet or cell just prior to port egress.

In DSLAM applications, the same VPI/VCI is used on all DSL ports. This removes the need for any dedicated CPE configuration. The DSLAM translates the users PVC to a unique VPI/VCI on the network port and vice versa. The actual translation used depends on the service and network operator, this is discussed in more detail in Section 10.5.1. DSLAMs in deployment today are nearly all PVC based and hence statically configured.

Ethernet-based IP DSLAMs can be self-learning or statically provisioned. In pure IP-based DSLAMs, the IP address can be statically assigned through provisioning or dynamically assigned network during log-on.

10.4   DSLAM Architecture Considerations

DSLAMs, like all access products, tend to be very cost sensitive. In telecommunications networks, the access portion is the only part where the equipment cost is directly incurred on a per user basis. There are several considerations in the design of a DSLAM, but two very important architectural elements need to be considered early on. These are as follows:

1. Whether to use a bus or a point-to-point star link backplane.
2. Whether to use distributed or centralized traffic processing.

10.4.1   *Bus versus Star Switching*

DSLAMs differ from a generalized switch, in that a switch typically has a limited number of ports, a high-speed cross-connect between all the ports, and where the ports normally have similar rates. By contrast, a DSLAM is a multiplexer that connects a large number of slower-speed ports to a single higher-speed network port used for the uplink. In addition, local switching between the DSL ports is disabled and all traffic flows between the DSL ports and the uplink port. Dual network link ports may be implemented for redundancy purposes.

Both bus- and star-based switching have been used to connect the DSL port cards to the network port card.

10.4.1.1   *Bus Switching*

The first DSLAM implementations took advantage of the multiplexer function and used a bus architecture shown in Figure 10.11 to keep the design very simple and low cost.

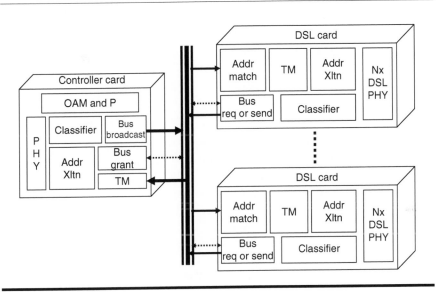

Figure 10.11   Bus-based switching.

In a bus-based architecture, the downstream traffic entering the DSLAM from the network link is classified, tagged, and policed, and then simply broadcast to all the port cards. Each DSL channel on the port card listens to the bus and if it sees a packet with its own tag, it drops the packet into its own port buffer. Upstream traffic is buffered on the port card. A bus arbitration method is used to give each port card equal access to the bus. The upstream packets are received by the network port card and forwarded onto the network link.

Upstream arbitration is not always necessary, in certain cases owing to the asymmetric nature of ADSL and the fact that the network link is symmetric and sized according to the downstream traffic, the upstream traffic could be non-blocking and a simple round robin mechanism can be used. Otherwise, standard arbitration techniques such as polling, priority encoding, request or grant, etc. can be used.

In the first ATM-based DSLAMs, the design was even simplified further. The network link was a maximum of 155-Mbps STM-1c/OC-3c. The VPI address part was used as the DSL port address while the VCI part of the address, corresponding to a service type, is passed through. With a UNI (User Network Interface), 252 DSL ports could be supported on a single DSLAM without any additional address translation. On the DSL port, the VPI was cleared so each DSL CPE modem always saw VPI = 0. In the initial video streaming application, line card buffering was minimal; it was subsequently increased for data. On the upstream, a simple round robin approach sufficed; for as long as the DSL rate was less 600 kbit/s per link, the network link would not saturate.

A bus-based system can be very simple and low cost. Its main limitation arose as link speeds were increased, making it very difficult to meet timing and skew requirements across a fully loaded backplane. The difficulty was further increased if redundant busses were a requirement.

10.4.1.2 Star Switching

The difficulty in implementing high-speed backplanes led to the adoption of star-switched systems. At first, it is counter-intuitive that higher through-put can be achieved over point-to-point links, than via a wide bus. Indeed, a bus is often several octets wide, while the point-to-point links have typically only between one and four parallel signals, each sent with a differential transmission technique. However, over these links the signals are trans-mitted at relatively much higher speeds than is possible over a bus. The achievable higher throughput led to the development of star switching sys-tems, which are based on a centralized switch fabric with point-to-point serial lines radiating out across the backplane as shown in Figure 10.12.

The switch fabric is generally mated with a companion line card device. For redundancy, the line card device connects to two switch fabrics. Incoming data is sent to both switch fabrics. The line card device receives transmit data from both switch fabrics and automatically selects the one without errors. To prevent loss of data, buffering is required on the fabric device to cater for the situation when two ingress ports simultaneously send data to the egress port. Most switches today use an input queue, combined with an output buffered approach in which data entering the switch is

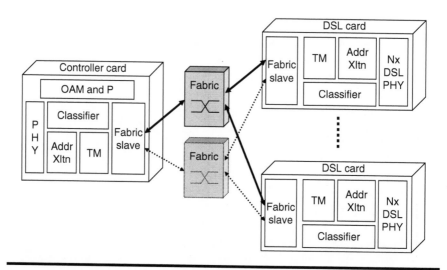

Figure 10.12   Star-based switching.

minimally queued and forwarded to the output buffer as soon as possible to prevent packets destined for another egress port from being blocked. When the switch has only a small number of ports, each ingress port may have a separate queue for each output port. Backpressure can be applied from the output port to the ingress queue to force it to hold the packet in the event that the output buffer is nearly full.

The switch fabrics also support multicast, so that data from one port can be sent to all the other ports. Simplified, multiplexer versions of switch fabrics that are both cheaper and prevent user port-to-port communication have also been developed and used extensively in Ethernet access systems.

10.4.2 Centralized versus Distributed Traffic Management

In a distributed system, TM, QoS buffering, and rate shaping is implemented on the port cards. Bus architectures are inherently distributed in that the port ingress data is broadcast to all output ports, so each port must buffer and process its own egress traffic.

In star backplane architectures, generic packet switches also tend to use a distributed traffic management approach. However, as previously indicated, DSLAMs are a special subclass of generic switches with the property that they have a high-port count, a low capacity, and that data is multiplexed, i.e., only flows between the line side ports and the network port, instead of being fully switched between any two ports. By taking advantage of this characteristic, all the traffic processing functions can be implemented on a single network-side switch card. This architecture is shown in Figure 10.13 and is the basis for PMC-Sierra SUNI Vortex and SUNI-Duplex devices, which have been successfully implemented in many DSLAMs. Some DSL chipset vendors have extended this concept by including a management channel endpoint and processing capability on the DSL transceiver itself, eliminating the requirement for a separate DSL port card processor.

To maintain complete per DSL port TM functionality in the centralized traffic processor, there must be a dedicated virtual port within the centralized traffic processor per physical DSL port, and the mechanics of the star backplane links must be such that each virtual port has a one-to-one relationship with the physical port. In addition, the port must be able to signal back via some form of handshake mechanism when it can accept new packets for transmission on the DSL links, or has packets ready, which it received from a DSL link. Each port has minimal buffering, with the switch only sending new, pre-prioritized data to the port when it knows it is ready to transmit it. On the upstream, the switch must be able to receive new data from each port card as quickly as possible. To prevent blocking, the link

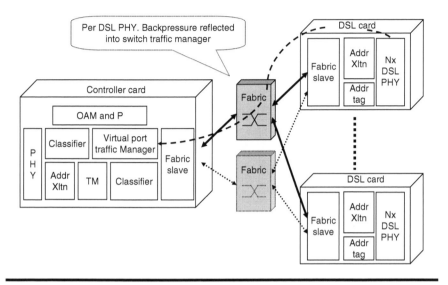

Figure 10.13   Centralized star-based switching.

speed between the switch card and the ports card must, in each direction, exceed the sum of the physical line rate for each of the ports on the card, e.g., if there are 24 ports on the DSL card, each running at 10 Mbit/s downstream, the port-to-switch card downstream link must be greater than 240 Mbit/s, if all ports are to be capable of running at full speed.

10.5   ATM DSLAMs

As discussed in Section 10.2.1, incumbent network operators were required to separate their layer 3 service provider networks from the layer 2 access and transport networks. This, together with ATMs ability to carry multiple service protocols over a single bearer, led to the widespread adoption of ATM as the preferred carrier class DSLAM technology.

The overall network architecture has been extensively reviewed in Chapter 9. In this section, we review the internal structure of a modern ATM DSLAM.

As can be seen in Figure 10.14, the basic functionality is very simple. The DSLAM node controller manages the overall operation, and performs the operations and maintenance provisioning (OAMP) function. The DSLAM is normally connected to its EMS using an inband OAM AAL5 VC that terminates on the management processor. In some cases, a separate network for the out-of-band management may be used, typically connecting to the

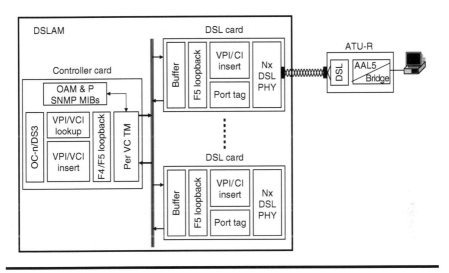

Figure 10.14   ATM data DSLAM.

DSLAM on an Ethernet port. The management protocol is normally SNMP using DSL Forum defined DSL MIBs (management information bases) and the ATM Forum defined ATM MIBs. The DSLAM routing table is usually statically provisioned. ATM F4 and F5 OAM cells are used for additional diagnostics and link status.

In the upstream direction, the DSL CPE modem, i.e., the ATU-R in the figure, converts the user data traffic to ATM cells using AAL5 adaptation. To avoid having to provision the CPE modem, all modems use the same VC addresses, typically VPI = 0 and VCI = 32. The DSL port card performs minimal ingress TM and then simply forward the cells to the network port card. A port ID is usually added either to the front of the cell or in the VPI/VCI fields, to identify which user the traffic came from. The network card buffers the traffic from each of the port cards, applies the appropriate address translation using the port ID (if the VPI/VCI field was not already filled-in at the DSL card), and then statistically multiplexes the cells onto the network link.

In the downstream direction, the data is queued on the network card. Depending on the architecture, centralized versus distributed, the address translation and QoS buffering may happen on the switch card or on the port card. The figure shows the centralized case. In general, a VPI/VCI address lookup is used to determine the correct DSL port ID. The traffic data is then sent to the DSL port card for transmission. Prior to egress, the cell VPI is normally set to 0 and the VCI to a predefined value based on the service.

In addition, a "Tuner" function is required if the DSLAM is being used to support switched digital video (SDV). Broadcast TV data is received on the network port card, with the VPI indicating that it is a SDV service VC, and the VCI containing the channel identifier. The DSLAM switch multicasts each channel to all DSL ports that are currently tuned to that channel.

For tuning, the DSLAM processor terminates or monitors a dedicated "Tuner Control Channel" VC on each DSL. A separate table is maintained for each DSL that indicates the "TV channels" the user on that line may receive. If a user selects a valid TV channel, the processor updates the fabrics SDV routing table to forward that channel to the DSL port for final address translation prior to transmission.

An alternative, more commonly implemented approach is to have the BAS implement the TV Tuner function, which is also well suited to TV distributed using IP multicast. The functionality is similar to that described above. IGMP is becoming standardized for the Tuner protocol. The only disadvantage is that, if the multicasting per user is occurring at the BAS, the link between it and the DSLAM needs to have a high capacity as users "tuned" to the same TV channel will still each receive separate traffic streams.

10.5.1 Network Address Translation

Aside from buffering, an important role of the DSLAM is address translation and its role in overall service delivery. The initial concept was that the DSLAM would follow an all ATM model. The DSLAM performs all the user address translation on a VP model as shown in Figure 10.15. Each ISP has a VP with a unique VPI between its BAS and the DSLAM. The VCI address corresponds directly or indirectly to the DSL port of a user. The DSLAM holds a user-to-ISP routing connection table. Once the cells have been tagged or switched to the correct DSL port, the VPI is cleared and the VCI set to a service value. By having different service values for each ISP, it is possible for the user to connect to multiple ISPs simultaneously. The VP is typically rate limited and policed by the network operator's intermediate ATM switches. This allows ISPs to buy VP bandwidth based on their own service concentration ratios and the number of users they have served off any one DSLAM.

The downside of the ATM VP approach is that a VC must be managed between the network operator and ISP for every user. The ISP may have to pay proportional to the number of ATM or FR VC they lease. With a VC running directly from the user to the ISP, the ISP has to pay—a setup charge and monthly rental—for one or more connections per user. In addition, as users churn, i.e., connect and disconnect on the network, the ISP will be charged additionally for all these re-incurring setups. Also the maximum number of VCs a single port on a FR switch, ATM switch, or a router

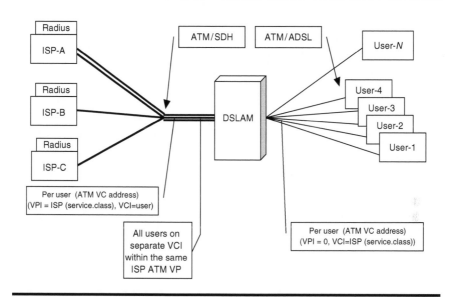

Figure 10.15   ATM virtual path (VP) grooming.

can support is limited. All together this results in a high user management overhead. Specifically, to overcome this problem the BAS was developed.

The BAS is a device, which grooms many users from multiple DSLAMs each with their own VC (thin pipes) onto a much smaller number of links (a few per ISP) each carrying multiple users (fat pipes). The BAS eliminates the per user VC issue, by grooming all the users of the same ISP and the same service onto a single VC, and multiplexes and demultiplexes them correctly before handoff to the DSLAM. The BAS is actually a layer 3 device and uses PPP and domain name addresses extensively to route traffic. It can also in effect act as a switched circuit device as shown below in Figure 10.16.

The network is set up so that each ISP has only one fixed VC or only a few VCs between their POP and the BAS, also called "fat pipes." The BAS may be located in each CO, or for smaller COs a BAS may be shared between COs. The bandwidth of the ISP VC is determined by the ISP's desired service concentration ratio. In contrast to the ATM VP switching, the only time this VC ever needs to be changed is if the link details (size or route) change.

Note that ISP may have several links with different domain name qualifiers to offer a differentiated service. The connection between the user and the ISP is made when the user's PPP login is authenticated. The users specify their name as user@service_class.ISP.com. The BAS uses the "service_class.ISP.com" domain name to identify which ISP VC to map the user's PPP link to. The ISP validates the user login using the standard

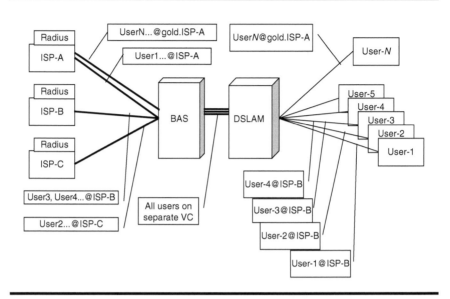

Figure 10.16   Internet Service Provider (ISP) user grooming.

radius authentication mechanism. This process means that the network provider does not need to keep configuring the DSLAM routing table once the path between the DSLAM and the BAS is in place.

A point to note is that this approach is very IP-centric and presumes a PPP/IP-based transport and application set. The connection between the BAS and the ISP POP does not have to be ATM, it can be FR, L2TP (Layer 2 Tunneling Protocol), or MPLS. This model has been extended to support multiple users on one DSL link, by using the PPPoE (PPP over Ethernet) protocol to extend PPP sessions to the users-PC. Here the ADSL-CPE modem acts as a simple bridge and the PPP session set up directly from the user-PC to the ISP, as relayed by the BAS.

10.5.2   ATM DSLAM Evolution

DSLAMs started out as simple devices, focused on simple multiplexing of ADSLs at low cost. The most visible enhancement has been the improvements of TM and of supported bandwidths. DSLAMs are slowly evolving into multi-service platforms. The first change has been the support for G.HDSL transceivers aimed at providing symmetric business class services. ATM inverse multiplexing (IMA), now standardized in the latest ADSL2+ specifications is being used both for ADSL2+ and SHDSL to increase the link speed by bonding several loops together. ADSL2+ is also bringing higher speeds leading the way to widespread adoption of digital video. Modern DSLAMs, with their gigabit network interfaces and back-plane capacity easily integrate Very High Bit-Rate DSL (VDSL) and ADSL2+ technologies.

On the physical side, DSLAMs of all capacities are now deployed: from highly integrated mini-DSLAMs, with a small number of ports, to gigabit DSLAMs supporting over 2000 ports and offering line cards with a port density of 96 and more. Network interfaces range from IMA bonded E1/T1 links to OC-48c optical connections. Hardened versions have been developed that can be pole-mounted and function in severe weather.

Of all the developments, the most ironic one has been the resurgence in derived-voice support. With falling costs and the rising adoption of packet voice, DSLAMs are in prime position to become a dominant platform for voice services. The DSLAMs will either transmit it over the twisted pair as a POTS signal, with the DSL still in overlay, or it will transmit the voice over the DSL, ultimately as Voice-over-IP (VoIP).

10.5.3 Integrating Voice into DSLAMs

Taking advantage of DSL's bandwidth to provide extra voice channels in addition to the baseband ASDL/POTS channel has been an enticing prospect from the outset. Over time, a number of approaches have been devised and standardized as discussed in Chapter 14. The various DSLAM architectures to implement the different Voice over DSL (VoDSL) approaches are shown in Figures 10.17 through 10.20.

The ANSI T1E1.4 standard defined separate TDM channels for voice support. An early TDM DSLAM architecture is shown in Figure 10.17. The broadband unidirectional channels are delivered over a T3 PDH or OC-3/STM-1 or higher SONET/SDH interface and demultiplexed onto the individual ADSL links. The bidirectional control channels are multiplexed into an X.25 connection. The LSx voice channels are multiplexed together into a GR.303 T1/V5.3 E1 switch interface. Note that for backhaul the X.25 T1

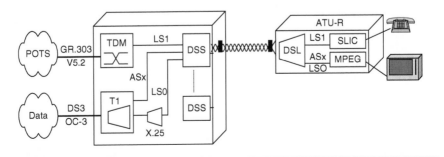

Figure 10.17   DSLAM with synchronous Voice over Digital Subscriber Line (VoDSL).

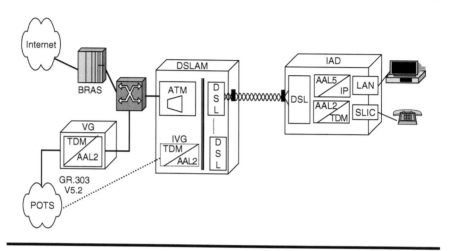

Figure 10.18   DSLAM with BLES VoDSL.

and switch T1 interfaces could also be carried over the SONET/SDH link. The benefit of dedicated, low-latency DSL bearer channels for voice was that no echo cancelation or DSP processing was required. The operation was similar to standard ISDN.

With the adoption of ATM as the primary convergence layer, TDM voice fell out of favor, but the fundamental driver persisted. Derived Voice over ATM (VoATM), based on AAL2 encapsulation was developed and specified as the BLES. A new form of DSL CPE, the integrated access device (IAD)

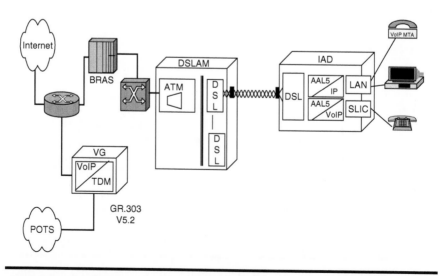

Figure 10.19   DSLAM with Voice-over-Internet Protocol (VoIP) VoDSL.

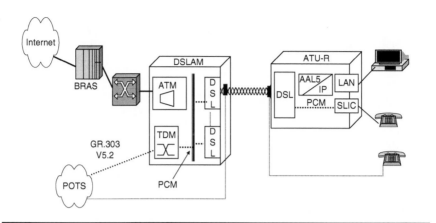

Figure 10.20   DSLAM with Channelized VoDSL (CVoDSL).

was developed that supports both AAL2 BLES voice and AAL5 data over a single DSL. The voice gateway was developed to terminate the BLES traffic and convert it to TDM. To take advantage of the installed base of DSLAMs, the voice gateway was initially an adjunct product, although it could be a separate card integrated into the DSLAM. The DSLAM aggregates the AAL2 voice into one ATM PVC linking the DSLAM and voice gateway, as shown in Figure 10.18. To avoid increased latency and cell delay, the IAD and DSLAM must implement QoS buffering in both upstream and downstream directions so that voice cells always have highest priority.

For business customers who have PBXs (Private Branch exchanges) rather than separate phones, and are usually connected over Symmetric DSL (SDSL), AAL1 encapsulation is sometimes preferred. With a PBX, the bandwidth reduction achieved by using AAL2 statistical multiplexing is reduced, as the PBX is already concentrating the voice traffic, so a higher percentage of trunk timeslots are always active. In AAL1, the entire T1/E1 trunk is assembled at a time, in the case of a full T1 with no voice compression, it takes just under two frames or only 250 µs to fill a cell. As long as the total end-to-end delay is low enough, this may avoid the need for any echo cancelation. Where AAL2 with its inherent statistical multiplexing retains an advantage over AAL1 is time-of-day bandwidth usage. For example, in a remote office application, during the day the DSL connection

---

Note: *On the need of echo canceling in VoDSL*

When there is a long delay between the both ends of a voice link, echo canceling is needed. Indeed, on commonly found, traditional telephones, a non-negligible part of the arriving signal is bounced back to the transmitting side. This is perceived by the sending end as an echo, which needs to be canceled somewhere. Indeed, when the echo is strong and arrives with an extreme delay, it hampers the speaker up to the point that it makes conversation impossible.

may have a high percentage of voice traffic, whereas at night the full link capacity is utilized by the data traffic to update the local databases and transfer data back to the corporate servers.

An alternative to using VoATM is VoIP. VoIP can be used over either an ATM or a packet DSL loop. The network diagram is shown in Figure 10.19.

One disadvantage of VoIP relative to VoATM is that more network elements must be traversed between the IAD and the voice gateway, which can complicate provisioning and troubleshooting. Another problem is that DSL user IP addresses are owned by the ISP and usually dynamically assigned, whereas VoIP ideally requires static IP endpoint addresses. This will become less of an issue as IPv6 deployment increases. A final issue for incumbent network operators is that the IP network is part of a separate deregulated entity, i.e., separate from the TDM voice business unit. The regulatory situation has caused incumbent operators to initially have favored VoATM while competing network operators, campus networks, and ISPs have favored VoIP.

The downside of both VoATM and VoIP is the latency resulting from assembling the voice into cells or packets. For example, in 64-kbps PCM encoded voice, a byte is generated every 125 µs. If 40 voice bytes are assembled per AAL2 cell, then it takes 5 ms to fill one cell. If the voice is compressed, as is often the case in ADSL to account for upstream bandwidth limitations, then the delay can be easily twice as long in one direction. If another few milliseconds are allowed for cell delay variation, the delay can be sufficient to require the use of echo cancelers. The result is a more expensive and power intensive IAD.

To address the IAD complexity, DSL standards came full circle and the most recent ADSL2+ standard has added a Channelized Voice over DSL (CVoDSL) option. With CVoDSL, the DSL framing structure supports up to four TDM channels that are allocated on a dynamic basis, meaning they do not take up any link bandwidth when voice channel is idle. A possible DSLAM architecture is shown in Figure 10.20.

Looking ahead, the TDM telephony network is slowly being replaced with a packet network. International segments were the first to be converted, followed by the long distance trunk network. Now it is the turn of the access network. One consequence of this is that voice traffic may undergo several TDM to packet conversions, which might add latency and could even be using a different voice compression codec at each instant. The result is greater network complexity and degraded voice quality as shown in the top part of Figure 10.21.

To overcome this, networks are evolving to a Softswitch model, where the call control is separated from the bearer network and the voice is packetized only once at the edge of the network. To avoid the complexity of IADs, the DSLAM is mutating into a multi-service access node and is becoming

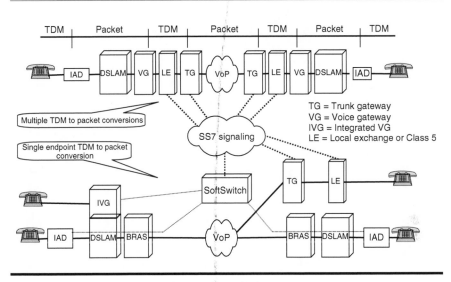

Figure 10.21   Voice over Packet (VoP) evolution.

the primary node where the voice is packetized. A DSLAM architecture that captures all the derived voice options is shown in Figure 10.22.

In places, where derived voice is used only in the local loop, the voice gateway can be integrated into the DSLAM. AAL2 traffic from IAD are converted back to TDM and processed through the GR.303/V5.x interface. CVoDSL is fed directly into the GR.303/V5.x block.

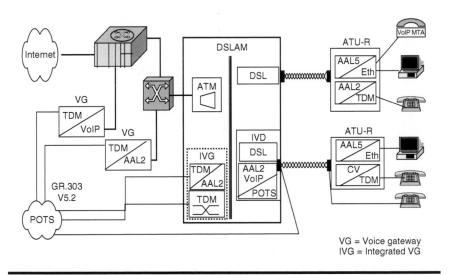

Figure 10.22   Fully featured VoDSL DSLAM.

For packet backbones, IAD packet voice is not terminated on the DSLAM; instead it is passed directly onto the voice trunk. CVoDSL is adapted on the DSL card and the resulting packet voice passed through. An option that is just starting to be deployed is the integrated voice data (IVD) card. The IVD card incorporates both a DSL transceiver and a voice SLIC (subscriber line interface circuit) for the analog baseband POTS. The analog POTS signal is converted to packet voice and multiplexed onto the voice trunk. Note that in an advanced implementation, an ADSL IVD card will support data, lifeline telephony over the baseband POTS, and additional telephone lines using CVoDSL. All the voice traffic, irrespective of source, will be converted to a common packet voice format. From a signaling perspective, the DSLAM is now acting like a media gateway and is controlled by the Softswitch using the H.248 or Media Gateway Control Protocol (MGCP).

DSLAM supported voice has been a constant theme or even a hot topic, of which the adoption was always "almost" about to happen. Now with the evolution of the public switched telephone network (PSTN) from TDM to packet, the DSLAM may become the primary focal point for voice conversion that bridges both derived voice on the local loop terminating on legacy TDM switches and network packet voice hosted off softswitches.

10.6   IP DSLAMs

Most applications that run over DSL today are, or are becoming IP-based. Data has moved from X.25 and FR to IP, voice is moving from TDM and VoATM to VoIP and video is moving from MPEG or ATM to video over IP. This raises the obvious question of, Why continue to use ATM as an intermediate layer?

As discussed earlier, incumbent telephone operators were required to separate their network operations from their service provider business units and open access to their DSL network to competing providers, a process known as unbundling. In many cases, they were specifically barred from using IP in their access or transport networks. As ATM was designed to transport many different services over a single network comprised of different physical media, it was a natural choice. In addition, many telecom operators already had extensive ATM backbone networks. ATM provided the means to create virtual leased lines between a DSL user and their ISP without having to have any awareness of the application traffic or type. In addition, the rate of early DSL uptake was uncertain, so multiple DSLAMs could be aggregated back to a single point that served as an IP POP. This reduced the amount of initial IP equipment required, as DSL penetration increased the IP nodes could profitably be pushed further toward edge.

For competing operators, who had their own networks and were not required to unbundle those networks, and campus networks including hotels, multi-tenant units, universities, business parks etc; ATM was considered an unnecessary overhead and complication. For these networks, the equipment could be simplified if ATM was eliminated, assuming another low-cost layer 2 multiplexing protocol could be identified.

Ethernet, which has a globally unique address for every endpoint, and contains both the source and destination address in every frame, proved an eminently suitable protocol, with a small implementation modification. Nearly, all DSL data traffic was already Ethernet-encapsulated as most of it emanated on Ethernet LANs. Even in ATM DSLAMs, the data mainly started or terminated in Ethernet. The ATM CPE modems bridged the traffic into ATM for transport over DSL.

In contrast, in an Ethernet DSLAM, the Ethernet traffic is bridged into HDLC (High-level Data Link Control) frames for transport over the DSL link. The DSLAM operates as a modified Ethernet layer 2 switch. In a generic switch, any port can send traffic directly to any other port. In an Ethernet DSLAM, the traffic may only flow from the DSL port to the network port and vice versa. Upstream broadcast traffic is routed only to the network port, while downstream broadcast traffic is optionally blocked, allowed through to all ports, or switched only to selected ports based on the source address. The DSLAM routing table can be statically provisioned or self-learning. In self-learning, on each upstream DSL port, the DSLAM learns the CPE MAC (Message Authentication Code) addresses by monitoring the source address of each packet, and then creates a binding between the source addresses and the port number. In the downstream direction, the DSLAM, compares the destination MAC address with the tables for each port, and forward the packet to the correct port on a match, and discarding those packets with no match.

Beyond basic Ethernet switching, the generic architecture discussed in Section 10.3.2 applies. For a pure Ethernet design, Ethernet 802.1p can be used directly for QoS buffering. Classification is based on the source and destination address, and packet type and priority. With the selection of bonded-SHDSL for low-speed Ethernet first mile access, Ethernet DSLAMs are sure to prosper and evolve.

The first enhancement to the basic Ethernet design is to include the BAS functionality. The integrated Ethernet DSLAM now includes the Ethernet multiplexer, the BAS sign-on capability and the backbone router.

A more advanced enhancement that is just starting to come to market is the all-IP DSLAM. This operates directly on the IP packets, irrespective of

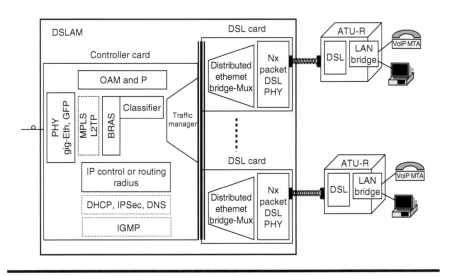

Figure 10.23   All-IP DSLAM architecture.

the underlying layer 2 protocol, which may be HDSL, Ethernet, or even FR or ATM. A key development enabling pure IP switches has been advances in classification technology. The ability to do wire speed Gigabit per second multilevel deep packet lookup and classification has eliminated the requirement for an underlying layer 2 protocol as a rapid switching mechanism. An all-IP DSLAM architecture is shown in Figure 10.23.

At the ingress ports, the IP packet is classified and tagged. Based on the tag, the packet undergoes admission control, switching, QoS buffering, and output rate shaping. Non-data packets are forwarded to the control processor. BAS functionality is fully integrated, as is the core routing function. Increasingly, MPLS is also integrated with the IP-DSLAM acting as a MPLS label edge router (LER). IGMP to implement the "Channel Tuner" for IP-TV may also be integrated.

10.7   DSLAM—Continuous Evolution

Looking ahead, DSLAMs and DSL continue to evolve on three main fronts:

1. Increasing speed with ADSL2+ line rates up to 24 Mbps and VDSL up to 100 Mbps.
2. Enhanced DSL capability and types including ADSL2+, CVoDSL, packet-mode operation, G.SHDSL, and EFM (Ethernet in the First Mile).

3. Further functional integration including BAS authentication, routing, MPLS, packet-voice, and broadcast TV.

The DSLAM has a significant role to play in broadband deployment and may yet prove to be the convergence platform for multi-service access.

Chapter 11

Operations, Administration, and Maintenance

Marko Loeffelholz, Thomas Haag, and Markus Freudenberger

CONTENTS

Abstract   This chapter addresses aspects of operation, administration, and maintenance (OAM) for Digital Subscriber Line (DSL) access networks. It includes descriptions of basic functionalities of OAM for DSL-based broadband services in dependence of the network elements and the usage for the "in service" end-to-end testing possibilities. Following this, line aspects of the xDSL related functions are described such as information exchange between the xDSL elements and the xDSL-type dependent functionalities (performance monitoring, configuration parameters, primitives, and failures). This chapter also includes basic descriptions of network management systems (NMS) and element management systems (EMS) with the commonly used interfaces and protocols.

11.1   Introduction

The purpose of operations, administration, and maintenance (OAM) in connection with a telecommunications network is to provide appropriate mechanisms for establishing and monitoring link operation. OAM provides network operators the ability to monitor the health of the network and quickly determine the location of failing links or fault conditions.

From the functional perspective, DSL systems are commonly standardized. From the OAM point-of-view there is a multiple set of single functions that are defined in several DSL-, Asynchronous Transfer Mode (ATM)-, and IP (Internet Protocol)-related standards. Examples of such mechanisms are remote fault indication and remote loopback control. Each set of these standards provide a rather complete view from a perspective which is layer dependent. The goal furthermore must be to describe the layer-specific functionalities in a perspective of layer-related interworking mixed with requirements from a network operator's perspective. Also, it is important to consider and describe the interfaces for reporting and control, mostly referred as network management systems (NMS) or element management systems (EMS) interfaces.

11.2   OAM Concepts

From a network operators point of view the aspects of operation and maintenance play a key role for cost-effective operation. Given this and considering the huge number and variety of network elements along the data path in a modern broadband telecommunications network, OAM needs a consistent concept. Such an OAM concept considers not only equipment, interface, and network requirements but also various issues for service and provisioning. The development of such a concept usually starts with the planning phase for deploying a system, platform, network, and services and will be adapted to actual requirements received during operation over the whole life circle. Once a good OAM concept is established, requirements for new systems or network components very often are derived from it because the other way around is commonly considered to be more cost intensive. The following aspects mainly constitute an OAM concept or should at least be considered for OAM:

- Equipment provisioning
- Network provisioning
- Service provisioning
- Billing aspects, content management, and if needed access control
- Service configuration
- Inventory management (discovery and configuration, updates)
- Service quality management, including performance and capacity management
- Fault management

One can see that OAM for network operators is not only related to particular network elements or technologies but also needs a broad base of considerations.

The following sections give an overview over the general technical means and means provided by xDSL access network elements, which constitute the basis for the above listed aspects for xDSL platforms.

11.3 General System Architecture

For consideration of OAM aspects, it is very useful to define a general model to describe the specific individual layer-related functionalities provided by OAM. The description is based on common standardized functions. Platforms supporting broadband services have a layer structure. Layer 1, which represents the physical layer, is dependent on the used physical medium. For the access section between customer premises equipment (CPE) and a DSLAM across the U-interface, xDSL variants like ADSL (2/2plus), SDSL, or VDSL can be used.

For the connection between the DSLAM and the broadband network SDH/SONET (synchronous digital hierarchy/synchronous optical

Table 11.1   Operations, Administration, and Maintenance (OAM) Flows and Corresponding Layers and Technologies

Flow	Layer	Technology
1	1	SDH/SONET, PDH, xDSL
2	1	SDH/SONET, PDH, xDSL
3	1	SDH/SONET, PDH, xDSL
4	2	ATM (VP-Level)
5	2	ATM (VC-Level)

networking) or PDH (plesiochronous digital hierarchy) is mainly distributed as layer 1 technology. From the end-to-end point-of-view, ATM is the dominant technology for layer 2.

Therefore, we distinguish the flow of OAM information (OAM flows) that will be considered into five hierarchical flows, named as F1–F5 (see Table 11.1 and Figure 11.1), according to the terminology used in recommendations from the ITU-T and other standardization bodies. In detail, each flow belongs to a certain layer.

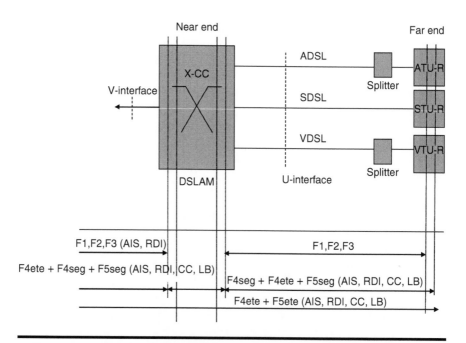

Figure 11.1   General system architecture and operations, administration, and maintenance (OAM) information flows.

For each flow three types of OAM functionalities exist:

1. Fault management
2. Performance management
3. Configuration management

The configuration management is mainly used for layer 1 configuration purposes and is described below. The configuration management for layer 2 or 3 aspects is not further considered here because this is mostly done by additional means besides the xDSL-related OAM flows described here.

11.4   Layer Model

11.4.1   Layer 1 Aspects

11.4.1.1   Layer 1 OAM Flow

The layer 1 represents the physical layer which contains, referred to the common xDSL reference models, the physical medium dependence layer (PMD) up to the TPS-TC-Layer (see also Chapter 13). The OAM functions are described in the various xDSL recommendations from ITU-T and other standardization bodies.

The physical layer contains the three lowest OAM information flow levels and allocates as follows (see also Figure 11.2):

1. F1: Regenerator section level (regenerators are only intended for Symmetric High Bit-Rate DSL [SHDSL])
2. F2: Digital section level
3. F3: Transmission path level

Each flow F relates to the respective superordinate flow level (see Figure 11.3).

In contrast to legacy SDH/PDH connections for xDSL using a layer 2 adaptation to protocols like ATM or Ethernet, the flows F1 or F2 or F3 alone do not ensure an end-to-end OAM for the entire xDSL connection because they are usually terminated within the xDSL span. End-to-end OAM can only be provided by using means of F4 or F5 (see layer 2 or 3 aspects).

This leads to a dependency between the different flows, particularly with regard to the fault management. A loss of signal for instance at the physical layer (F1 to F3) will be reported to the NMS but also a fault at F4 or F5 corresponding to [ITU-T I.610] will be generated as well. In most cases,

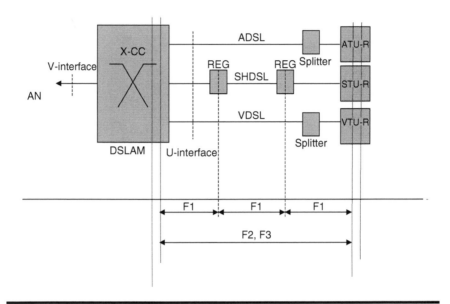

Figure 11.2   xDSL physical layer OAM flows.

only the F4 or F5 fault is available beyond the NMS-controlled elements, which is important for NMS independent end-to-end OAM.

The F1 and F2 flows are often combined into the F3 flow to which a management interface via an InterWorking Function (IWF) can have access

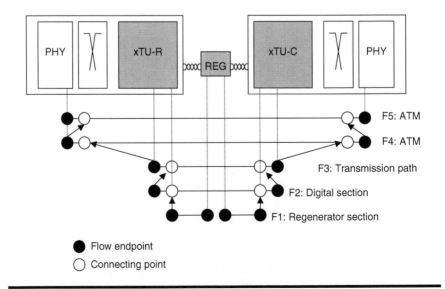

Figure 11.3   Interconnection between xDSL physical layer OAM flows.

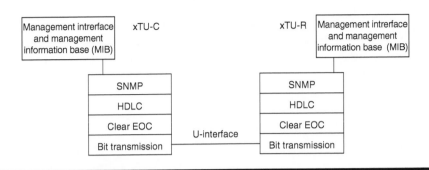

Figure 11.4   Management interface access to OAM channel.

to operate the physical layer. An example according to [ITU-T G.997.1] is given in Figure 11.4.

The transport of the F3 flow information between the xTU-R at the customer side and the xTU-C at the central office (CO) side can be handled in different ways, but always across the U-interface:

- Via a dedicated OAM channel in a clear embedded operations channel (EOC) transmission
- Via the EOC
- By using indicator bits from the xDSL overhead

The xDSL-specific use of these transport means are specified in the xDSL recommendations and standards. Each way uses appropriate link control mechanisms to protect the OAM information flow. The appropriate way can be identified during the handshake or initialization phase.

11.4.1.2   Layer 1 OAM Contents

A set of information is defined to be communicated for layer 1 OAM. They can be divided into far-end and near-end primitives, and corresponding xTU-R and xTU-C generated information, respectively. The definitions of the below listed anomalies, defects, and failures are given in the xDSL recommendations for ADSL (2/2plus), SHDSL, and VDSL. Some examples are listed in Table 11.2. Depending on the transmission type subsets of these lists are used.

Apart from the line-related parameters data path-related parameters depending on the transmission layer type (ATM, STM, packet transfer mode, etc.) are also used. More details are specified in the ITU-T G.997.1 and the individual xDSL recommendations (ITU-T [International Telecommunication Union-Telecommunication], ETSI [European Telecommunications Standards Institute], ANSI [American National Standards Institute]).

Table 11.2   Examples of Far-End and Near-End Primitives to Be Communicated to Layer 1 OAM

Line-Related Primitives	Configuration Parameters	Diagnostic and Status Parameters
Anomalies Information about corrected errors (counter) Information about code violations (counter, e.g., for CRC errors) **Defects or failures** Examples for and defects failures are as follows: Loss of Signal (LOS) Loss of Frame (LOF) Loss of Power (LPR) Loss of Sync Word (LOSW) SNR Margin Defect (LOM) Loop Attenuation Defect	Line type Transmission rates Noise margin data (target, maximum, minimum, etc.) Operation mode (rate adaptive, etc.) Power modes (Power down or power saving modes) Spectrum usage (PSD configuration)	Transmission capabilities Line attenuation SNR margin Current rate Attainable rates Output power Line delay Line states Line diagnostics including vector formatted data over the frequency Inventory information (vendor ID, version number, serial number)

The configuration parameters can be organized in three profiles, dedicated to a line (xDSL port), a channel, and a data path. Whereas line configuration profiles combine parameters related to physical characteristics like operation modes, power management, spectrum usage, and noise margin thresholds, channel configuration profiles for each bearer channel combine data rate and delay related configuration parameters. The data path configuration profiles are transport layer dependent and cover all parameters needed to define the layer 2 adaptation in terms of operational modes and failure thresholds.

11.4.1.3   Performance Monitoring

Generally, the xTU-C acts as the near end and provides a monitoring function. Therefore, it collects the near-end and the far-end information and provides it to an EMS via a standardized management interface (Q-interface).

Optionally, according to demand the xTU-R can also provide the monitoring function, e.g., for applications in the customer environment or for

further use in additional management concepts as described in recommendation [DSL-Forum TR-069].

The performance monitoring (PM) contains collection and storage of
current and history line and often data-path-related parameters. Apart from
the anomaly counters, the defects or failures or anomalies are presented as
event seconds, which means that all seconds are counted in which a defect
or anomaly was recognized.

Examples for PM parameters are Errored Seconds (ES) (seconds with
code violation anomalies), Severely Errored Seconds (SES) (seconds
with LOS or a certain number of anomalies exceeded), Unavailable Seconds
(UAS) (seconds, where the xDSL transmission is not available), and the
counters for anomalies and defects.

The PM storage is typically organized in time intervals of 15 minutes
and 24 hour. The current data is presented in one 15 minute interval and
one 24 hour interval. The history entries contain the last $N \times 15$ minute
intervals (e.g., $N = 96$) and $M \times 24$ hour intervals (e.g., $M = 7$). The amount
of parallel performance monitoring is strongly dependent on the internal
memory of an xDSL system (DSLAM) because all data normally has to be
collected and stored inside the xTU-C network elements. So, the system
integrator is responsible to use this feature carefully in terms of memory
consumption and processor load.

Appropriate EMS or NMS functionality is necessary to handle the further
processing of the collected data. With this information, a detailed analysis
of the xDSL performance over a certain time interval is possible.

11.4.1.4 PDH/SDH Layer

In case of a PDH/SDH, at the interface at the V-reference point at the
DSLAM, which is basically the network interface, the appropriate OAM
flows for PDH/SDH are used. Usually, at this interface, the ATM traffic
from the DSLAM is mapped into SDH/PDH because in an ATM transport
network the interfaces are SDH based. The physical layer also contains the
three lowest OAM levels and allocates as follows (see Figure 11.5):

1. F1: Regenerator section level
2. F2: Digital section level
3. F3: Transmission path level

For SDH-based transmission systems, the ITU-T recommendations
G.707–709, G.782, and G.783 are relevant.

Flows F1 and F2 are carried on bytes in the Section OverHead (SOH),
and flow F3 is carried in the higher-order Path OverHead (POH) of the
transmission frame. ATM transport over SDH may also be supported on

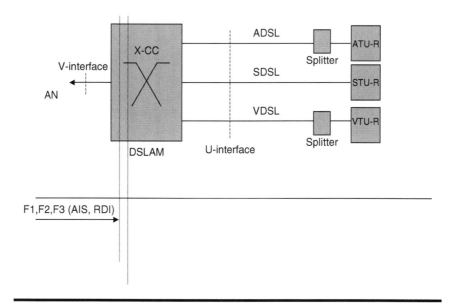

Figure 11.5   PDH/SDH (plesiochronous digital hierarchy/synchronous digital hierarchy) physical layer OAM flows.

lower-order paths, which support POHs. These lower-order path OAM flows are subsets of the F3 flow. But note that the SDH path is different than the ATM path. It is essential to distinguish between ATM path and SDH/PDH path, because the SDH/PDH path is covered by OAM F3 while the ATM path is covered by F4.

For PDH-rate frame-based transmission systems the ITU-T recommendations G.702, G.804, and G.832 are relevant.

Specific means to monitor the section performance (e.g., code violation counting, CRC, etc.) are defined within the ITU-T recommendations [ITU-T G.702], [ITU-T G.804], and [ITU-T G.832]. If the ATM layer is supported only on a PDH section then the G.804, overhead constitutes both F1 and F3 flows (but not F2).

F1 is supported by frame alignment bytes and the remainder of the overhead constitutes the F3 flow.

11.4.2   Layer 2 ATM Aspects

11.4.2.1   Overall

Mainly, the distributed DSL architecture for layer 2 aspects is the ATM-based architecture. Alternatively, Ethernet will be used for transport. Independent from the used transport medium a control plane needs management functions, which can be addressed by the management plane. A control plane

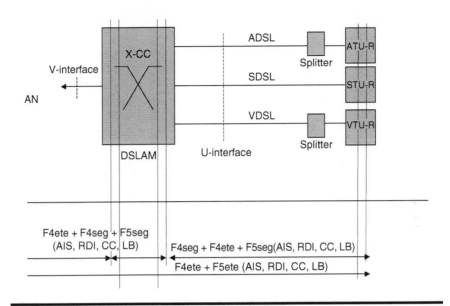

Figure 11.6   Layer 2 OAM flows.

means the entity to setup and disconnect connections. A management plane represents the IWF between the control plane and the management interface. In other words, the control plane defines the action while the management plane defines how to address and control the action of the control plane.

Usually, these functions are described in the overall model as flow F4 and F5 and each is clustered in the following areas:

- Fault management
- Performance management

To derive the OAM model for DSL equipment from existing models that are described in the recommendation [ITU-T I.610] for transport networks, the details of network architecture and the impact to specific OAM functionality have to be considered (see Figure 11.6).

11.4.2.2   OAM Types

The ATM approach according to [ITU-T I.610] is based on fault and performance management issues. That means that fault management functionalities like alarm indication signal (AIS), remote defect indication (RDI), loop back (LB), and continuity check (CC) are applicable. Regarding performance management issues, functionalities like forward performance monitoring (FPM) and backward reporting (BR) will be required.

Table 11.3   Overview of the OAM Functions of the ATM Layer

OAM Function	Main Application
AIS (alarm indication signal)	For reporting defect indications in the forward direction
RDI (remote defect indication)	For reporting remote defect indications in the backward direction
CC (continuity check)	For continuously monitoring continuity
LB (loop back)	For on-demand connectivity monitoring For fault localization For preservice connectivity verification
FPM (forward performance monitoring)	For estimating performance in the forward direction
BR (backward reporting)	For reporting performance estimations in the backward direction
Activation or deactivation	For activating or deactivating PM and CC
System management	For use by end-systems only
APS	For carrying protection switching protocol information

For maintenance purposes, F4 and F5 flows are defined at the ATM layer covering the virtual path (VP) and virtual channel (VC level), respectively. Both flows are bidirectional and follow the same physical route as the user-data cells, thus constituting an in-band maintenance flow.

Besides the vertical subdivision into F4 and F5 levels a "horizontal" partition also exists; both flows can either cover the entire virtual connection (end-to-end flow, "ete") or only parts of the virtual connection (segment flow, "seg").

Table 11.3 gives an overview of the OAM functions of the ATM layer. Additional functions for testing, fault localization, and performance measurement are not mentioned here because they are not finally defined.

11.4.2.3   OAM Transport Topology

The main difference between traditional transport networks and DSL-based access networks is the usage of one copper pair for receiver and transmitter simultaneously. This aspect influences the usage of the specific OAM functions specified by the standardization body ITU-T. For traditional transport networks, one segment consisting of transmitter and receiver for separated transport medium is shown in Figure 11.7.

In DSL-based networks, the receiver and transmitter use the same medium (see Figure 11.8).

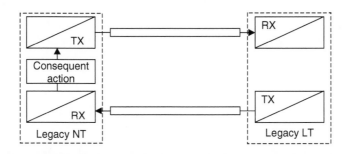

Figure 11.7   OAM transport for legacy transmission.

11.4.2.4   Consequence

For ATM-based DSL systems, the standard ITU-T I.610 describes fault management purposes for F4 and F5 flows in two basic messages, the AIS and the RDI. Both are transported together with data (inband) through the network. They usually represent an action or reaction on certain conditions (faults) regarding the transmission path and can therefore also be called as "consequent actions." It must be noted that in an xDSL environment an RDI signal cannot be transported while the connection is interrupted. In such a case, the xTU-R receiver detects a loss of cell delineation (LCD)* defect which should be reported as RDI to the xTU-C. The xTU-C would then send for its part as consequent action an AIS. But the RDI signal will never reach the receiver of the xTU-C because the line already was interrupted. Therefore, a single AIS/RDI signaling on layer 2 for F4/5 does not make

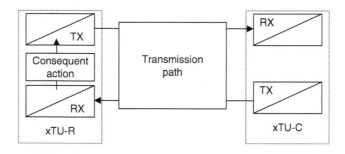

Figure 11.8   OAM transport for xDSL transmission.

---

*   Cell delineation, based on the HEC (Header Error Control) field in the ATM cell header, allows identification of the cell boundaries. When consecutive errors affect the HEC, an LCD defect may be observed.

sense in terms of a signaling between DSLAM and xTU-R only. However, for the supervision of segments across the V-interface, F4/5 is suitable for signaling as per end-to-end (ete) or segment (seg) connection.

11.4.3 Layer 2 Ethernet Aspects

11.4.3.1 Overall

Broadband access transport networks are commonly evolving into layer 2 Ethernet transport technologies. There are several worldwide migration or even replacement activities to move from ATM to Ethernet and maintaining at the same time the amount and handling of OAM functionalities known from ATM OAM. Because Ethernet transport technology was originally developed for MAN and LAN applications there is still some need to develop and standardize OAM functions suitable for existing xDSL broadband access transport and aggregation networks. Actually, there are several standardization activities, in the ITU, Institute of Electrical and Electronics Engineers (IEEE), Internet Engineering Task Force (IETF), and Metro Ethernet Forum (MEF), aiming on the introduction of Ethernet in carrier networks. The intent of introducing OAM mechanisms for Ethernet is to maintain similar layer 2 end-to-end OAM capabilities that were available when the aggregation was ATM based, via ATM OAM as specified in ITU-T (I.610).

[IEEE 802.1ag] and [ITU-T Y.1731] define the concept of a maintenance association end point (MEP) and maintenance association intermediate point (MIP), which are configured on a per port, per virtual local area network (VLAN), and per maintenance domain level. MEPs initiate connectivity fault management (CFM) OAM messages and are configured at the far end of the service perimeter or S-VLAN (e.g., in the service node and DSLAMs). MIPs are configured across the path of the S-VLAN (e.g., in Ethernet aggregation nodes). Various domain maintenance entity (ME) levels can be configured, allowing the network administrator to divide the network into multiple administrative OAM domains and to allow nesting of OAM domains, where an ME level corresponds to an OAM domain.

Three domains are defined such as "carrier domain," "intra carrier domain," and "Customer Domain."

At the time of publishing, this work was being performed in ITU (ITU-T Y.1731 Ethernet OAM mechanisms) and IEEE (802.1ag). The architectural framework is described and published in [DSL-Forum TR-101].

11.4.3.2 Ethernet OAM Types

Table 11.4 lists the OAM message types, which are defined for CFM Ethernet OAM.

Table 11.4 OAM Message Types Defined for Connectivity Fault Management (CFM) Ethernet OAM

OAM Function	Main Application
LoopBack Message (LBM)	Defined as a unicast message sent to an MP's MAC address.
LoopBack Reply (LBR)	Originated by the LBM, the LBR is sent as a unicast message from an MP to the MAC address of the MEP.
LinkTrace Message (LTM)	This is a multicast message that is relayed to a given target MEP and inspected by every MIP along the data path to determine whether the target MAC is known by the MIP.
LinkTrace Reply (LTR)	An LTR message is a unicast message sent from a MEP or MIP upon receiving and forwarding an LTM.
Continuity Check Message (CCM)	Defined as a multicast message from a MEP that is received by all MEPs in the same service instance on the same ME level.

11.4.3.3 *Interworking between Ethernet and ATM OAM*

Testing end-to-end connectivity between a service node and xTU-R requires an end-to-end visibility between both network elements. Using ATM between service node and xTU-R fulfills this requirement by providing an end-to-end visibility in the data path and is defined in [ITU-T I.610].

The actual scope of functionality of I.610 within this document's context is limited to ATM only. It is expected that I.610 functionality will only have utility in the absence of tools based upon the emerging Y.1731 or IEEE 802.1ag standards and will merely exercise more of the NT than the DSL PHY.

When part of the ATM is replaced by Ethernet the end-to-end visibility between service node and residential gateway (RG) is lost. Restoring the lost visibility can be performed by a data path or control path mechanism, e.g., CLI (Command Line Interface), EMS, or layer 2 control. All solutions require the presence of an approach-dependent IWF, which provides the translation of a trigger message to DSL port specific ATM OAM messaging.

This requires an interim mechanism at the DSLAM to remotely invoke I.610 LB (either CLI or control plane action).

There is a subset of DSLAM functions, which can be initiated by the DSLAM management system request, which would be more conveniently initiated by the service node in some networks. To implement such an approach, a communication mechanism is needed between the service node and DSLAM (see Figure 11.9).

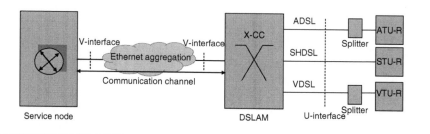

Figure 11.9   Layer 2 control architecture.

There are several options, e.g.,

- A management channel, as described in Y.1731 (ETH-MCC)
- A layer 2 control mechanism

Messages on the selected communication channel need to support functions including the following:

- Ability for the service node to check a DSL's status.
- Ability for the service node to trigger the DSLAM to initiate an ATM LB to an xTU-R, on a specified VPI/VCI on a specific access line.
- Ability for the service node to trigger the DSLAM to initiate an ETH-LBM to an xTU-R, at a specific access line.
- Ability for the DSLAM to report the results to the service node.

At the time of publishing, this work was still in progress.

11.5   Controlling and Reporting OAM

11.5.1   *Main Principle*

As an operator in the network control center (NCC), it is necessary to have all the OAM information reported to the EMS or NMS at the time an anomaly or defect in the network is detected. The most common model on the network level is the manager agent model. This model is based on the client server principle. Usually, an EMS or an NMS is placed in an IP-based DCN (Data Communication Network) and it handles the management of the whole network. The manager (client) sends a request to the agent (server) and the agent replies back to the manager. The difference to the well-known client server model is that several managers can reach one agent and the agent is able to send spontaneous notifications to several managers. The manager is often referred to as the EMS, to emphasize that it is usually realized as a user interface that presents data to an operator. The agent is an entity that executes within the network element (NE, e.g., the DSLAM). The DSLAM usually can be reached over a separated management interface by one or by several IP addresses using a standard Ethernet stack. It is preferred to use an implementation with one agent per NE and not per DSL, so each NE can be addressed by one IP address only. This solves problems regarding the required amount of distinctive IP addresses between different DSLs with different agents for each. It is suitable to access one network element with one IP address and to reach a couple of lines with one agent.

The management information is defined in the so-called management information base (MIB). The MIB is a structured collection of managed objects that represent resources in the network elements. It is a formal definition of the kind of information the agent makes available to the manager. The manager accesses the MIB through a management protocol, such as SNMP, CMIP, HTTP, or CORBA. Each of these protocols has its own MIB definition language.

In Simple Network Management Protocol (SNMP), it is a subset of Abstract Syntax Notation 1(ASN.1); in Common Management Information Protocol (CMIP), it is Guidelines for the Definition of Managed Objects (GDMO); in Hypertext Transfer Protocol (HTTP), it is implicit; and when using Common Object Requester Broker Architecture (CORBA), it is Interface Definition Language (IDL). Usually, the entities defined in the MIB are called managed objects, although these objects do not have to be objects in the Object-Oriented (OO) way, e.g., a simple scalar variable defined in an MIB is called a managed object. The managed objects are logical objects, not necessarily with a one-to-one mapping to the resources. Each managed object, defined in the MIB, is uniquely identified by an Object Identifier (OID) that is located in the lexicographically ordered list of the names of all the variables.

11.5.2 SNMP Fundamentals

It is a matter of fact that the SNMP is the most popular network management protocol and provides interoperable solutions to the problem of managing networks, enabling effective monitoring and control of heterogeneous devices.

Almost all manufacturers of DSL equipment implement the SNMP as the most common protocol for network devices for communication with the appropriate EMS. Hence, the following chapter is based on this management protocol, as defined in the IETF documents RFC 1157 ("A Simple Network Management Protocol (SNMP)") [RFC 1157] and RFC 1215 ("A Convention for Defining Traps for Use with the SNMP") [RFC 1215].

During the ongoing deployment of the SNMP protocol several updates and enhancements were agreed upon. These were reached in Version 3 of SNMP (RFC 3411–3418) [RFC 3411] through [RFC 3418] which is recommended as the mandatory standard protocol for managing broadband networks. It fixes several security issues in the previous versions of the protocol like authentication and privacy. More details are specified in the appropriate RFC recommendations.

The SNMP management framework includes four components:

1. Managed node, respectively the Network Element, each containing a processing entity, termed as an agent
2. Network management station, also known as the manager, on which the management application reside
3. Network management protocol that is used to exchange management information between the management station and the agents
4. Management information, which is specified in the structure of management information (SMI) and the definition of managed objects, contained in the MIB

Note: The IETF is a large open international community of network designers, operators, vendors, and researchers concerned with the evolution of the Internet architecture and the smooth operation of the Internet, open to any interested individual.

Figure 11.10 describes the main principle of SNMP.

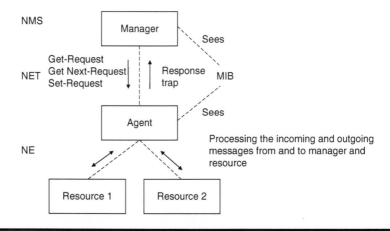

Figure 11.10   Simple Network Management Protocol (SNMP) Principle.

As one can see here, the main task of the SNMP agent, placed in the NE, is to process and correlate the messages it gets from the several resources (Resource 1, Resource 2) inside the NE, to map them into a specified, readable format, and to make them available for a manager. The five fundamental functions available to realize the communication between manager and agent are represented as SNMP-PDUs (Protocol Data Unit). Their denotations are as follows:

1. Get-Request-PDU: Generated and transmitted as the request of the manager to the agent.
2. Get next-Request-PDU: Almost the same as the Get-Request with the difference of viewing data without requiring prior knowledge. It is possible to traverse the management information in the lexicographical order.
3. Set-Request-PDU: Command to change the value of a specified managed object in the agent.
4. Response-PDU: Generated by the agent as a reaction to a Get-, Get next-, or Set-Request-PDU. In the case of a Get- or Get next-Request, the agent delivers the value of the concerning objects. If an error occurs, an error code with the corresponding description is delivered.

   In the case of a Set-Request, the Response-PDU delivers the configured value or in the error case an error code with the corresponding description.
5. Trap-PDU: A Trap is a spontaneous message from the agent without intervention from the manager to notify the manager that an extraordinary event has occurred in the agent.

   It can be divided into two categories:
   a. Alarm
   b. Notification

For reporting several OAM messages to the EMS, the above described Trap-PDU is used. High priority messages with service interruption must be reported as an alarm. In this context, it is absolute necessary to prohibit a high amount of alarm owing to a primary alarm that leads to a huge amount of spontaneous traps, also called trap storm. Only the origin primary alarm has to be notified to the EMS.

For example, if there is an AIS detected on the transmission layer, this leads to a multiplicity of F4 and F5 AIS messages. So the spontaneous notification of these alarms must be suppressed by appropriate filter mechanisms in the NE and only the primary alarm is notified to the EMS.

Additionally, the NE has to provide possibilities to assign appropriate severity levels (minor, major, etc.) for each alarm.

11.5.3 *Management Information Bases*

The MIBs [RFC 2515], [RFC 3635], [RFC 4363], [RFC 2108], [RFC 4188], [RFC 2662], [RFC 4319], [RFC 3592], [RFC 2863], [RFC 4087] and [RFC 2933] are specified and are available from the IETF. They contain the fundamental functions to report OAM. Tables 11.5 and 11.6 list the applicable RFC documents for an ATM and Ethernet approach, respectively.

ATM-Based Approach

Table 11.5   RFC (Request For Comments) Corresponding to ATM Management Information Base (MIB)

RFC No.	Meaning
2515	ATM MIB

Ethernet-Based Approach

Table 11.6   RFCs Corresponding to Ethernet MIBs

RFC No.	Meaning
3635	Ethernet overall
4363	Multicast, Filtering, VLAN
2108	Repeater
4188	Bridge

Physical Interface

The MIB of the physical interface is independent of Ethernet or ATM. Therefore, Table 11.7 contains the relevant references separately.

Table 11.7   RFCs Corresponding to the MIBs of the Physical Interface

RFC No.	Meaning	Comments
2662	ADSL	MIBs for ADSL2 or 2plus and VDSL2 are still in the standardization process in the IETF.
4319	SHDSL	
3592	SONET/SDH	
2863	Interface MIB	Relevant for both ATM- and Ethernet-based systems

Note: Because of the focus on layer 1 and 2 specific investigations, the references for IP are not shown. Important RFCs may be [RFC 4087] for IP Tunneling and [RFC 2933] for IGMP.

Bibliography

[DSL-Forum TR-069] DSL-Forum Technical Report TR-069. *CPE WAN Management Protocol.* May 2004.

[DSL-Forum TR-101] DSL-Forum TR-101. *Migration to Ethernet-Based DSL Aggregation.* April 2006.

[IEEE 802.1ag] IEEE Standard 802.1ag. *Connectivity Fault Management.*

[ITU-T G.997.1] ITU-T Recommendation G.997.1. *Physical Layer Management For Digital Subscriber Line (DSL) Transceivers.* June 2006.

[ITU-T I.610] ITU-T Recommendation I.610. *B-ISDN Operation and Maintenance Principles and Functions.* February 1999.

[ITU-T G.702] ITU-T Recommendation G.702. *Digital Hierarchy Bit Rates.* November 1988.

[ITU-T G.804] ITU-T Recommendation G.804. *ATM Cell Mapping into Plesiochronous Digital Hierarchy (PDH).* June 2004.

[ITU-T G.832] ITU-T Recommendation G.832. *Transport of SDH Elements on PDH Networks—Frame and Multiplexing Structures.* October 1998.

[ITU-T Y.1731] ITU-T Recommendation Y.1731. *OAM Functions and Mechanisms for Ethernet-Based Networks.* May 2006.

[RFC 1157] IETF RFC 1157. *Simple Network Management Protocol Version 1.*

[RFC 1215] IETF RFC 1215. *Convention for Defining Traps for Use with the SNMP Version 1.*

[RFC 3411] IETF RFC 3411. *An Architecture for Describing Simple Network Management Protocol (SNMP) Management Frameworks.*

[RFC 3412] IETF RFC 3412. *Message Processing and Dispatching for the Simple Network Management Protocol (SNMP).*

[RFC 3413] IETF RFC 3413. *Simple Network Management Protocol (SNMP) Applications.*

[RFC 3414] IETF RFC 3414. *User-Based Security Model (USM) for Version 3 of the Simple Network Management Protocol (SNMPv3).*

[RFC 3415] IETF RFC 3415. *View-Based Access Control Model (VACM) for the Simple Network Management Protocol (SNMP).*

[RFC 3416] IETF RFC 3416. *Version 2 of the Protocol Operations for the Simple Network Management Protocol (SNMP).*

[RFC 3417] IETF RFC 3417. *Transport Mappings for the Simple Network Management Protocol (SNMP).*

[RFC 3418] IETF RFC 3418. *Management Information Base (MIB) for the Simple Network Management Protocol (SNMP).*

[RFC 2515] IETF RFC 2515. *Definitions of Managed Objects for ATM Management.*

[RFC 3635]	IETF RFC 3635. *Definitions of Managed Objects for the Ethernet-like Interface Types.*
[RFC 4363]	IETF RFC 4363. *Definitions of Managed Objects for Bridges with Traffic Classes, Multicast Filtering, and Virtual LAN Extensions.*
[RFC 2108]	IETF RFC 2108. *Definitions of Managed Objects for IEEE 802.3 Repeater Devices using SMIv2.*
[RFC 4188]	IETF RFC 4188. *Definitions of Managed Objects for Bridges.*
[RFC 2662]	IETF RFC 2662. *Definitions of Managed Objects for the ADSL Lines.*
[RFC 4319]	IETF RFC 4319. *Definitions of Managed Objects for High Bit-Rate DSL—2nd generation (HDSL2) and Single-Pair High-Speed Digital Subscriber Line (SHDSL) Lines.*
[RFC 3592]	IETF RFC 3592. *Definitions of Managed Objects for the Synchronous Optical Network/Synchronous Digital Hierarchy (SONET/SDH) Interface Type2.*
[RFC 2863]	IETF RFC 2863. *The Interfaces Group MIB.*
[RFC 4087]	IETF RFC 4087. *IP Tunnel MIB.*
[RFC 2933]	IETF RFC 2933. *Internet Group Management Protocol MIB.*

Chapter 12

Security

Randy Turner

CONTENTS

Abstract   This chapter discusses the current issues and best practices regarding the security of Digital Subscriber Line (DSL) networks. The last sections of this chapter highlight the direction several standards organizations are taking for enhancing and improving on security methods for broadband networks, including DSL. This chapter begins by describing the types of threats that face DSL and broadband Internet users in general. These descriptions are then followed by some elaboration and detail regarding the impact that these types of security threats can impose on Internet users. The discussion then proceeds with current and future solutions to each one of these types of threats.

12.1 Scope

The scope of Digital Subscriber Line (DSL) network security discussed in this chapter concerns both higher-layer and lower-layer families of network protocols. The lower layers discussed center on layer 2 and layer "2.5" of the OSI (Open Systems Interconnection) networking model. Within the context of this text, layer 2.5 is referred to as the "nether region" of protocol that spans both normative layers 2 and 3. An example of layer 2.5 protocol would be Point-to-Point Protocol/Internet Protocol Control Protocol

(PPP/IPCP), Dynamic Host Configuration Protocol (DHCP), or any protocol that is used to bootstrap an internetworking layer like Internet Protocol (IP). We call this layer 2.5, because the protocol carries, and more importantly is "aware of," information from layer 2 and 3 in its payload. The security mechanisms utilized to protect transport and application layer protocols are common to any broadband Internet access medium; however, some high-level coverage of transport and application security is included. Later sections in this document include future security methods for DSL network provisioning that involve transport and application-level security.

Also, regarding customer premise DSL deployment scenarios, the majority of the discussion will focus on security issues and vulnerabilities that can be encountered by individuals that access the Internet from their home computer, or home network. Office deployments are considered to be secure through the common use of medium to high-end firewall appliances, and are usually managed by competent network administrators. However, in the case of small businesses without adequate network administration staff and perimeter defenses, the issues and possible solutions examined in this section would apply equally to these environments.

12.2   What Are We Worried About?

Before embarking on a description of security solutions for DSL networks, it is important to understand just what needs to be secured or protected. Just what is the problem? Security threats can be reduced to three basic categories:

1. Unauthorized access to information
2. Denial of service
3. Theft of identity

The intent of every known security attack or breach can be grouped into one of the above categories. For a typical broadband Internet user, the following text provides examples of the threat categories.

12.2.1   *Unauthorized Access to Information*

Most individuals access the Internet directly from their home computer, or indirectly through a home network. If the computer being used for Internet access is also used for any other household activity, confidential information could be stored on this computer in the form of banking account information, work-related intellectual property, usernames or passwords to other services, or other types of information that the user would not want exposed without explicit authorization. Virus and "Trojan horse" software programs are designed to be unintentionally installed on a user's computer with the goal of "mining" the resources of the computer for confidential or otherwise valuable information.

In addition, through e-mail or other types of file transfer programs, users can unintentionally expose confidential information as this e-mail or file content is transported over the Internet. Rather than allowing a virus or Trojan horse software program to access the confidential information on a Personal Computer (PC) hard drive, unauthorized parties just "snoop" the network and extract the information from the "wire" as the information flows from authorized entity to authorized entity. This is one form of an attack called a "man in the middle" attack, where an unauthorized party is unknowingly monitoring the traffic between two trusted parties in a communication. This is analogous to a "wire-tap" of the public telephone network, a traditional technique used by the law enforcement community.

12.2.1.1 Theft of Information

Snooping: This type of threat is exhibited by untrusted or unauthorized third parties snooping or eavesdropping on a communication between two entities that have a trust relationship with each other. For the communication session to be snooped, the network traffic between the two trusted parties must pass through network facilities that are unsecured or untrusted, such as the public Internet. However, snooping has been found to be a problem even within the so-called private networks such as internal corporate intranets and other secured environments.

Trojan Horse Software: Trojan horse software programs are often confused with virus software programs, but they are not necessarily designed to proliferate with the rapidity and aggressiveness of viruses. They do share the property with viruses wherein the software is meant to be distributed indirectly to victims, usually in e-mail or other types of downloadable content, primarily without the user's knowledge that the software has made its way to their PC.

The Trojan horse software can analyze and collect private files from a user's PC and deliver it to other interested, but unauthorized, parties.

Spyware: Spyware is a form of Trojan horse software that is designed to be unwittingly installed on a PC. There is also another form of spyware that is unknowingly included as a "feature" in what would be, ordinarily, an authorized software program installation.

Spyware is designed to interrogate the computer's system and application-specific management files to determine the types of Web sites, newsgroups, and network activity that a user has accessed in past Internet activity. Some spyware software can also install itself at a very low layer within the PC system software, and actually log a user's keystrokes from the keyboard, as they type. This is often used to obtain usernames and passwords to valuable services, as well as other personal information.

Phishing: Phishing attacks are a form of fraudulent activity that involves sending e-mail to unsuspecting recipients. The e-mail messages are typically counterfeit, to appear as if they originate from a business or organization that is trusted by the recipient. These types of messages attempt to lure the recipient to access a Web site that will subsequently ask the recipient for confidential or personal information. Most phishing attempts pose as queries from a financial institution or Internet Service Provider (ISP), asking the user for their ISP login credentials or asking the user to "reverify" their banking information.

12.2.2   Denial of Service

Unlike Trojan horse or virus software, a denial-of-service (DoS) attack is not designed to access information on a device. DoS "attacks" do not necessarily have to be initiated by software on the PC. The DoS attack is usually initiated by a remote computer somewhere on the Internet with the intent of rendering one or more application services on a "target" computer or host unusable. For example, a target computer may serve as the primary e-mail forwarder for a corporate network. A DoS attack on this machine could render the e-mail service unusable, thus disabling individual users on the corporate network from sending or receiving e-mail. There are numerous other forms of DoS attacks, depending upon the type of "service" being attacked, as described in the following sub-sections.

12.2.2.1   Corruption of PC

Corruption of computer resources such as files on a hard disk, or the operating system itself is a classic attack employed by virus and Trojan horse software programs. Luckily, a number of recent types of viruses have concentrated on proliferation and rapid distribution of the virus, through e-mail and other means. This can create a DoS attack on the Internet at large, because of the volume of traffic this activity creates, as well as clogging up enterprise mail servers. However, for the average home computer user, these types of viruses have much less destructive impact than what could be designed.

12.2.2.2   Distributed Denial-of-Service Attacks

As described in Section 12.2.1.1, a Trojan horse software program is unknowingly installed onto a PC, for the ulterior purpose of information gathering. There is another kind of Trojan horse program that is also distributed in the same fashion, but the goal of the software distribution is not for gathering private information from a PC. Instead, these programs are designed to be "remotely controlled" across the Internet, usually by their creator, or someone with the knowledge of how these programs operate.

When so ordered, these Trojan horse programs can initiate high-volume network traffic to a particular destination on the Internet, flooding the Internet victim with so much invalid traffic that the Internet site is unable to provide enough bandwidth or computing resources to serve normal, valid user requests. If the Trojan horse software has successfully been installed on enough PC "collaborators" (say, thousands of PCs), then the volume of traffic the combined attack can create can easily overwhelm typical Internet Web sites or file or download repositories, typically causing these sites to fail.

This software was created to overcome the inherent problems with one user at a single attacking PC causing a problem for a major Internet site. Ordinarily, most attackers initiate attacks from computers that are not connected to the Internet with a very high-speed connection. Indeed, in the past, hackers were dialed in to the Internet and could only send data to the network at a speed of 2800 bytes/s. This rate would hardly cause a problem for a Web site the size of Yahoo or Amazon.com, with their multiple connected Gigabit Internet links, each circuit capable of 45 Mbits of bandwidth.

However, if a remotely controlled Trojan horse attack program can be disseminated to thousands of such dial-up users' PCs, then the total aggregate amount of traffic that can be generated to target the Internet site can have a much greater impact.

12.2.3 *Theft of Identity*

Theft of identity attacks are chiefly used by unauthorized parties to impersonate a particular user or group of users. The specific use of impersonation is twofold: (1) To gain access to a user's identity credentials for the purpose of gaining access to a value-added service to which the user has subscribed (theft of service), or (2) to forge a user's credentials on a particular transaction, whether this be an e-mail message, or a digital signature on some type of digital content.

12.2.3.1 *Theft of Service*

The most common form of theft of identity involves impersonating a user to obtain access to a service to which the user has subscribed. This may be in the form of stealing a subscriber's username and password to obtain access to the subscriber's Internet account. If this is a broadband Internet access account, these types of accounts support unlimited access to the Internet for a flat monthly fee. However, if this is a "metered" service where the customer is charged according to usage, the theft of the subscriber's username and password can be much more damaging. Worse, if the service for which credentials have been acquired happens to permit arbitrary credit transactions, or other types of banking activities, the ceiling on damage could be quite high.

As described earlier, certain types of Trojan horse software designed to monitor keystrokes from a keyboard can be used to steal a user's credentials and transmit these credentials to the owner or operator of the Trojan horse software. This is one example of theft of service.

12.3 The "Always-On" versus "Dial-Up" Network Connection

Until recently, the threat of security issues for individual consumers was never a major point of concern. Although some consumers have been aware of possible security risks associated with online "E-commerce" transactions, where their credit card numbers were being transmitted, the industry had developed a de facto (and now official) standard for protecting the confidentiality of these transactions. There was too much business at stake for this problem not to be solved. However, ever since the proliferation of the Internet, security has always been a concern of corporations and large enterprises that were connected to the Internet. Financial records, personnel information, and core intellectual property were at risk if these firms did not provide sufficient security for their network.

More and more consumers are beginning to use their PCs at home to store confidential information such as medical records, personal letters, and other correspondence, and also tax records and banking information. To date, this information has not been exposed "online" all the time, because of the fact that consumer Internet access has traditionally been relegated to transient dial-up connections. The computers that contained the users' sensitive data were not generally available at all times. Instead, the computers were only open to attack during a particular user's dial-up session. Even when the user had chosen to establish a dial-up connection to the Internet, the IP address of the PC would be different each time a connection was established, making it difficult for a potential attacker to sustain any kind of robust attack strategy.

With the advent of broadband Internet connections like DSL, the security considerations of the consumer or end-user population were broadened considerably. With broadband connectivity, a user's residential home network or PC was looking more like a corporation or large enterprise, from an Internet connectivity perspective. The Internet connection is always available, and the IP address rarely changes, because of the always-on connection. This new always-on profile for Internet connectivity is a major factor in the heightened awareness of security-related issues for the connected household.

12.4 High-Level Security Requirements

Given the security threats and related issues described in previous sections, it is possible to derive a set of high-level security requirements for broadband DSL connectivity. The reality of broadband connectivity is that there are a myriad of business models that can be used to deliver Internet access to a subscriber. For reasons of simplicity, the high-level requirements are broken down into three different perspectives:

1. Network Access Provider (NAP) security requirements
2. Internet Service Provider (ISP) security requirements
3. End-user (subscriber) security requirements

Looking at Figure 12.1, there are a number of network links over which a user's traffic can transit while using the Internet. These reference points are labeled N1 through N7 and are to be referenced later in this chapter. In addition, the integrity of each device over which the traffic moves needs to be protected as well. Some of these links and associated devices are clearly in the domain of specific business entities that cooperate to deliver Internet connectivity to a consumer. By dividing security requirements into specific categories, it becomes clear where some security solutions are more attractive than others. For example, from an ISP's perspective, the links as indicated by N5 and N6 may not need to offer confidentiality of data traffic, because the internal ISP network connection points would ordinarily be trusted by the ISP, as would be the physical media connecting the network elements comprising the ISP intranet. However, ISP end users are typically blind to network topology and would tend to not trust the network links that exist outside their home or home network.

The individual security requirements were previously separated into three categories: NAP, ISP, and end-user requirements. The end user owns the N1 interface, because this physical link resides at the customer premise. The NAP might own the N2, N3, and N4 interfaces as in Figure 12.1. The ISP manages the remainder of the interfaces and the devices that implement these interfaces. There is no strict model for broadband access network topology, but the diagram in Figure 12.1 is meant to

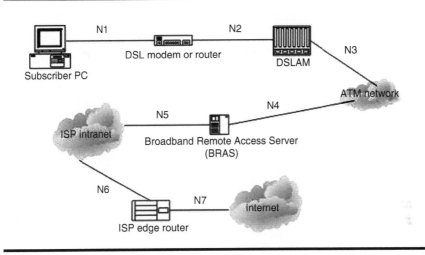

Figure 12.1   Broadband access model.

capture the most common architecture so that a sufficient framework can be referenced in the remainder of the security discussion.

The remaining sub-sections discuss the high-level security requirements for the network described in Figure 12.1.

12.4.1   *Confidentiality*

Confidentiality in the context of this discussion involves encryption of information that is transported over the DSL network. What is confidential is in the eye of the beholder. Information is classified as confidential from the point-of-view of the parties exchanging the information. This security discussion focuses on two classifications of data, and is dependent upon the parties exchanging the data, as just described. There is information that is confidential with respect to an end user of an Internet application, such as when an end user interacts with an E-commerce Web site, and this is considered "user confidentiality." There is also information that is transported over the broadband network that is confidential to one or more of the business entities that collaborate to provide service to the end user. This classification is called "network management confidentiality," or more briefly "management confidentiality."

These two classifications of confidential data are, for the most part, disjoint. Neither of the parties of either type of communication have visibility to the others' data. The transmission of network management data throughout a DSL network is transparent to an end user, and likewise, an end user's Internet data session is just "bits on the wire" to the network, and the network knows very little, if any, about the semantic contents of the traffic being transported, especially if it is encrypted.

User confidentiality is achieved through a common set of technologies that are typically access-network agnostic. Two popular techniques for solving the user confidentiality problem are Transport Layer Security (TLS 1.0), and IP Security (IPSec). TLS is the encryption solution employed by the vast majority of Web-based, or HyperText Transfer Protocol (HTTP), applications. TLS is the IETF (Internet Engineering Task Force) standard version of the de facto SSLv3 (Secure Sockets Layer Version 3.0) solution, originally designed by Netscape, and adopted by the industry in the late 1990s. IPSec is becoming the dominant technology for secure remote access to enterprise networks for corporate workers who spend the majority of their time on the road (road warriors), and also by employees that work from home, either full-time or occasionally (telecommuting).

Management confidentiality involves the secure transport of provisioning and configuration data throughout a network, and can include the secure transport of proactive network monitoring (performance management) of specific elements in the broadband network. One example of management data is data that is associated with the sale of new DSL Internet service. IP routing tables, IP address assignment, usernames, and passwords have to be propagated to all of the appropriate network elements in the broadband network before the user can successfully establish Internet connectivity.

As indicated, the user confidentiality problem is solved using Transmission Control Protocol/Internet Protocol (TCP/IP)-based end-to-end technology that is access-network independent, and therefore the main focus of the confidentiality discussion in this chapter is oriented toward management confidentiality solutions. Management confidentiality is currently being applied in an access network-specific fashion.

12.4.2 Authentication

Authentication is essentially one entity "proving" its identity to another entity. However, rather than providing its identity absolutely, the scope of authentication amounts to one party providing "sufficient evidence" regarding its identity to allow the other party to conclude that the identity being presented can be trusted. What is "sufficient" in the above description is usually what is subjective in any assessment of authentication technology, and usually depends upon the robustness of the technology used, as well as the probability of specific identities on the communication channel being forged or otherwise "hijacked."

As will be illustrated in later sections, authentication is the weak link in currently deployed management and provisioning communications, as well as end-to-end IP-based services.

In client–server environments, such as subscriber–ISP relationships, authentication can be either one-way or two-way authentication. In one-way authentication, the server normally requires a client to authenticate itself prior to the server allowing the client to access the service. In two-way authentication (mutual-authentication), the client authenticates itself to the server, and the server is also required to authenticate itself to the client.

12.4.3 Authorization

The process of authorization occurs after an entity has successfully authenticated to a service. Authorization of an identity is a procedure that determines what type of access will be granted to an authenticated identity (or user). There are a number of authorization methods that are employed by service providers to enforce access control to service-related resources. Unfortunately, authorization is the one security-related area where standards organizations have not yet addressed this problem with a complete solution. Therefore, the authorization technology space is very fragmented. Most equipment vendors are converging on a centralized directory where all authorization information is located. These directory infrastructures are typically accessed through a Lightweight Directory Access Protocol (LDAP) front-end that provides access to specific schemas (sets of attributes) where the authorization information is kept on a per-entity basis. Microsoft employs proprietary directory attributes in their active directory product. These attributes enable either an access-control-list (ACL) model, or a role-based access control (RBAC) model. ACLs allow very fine-grained access control policies to be enforced and can degenerate to a per-user access control policy. However, these techniques generally do not scale well in the face of very large user communities. One solution to the ACL scalability problem is using RBAC. In an RBAC model, there is a finite set of "roles" into which users or entities are mapped. Examples of roles are "normal user," "manager," "administrator," or "operator." These abstract role names imply levels of access to resources. Roles are typically defined on a linear scale from "no-access" to "full-access" to all service resources.

12.4.4 Data Integrity

Data integrity involves two parties trusting that the data that is being exchanged over a communications channel has not been modified or otherwise tampered with in route or in transit from one party to another. The mechanisms for data integrity have historically employed techniques such as checksums or cyclic-redundancy checks (CRCs). However, data integrity has evolved to mean more than answering the question "Has the data reached its destination intact ?"

The use of checksums and CRCs was primarily used to detect data corruption by unintentional modification of the data bits as they pass over certain types of unreliable network equipment or network media. The latest technology used for data integrity can still protect against this type of data corruption, but new methods also provide integrity protection in the event of active third party manipulation of the data while in transit from source to destination.

For example, a CRC would probably protect against data corruption because of bad network wiring, but would not protect against an untrusted party intercepting a transmission, modifying the data, calculating a new CRC, and continuing the transmission to its intended destination.

12.5   Currently Deployed Security Solutions

12.5.1   *Authentication*

12.5.1.1   Layer 2

Layer 2 network traffic in a DSL access network is typically a collection of Asynchronous Transfer Mode (ATM) permanent virtual circuits (PVCs), where each PVC is transporting "Ethernet-like" frames. The relationship to Ethernet is that the bulk of packet header information carried by these layer 2 frames looks like Ethernet. The specification that describes these frames in detail is RFC 2684. For all practical purposes, the remainder of this chapter considers layer 2 DSL network traffic as Ethernet frames.

DSL service providers rarely impose any authentication requirement on DSL consumer or customer premise equipment (CPE) for injecting packets onto their access networks at layer 2. Allowing any DSL equipment to generate traffic on a portion of the access network has varying levels of security implications, depending upon the topology of the provider's network.

Referring to Figure 12.1, the impact of unauthenticated traffic on the access network involves links N1 through N4. N4 is the first "hop," or packet-forwarding device in the network, that will attempt to address the higher-layer protocol specifier in the Ethernet frames originating at the customer premise. Typically, this higher-layer protocol specifier in the Ethernet header will either specify IP or PPPoE (PPP over Ethernet) (described in the next section).

The risk of allowing any CPE device to inject packets onto the portion of the access network identified by N1 through N4 (or more importantly to the provider, N2 through N4), depends upon the upstream bandwidth that has been provisioned, at layer 2, by the access network provider. The more traffic that an individual customer or subscriber can generate per unit of time, the more impact the device could have on the availability of bandwidth on the access network. The impact of one subscriber's traffic on the access network is only important when the subscriber's traffic is aggregated with other subscriber's traffic onto a shared link. In Figure 12.1, this aggregation point is the Digital Subscriber Line access multiplexer (DSLAM), and the impact of an individual's upstream traffic becomes an issue on link N3 (and beyond). The DSLAM is the equipment in the access network that routes DSL signals from individual subscriber phone lines and multiplexes these data channels onto ATM virtual circuits (VCs).

12.5.1.2 Layer 2.5

PPP over Ethernet (PPPoE): The primary subscriber authentication mechanism employed by Internet service providers over DSL is PPP (Point-to-Point Protocol) over Ethernet, or simply PPPoE. The currently deployed use of PPPoE requires a username and password to be associated with each subscriber. The actual authentication mechanism used is PPP-specific and is independent of the specific mapping used to transport the PPP over Ethernet. In other words, the same authentication mechanisms available to PPPoE sessions are available to any implementation of PPP, independent of what type of media is being used to transport the PPP session. Available PPP authentication mechanisms are documented in an evolving set of RFC (request for comments) specifications in the IETF.

The dominant form of PPP authentication in use today is Password Authentication Protocol, or PAP. This method involves a simple transfer of username and password credentials from client to server. The credentials are passed in plaintext (unencrypted form) and are vulnerable to third parties that can intercept or otherwise monitor the network traffic between client and server. Some service providers are beginning to provision their networks for a more secure form of username and password PPP authentication called the "Challenge Handshake Authentication Protocol, or CHAP. In this method, the username and password are hashed with a random "nonce" value supplied by the server so that the subscriber's credentials are not exposed as plaintext to third party snooping.

Note that only "subscriber" authentication is utilized in current DSL deployments for accessing the provider's layer 3 (IP) network. This type of authentication refers to "client-only" authentication discussed previously, as opposed to mutual authentication.

Some providers allocate IP addresses to subscribers using DHCP; however, this form of layer 3 connectivity is not authenticated by the ISP and is the exception rather than the rule in current DSL deployments.

12.5.2 Confidentiality

As indicated previously, the discussion on confidentiality will reflect only the confidentiality of management and provisioning information that is communicated between subscriber and service provider. In current DSL deployments, there is little, if any, remote management or remote provisioning occurring between subscriber CPE and service provider network management systems. Instead, most providers are supplying PC-based application installations that connect directly to the subscriber CPE or connect to the CPE over a customer premise LAN (local area network). Some other types of subscriber CPE are advanced routers or residential gateways that contain Web servers for configuration by customers through a Web browser on their PC. All of these methods generate management and provisioning network traffic that only flows over LANs on the customer premise, or via direct connections (USB [universal serial bus], serial, etc.) to the CPE. No management or provisioning traffic flows over the public Internet, so the requirement for confidentiality of this traffic has never been an issue.

12.5.3   Data Integrity

As stated in the previous section, with very few exceptions, management and provisioning operations have not been performed remotely, where the data could be intercepted or otherwise corrupted. Therefore, the requirement for integrity of network management operations has never been an issue. There are other types of data besides network provisioning and configuration operations that might require a very high degree of integrity. One such datum might be a software or firmware image that is either locally or remotely transferred to the CPE. The CPE (DSL router, DSL modem, or residential gateway) should be able to perform some type of integrity check on the received software image file, prior to actually using the update. This kind of integrity check on bulk data should prevent at least two potential problems: (1) the image file could have been inadvertantly modified because of either a communication error or some other type of unexpected, accidental modification of the image, and (2) the image file should be "trusted" as a software image file appropriate for the service provider, subscriber, and CPE. This check ensures that a "rogue" image file constructed by an untrusted third party was not received. This type of bulk integrity check should be applied to any type of bulk management data transfer from the service provider to the CPE.

12.6   Coming Attractions

A number of technologies are being developed to address the perceived gap in securing DSL service deployment. Some equipment vendors and network operators have attempted to deploy proprietary solutions and have met with some success, although network operators would rather see interoperable standards for a security solution so that their fundamental requirement for multiple suppliers of a given technology can be met. This requirement is most important for equipment at the outermost edge of the provider's network where volumes are highest and so is the risk of maintaining a single supplier. The industry standards organizations that are working on security solutions applicable to DSL include the following:

- DSL Forum
- ATM Forum
- Internet Engineering Task Force (IETF)

The DSL Forum is working on technology that targets DSL networks specifically, and there is no requirement in DSL Forum specifications for being able to apply this technology in an access network-agnostic way. However, some of these standards do not preclude their application on other types of broadband connectivity.

Because the fundamental layer 2 switching technology being utilized in DSL networks today is based on ATM PVCs, the DSL Forum leverages one or more ATM Forum standards for solving specific security issues. The ATM Forum maintains a set of network security recommendations for the integration and deployment of standards that are published through the forum.

The IETF focus is on solutions specific to IP-based networks. Because IP mappings exist for all current and future broadband technologies, the work of the IETF with regards to security solutions is fundamentally access network agnostic.

In the following sections, standards from one or more of these organizations will be highlighted as solutions to remaining security issues in DSL networks.

12.6.1 Access Network Independence

It should be noted that one of the fundamental technology directions being pursued by network operators for future broadband solutions is a focus on keeping access network-specific operations as close to the edge of the network as possible. Having a common management interface to all network elements would allow operators to avoid the deployment of a fragmented vendor-proprietary management capability for each vendor and every kind of network element in a topology. Because practically every broadband media technology such as DSL, cable, or wireless has a corresponding method for transporting IP packets, the network operators are focusing their attention and formulating requirements toward sophisticated network management and provisioning infrastructure that is IP-based.

12.6.1.1 IP-Focused

The emphasis for deployment of future management systems on IP implies that access-specific technology will only be employed to "bootstrap" a particular network element into the IP network management framework. Traditionally, protocols such as ATM or Frame Relay (FR), or ITU (International Telecommunication Union) and ETSI (European Telecommunications Standards Institute)-based telephony protocols have been distributed with their own OAM (Operations and Management) protocol family. With the adoption of IP-based management methods, much of this infrastructure will become obsolete, and would only be utilized during a transition strategy toward an IP management "backbone."

12.6.2 Authentication

A number of authentication technologies will be deployed to fill the perceived gap in the security that is currently being used by DSL network service providers (NSPs). The umbrella for these authentication technologies is the Extensible Authentication Protocol, or EAP, which is described in the next sub-section.

12.6.2.1 Extensible Authentication Protocol

PPP-related technology will continue to be used for subscriber authentication by DSL service providers. However, as described PPP is a framework wherein a number of PPP-specific authentication methods have been specified in IETF-related standards. The IETF is no longer entertaining proposals for extending PPP for authentication. Instead, using PPP for authentication will be replaced with a more extensible method of authentication protocol selection. The EAP was originally designed to allow multiple different authentication mechanisms to be deployed without regard to how the protocol was transported. EAP augments the negotiation capability of the PPP by allowing PPP services to support an evolving set of authentication methods that have recently been specified, or for which work is under development. A side benefit of being able to negotiate multiple levels of authentication capability is that one or more EAP mechanisms support mutual authentication, as opposed to the PPP-specific client-only authentication mechanisms currently deployed in DSL networks.

Examples of how EAP can be reused across multiple transports can be found in IEEE 802.1x and PPP use-cases. EAP works equally well for both of these standards. In addition, another IETF standard (PANA [Protocol for carrying Authentication for Network Access]) also carries EAP as its authentication payload.

EAP Methods: As described in Section 12.6.2.1, EAP supports a number of different types of authentication algorithms. A subset of the evolving EAP methods applicable to DSL network security is provided below.

Method	Description	Mutual Authentication Support
EAP-MD5	Digest "challenge-response"	No
EAP-TLS	X.509 certificate exchange	Yes
EAP-TTLS	Certificate-based server authentication, EAP-based client authentication	Yes

12.6.2.2 Layer 2

802.1x: 802.1x is an IEEE standard for layer 2 authentication and access control. The standard is primarily aimed at Ethernet switch vendors and wireless infrastructure equipment vendors to meet the perceived requirement for authenticated access to a layer 2 network domain. The model is based upon a per-port access control policy and authentication model, which is ideal for equipment like an Ethernet switch. The model is also ideal for Ethernet termination equipment that implements the idea of multiple "virtual" ports. Wireless access points, such as those based on 802.11, emulate a multi-port Ethernet switch for all wireless clients to which it is logically connected. Likewise, a DSLAM also emulates a multi-port Ethernet switch, where each "port" is associated with a particular ATM PVC that terminates at the DSLAM. As stated previously, RFC 2684 utilizes Ethernet (802)-like framing for its transport. These two qualifiers make it possible for DSLAMs to implement 802.1x security for each port over which a particular subscriber is trying to connect.

As the layer 2 provisioning of upstream bandwidth increases, it is very likely that layer 2 access control will become a requirement by network access providers, and 802.1x is the most likely candidate for meeting this requirement.

PANA: The IETF has produced a standards-track mechanism for network access authentication. This specification is called PANA (Protocol for carrying Authentication for Network Access), and provides a network access authentication mechanism similar to 802.1x, except at layer 3, using UDP (User Datagram Protocol). The one advantage that PANA offers is that it is layer 2 "agnostic." Although 802.1x is defined over links that utilize 802-style packets, PANA can theoretically be employed over any network media for which an IPv4 or IPv6 link-specific transport method has been defined.

PANA uses a temporary IP address assignment to obtain access to the "limited scope" network until the user successfully authenticates with a NSP.

Like 802.1x, the PANA protocol is a transport for EAP, enabling an extensible set of authentication protocols to be employed, depending upon the requirements of the access provider. Unlike 802.1x, PANA is not link-specific. The layer 2 agnostic nature of PANA makes it especially attractive for complex NAP, NSP, ISP, or ASP (Application Service Provider) topologies. Complex interconnects between business partners (like NAPs, NSP, and ISPs) can consist of a number of link-layer technologies. In most cases, these links all share a common attribute: they support the transport of IP packets using a common link-layer-specific standard. Relying on this end-to-end connectivity across arbitrarily complex networks allows PANA to be used in a multi-layered authentication scenario.

PANA is being adapted and deployed by a number of equipment vendors and network operators, and is expected to become a solution used in a multi-service, multi-tier, broadband network.

12.6.2.3 Layer 2.5

Authenticated DHCP: In the previous discussion, PPP authentication mechanisms were described in the context of the NSP requiring successful authentication of a client (subscriber), prior to allowing the client access to the Internet. The PPP is also used to automatically configure the client with key configuration parameters that the client will need for proper Internet connectivity. A more robust method for dynamic network configuration of a client is through the use of DHCP. As stated previously, PPPoE has been the dominant form of authentication in use for DSL. One of the reasons for this lack of popularity with DHCP is that DHCP had no corollary for username and password authentication of a client.

RFC 3118, published in 2001, is an effort to create a version of DHCP that is more secure than the base DHCP defined by RFC 2131. However, there were numerous security vulnerabilities identified against the security measures proposed by RFC 3118. This fact, combined with the introduction of 802.1x, slowed the deployment and take-rate of authenticated DHCP, and it has never been widely deployed.

PPP/EAP: The current PPP technology deployed utilizes PPP-specific mechanisms such as PAP, CHAP, and MS-CHAP, which are legacy forms of authentication. These legacy forms of authentication are being replaced by EAP mechanisms, which provide a future roadmap for extensibility. The IETF is no longer entertaining proposals for extending PPP authentication methods. PPP/EAP is essentially EAP protocol carried within PPP frames on the network. The EAP protocol is engaged during the authentication portion of the PPP state machine by the PPP endpoints.

12.6.3 Confidentiality

Confidentiality with respect to the management and provisioning of DSL networks has recently become a feature of interest in requirements documents from network operators. Even when the management and provisioning networks are internal network links within the service providers' operations networks, confidentiality must be available at the provider's discretion. The growing threat from both external and internal sources is shifting advanced security requirements from the "desired" column into the "mandatory" column in most of these requirements specifications.

There are two standards-based mechanisms that are being employed to solve the confidentiality problem: a transport-level (layer 4) solution, and a network-level (layer 3) solution. These two methods are discussed in Sections 12.6.3.1 and 12.6.3.2, respectively.

12.6.3.1 SSL 3.0/TLS 1.0

The dominant form of confidentiality protection used on public networks like the Internet is the application-level Secure Sockets Layer (SSL) Version 3.0 protocol, a de facto standard developed by Netscape in 1997. The IETF has produced a sanctioned industry standard version of this protocol and renamed it Transport Layer Security, or TLS Version 1.0. TLS supports mutual authentication of both client and server in a client/server application session, and also supports confidentiality (encryption) protection of the data traffic occurring on the network connection between client and server. The majority of TLS deployment is in standard HTTP/1.x Web browsers such as Microsoft Internet Explorer or Mozilla Firefox, and Apple Safari. TLS imposes a requirement on servers that they be a member of a public key infrastructure (PKI), and have an associated public key certificate issued by the PKI certificate authority. The TLS-capable server will send a copy of this certificate to a TLS client application during TLS connection establishment. The server certificate can be used by the TLS client to robustly authenticate the identity of the server. TLS is capable of mutual authentication through a procedure involving the server requesting a client's public key certificate, and subsequently performing the same certificate authentication and verification on the client's certificate that the client performs on the server's certificate.

However, distributing public key certificates to a population of potential clients has historically been much more problematic because of the trust relationships and infrastructure required to make certificate distribution easy and scalable. The issues surrounding PKI deployment are being addressed by evolving IETF and other industry standards to make the burden of a PKI more tolerable.

For purposes of management and provisioning between a DSL service provider and a subscriber, the scope of PKI can be reduced, and the associated trust model does not have to support a very heterogeneous population of subscribers. Instead, certain assumptions can be made regarding the client population of subscribers, especially if the service provider has particular customer premise network elements that are installed for accessing the service.

TLS is gaining popularity in the service provider arena for transporting management and provisioning information to and from the service provider network and the customer premise. Even in the absence of a full PKI with certificate distribution capability, network operators can still employ the encryption features of TLS by leveraging noncertificate-based client authentication within an encrypted TLS connection. With this method, only the server-side component of the provider's service needs to have a PKI certificate installed. This is sufficient to allow the subscriber or client to authenticate the server, as well as establish an encrypted communications channel between the subscriber premise and the provider's server systems. Once this encrypted channel is established, any form of simple plaintext username and password authentication can be used for allowing the server to authenticate the client. The plaintext username and password is protected from snooping by transporting the plaintext credentials within the encrypted TLS connection. This server-side operation of TLS is the foundation for all E-commerce transactions now employed by Web sites such as Amazon, PayPal, as well as major financial institutions that provide Web-based access to bank accounts.

The server-side form of TLS deployment allows network operators to evolve to advanced PKI functionality through a step-wise transition mechanism.

12.6.3.2 IPSec

In the SSL/TLS discussion in Section 12.6.3.1, the management and provisioning connections being established between the customer premise and the service provider's systems were between specific applications, with the secure connection being established at layer 4. This implies that if more than one application is needed to perform management and provisioning for the subscriber, or if another service provided by the service provider also requires the same security as that provided by TLS, then multiple TLS connections must be established and maintained between the customer premise and the service provider systems. Each TLS connection requires a certain amount of network resources on both client and server, and multiple applications requiring multiple TLS connections will have a definite impact on the resources available on both client system, and especially the server systems in the network.

IPSec is another way for two systems to maintain a secure, confidential connection. The benefit of IPSec over SSL/TLS is in the amount of system resources required to offer security for multiple concurrent applications with confidentiality requirements. With SSL/TLS, each application is required to establish and maintain a security association with a peer application. Using TLS, if four client applications on system "X" want to establish an encrypted channel of communication between four server applications on host "Y," then four independent security associations would have to be established, negotiated, and maintained. With IPSec, the layer 3 (IP) communications link between system X and system Y would be protected, and the applications could share and benefit from the single security association that is established at layer 3. The most powerful benefit derived from IPSec in most end-user scenarios is that it provides a secure communications path for applications and services that may not have been originally designed with "built-in" security mechanisms or features.

12.6.4   Integrity

The extent of requirements for integrity protection on DSL networks is currently under investigation by a number of service providers. As described earlier, integrity involves ensuring that the data passed from one party to another party over a network has not been modified, either accidentally or with malicious intent. The TCP/IP family of network protocols provides some degree of end-to-end integrity between the network node that is sending the data and the network node that is receiving the data. This type of protection is meant to detect transmission errors that occur over different types of physical media (cable, DSL, wireless, etc.) as the data moves from the sender to the receiver. Figure 12.2 below shows another potential for integrity violation that would not be detected by network communications integrity checks.

In Figure 12.2, there is a content producer that has produced some type of content. The content producer is tasked with delivering this content to a content consumer. The content producer would like to transfer this content to the content consumer, with integrity, i.e., without incurring any modification either accidentally or intentionally. The path from content producer to content consumer is indicated by the arrows in the illustration. The path can consist of any number of intermediate "hops" on the way from producer to consumer. The content consumer operates a client node that is connected, presumably over a public network like the Internet, to the server. Once the content is published on the server,

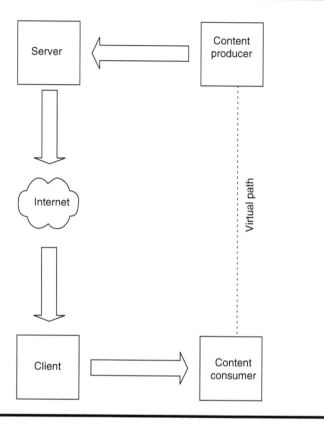

Figure 12.2   Content integrity.

or at any intermediate point along the way from producer to consumer, the content could be modified or forged, without the producer's knowledge. The requirement for integrity is that the consumer be able to detect whether or not the content has been modified since the transfer began. In simplest terms, this means to ignore all of the intermediaries that are involved in transmission of the content over a network; the goal of providing integrity is to try and approximate the virtual link that is illustrated by the dashed line in Figure 12.2. Individual physical network links along the path from producer to consumer are protected against transient link corruption by facilities provided by protocols like TCP or UDP. However, there needs to be a mechanism that offers integrity protection specifically between the content consumer and the content producer. Throughout this chapter, this type of integrity will be referred to as end-to-end integrity. This type of integrity protection is offered through a unique "security association" that exists between the parties involved in the content exchange. The idea of a formal security association is discussed further in Sections 12.6.6 and 12.7.

12.6.4.1 Cryptographic Hash

Most end-to-end integrity methods now being employed or considered for deployment utilize cryptographic hash algorithms for computing a unique value for a specific stream of digital content. This value is similar to a checksum because it is calculated against the entire length of content to be protected. The two most popular hash algorithms in use today are MD5 and SHA-1, which are 128-bits and 160-bits in length, respectively. Both algorithms compute a unique multi-bit value for a given sequence of bytes. Changing one or more bits in any byte or bytes of the content to be hashed would produce yet another unique value. Herein lies the utility of this type of algorithm with respect to integrity checking. It is not enough, however, to use this algorithm, unadorned by any other type of security. If a content producer was to compute a hash value on a particular content, and supply this hash value to the consumer as an integrity check, any party having access to an intermediary network node between producer and consumer could intercept the data, modify it, and recompute the hash value.

12.6.4.2 Message Authentication Codes

The most common way of augmenting the utility of cryptographic hashes on content is to utilize a shared secret key during the hash computation. The addition of a shared secret into the hash computation allows some level of "trust" or authentication to be included in the integrity check. Instead of any party being able to compute a generic hash on the content being transmitted, only parties that know the shared secret can compute the correct hash value. Message authentication codes (MACs), or "Keyed" MACs, represent the family of algorithms that combine traditional hashing with shared secrets. The IETF has published an RFC that describes one of the most popular forms of MAC algorithms, call the HMAC.

12.6.4.3 Digital Signatures

Similar to MACs, digital signatures are another form of incorporating the authentication of trust with traditional cryptographic hashing. Rather than utilizing a shared secret or password as in Section 12.6.4.2, digital signatures rely on the existence of a PKI. Referencing Figure 12.2, the content producer would calculate a hash over the content and then subsequently encrypt the resulting hash value with a public key encryption operation using the producer's private key. The content consumer would be a member of a PKI, or trust relationship, that included the content producer as well. The consumer would obtain the public key certificate from the content producer (it could be delivered with the content), and use the information from the public key certificate to decrypt the hash value. As in MACs, digital signatures imply either trust or specific authentication (and authorization) of the content producer.

12.6.4.4 Applicability

The advanced integrity protection mechanisms described in this chapter are being planned for deployment in a number of broadband service scenarios. Some scenarios for the applicability of integrity protection are outlined in the following sub-section.

Software Updates: Many types of broadband CPE can have their resident software upgraded over the network (either LAN or WAN). Sometimes, this feature requires end-user intervention, and some CPE software can be updated over the broadband network with no end-user intervention at all. Without regard to the actual mechanism of transporting the software image to the device, the device needs to verify that the image was generated by a trusted source, and that other information such as compatibility and versioning is taken into account.

12.6.5 Signaling Security (QoS)

Within the DSL Forum and other standards organizations, including the IETF, there is a requirement for the dynamic negotation of network "quality of service," based on application type. Historically, broadband networks have been preprovisioned with a particular level or quality of service (QoS) (bandwidth, latency), with no way to adapt to the requirements of multiple types of network applications that are being transported over the network. A subscriber to a broadband service that required very low latency and high bandwidth would have a network connection which is preprovisioned for these requirements during the installation process. If this connection also offered Internet connectivity, and the subscriber established "best-effort" connections over this network for applications like e-mail and Web surfing, then the subscriber would be paying considerably more for using these types of simple Internet applications than someone else who only subscribed to the basic best-effort class of Internet service. The key point here is that the cost of layer 1 and 2 provisioning is a significant percentage of a NSP's cost of delivering the service to the subscriber. This is why a subscriber has historically been statically provisioned for a particular class of service (CoS). Once the circuit was provisioned, the CoS was fixed for all data traffic flowing over the circuit.

The requirements now being specified by network operators are for a single mechanism to be employed, at layer 3, for a dynamic signaling protocol that can offer a varying number of CoSs depending upon the application or service type that is in use, as well as a potential ceiling on QoS that is authorized to a particular subscriber.

For DSL, this implies that all DSL layer 2 provisioning will be identical, which also implies that the layer 2 provisioning of the subscriber's connection will be equivalent to the highest CoS offered by any application or service that is offered by the NSP. The provider may also elect to simply provision their layer 2 DSL circuits at the DSLAM for the full rate possible. As an example, ADSL broadband service to all customers would be provisioned at the DSLAM to train at 8 Mbits/s downstream, and 1.5 Mbits/s upstream. The actual network resources that are consumed by a particular subscriber would depend upon the types of broadband services which the subscriber has been authorized to use by the NSP. During the startup of each application or service connection, the application would use dynamic signaling protocols (similar to ATM SVCs), to reserve the network resources necessary to realize the particular application. When the service connection is disconnected, or the application terminates, the network resources previously used would be released for other subscribers to use.

The particular strategy behind placing this type of signaling and QoS management at layer 3 is to allow the NSPs to deploy one single QoS management and provisioning system for IP services, and have this system be reused for any and all media-specific broadband technologies that are used now or in the future. This reuse is possible because of the fact that, as new networking technologies are created and standardized, there is always an effort to standardize an associated mapping of the IP over this new technology. This should allow the NSPs to more quickly adopt new broadband technologies as they become available.

There is, however, a risk in allowing the CPE to dynamically allocate broadband network resources. Any unauthorized resource allocation would have detrimental effects on the availability of network resources for other broadband customers. To reduce the risk of unauthorized traffic taking up resources on the network, there must be a way for applications and application users to be formally authorized to signal for higher levels of service from the network. Section 12.6.5.1 discusses an IETF working group effort that addresses the problem of signaling, as well as the security requirements for such a mechanism.

12.6.5.1   IETF NSIS Working Group

The IETF NSIS (next steps in signaling) working group was formed to address the on-demand requirement for network resource reservation. The output of the working group is also meant to address a perceived lack of scalability on the part of a previous signaling protocol, RSVP (Resource Reservation Protocol). Given the amount of augmentation required to address the issues with RSVP, the IETF thought the better way to go was with the creation of a new protocol. At the time of writing of this chapter, the working group had published initial drafts of their work, including a document on security mechanisms for the protocol.

The initial work on security for signaling has focused on three different scenarios:

1. PKI/X.509
2. EAP/AAA
3. 3GPP

The authentication and authorization systems chosen by service providers in the future should be reused for security (and especially authorizing) NSIS dynamic QoS requests. At the time of writing of this text, the NSIS working group has not ratified a final decision on security, so service providers with future dynamic QoS requirements should monitor activity in this area.

12.6.6 Security for DSL Service Provisioning and Management

In 2004, the DSL Forum published Technical Recommendation (TR) 069, specifying an XML (eXtensible Markup Language)-based network management and provisioning protocol for DSL networks. TR-069 provides for management and provisioning of DSL CPE through secure connections between the CPE and a head-end management system operated by the service provider. The protocol is designed as an XML specification with support for provisioning of fine-grained layer 1 DSL parameters, all the way through high-level service assurance and broadband service billing options. The transport layer for the XML protocol is SOAP (Simple Object Access Protocol) 1.2, but with DSL Forum-specific modifications to the standard request or response nature of SOAP.

From a confidentiality perspective, most deployments are choosing to protect TR-069 communications with TLS, leveraging either network provider PKI management systems, or a "managed" PKI service for handling distribution and maintenance of certificates. In some cases, there are even "hybrid" deployments that utilize managed PKI services in the back-end, and provider-issued certificates to individual CPE devices.

There are many aspects in the design of TR-069 that foster reusability in other types of broadband network topologies. The data model expressed in the XML schema for TR-069 is very rich in its ability to represent varying broadband services business models (NSP, ISP, ASP, etc.) and is therefore a candidate for reusability. A short-list of TR-069 functionality is included in the following sub-sections.

12.6.6.1   TR-069 Functional Highlights

Broadband Service Provisioning: TR-069 supports dynamic provisioning of services to CPE that is "multi-service" capable, such as set-top boxes and residential gateways. The dynamic service provisioning process can be instantiated in one of two ways: (1) the subscriber can request new service over the phone through a customer service representative (CSR), and the CSR can subsequently request the head-end provisioning system to "turn on" the service at the customer premise, or (2) the subscriber can visit a service provider Web page and request the service, initiating the same dynamic provisioning request to the head-end systems.

CPE Firmware Update: If the firmware residing in the CPE at the customer premise does not currently support the new service being ordered, the management system can automatically initiate a firmware update for the device to allow the device to support the new service.

Remote Diagnostics: At the request of the subscriber, or periodically as a part of a proactive maintenance strategy, CSRs can schedule remote diagnostics to be performed on CPE, with detailed reports from the diagnostics returned to the CSR for analysis.

12.6.6.2   TR-069 Futures

The DSL Forum will no doubt extend and improve upon the data model used by TR-069 as the use-case scenarios and business models for broadband services evolve over time. For DSL service providers, it is expected that TR-069 support will become a requirement, as well as a springboard for newer models of broadband service provisioning and content delivery networks.

12.6.7   Security Frameworks

A DSL NSP which also offers Internet connectivity is much like a very large corporation, except with many more employees. Large corporations tend to formalize all of their security instruments into a manageable framework. The framework is built on a number of interrelated technologies and systems that, together, provide the fundamental requirements that have been discussed in previous sections, including one or more additional features. For review, these fundamental requirements are enumerated below.

- *Authentication (and subsequent authorization):*   As stated earlier, an organization must be able to identify who is requesting or accessing network resources. Using the confirmed identity, the network can arrange for network resources to be allocated to the requestor based on prearranged authorization.

- *Confidentiality:* Organizations would like to protect both online and offline information, as well as confidential information transferred over the network.
- *Integrity:* An organization would like to maintain all important content with some degree of integrity, whether the content is online or offline.

12.6.7.1 Usernames and Passwords

Traditionally, ISPs have built a security infrastructure around the idea of a username and password. This model has evolved from legacy operating system authentication methods dating back to the origins of the multi-user computer system. The model is well understood by network operators, as well as their broadband customers. For users migrating from dial-up to broadband, the network authentication procedure is the same, and the service provider does not have to replace their authentication procedures and systems.

The subscriber, or the CPE device on behalf of the subscriber, logs into the network as described in Section 12.5.1.2. The credentials are passed to the broadband remote access server (BRAS, depicted in Figure 12.1). The BRAS can either consult an internal file or database of known users to verify the username and password, or access a remote RADIUS (Remote Authentication Dial In User Service) server for username and password validation. The RADIUS server accesses the repository for the service provider's known subscribers. This repository can be exposed to other applications and business processes by publishing an LDAP schema for the subscriber records in the database, and using a database that supports an LDAP front end, providing generalized access to the database using the LDAP. However, it is here that any industry standard solution for a security framework ends. Most standardized LDAP "user" schemas are based on user certificates, issued by an available certificate authority, which maintains the validity of the certificates, as well as the database records that are exposed through LDAP. Also, if network confidentiality is required for connections to the LDAP server, the only confidentiality mechanism that exists for LDAP servers today is based on SSL/TLS, which also implies support for PKI certificates on the database servers.

The central issue with evolving a legacy username and password architecture is the fact that no interoperable standards exist for a complete security framework, offering confidentiality, integrity, and authentication features, that are based upon only username and password credentials. Organizations and enterprises that follow this path, quickly find themselves developing an architecture that cannot be leveraged to integrate new security technologies as they come along. The architecture is also not interoperable with any other sophisticated security frameworks that might be used by business partners and other security-related service providers.

12.6.7.2 Public Key Infrastructure

A PKI provides all of the basic mechanisms required of a robust security framework: confidentiality through public key encryption, integrity through digital signatures, and authentication through certificate exchanges and subsequent validation. However, this robust framework comes at a high infrastructure cost, because of the requirement for supporting a certificate authority (CA). Hosting a CA implies that the system is capable of X.509 certificate distribution to clients of the PKI (broadband subscribers in this case). The CA must also supply online capabilities for verification and validation of certificates that are issued by the CA. There is also the mission critical 24×7 uptime requirement that must be maintained for the system. Because the CA is responsible for digitally signing all of the certificates that are issued, extreme security measures must be in place to protect the CA's "root" signing key, or its private key that is associated with the CA's root public key certificate. If the CA's private key is compromised, the entire PKI is considered corrupted and must be reinitialized. All certificates previously signed by the compromised key must be revoked, and new certificates reissued.

Luckily, there are specifications and recommendations in place that unambiguously lay out both de facto and industry standard techniques for certificate validation, revocation, and automatic distribution. Although recovering from CA security breaches is still painful, these standard mechanisms offer an open and straightforward path to PKI recovery.

These recovery procedures, and the introduction of security breaches into any security framework, comprise the key advantage that PKI deployment maintains over other potential framework solutions. The deployment of a PKI is complex because the underlying problem of maintaining a robust security framework is complex. PKI-related standards cover a broad range of security framework problems, such as

- How do I enroll a new user into my security framework with little or no end user involvement (i.e., via automatic mechanisms)?
- How do I revoke a user's credentials, once they are issued?
- How can I indicate to the network that a user's credentials are only valid for a specific period of time?
- How can I indicate to the network that a user's credentials only allow access to a specific set of services or capabilities?

- How can I move a user's credentials from system to system, and do this securely?
- How do I enable standard IPs to use the security instruments available in the framework?
- How do I leverage my security framework to create a "circle of trust" between my framework and those of my business partners?

If one were to attempt to evolve a simpler security framework, like one based on usernames and passwords, to address the issues outlined above, and do so in a standards-based fashion, it is reasonable to assume that the output of such an evolution would be a system at least on a level of complexity suggested by a PKI—if not more. There is a great deal of work going on in the PKI community about how to simplify the complex issues surrounding operational deployment of a CA, without sacrificing robustness. Some organizations have chosen to outsource the operation of a CA to third parties to expedite the availability of PKI services for their applications. For some large enterprises, this may be the right way to go, but for other organizations like broadband service providers, the costs of outsourcing the authentication (PKI) infrastructure may be prohibitive. Most third party CA providers (those that actually operate the CA), capacity license their fees according to the number of clients (certificates) managed by the CA. If a service provider numbers their end users in the millions, then the cost of outsourcing the infrastructure may cut a little too deep into the revenues the provider collects on their service. Of course, the more services that are deployed that can leverage the PKI (ROI), the better. However, the clear advantage would be with a service provider with a large palette of broadband services that has figured out a way to support their own PKI, with as much automated certificate management as possible for their clients.

The fundamental idea behind identity federation is to create a common framework for the management of identity across a set of normally disjoint authentication domains. Figure 12.3 illustrates the federation interface between independent domains. With respect to broadband services, each identity provider would be the authentication domain for a particular service provider. Some service providers offer a closed set of services to their subscribers, while at the same time, offering compelling, value-added services from other service providers as well. This is shown in Figure 12.3 where identity provider #1 is an ISP that offers basic Internet service and a digital video or video-on-demand (VoD) service. The ISP has also established an identity alliance (federation) with identity provider #2, an Internet gaming site, and wishes to offer one or more of the network-based games that the gaming site has available.

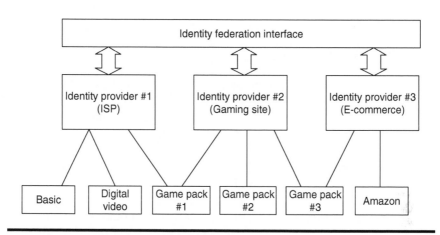

Figure 12.3   Federated identity infrastructure.

One attractive feature regarding the sharing of identity information is that each identity provider (in this case, service provider) can tailor a particular service offering for a subscriber according to the "context" information that is available about the subscriber. This subscriber context information is available through the Identity Federation Interface, and allows service providers to offer a more custom set of services depending upon the perceived attributes or preferences associated with a subscriber.

Liberty Alliance: The Liberty Alliance (www.projectliberty.org) is comprised of several large corporate entities from every corner of the business community, and its goal is to establish a standard for identity federation, as described above. The Liberty architecture allows for a decentralized approach to identity management and associated infrastructure. The alliance between identity providers is loosely coupled and can exist in any form, as long as the interfaces between identity providers are compliant with the Project Liberty recommendations.

Microsoft Passport: The .NET initiative from Microsoft is similar to the Liberty Alliance architecture, providing many of the same features and benefits. The main difference between the two solutions is that, while the Liberty Alliance architecture is loosely coupled and decentralized, the .NET model is tightly coupled with its infrastructure components, and is completely centralized. Rather than a collection of peer identity providers as described in Figure 12.3, there is a "master" identity provider, deployed and managed by Microsoft, and a sub-ordinate identity provider systems that collaborate with the centralized server to achieve the identity federation.

12.7 Sample Next-Generation DSL Security Framework

From the collection of technologies described in Section 12.6, the framework for a next-generation security model for DSL can be derived. This section illustrates an example of how these technologies can be combined to offer a complete security association between multiple service providers and their subscribers.

The next-generation DSL service network illustrated in Figure 12.4 interconnects subscribers and broadband services with a federated identity infrastructure. The instrumentation "glue" that binds these layers together is a PKI. The preferred identity infrastructure should be decentralized as much as possible because of the different types of broadband service business models that are yet to be defined. Figure 12.4 illustrates one complete autonomous broadband services organization, each with its own customer base, network, and security framework.

12.7.1 Identity Infrastructure

As described in Figure 12.3, multiple instances of these autonomous service organizations can be federated through the identity infrastructure to allow subscribers from one domain to subscribe to services in another domain. There are some classes of broadband services that can make optimal use of the identity infrastructure, and there also exist other classes of broadband services that do not necessarily take advantage of the existence of a federated identity architecture. Most Web-based services would see a direct benefit from the exchange of identity information with other members of the federation. However, other types of services could be evolved or otherwise

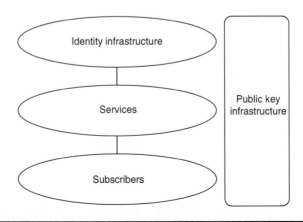

Figure 12.4   High-level next-generation overview.

encapsulated to enable the benefits of the infrastructure. Because of the decentralized nature of the identity infrastructure, and the clean interface between the identity management components and the actual broadband services, the design would allow for a flexible transition to offering third party broadband services, through federation, at the discretion and pace with which the primary service provider is comfortable.

12.7.2 PKI Deployment

The PKI supports all of the fundamental aspects of a robust security framework enumerated earlier: confidentiality, authentication, integrity, and, if required, non-repudiation. The service provider would deploy a CA capable of handling the subscriber population, as well as the collection of hosted broadband services. The CA could be outsourced completely to agencies like Verisign, Baltimore Technologies, or a host of other third party CA vendors, or the network operator could decide that the leverage and return on investment is high enough to deploy their own CAs.

12.7.3 Subscribers

This example proposal assumes that service providers will migrate to a multi-service-capable model, requiring intelligent CPE to be deployed in all customer environments. Given the requirement for intelligent CPE, the CPE device will evolve to become a proactively managed network element, much like any other device in the DSL access network. The CPE will also be integrated into the security framework that is maintained by the service provider. When communicating with the service provider network, the CPE device will proxy authenticate on behalf of the subscriber; in other words, a particular instance of DSL CPE will be associated with a particular subscriber in the service provider's security framework. Each DSL subscriber's CPE will be assigned a public key certificate within the PKI being employed by the service provider. The assignment of the certificate to the CPE will be performed automatically by the network during the CPE installation process; subsequent maintenance of the certificate's validity would also be performed automatically by the network. This device certificate could be used for the following purposes:

- Secure provisioning of the CPE using SSL client (pull) or server (push) capability
- IPSec VPN (virtual private network) support (client or server)
- Secure Web-based access to CPE-resident HTML (HyperText Markup Language) pages
- Broadband service content integrity checking via digital signatures
- Software update image verification via digital signatures

Chapter 13

Data Packets Transport over DSL

Vladimir Oksman

CONTENTS

Abstract   This chapter describes how various data streams formed into sequence of packets are transported over Digital Subscriber Line (DSL). The presented material includes principles and requirements used by standardization bodies to define packet transport for various types of packets, and describes important aspects of data packets transport technique used in recent DSL.

13.1 Overview

Packet-based services use protocols such as Point-to-Point Protocol (PPP), Ethernet, and frame relay. These same protocols have also been the most popular packet-based service protocols in the Digital Subscriber Line (DSL) world from the very beginning. However, for a long time no special protocol for packet transport over DSL was specified, and transport protocols such as Asynchronous Transfer Mode (ATM) and Synchronous Transport Mode (STM) were used to carry packets.

ATM is still the main packet carrier. ATM can deliver any packet-based service by segmentation of original data packets into 48-octet segments and encapsulating them into standard ATM cells [ITU-T I.432.4]. The primary advantage of this method is that it is universal and can be used to accommodate any packet protocol. The disadvantage is high overhead because of a five-octet ATM cell header attached to every ATM cell at the physical layer [ITU-T I.432.4]. Additional overhead is also added at the segmentation sublayer to reconstruct the entire packet properly from the received segments. As a result, the total overhead can be as high as about 15 percent, which wastes valuable capacity of the DSL transmission core.

STM is also used for packet transport, such as for frame relay transport and Ethernet transport over T1/E1 or HDSL/SHDSL (High Bit-Rate DSL/Symmetric HDSL), for example. The drawback of this solution is the unjustified complexity, mostly because STM was originally intended to carry continuous, delay-sensitive applications, and thus it uses complex techniques based on accurate time stamps to recover application data timing. For packet transport, this functionality is unnecessary because the connected packet data networks run with independent data rates and do not require common timing.

Recently, working group IEEE 802.3 and study group 15, Question 4 (SG15/Q4) of the ITU-T (International Telecommunication Union-Telecommunication) specified two innovative methods for packet transport over DSL. These methods, which are described in this chapter, are free from the disadvantages of the ATM and STM protocols. They allow transport of packets of any size and type with high efficiency and minimum implementation complexity. Furthermore, these methods include a traffic management technique called preemption, which allows priority access of assigned packets to the transmission media.

13.2 Key Requirements for Transport of Data Packets

Modern data networks use a variety of different protocols and formats. The most popular is Ethernet, although many others are used as well, and new protocols are expected in the future. Therefore, the first requirement considered by ITU-T is to specify a packet transport mechanism that is independent of the user's data protocol. In other words, the specified method should be capable of transporting data packets of arbitrary length and without any knowledge of their content (i.e., whether they have headers, how the payload is formatted, or whether each packet has a trailer). Such a requirement implies two features that are necessary:

1. Any packet of data needs to be transported transparently, regardless of its length or type, or of the content of its header, payload, or trailer, except that the packet should contain an integer number of bytes. The packet integrity, i.e., the number of bytes in the packet, their contents, and their sequential order, needs to be maintained.
2. Method should not require any reference to the data bit rate of either the source or the customer. In other words, the DSL should be able to connect the source and customer even if the bit rates of the source and customer networks are set independently of each other. It is usually a sufficient condition for packet transport that the average bit rate of the source is less than the (typically constant) bit rate of the DSL link.

An additional goal of the new packet transport mechanism is that the method should be more efficient than the conventional packet transport over ATM. Thus, the overhead of the new method should be less than 5/53, or less than about 9.5 percent.

Another desired feature for the new method is error-monitoring capability. Error monitoring at the packet level is required for proper management of the physical layer operation. Errored packets need to be identified, processed, and sent to the upper layers accompanied by an indicator of a packet error.

Finally, other issues concerning specific protocols for packet transport need to be considered. Some of the protocols impose restrictions on the undetected error probability in the access network, and others are sensitive to changes in the transmission overhead, and still others are just too complex for implementation (e.g., generic framing procedure (GFP) [Castagnoli 1993]). These issues were also considered by the IEEE 802.3 and the ITU-T.

13.3   Functional Model of Packet Transport

The functional model of packet transport is shown in Figure 13.1. In the transmit direction, the packet transport mode (PTM) entity obtains data

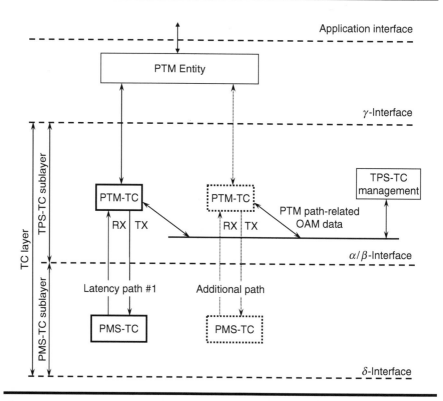

Figure 13.1   Generic functional model of packet transport over Digital Subscriber Line (DSL).

packets to be transported over the DSL connection from the application interface. The application interface might be, for example, the media ind- ependent interface (MII) if Ethernet transport is in use. The PTM entity processes each packet and submits it, based in part on the latency require- ments of the service delivered by the packet, to the γ-interface of the app- ropriate latency path of the DSL transceiver. At the γ-interface, the packet transport mode transmission convergence (PTM-TC) function encapsulates the packet into a PTM-TC frame and sends the encapsulated packet to the physical medium specific transmission convergence (PMS-TC) function for transmission over the DSL link. In the receive direction, PTM-TC frames are extracted from the PMS-TC. The PTM-TC recovers received packets from the PTM-TC frames and submits them to the γ-interface. The PTM entity processes the packet and delivers it to the application interface. Information on errored packets and other violations in the packet transport mechanism is collected by the transport protocol specific transmission convergence (TPS-TC) management entity to assist the related performance monitoring functions.

A DSL modem in PTM usually uses only one latency path, of which the latency is adjusted to meet the requirements of all delivered services. However, one or more additional latency paths may be used if no common latency setting is appropriate for all services. An optional second latency path is shown in Figure 13.1 by dotted lines and differs from the first path by its forward error correction (FEC) settings and interleaver settings in the PMS-TC. The PTM-TCs of both latency paths have the same characteristics, and each provides a fully transparent packet transfer between the γ-interfaces of the DSL transceivers at both ends of the line. The upstream and downstream bit rates of the DSL connection may be set independently of each other and independently of the bit rates used by the data networks at the service provider end and the customer end, which are connected by the DSL. However, the DSL net bit rate in each direction of transmission must be slightly higher than the average bit rate desired for transport of data packets in that direction to accommodate the overhead added because of PTM-TC framing.

13.3.1   Data Processing by the PTM-TC

Two types of PTM-TC will be considered: a generic PTM-TC (GPTM-TC) specified by the ITU-T [ITU-T G.992.3], and the Ethernet PTM-TC (ETM-TC) specified by the IEEE [IEEE 802.3]. The GPTM-TC provides basic transport of packets of arbitrary size and with arbitrary contents. The ETM-TC has a limitation on the minimum size of the transmitted packet, per requirements of Ethernet transport. Both methods have a very low probability of undetected error, but they differ in possible fluctuations of encapsulation overhead.

To identify the beginning and end of each packet sent over the line, the PTM-TC encapsulates each packet into a PTM-TC frame. The GPTM-TC defined for ADSL2 [ITU-T G.992.3] and VDSL1 [ITU-R G.993.1] uses for encapsulation the well-known high-level data link control (HDLC) frame (GPTM-TC$_{HDLC}$), which is ISO-standardized [ISO/IEC 3309] and simple to implement. HDLC does not impose any specific limitation on the length of the packet (although packets cannot be too long to meet the latency requirements), but it adds a variable amount of overhead that depends on the contents of the transmitted data. To reduce the variable overhead of the PTM-TC, IEEE 802.3 developed an encapsulation protocol named 64B/65B. The 64B/65B encapsulation is free of the overhead dependence on the packet contents, but it restricts the minimum length of the transmitted packet to 64 bytes. ITU-T generalized the 64B/65B encapsulation method by adding the means to allow transmission of packets of all sizes. This type of GPTM-TC is specified in the latest ITU-T Recommendations for all types of DSL, including VDSL2 [ITU-R G.993.2]. The details of the 64B/65B encapsulation variants are described in Section 13.3.2.

Table 13.1   Signals at the γ-Interface

Signal	Description	Direction
	Transmit direction	
Data	Transmit data packets	PTM ⟶ PTM-TC
SoP, EoP	Start and end of the transmit packet	PTM ⟶ PTM-TC
Tx_Enbl	Set by the PTM-TC, allows the PTM entity to send data	PTM ⟵ PTM-TC
Tx_Err	Transmit signal error indicator	PTM ⟶ PTM-TC
	Receive direction	
Data	Receive data packets	PTM ⟵ PTM-TC
SoP, EoP	Start and end of the receive packet	PTM ⟵ PTM-TC
Rx_Enbl	Set by the PTM-TC, allows the PTM entity to retrieve data from PTM-TC	PTM ⟵ PTM-TC
Rx_Err	Received signal error indicator	PTM ⟵ PTM-TC

In addition to encapsulation, the PTM-TC provides adaptation between the data bit rate of the application (from the perspective of the PTM entity) and the data bit rate provided by the DSL link. This adaptation uses a simple flow-control mechanism at the γ-interface (see Figure 13.1 and Table 13.1). In the transmit direction, overflow of the PTM-TC data buffer is prevented by the transmit enable (*Tx_Enbl*) signal, asserted by the PTM-TC, which stops the packet flow from the PTM entity to the PTM-TC. In the receive direction, the receive enable (*Rx_Enbl*) signal indicates to the PTM entity that the packet has been received and is ready to be retrieved from the PTM-TC. As the packet is retrieved, the PTM-TC buffers are promptly flushed, allowing space for a new incoming frame. Note that the described flow-control mechanism relates to the physical (PHY) layer only and does not address the application-layer flow-control issues, such as backpressure. The latter are facilitated using in-band operations and management (OAM) messages of the higher-layer protocol (e.g., Ethernet or Transmission Control Protocol [TCP]).

The boundaries of each transmitted packet are indicated by two signals at the γ-interface. The *SoP* and *EoP* signals identify, respectively, the beginning and end of each transmitted packet. The signals are asserted by the PTM entity in the transmit direction and by the PTM-TC in the receive direction.

Packets received with errors are accompanied by the receive error (*Rx_Err*) signals. The PTM-TC, being a PHY component, does not make a decision whether the errored packet should be discarded. Instead, the PTM-TC passes all packets, including errored, to the PTM entity, which typically uses a more sophisticated mechanism to process errored packets.

A crude mechanism of preemption is effected by the transmit error (*Tx_Err*) signal, which is used in some applications to abort the packet currently being transmitted. Upon receiving *Tx_Err* from the PTM entity, the PTM-TC flushes the buffer and terminates transmission of the packet. The transmitted fragment of the packet is expected to be recognized and discarded at the receive side during frame check sequence (FCS) verification. Using this mechanism, a long packet can be discarded to release the channel for transmission of a packet with higher priority. The disadvantage of this method is that it requires retransmission of the aborted packet at a later time, which is inefficient. The latest version of the GPTM-TC (in ITU-T Recommendation G.992.3 [ITU-T G.992.3]) specifies a more efficient method of preemption, which is described in Section 13.3.2.

Table 13.1 summarizes the signals at the γ-interface.

The data clock signals at the γ-interface, in both directions, are sourced by the PTM entity. In practical applications, these clocks are usually set to a much higher rate than the expected DSL bit rate, so the latency of the packet transfer over the γ-interface is reduced by fast loading and draining of the PTM-TC buffers. For 10 Mbit/s Ethernet applications, for instance, the typical data clock rate is 100 Mbit/s, whereas the line rate might be less than 1 Mbit/s in the upstream direction (e.g., for ADSL).

13.3.2   Encapsulation

As mentioned previously, DSL standards specify two types of packet transport that are based on, accordingly, two different types of encapsulation. The basic 64B/65B encapsulation is used in the ETM-TC, HDLC is used in the GPTM-TC$_{HDLC}$, and the newest ITU-T DSL Recommendations specify GPTM-TC encapsulation based on 64B/65B.

13.3.2.1   HDLC Encapsulation

An ISO-standard HDLC frame used in the GPTM-TC$_{HDLC}$ is shown in Figure 13.2. The start and end of each frame are identified by flag sequences (with a hexadecimal value of $7E_{16}$). The address field and control field may be used for proprietary information, or they may be set to their default values. The data field carries the transported packet, and the two-octet FCS is used for packet error monitoring. The FCS is calculated over all fields of the frame, except flags, using the 16-bit ISO/IEC 3309 cyclic redundancy check (CRC-16) algorithm [ISO/IEC 3309]. It should be noted that calculation of the FCS does not necessarily require buffering of the entire HDLC frame, and thus it does not restrict the length of the transmit packet.

Time gaps between transmitted HDLC frames are filled with additional flags, but only one flag is required between two consecutive frames (i.e., a single flag can serve as both the closing flag of the previous frame and the opening flag of the next frame). Additional flags fill the inter-packet gaps to equalize the average bit rate of the transmitted data packets to the bit

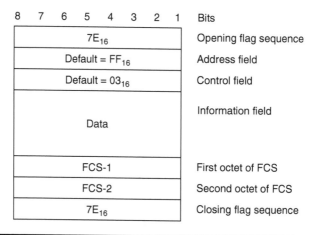

Figure 13.2 (GPTM-TC$_{HDLC}$) frame format.

rate provided by the DSL connection. All flags and all HDLC fields, except the data field, are discarded by the receiver.

At the receiver, HDLC frame boundaries are determined by identifying the flags. Clearly, if any "information" bytes, such as those in the data or FCS fields, have the value of $7E_{16}$, they could be mistaken for flags. Therefore, to avoid false identification at the receiver, the transmitter replaces all octets in the data field or in the FCS field that are equal to $7E_{16}$ using the ISO/IEC 3309 octet stuffing algorithm [ISO/IEC 3309]. The transmitter identifies those non-flag octets equal to $7E_{16}$ and replaces them by the two-octet sequence $7D_{16}5E_{16}$. Because the meaning of the value $7D_{16}$ is changed by this procedure, any instances of the octet $7D_{16}$ in the data or FCS fields must also be replaced by a special two-octet sequence, which is $7D_{16}5D_{16}$. The receiver identifies the replacements by looking for octets with the value of $7D_{16}$ (called the control escape octet) and reconstructs the original packet based on the values of the octets that follow $7D_{16}$. The FCS is calculated prior to the stuffing procedure, i.e., over the original packet. Any received HDLC frame with a mismatch in the FCS calculation is qualified as errored, and the packet is passed to the PTM entity accompanied by an *Rx_Err* signal.

Because of the octet stuffing, the length of an HDLC frame changes depending on the number of data octets equal to $7E_{16}$ or $7D_{16}$ it includes. In the extreme case, the length of the transmit frame can be almost twice as long as the original length of the packet, thus increasing the overhead of the HDLC encapsulation and reducing the effective bit rate of the DSL channel. Fortunately, the probability of this event is very low if the content of the packet is sufficiently random. If this is not the case, a data randomizer with a generating polynomial of power greater than 16 and not divisible by 8 usually provides a good result. Current DSL standards do not require the use of a randomizer assuming that 64B/65B encapsulation will be used in cases which concern about the mentioned overhead expansion.

13.3.2.2 Basic 64B/65B Encapsulation

The 64B/65B encapsulation is defined by the IEEE 802.3 standard [IEEE 802.3]. This technique was developed to reduce the variability of the encapsulation overhead, such as that caused in HDLC by the octet stuffing mechanism. The transmit 64B/65B signal is a continuous sequence of 65-byte codewords. Each codeword starts with a sync byte, followed by a 64-byte data field, where each octet is either a data byte, or one of the valid control bytes (S-byte, C-byte, Y-byte). A detailed description of 64B/65B coding rules is presented in Table 13.2. The transmit packet is mapped with no alignment with the boundaries of the 64B/65B codeword. The S-byte and the C-byte mark the start and the end of the packet, respectively, inside the codeword. The average bit rate of the transmit data is equalized to the bit rate provided by the DSL connection by insertion of Z-bytes (idle bytes) between packets. At the receiver, packets are delineated based on the received control bytes and coding rules presented in Table 13.2.

The values of C_k, Z, and S and Y are defined as follows:

- $C_k = k + 10_{16}$, with the MSB (most significant byte) set so that the resulting value has even parity, i.e., $C_0 = 90_{16}$, $C_1 = 11_{16}$, $C_2 = 12_{16}$, $C_3 = 93_{16}, \ldots, C_{62} = 43_{16}, C_{63} = CF_{16}$.
- $Z = 00_{16}, S = 50_{16}, Y = D1_{16}$.

The control byte Y followed by Z-bytes is intended to signal the remote side that the receiver in PTM-TC lost synchronization (multiple violations of sync byte).

Prior to encapsulation, a two-byte or four-byte FCS is appended to the packet to enable error monitoring in the transmission path. The two-byte (four-byte) FCS is calculated using the ISO/IEC 3309 CRC-16 (CRC-32) standard algorithm. The four-byte FCS is used for those types of DSL that do not use forward error correction (FEC).

A low and stable overhead is the primary benefit of 64B/65B encapsulation. The 64B/65B coding table (Table 13.2) shows that the overhead is one byte per codeword containing data and either two or three bytes in codewords at the beginning or end of the packet. Unlike HDLC, the overhead in 64B/65B has small variations of about two bytes per packet (the S-byte at the beginning and the C_k-byte at the end).

Table 13.2 Basic 64B/65B Coding Rules

Codeword Type	Codeword Contents	Sync Byte	Codeword Fields, Bytes 1–64										
Data	Data octets (D) only	$0F_{16}$	D_0	D_1	D_2	D_3	D_4	D_5	$\cdots$		D_{61}	D_{62}	D_{63}
End of packet	Control byte (C_k) followed by k bytes of data, others are idle (Z)	$F0_{16}$	C_k	D_0	D_1	D_2	D_3	$\cdots D_{k-1}$	Z		$\cdots$		Z
Start of packet while idle	First k-bytes are idle (Z), last 64-k-1 bytes are data. S-byte indicates start of the packet	$F0_{16}$	Z	Z	S	D_0	D_1	$\cdots$	$\cdots$		D_{k-3}	D_{k-2}	D_{k-1}
End of packet and start of the next packet	Control byte (C_k) followed by k data bytes. Last j bytes are data (next frame). S-byte indicates start of the packet	$F0_{16}$	C_k	D_0	$\cdots D_{k-1}$	Z	$\cdots$		S	D_0		$\cdots$	D_{j-1}
Idle	Idle octets (Z) only	$F0_{16}$	Z	Z	Z	Z	Z	Z	$\cdots$		Z	Z	Z
Idle, out-of-sync	Y-byte followed by Z–octets only	$F0_{16}$	Y	Z	Z	Z	Z	$\cdots$	Z		Z	Z	Z

Fast and robust synchronization is another feature of 64B/65B. Because sync bytes are transmitted at predefined positions, codeword synchronization is robust to errors, regardless of where in the codeword those errors appear. Synchronization can be maintained even if the sync byte has errors. The process of synchronization and codeword delineation is similar to the one used in [ITU-T I.432.4]. The IEEE 802.3 standard recommends to declare synchronization after at least four correct sequential sync bytes. Conversely, a lack of synchronization is declared following at least eight errored sequential sync bytes. These definitions allow a system to maintain synchronization under very severe error conditions, such as with an average bit error ratio (BER) of up to 10^{-2} and error bursts (erasures) that are up to several milliseconds long.

13.3.2.3 ITU-T Enhancements of 64B/65B

Support of short packets: The encapsulation format of basic 64B/65B does not allow a packet to start and end inside the same 65-byte codeword. Thus, the size of encapsulated packets cannot be shorter than 63 bytes, and the size of transmitted packets, respectively, cannot be shorter than 59 or 61 bytes, depending on the FCS type. These restrictions are fine for Ethernet transport, in which the shortest packet size is 64 bytes, but other data protocols may allow shorter packets. Therefore, ITU-T investigated various means to use 64B/65B encapsulation with packets shorter than 63 bytes. One way to accommodate short packets using the original 64B/65B codewords is to pad them (i.e., insert dummy bytes) to 63 bytes. Another way is to adjust the gap between the previous packet and a short packet so that each short packet starts in one codeword and ends in the next codeword. Both approaches require the insertion of additional (meaningless) bytes, which results in higher overhead. As a result, ITU-T defines an enhancement of 64B/65B to support short packets. With this enhancement, 64B/65B is suitable for generic packet transport.

Additional coding rules to support short packets are presented in Table 13.3. In the case of a short packet, an additional control byte C_j is inserted immediately before the S-byte for any packet that will finish before the end of the codeword in which it starts. The value of the C_j-byte marks the end of the short packet. The value of C_i is defined in the same manner as the control byte C_k specified in Table 13.2 in regard to the position j in the codeword where the packet ends. If no C_j-byte precedes the S-byte, then the data bytes continue to the end of the codeword as per the definition in Table 13.2. The number of short packets in a codeword is not limited by coding rules, but only by the codeword size (64 bytes). For simplicity, examples in Table 13.3 show only one short packet per codeword.

This enhancement to support short packets changes the basic coding rules; however, if no short packets are transmitted, the enhanced 64B/65B operates identically to the basic 64B/65B. Furthermore, the overhead variation for enhanced 64B/65B is two bytes per packet, which is the same as for basic 64B/65B.

Support of preemption: Preemption is a traffic control mechanism on the physical layer that minimizes propagation delay of high-priority packets. Preemption can be useful for services such as Voice-over-Internet Protocol (VoIP) and videoconferencing to comply with requirements for delay variation. A simple preemption mechanism, using the Tx_Err signal, is available at the γ-interface, but it requires retransmission of preempted packets. The preemption mechanism developed by ITU-T for 64B/65B encapsulation is free of this disadvantage.

Table 13.3   Enhanced 64B/65B Coding Rules for Short Packets

Codeword Type	Codeword Contents	Sync Byte	Codeword Fields, 1–64
Start short packet after end of packet	Last k data bytes of the first packet and $j1$ data bytes comprising the short packet	$F0_{16}$	C_k D_0 $\cdots$ D_{k-1} Z $\cdots$ C_{j1} S D_0 $\cdots$ D_{j1-1} Z, S or C_{j2} $\cdots$
Start short packet after idle	Up to $(62-j1)$ idle bytes and j data bytes comprising the short packet	$F0_{16}$	Z $\cdots$ Z $\cdots$ C_{j1} S D_0 $\cdots$ D_{j1-1} Z, S or C_{j2} $\cdots$
Start short packet immediately after sync byte	$j1$ data bytes comprising the short packet and $(62-j1)$ idles	$F0_{16}$	C_{j1} S D_0 $\cdots$ $\cdots$ $\cdots$ D_{j1-1} Z, S or C_{j2} $\cdots$

According to the ITU-T preemption specification, when a high-priority packet is ready for transmission, under control of the PTM entity, the transmission of a low-priority packet is paused, releasing the channel for a packet with higher priority. After the high-priority packet has been transmitted, transmission of the low-priority packet resumes. Thus, preemption minimizes the delay in transmitting high-priority packets at the expense of increased delay of the low-priority packets.

A functional diagram of transmission using the preemption mechanism is presented in Figure 13.3.

The high-priority (preempting) and low-priority (preempted) packets enter the PTM-TC through two separate γ-interfaces. The PTM entity indicates to the PTM-TC that one or more high-priority packets are ready for transmission by asserting the Tx_Avbl signal at the γ-interface for the preempting packet. Upon assertion of this Tx_Avbl signal, the PTM-TC suspends transmission of the low-priority packet and encapsulates the

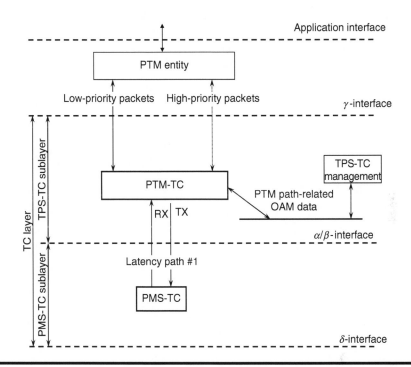

Figure 13.3 **Functional model of packet transport using preemption.**

preempting packet according to the preemption rules of enhanced 64B/65B (see Table 13.4). As the last preempting packet is sent, the *Tx_Avbl* signal is de-asserted at the γ-interface of the preempting packets, and transmission of the low-priority packet continues after the interruption caused by the preemption.

Additional 64B/65B coding rules that facilitate preemption are presented in Table 13.4. When a preempting packet is to be sent, the transmission of the preempted (low-priority) packet is suspended at the end of the code-word, and the preempting packet is mapped into the next codeword imme-diately after the Sync byte (see the first line of Table 13.4). This codeword is identified by a special Sync byte value of $F5_{16}$. The subsequent code-words carrying the data of the preempting packet are indicated by a Sync byte value of AF_{16}, and the last codeword carrying a preempting packet is again marked by Sync byte value of $F5_{16}$. Transmission of the preempted packet can resume starting from the next codeword after transmission of the preempting packet is complete. This codeword is marked by a regular Sync byte value, which is $F0_{16}$ or $0F_{16}$, depending whether the preempted packet ends or not in the codeword (see Table 13.2).

Table 13.4   Enhanced 64B/65B Coding Rules for PreEmption

Codeword Type	Codeword Contents	Sync Byte	Codeword Fields, 1–64								
Start of new pre-empting packet from idle	S-byte followed by 63 data bytes	$F5_{16}$	S	D_0	D_1	D_2	D_3	$\cdots D_{k-4}$	D_{k-3}	D_{k-2}	D_{k-1}
Start of new pre-empting packet after the end of a pre-empting packet	Last k data bytes of the first packet $(0 \leq k \leq 62)$, $(62 - j - k)$ and first j data bytes of the second packet $(0 \leq j \leq 62 - k)$	$F5_{16}$	C_k	$D_0 \cdots D_{k-1}$	Z	Z	S	D_0	$\cdots$		D_{j-1}
All pre-empting data	64 data bytes	AF_{16}	D_0	D_1	D_2	D_3	D_4	D_5	$\cdots$	D_{61}	D_{62} D_{63}
End of pre-empting packet	k data bytes $(0 \leq k \leq 63)$ followed by $(63 - k)$ idles	$F5_{16}$	C_k	D_0	D_1	D_2	D_3	$\cdots D_{k-1}$	Z	$\cdots$	Z

A comparison of Tables 13.2 and 13.4 shows that the rules of coding for preempting and preempted packets are the same, except for a different Sync byte value (AF_{16} or $F5_{16}$ for preempting packets, and $0F_{16}$ or $F0_{16}$ for preempted packets). The similarity of coding rules greatly simplifies the implementation of preemption. Switching from a low-priority packet to a high-priority packet and back is also simple: transmission of both types of packets can begin only in the next codeword that follows the request for high-priority packet, or next codeword after transmission of high-priority packet has been completed.

At the receiver, preempting and preempted packets are identified by the value of the Sync byte. Preempting packets are routed to the PTM entity through the γ-interface for high-priority packets. The received fragments of low-priority interrupted packets are stored while transmission of high-priority packets takes place. After transmission of low-priority packets resumes and all parts of the preempted packet are received, the completed packet is routed to the PTM entity through the γ-interface for low-priority packets.

13.3.3 Performance

The performance of a PTM-TC is characterized by its effectiveness, which is usually determined by the ratio between the number of overhead bytes and the length of the transmitted packet.

13.3.3.1 GPTM-TC$_{HDLC}$

In GPTM-TC$_{HDLC}$, overhead is introduced due to the HDLC encapsulation requirements, and by the octet stuffing. The six octets of HDLC encapsulation overhead shown in Figure 13.2 result in an overall overhead of $\frac{5}{L} \times 100$ percent, where L is the length of the transmitted packet in bytes (including the overhead, and assuming the closing flag is counted as an opening flag of the next frame).

The overhead introduced by octet stuffing depends on the content of the data to be transmitted. For random data, the overhead owing to octet stuffing is determined by the probability $P(k)$ that k bytes, each equal to either $7E_{16}$ or $7D_{16}$, will appear in a L-byte packet. For $k \geq 1$ the probability is

$$P(k) = C_L^k \cdot p^k (1-p)^{L-k} \tag{13.1}$$

where p is the probability that the value of the byte is either $7E_{16}$ or $7D_{16}$. Assuming that the transmitted data stream is random, $p = 2 \times 2^{-8} = 2^{-7}$.

The probability that the number of stuffed bytes will be any from 1 to N is

$$P(0 < k \leq N) = \sum_{k=1}^{N} C_L^k \cdot p^k (1-p)^{L-k} \tag{13.2}$$

and therefore the probability that no stuff bytes will be necessary is

$$P(k=0) = 1 - P(0 < k \leq L) = 1 - \sum_{k=1}^{L} C_L^k \cdot p^k (1-p)^{L-k}$$

Consequently

$$P(k \leq N) = P(k=0) + P(0 < k \leq N)$$

$$= 1 - \sum_{k=1}^{L} C_L^k \cdot p^k (1-p)^{L-k} + \sum_{k=1}^{N} C_L^k \cdot p^k (1-p)^{L-k} \tag{13.3}$$

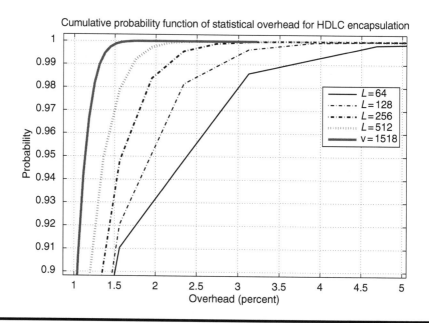

Figure 13.4   Cumulative distribution function of the high-level data link control (HDLC) statistical overhead.

is the cumulative probability function (CDF) of the statistical overhead introduced by octet stuffing. Equation 13.3 shows the probability that the overhead owing to octet stuffing will not exceed $\frac{N}{L} \times 100$ percent and the total overhead will not exceed $\frac{N+5}{L} \times 100$ percent. The CDF clearly depends on the length of the packet L. The results of calculations for different L using Equation 13.3 are presented in Figure 13.4. The curves show the maximum overhead in terms of a percent that will not be exceeded with the given probability. For instance, considering $L = 1518$, the probability that the overhead will not exceed 1.3 percent is 99 percent, and the probability that the overhead will not exceed 1.58 percent is 0.999 percent. In terms of bytes, the absolute value of the overhead in the first case can be calculated as round $(1518 \times 0.013) = 20$ bytes.

The 99 percent and 99.9 percent worst-case values of overhead, in percentages, are listed in Table 13.5 for various packet lengths L.* The table shows that the total overhead strongly depends on the packet length L. For short packets, the encapsulation overhead is the issue, whereas for long packets the contribution of stuffing overhead becomes more significant.

---

* In other words, the overhead will exceed the given values in only 1 percent and 0.1 percent, respectively, of cases.

Table 13.5 High-Level Data Link Control (HDLC) Overhead

Overhead	Packet Length (L), Bytes					
	64	128	256	512	1024	1518
Encapsulation overhead (percent)	7.25	3.76	1.92	0.97	0.49	0.33
Stuffing overhead 99 percent case (percent)	3.56	2.75	2.20	1.75	1.49	1.30
Stuffing overhead 99.9 percent case (percent)	6.25	3.90	3.13	2.15	1.75	1.58
Total overhead 99 percent case (percent)	10.81	6.51	4.12	2.72	1.98	1.63
Total overhead 99.9 percent case (percent)	13.50	7.66	5.05	3.12	2.24	1.91

If many octets of the transmit packet are equal to either $7E_{16}$ or $7D_{16}$, the overhead may increase greatly, approaching 100 percent if all data octets are $7E_{16}$ and $7D_{16}$. This effect is called overhead explosion. Explosion delays the arrival of the exploded packet and also several packets after it, thus causing a temporary slowdown of the packet transport. For random data, the probability of explosion is very low (see Figure 13.4), and the average total overhead is usually less than 2 or 3 percent. In practical cases, data is sufficiently random, and current DSL standards do not require additional randomization.

In some specific applications, however, even a rare overhead explosion needs to be considered. The goal is to ensure that the explosion causes a very short slowdown in the packet transport, ideally delaying only a few subsequent packets. Usually, mitigation of the effects of explosion happens naturally because of the idle intervals between transmitted packets. However, explosion may be more of a problem if the application can send long sequences of packets without any idle intervals or with very short idle intervals. The simplest way to mitigate the impact of overhead explosion is to introduce additional idle intervals (inter-packet gaps) by slightly decreasing the average bit rate of the packet transport, or by slightly increasing the bit rate of the DSL connection. (In practice, throttling the packet transport rate is usually the only available option.) This method is illustrated in Figure 13.5.

Assume that there are I idle octets between the packets transmitted by the PTM entity. The overhead explosion of E octets will delay the exploded frame and J following frames. If the average overhead per frame equals OH, the number of frames J that are slowed down can be calculated as

$$E + J \cdot \text{OH} = J \cdot I \quad \text{and} \quad J = \frac{E}{I - \text{OH}} \tag{13.4}$$

518 ■ *Implementation and Applications of DSL Technology*

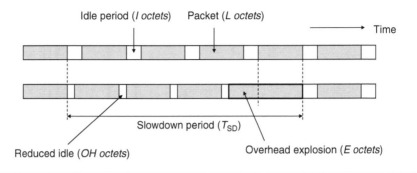

Figure 13.5 Illustration of the impact caused by overhead explosion.

Equation 13.4 shows that if $I = OH$, i.e., if the number of idle octets can only cover the average overhead per frame (see Table 13.5), any explosion will slow down all of the traffic ($J = \infty$). For any $J > OH$, the slowdown will affect a limited number of packets. For example, if the explosion is 20 percent of the packet length, $I = 2.3$ percent, and $OH = 1.3$ percent of the packet length, the number of delayed packets will not exceed $\frac{20}{2.3-1.3} = 20$ packets. If the transmission bit rate is 1 Mbit/s and the average packet size is 500 octets, the slowdown will only be during $10^{-6} \times 8 \times 500 \times 20 = 0.08$ s, which is usually negligible.

The analysis above shows that impact of the overhead explosion can be controlled by appropriate selection of the transmission bit rate. A computation similar to the one above proves that increasing the bit rate by about 1 percent is sufficient to avoid any noticeable traffic slowdown owing to overhead explosion, which is an insignificant additional redundancy.

13.3.3.2 ETM-TC

Every 64B/65B codeword includes at least one overhead byte (the Sync byte). Additionally, a C_k-byte is transmitted to indicate the end of the packet, and an S-byte is transmitted to indicate the start of the packet. Assuming that packets are being transmitted without gaps (no Z-bytes), there should be at least two bytes added to each packet and one byte after every 64 data bytes. Additionally, two or four FCS bytes are added. As a result, the total overhead per packet may be calculated as

$$OH = \left\lfloor S + C + FCS + \left(L + FCS\right)/64 \right\rfloor, \text{[bytes]} \qquad (13.5)$$

where
    L is the length of the transmit packet
    FCS is the number of FCS bytes
    $\lfloor . \rfloor$ represents rounding down

Table 13.6   64B/65B Overhead

	Packet Length (*L*) (in Bytes)					
	64	128	256	512	1024	1518
Maximum overhead (percent)	9.38	5.47	3.52	2.54	2.05	1.84

The results of calculations using Equation 13.5 and *FCS* = 2 bytes are presented in Table 13.6. They show that, similar to HDLC, the encapsulation overhead of 64B/65B changes with the length of the transmit packet. However, with 64B/65B, the overhead is less than with HDLC, as shown by the values presented in Table 13.5. In addition, 64B/65B does not suffer from the overhead explosion problem.

13.3.4 *Undetected Errors*

Impairments in the transmission line may cause bit errors in the PTM-TC frame. Most of the errors are detected by the FCS, but some may be missed due to the limited capabilities of the FCS algorithm. The residual errors are called undetected errors. The probability of undetected error strongly depends on the probability of errors caused by the DSL loop, and also on the statistics of these errors. The robustness of the system regarding undetected errors is usually specified by the mean time to false packet acceptance (MTTFPA). Requirements for the MTTFPA vary between 10^6 and 10^{12} years for different applications. MTTFPA values this high are provided by the error detection capabilities of both the physical layer and the application protocol. (Standard Ethernet frames, for instance, have an embedded CRC-32 which greatly improves error detection capabilities.)

The focus of this chapter is on the MTTFPA provided by the physical layer (by the PTM-TC), which is estimated using the probability P_{FP} that an errored packet is passed to the PTM entity. For simplicity, this probability is referred to as the probability of a false packet. To comply with a specific MTTFPA value, the probability of a false packet must satisfy

$$P_{FP} \leq \frac{1}{K \cdot MTTFPA}$$

where *K* is the total number of packets transmitted over the line during the year.

The total number of packets *K* sent in a year depends on the utilization of the line (the percentage of time the line is used), the bit rate, and average packet length. For example, a 10 Mbit/s line bit rate with an average packet length of 750 bytes and a utilization time of 20 percent results in

$$K = 0.2 \times 10^7 \times (365 \times 24 \times 3600)/(750 \times 8) \approx 10^{10} \text{ packets/year}$$

Assuming the DSL is configured for a symmetrical connection and provides 10 Mbit/s in both the downstream and upstream directions with the parameters given above, $K = 2 \times 10^{10}$ packets/year. To provide a value of $MTTFPA = 10^9$ years over this line, the probability of a false packet cannot exceed 5×10^{-20}.

The ISO-standard 16-bit FCS (CRC-16) is capable of detecting all frames with an odd number of errored bits, all cases of two-bit errors, and some cases of four-bit error combinations if the length of the frame is less than 2048 octets. The standard CRC-32 is capable of capturing almost all four-bit errors if the length of the frame is less than 2048 octets [Baicheva 2000] and [Castagnoli 1993]. Thus, the probability of a false packet may be estimated as the probability of all m-bit error combinations, where m is even and exceeds the error detection capability of CRC. In the case of randomly distributed errors, this probability can be computed as

$$P_{\text{FP\_R\_Err}} = P_{m>E}(L) = \sum_{m=E+1}^{L \times 8} C_{L \times 8}^m \cdot p^m \cdot (1-p)^{L \times 8 - m} \qquad (13.6)$$

where
p is the probability of a bit error caused by the DSL channel
L is the length of the transmitted packet in bytes
E is the CRC error detection capability, which is equal to 3 errors in the case of CRC-16 and 5 errors in the case of CRC-32.

In many types of DSL, FEC coding is used. With FEC coding, errors are not randomly distributed, but instead are structured into multi-byte error bursts, with bytes containing multiple bit errors. These types of error bursts are poorly detected by CRC. Thus, the probability of a false packet in this case can be estimated as same as in the case of a frame with random contents:

$$P_{\text{FP\_B\_Err}} = 2^{-M} \times P_{\text{eb}} \qquad (13.7)$$

where
M is the number of bits in the FCS field (16 or 32)
P_{eb} is the probability of the error burst

For the popular FEC of Reed–Solomon type RS(255, 239) with 8-byte error correction capability (see chapter 9 of [Golden 2006]), the most probable number of errored bytes per codeword is 9, and the average number of bit errors in the burst is approximated as $9 \times 4 = 36$. Thus, P_{eb} can be estimated as $\sim p \times 36$.* Other FEC types will result in slightly different values of P_{eb}, which depend on their error correction capability. Trellis coding (see chapter 8 of [Golden 2006]) also causes burst errors, and thus the same behavior is expected.

Similarly, estimation using a frame with random contents can be used in the case of false frame delineation, which may happen due to an error in the bytes indicating either the start or the end of the frame. The probability P_{FP-FL} of a false packet owing to loss of frame delineation can be found using Equation 13.7 if the probability of an error burst P_{eb} is replaced by the probability of a loss of frame delineation P_{ff}.

13.3.4.1   False Packet Probability with GPTM-TC$_{HDLC}$

The probability of a false packet can be found as a combination of probabilities of a false packet owing to FCS misdetection (P_{FP_Err}) and that owing to a false HDLC frame (P_{FP_ff}):

$$P_{FP} = P_{FP\_Err} + P_{FP\_ff} \tag{13.8}$$

In the GPTM-TC$_{HDLC}$, a false frame may occur when errors create flag-like octets inside the data field of the HDLC frame, or corrupt the opening or closing flags. The second event is much more probable, because a single bit error is enough to cause it. Approximating the probability of a corrupted flag is

$$P_{ff} = \sum_{k=1}^{8} C_8^k \cdot p^k \cdot (1-p)^{8-k} \approx 8 \cdot p$$

The probability of a false frame (losing a frame) is estimated as

$$P_{FP\_ff} = 2 P_{ff} \cdot 2^{-16} = p \cdot 2^{-12} \tag{13.9}$$

Substituting Equations 13.6 and 13.9 into Equation 13.8 yields the probability of a false packet for a DSL link with randomly distributed errors:

---

* It is assumed for simplicity that the Reed–Solomon decoder bypasses the codeword if more than eight errored bytes are detected.

$$P_{FP} = P_{FP\_R\_Err} + P_{FP\_ff}$$

$$= \sum_{m=E+1}^{L \times 8} C_{L \times 8}^{k} \cdot p^m \cdot (1-p)^{L \times 8 - m} + p \cdot 2^{-12}$$

$$\approx p \cdot 2^{-12}$$

$$\approx 2.44 \ p \times 10^{-4}$$

If the DSL link uses FEC, the probability of a false packet is

$$P_{FP} = P_{FP\_B\_Err} + P_{FP\_ff}$$

$$= P_{eb} \cdot 2^{-16} + p \cdot 2^{-13}$$

$$\approx p \cdot 32 \cdot 2^{-16} + p \cdot 2^{-12}$$

$$\approx 6p \cdot 2^{-13}$$

$$\approx 7.32 \cdot p \times 10^{-4}$$

DSL systems are typically specified to have a BER no greater than 10^{-7} (i.e., $p \leq 10^{-7}$). This results in $P_{FP} \leq 7.32 \times 10^{-11}$.

Assuming the MTTFPA is at its lower bound of 10^6, the maximum number of packets K that can be transported per year is

$$K = \frac{1}{10^6 \times 7.32 \times 10^{-11}} = 13661 \text{ packets/year.}$$

If the average packet length is assumed to be 750 bytes, then the maximum average rate at which bits can be transmitted over the link is 2.6 bits/s. Clearly, such a link is not a DSL. This result shows that the BER of a link operating at normal DSL speeds (typically, at least 1 Mbit/s aggregate in the upstream and downstream directions) is too high to enable the required MTTFPA values without additional error detection encoding included in the PTM entity and higher protocol layers. Fortunately, Ethernet frames contain an additional 32-bit CRC, which allows compliance with the MTTFPA requirements when using a GPTM-TC$_{HDLC}$.

13.3.4.2 False Packet Probability with ETM-TC and GPTM-TC

In the case of 64B/65B, a false packet occurs if a corrupted S-byte, or a Z-byte corrupted into an S-byte, occurs in a "Start of packet" codeword. This condition requires at least two bit errors. A similar effect occurs if a C-byte in an "End of packet" frame is corrupted. This can be caused by a single bit error. Because a single bit error is sufficient to corrupt the packet, the probability of a false packet is close to (but slightly lower than) the same value (P_{FP_ff}) for the GPTM-TC$_{HDLC}$ if a 16-bit FCS is used. The same conclusion is valid regarding the false packets occurring due to errors inside the packet (P_{FP_Err}). Therefore, the expected MTTFPA provided by the ETM-TC and GPTM-TC are only slightly higher than the values provided by the GPTM-TC$_{HDLC}$ and presented in the previous section. Using a 32-bit FCS, further reduces the probability of false packets by 2^{16}. Similarly, to the case of the GPTM-TC$_{HDLC}$, with the ETM-TC and GPTM-TC, the 32-bit CRC of Ethernet frames enables the MTTFPA requirements to be met.

Bibliography

[Baicheva 2000] T. Baicheva, S. Dodunekov, and P. Kazakov. *Undetected Error Probability Performance of Cyclic Redundancy-Check Codes of 16-bit Redundancy. IEEE Transactions on Communications*, Vol. 147, No. 253–256, October 2000.

[Castagnoli 1993] G. Castagnoli, S. Braeuer, and M. Herrman. *Optimization of Cyclic Redundancy-Check Codes with 24 and 32 Parity Bits. IEEE Transactions on Communications*, Vol. 41, No. 6, June 1993.

[Golden 2006] P. Golden, H. Dedieu, and K.S. Jacobsen, Eds., *Fundamentals of DSL Technology.* Auerbach Publications, Boca Raton, Florida, 2006, p. 457.

[ITU-T I.432.4] ITU-T Recommendation I.432.4. *Integrated Service Digital Network (ISDN). B-ISDN User-Network Interface—Physical Layer Specification.*

[ITU-T G.992.3] ITU-T Recommendation G.992.3 2006. *Asymmetric Digital Subscriber Line Transceivers 2 (ADSL2).*

[ISO/IEC 3309] ISO/IEC 3309 1993. Information technology, *Telecommunications and Information Exchange between Systems. High-Level Data Link Control (HDLC). Frame Structure.*

[IEEE 802.3] IEEE 802.3 2005. *Part 3: Carrier Sense Multiple Access with Collision Detection (CSMA/CD) Access Method and Physical Layer Specification: Amendment: Media Access Control Parameters, Physical Layers, and Management Parameters for Subscriber Access Network.*

[ITU-R G.993.1] ITU-R Recommendation G.993.1 2004. *Very-High-Speed Digital Subscriber Line Transceivers (VDSL).*

[ITU-R G.993.2] ITU-R Recommendation G.993.2 2006. *Very-High-Speed Digital Subscriber Line Transceivers 2 (VDSL2).*

[ITU-R G.114] ITU-T Recommendation G.114. *Generic Framing Procedure (GFP).*

Chapter 14

Voice over DSL: CVoDSL, VoATM, and VoIP

Ingo Volkening

CONTENTS

Abstract   We focus on techniques allowing to transport voice signals within the digital frequency band in such a way that they can still make use of the enormous investments of a network originally designed only for telephone services.

14.1 Introduction

The telecommunications industry is pursuing the provision of broadband connectivity with Digital Subscriber Line (DSL) technology, which allows the use of deployed copper pairs, rather than requiring the installation of fiber-optic cables everywhere. To use these pairs, it is vital to support the voice services these pairs already provide to the customer. An approach already discussed in previous chapters is to provide baseband voice: voice service carried as an analog signal in the lower-frequency band and separated, by a splitter, from the frequency band used for digital communications. Our focus in this chapter will be on techniques for transporting voice signals within the digital frequency band in such a way that they can still make use of the enormous investments of a network originally designed only for telephone services. We will denote the latter approach "Voice over DSL (VoDSL)."

The terms used in this chapter are as follows:

- Asymmetric DSL (ADSL): which applies to ADSL, ADSL2, and ADSL2plus
- Symmetric DSL (SDSL): which applies to SDSL and SHDSL (Symmetric High Bit-Rate DSL)
- Very High Bit-Rate DSL (VDSL): which applies to VDSL and VDSL2

14.2 VoDSL Reference Architecture

The reference architecture shown below in Figure 14.1 locates an Interworking function (IWF) at the customer's side of the T_B reference point. This IWF may be implemented as stand-alone equipment or it may be physically integrated into other equipment.

The customer premises distribution network can include switching or routing equipment to deliver voice and data signals to the customer

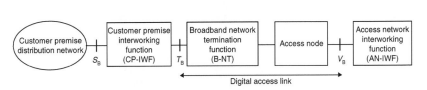

Figure 14.1 **Voice over Digital Subscriber Line (VoDSL) reference architecture.**

premises equipment (CPE). When both VoDSL and broadband data services are deployed, the switching and routing equipment within the customer premises distribution network perform the separation of these sessions, to their appropriate destinations. The customer premises interworking function (CP-IWF) performs the translation from signaling and bearer methods used by existing telephony equipment to the signaling and bearer methods described in this chapter. This can be a "null" function if the telephony equipment supports VoDSL services. The Broadband Network Termination (B-NT) performs the function of terminating the DSL signal entering the customer premises. For the different voice services such as Voice over ATM (VoATM), Voice-over-Internet Protocol (VoIP), and channelized voice, specific Transport Specific Transmission Convergence (TPS-TC) layer functions are defined. The access node (often referred to as the DSL access multiplexer (DSLAM)) provides concentration of multiple digital access links and concentration of bandwidth to the access network. The access network interworking function (AN-IWF) performs the complementing translation from signaling and bearer methods used by existing telephony equipment to the signaling and bearer methods described in this chapter. This may also be a null function when telephony equipment supports VoDSL services. The AN-IWF is not necessarily directly connected to the access node, as its location is specific to different VoDSL techniques, and is described in later sections of this chapter.

14.3 General Requirements for VoDSL

This chapter will give an overview what techniques for VoDSL exist and what kind of requirements have to be taken into account.

14.3.1 Approaches to VoDSL

Several approaches to VoDSL have been created, as illustrated in Figure 14.2.

As systems designed to carry high-speed digital communications, DSLs can carry any service that can be carried over IP or over Asyncronous Transfer Mode (ATM). Voice over Packet (VoP) technologies, which have been widely discussed in recent years, meet the challenge of harnessing

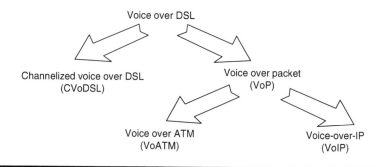

Figure 14.2   Relationship of VoDSL approaches.

legacy voice networks with packet networks by transporting both voice and signaling information over the packet network. This approach includes two specific approaches discussed in this chapter: VoIP and VoATM. VoP applications require real-time software and hardware modules that can be dynamically configured to provide flexibility and scalability in communications systems.

Under the VoP approach, the DSLAM will send its voice traffic over its packet-network interface; to interwork with the public switched telephone network (PSTN), the packet network must have a voice gateway to interface with a plain old telephone service (POTS) circuit switch, and to convert the stream of packets back into a continuous signal. The resulting complexity can be avoided by using a different approach: the channelized voice over DSL (CVoDSL) technique is to provide a channel to carry the voice bitstream unbroken from the CPE to the circuit switch. CVoDSL can be provided by next-generation DSLAMs with pulse code modulation (PCM) voice capabilities in addition to the current ATM functionality; these are sometimes called multi-service access platforms (MSAPs). Advantages of this approach include much lower latency, reduced cost of the CPE and central office (CO) hardware, and simplified provisioning and maintenance requirements.

14.3.2   Voice Quality

The overall goal of specific requirements placed on VoDSL capabilities must be set to ensure that voice connections provided over DSL are indistinguishable by the user from those provided by a conventional wireline connection to the PSTN. Maximum delay and echo cancelation requirements must be supported according to ITU-T Recommendations G.114 and G.131. Delay allocation objectives for national and international voice connections have been specified in Recommendations G.165 and G.168. The end-to-end one-way delay for an international connection must

Table 14.1   Mean Opinion Score (MOS) Used for Speech Quality Assessment

Rating	Speech Quality	Level of Distortion
5	Excellent	Imperceptible
4	Good	Just perceptible, not annoying
3	Fair	Perceptible, slightly annoying
2	Poor	Annoying but not objectionable
1	Unsatisfactory	Very annoying, objectionable

be less than 150 ms to be acceptable for most user applications. A one-way processing time of no more than 50 ms in each of the national systems and for the international chain of circuits is further recommended. This allocation includes the delay introduced in the CP-IWF, DSL network, AN-IWF, and the transport network between the call originator and the destination. It should be noted that the delay in the DSL network itself can be significant. For example, with ADSL (G.992.1 and G.992.2), it may be appropriate to either set the interleaver depth to minimum, or to avoid the interleaver entirely by using the "fast path" for the voice services. Also, when the ADSL payload is subdivided (e.g., using dual-latency or reserved channels), delay will be increased because the effective line rate for cell insertion will be reduced.

As a reasonable way to characterize the speech quality, the "mean opinion score (MOS)" is used. This rating distinguishes between different speech quality levels. The rating is defined in Table 14.1.

This rating is used in the industry to rate the speech quality, for example, a toll quality is a MOS of 4. Along with MOS, delay and delay variation, echo cancelation, background noise, and silence suppression are other metrics used to compare the quality of voice.

14.3.3   Clock Requirements on Voice Services

For voice services, three different techniques are used to ensure synchronization of the sender and receiving side of the voice call. These techniques are as follows:

1. Network timing reference
2. Deduce timing by deriving information out of the arriving cells or frames
3. None (free running clock)

All of the mentioned clock methods along with the transport protocol have different effects on the voice quality. To address this, the available clock methods are described by DSL voice protocol.

1. "Network timing reference" is the technique to ensure voice quality for VoDSL where the network timing is derived from the receiving data frame (ADSL or VDSL or SDSL plesiochronous mode: frame data rate + specific marker bits; SDSL synchronous mode: derived from the frame). At the DSLAM, the timing signal, which is synchronized to a primary reference source, is inserted to the transmit channel of the DSL transceiver. At the CPE, the local clock system is synchronized with the derived reference clock. For VoATM using the ATM Adaptation Layer 1 (AAL1), a specific method called Synchronous Residual Time Stamp (SRTS) uses the residual time stamp to measure and convey information about the frequency difference between a common reference clock derived from the network and a service clock. The same derived network clock is assumed to be available at both the transmitter and the receiver. If the common network reference clock is unavailable (i.e. when working between different networks which are not synchronized), then the asynchronous clock recovery method will be in a mode of operation associated with "Plesiochronous network operation." A detailed method of dealing with plesiochronous operation with AAL1 is standardized but subject to interpretation.

2. "Adaptive clock method" is a technique which derives the sender clock from the received data cells or frames. For all DSL technologies, supporting VoATM, this technique can be implemented on the receiving side. Cells are received and stored in a sample buffer. With respect to the predefined buffer depth, the voice samples contained in the ATM cells are played out at a certain clock. The output rate is adjusted to the mean buffer filling level. This ensures to prevent buffer under-run and overflow situations. A similar approach can be used for VoIP over DSL. For the sample buffer is built with the real-time packets (RTP). These packets have a time stamp which is used to synchronize the play-out of voice samples.

3. A "free running clock" DSL system supports the voice service without a special clock adaptation. These systems have the worst voice quality compared to all others.

14.3.4 Echo Cancelation

As described in the previous chapter, the subscribers use speech quality as the benchmark for assessing the overall quality in a network. Echo cancelation is a very important part of the voice service provided by the DSL technology.

The different types of echo are acoustic echo and hybrid echo. Acoustic echo is generated by the vocoders (voice-compressing encoding or decoding devices). Vocoders are typically used for VoP applications. Complex algorithm procedures are used to compute speech models. This involves generating the sum from reflected echoes of the original speech, then subtracting this from any signal the microphone picks up. The result is the purified speech of the person talking. The format of this echo prediction must be learned by the echo canceler in a process known as adaptation. The parameters learned from the adaptation process generate the prediction of the signal, which then forms an audio picture of the room in which the microphone is located. Other important performance criteria involve the acoustic echo canceler's ability to handle acoustic tail circuit delay. This is the time span of the acoustic picture roughly represents the delay in time for the last significant echo to arrive at the microphone.

Hybrid echo (known as line echo cancelation [LEC]) is originated from the A/D converter. From the hybrid circuit and the impedance mismatch in comparison with the connected phone, there is a line echo in the transmit path. Because echo causes diminished voice quality in packet networks it has to be canceled. This removing of the echo is done as the first instance in the upstream path by a LEC unit.

As a rough guideline, it can be stated that echo below 20 ms are typically not canceled. Values above are implemented in conjunction with the codec device and application. The following order gives an idea for what voDSL system the echo cancelation requirements are necessary:

1. Lowest: Channelized voice over DSL
2. Medium: Voice over ATM
3. Highest: Voice over IP

The above mentioned rating is based on the following delay relevant parts.

14.3.5 Encoding Delay

Analog voice input to the POTS port at the CPE-IWF is converted to digital and then encoded as a serial voice stream whose data rate depends on the encoding format. Examples for voice encoding delay are given in Table 14.2.

Table 14.2 Encoding Delay

Encoding Method (kbit/s)	Encoding Delay (ms)
G.711 PCM 64	0.75
G.726 ADPCM 32	1
G.728 LD-CELP 16	2

Table 14.3   Packetization Delay

Encoding Method (kbit/s)	Packet Size and Packetization Time (bytes)			
	20	36	40	44
G.711 PCM 64	2.5 ms	4.5 ms	5 ms	5.5 ms
G.726 ADPCM 32	5 ms	9 ms	10 ms	11 ms
G.728 LD-CELP 16	10 ms	18 ms	20 ms	22 ms

14.3.6   Packetization Delay

Another delay contributor is the packetization delay, which is present for VoP methods. The time taken to accumulate voice to fill a packet (packetization delay) is the first major element of transmission delay in the voice path. Examples for packetization delay are shown in Table 14.3.

14.3.7   Mapping Delay

The voice packet is then mapped to the underlying link layer, which is also only present for VoP methods. The delay introduced by this depends on the nature of this mapping.

Mappings that may be seen in VoDSL systems for VoATM include different delays. AAL1 and AAL2 mapping involve one AAL2 packet occupying the entire payload of a single ATM cell. This mapping involves negligible incremental delay. More delay is added, where sub-cell multiplexed AAL2 packets are packed into ATM cell payloads, with no fixed relationship between packet boundaries and cell boundaries. The delay involved in this mapping depends upon the value chosen for the "combined use" timer in the AAL2 common part process defined in I.363.2. A typical value of this timer is 1 ms.

In VoIP systems, the equivalent is IP trunking mapping, where multiple RTP are packed into a single IP packet payload. The delay involved in this mapping depends upon the degree of synchronization between the RTP generation processes on each voice channel. In addition to this, it must be distinguished between VoIP over ATM, where further processing is required for lower layers of the protocol stack. IP packets are typically carried over PPPoA (Point-to-Point Protocol over ATM) AAL5, which will be described in Section 14.7.3. An alternative to using this type of processing delay is using the Ethernet in the First Mile Transmission Convergence (EFM TC) for VoIP, which will be described in Section 14.7.4.

14.3.8   DSL Link Queuing

The DSL link queuing consists of the components sharing of the DSL link between voice and data and queuing of voice packets to be sent over the DSL link. Sharing of the DSL link heavily depends on this method. For CVoDSL, there is no influence, because voice will be sent over different frame payload bearers. VoATM takes care for quality of service (QoS) at the ATM layer. The ATM cell size of 53 bytes ensures that the delay for a voice cell is low. In VoIP applications, there are again two options. Option 1 is transmission over ATM via AAL5 and Option 2 is the Ethernet over DSL TC layer. Option 1 does not differ from VoATM except for the optional traffic QoS on IP/Ethernet level. Option 2 has been preferred because it uses QoS on IP/Ethernet level, whereby the link queuing may become significant. The reason for this is, for example, that the upper Ethernet frame size of 1512 bytes. If a data frame of this size is transmitted, the queuing must ensure that the blocking of voice packets is at a minimum.

14.3.9   DSL Transmission Delay

DSL transmission delay depends on the DSL technology and the choice of latency path, if applicable. In ADSL, two latency paths are available, fast and interleaved path. The lower bound of delay for ADSL encoding that is valid for fast path is 4 ms. The upper bound using interleaved path can be as much as 263.75 ms. The network itself adds a transmission delay which must be taken into account. At the receiving side, a voice decoding might be necessary as well. It also adds a delay similar to the encoding delay. It is also necessary for VoP methods to implement a de-jittering buffer. The buildup that is required in the jitter buffer is a function of the variability of delay in the entire voice path, for example, variation caused by DSL link queuing and transit through the packet network.

To summarize the mentioned factors influencing the voice quality, it can be stated that as more contributing factors for delay are added, there is less voice quality. Therefore, the ranking that takes only delay factors into account leads is as follows:

1. Lowest: Channelized Voice over DSL
2. Medium: Voice over ATM
3. Highest: Voice over IP

14.4   Lifeline Service

Traditionally, voice services have been powered from the CO and will work even during a local power failure. However, in a DSL system, power failure to the B-NT diminished most of the capabilities of the system.

If the voice services are provided using baseband voice, the situation is exactly as it would be for traditional POTS: because the analog service

534 ■ *Implementation and Applications of DSL Technology*

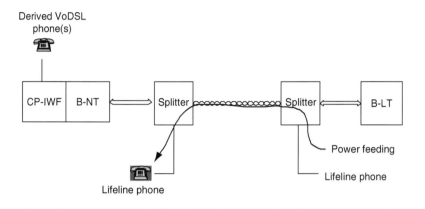

Figure 14.3 **Lifeline phone service for splitter-based Digital Subscriber Line (DSL).**

is always powered by the network side, it will continue working without power to the B-NT. This approach, illustrated in Figure 14.3, is applicable to ADSL and VDSL. The derived VoDSL services will be inoperative, but the lifeline phone will continue to work.

The lifeline phone provided over the splitter is totally independent from the DSL service. For SDSL, the use of baseband voice is not standardized, because the current architecture is "all-digital" and does not support a separate low-frequency band. The new ADSL standard, ADSL2, also supports an "all-digital loop" mode, which has the same issue as SDSL does. However, as described in the ETSI (European Telecommunications Standards Institute) specification of SDSL, lifeline POTS can be provided by using the CVoDSL approach with the POTS-TC layer or the Integrated Services Digital Network (ISDN)-TC layer. This approach is illustrated by Figure 14.4.

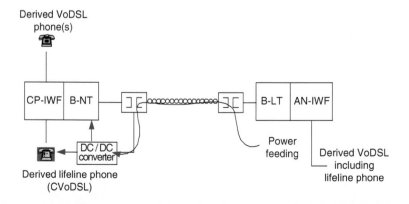

Figure 14.4 **Lifeline phone service for DSL operating in all-digital loop mode.**

For an all-digital loop, the lifeline phone service must be derived from the DSL frequency band. For SDSL, a maximum of 2.1 W of DC (direct current) feeding power is provided. This power is inserted at the hybrid of the access node and extracted at the hybrid of the B-NT. A DC/DC converter provides the lifeline phone, the B-NT, and the CP-IWF with power. Because of the relative simplicity of CVoDSL functionality, the power consumption of the B-NT and CP-IWF can stay within the power budget requirements. More details on the CVoDSL functions performed in the B-NT and CP-IWF are described in Section 14.5.

Other approaches for providing lifeline service in combination with VoATM or VoIP are in general supported by DSL.

14.5   Channelized Voice over DSL

CVoDSL technology has been defined for derived voice functionality in ADSL and SDSL. The CVoDSL approach transports voice without packetization, so the processing needs are reduced while the QoS characteristics are improved. The access network architecture of existing POTS and ISDN at the CO is used for this technique.

14.5.1   CVoDSL Network Architecture

For the CVoDSL network, it is required that the DSL CPE (B-NT and CP-IWF) and the DSLAM have the capability to switch the voice information and perform interworking for the basic POTS control functions.

The DSL CPE supports POTS and ISDN phones derived from the DSL band. At the DSL link, the voice is transported in parallel with the data packets, but in a separate part of the DSL payload. With this approach, a high QoS is ensured, because the voice transmission is not influenced by the data traffic. The DSLAM requires two interfaces at the V_B reference point

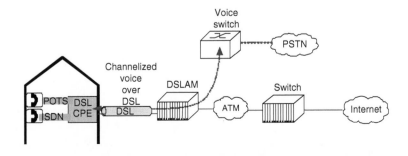

Figure 14.5   **Channelized voice over DSL (CVDSL) network architecture.**

(see Figure 14.5): one for the broadband data and one for the connection to the voice switch. The broadband data is transported to the ATM network, whereas the voice signal is already in the standardized format (GR-303, TR-08 or V5.x) for delivery to a class 5 voice switch. The class 5 switch is a telephony switch providing dial tone, call routing, and all other services required for connection to the PSTN.

14.5.2 CVoDSL Functionality and Protocols

From the functional point-of-view, the CVoDSL is targeted to providing multiple phone ports at the customer side by using the existing infrastructure of today's voice networks. The DSL CPE does not require additional processing functions, such as echo cancelation. At both sides of the DSL (DSL CPE and DSLAM), the voice samples are mapped into or extracted from the DSL transceiver payload, as specified for ADSLs in ITU-T G.9xx and for S(H)DSL in the relevant ITU-T and ETSI specifications. ADSL2 supports voice coding as specified in G.711 and G.726, while SDSL supports only G.711 encoding. Signaling and control functions and protocols specific to the type of DSL are described in the following sections.

14.5.3 ADSL-Specific Aspects

ADSL offers a POTS TC in which the voice samples (G.711) are either directly mapped into the ADSL frame or the voice is compressed to G.726 (Adaptive Differential PCM (ADPCM)) before the samples are mapped into the frame.

As shown in Figure 14.6, the methods for channelized voice in ADSL support several POTS phone ports at the customer premises. The codec and SLC (Subscriber Line Circuit) codes the analog voice to G.711 compliant format. Optionally, the voice samples can be compressed to G.726 format. The analog signaling is converted to network-specific required format. This format can differ from countries and is therefore described in this handbook.

ADSL PMS-TC (Physical Media Specific-Transmission Convergence) frame transports two bytes of data per active PCM channel (i.e., 64 kbps) and one byte of signaling information. If there are N PCM byte streams $P(i,j), i = 0 \cdots N-1$ $(N \leq 7)$, and j is the sample index, then the data bytes are ordered as $C(i + Nj)$, as shown in Figure 14.7.

One STM-TC bearer channel is used to carry both the PCM voice samples and the signaling information, which is multiplexed as shown in Figure 14.8. The signaling byte is appended to the end of the STM frame.

The bandwidth requirement for the Synchronous Transport Mode-Transmission Convergence (STM-TC) bearer is $(2N + 1) \times 32$ kbps. The

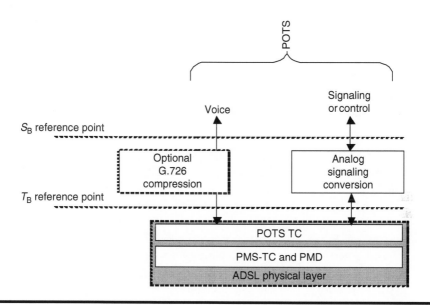

Figure 14.6   CVoDSL over Asymmetric DSL (ADSL) protocols (CPE (customer premises equipment) side).

bytes are transported from left to right. In each PCM byte as well as the signaling channel, the MSB shall be transmitted first.

The mapping of the CVoADSL is performed over the fast path. This path is chosen to ensure the lowest possible transmission delay for this service.

14.5.4   SDSL-Specific Aspects

SDSL offers three kinds of channelized voice services. The defined TPS-TC functions are the POTS TC (which is similar to the ADSL POTS TC), the ISDN TC, and the LAPV5 (Link Access Protocol Version 5) TC.

The methods shown in Figure 14.9 require a different SDSL frame structure for the voice transport of POTS and ISDN. The PMS-TC sublayer function is special for SDSL. SDSL offers a frame structure with payload subblocks operating at 125 μs. This is one-to-one equivalent with the frame sync clock (FSC) used at the S_B reference point.

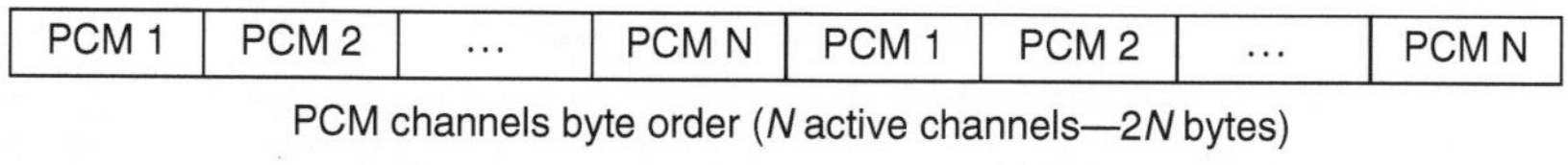

PCM 1	PCM 2	...	PCM N	PCM 1	PCM 2	...	PCM N

PCM channels byte order (*N* active channels—2*N* bytes)

Figure 14.7   CVoADSL pulse-code modulation (PCM) byte ordering.

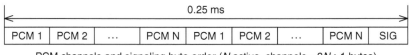

PCM channels and signaling byte order (*N* active channels—2*N*+ 1 bytes)

Figure 14.8   CVoADSL frame structure with PCM and signaling.

The POTS TC carries voice data, encoded per G.711, over a B-Channel which carries an 8-bit voice sample every 125 μs for the POTS phone. The signaling and control information is translated by the CP-IWF. The content of this information depends on the signaling protocol used: for example, the GR-303 specification uses channel-associated signaling (CAS). The frame map of the POTS TC is illustrated in Figure 14.10.

The voice samples are transparently forwarded by the CP-IWF from the S_B to the T_B reference point. Voice samples from the CVoDSL phone(s) are mapped every 125 μs to the respective payload block B-channel of the SDSL frame. The payload blocks and the CVoDSL service must be synchronized to a common clock source.

The signaling channel contains coded signaling or control information. This signaling information could be country specific and therefore it cannot be generalized.

The ISDN TC offers the basic characteristics to transport the ISDN data, which are as follows:

1. B-channels and D-channels are mapped onto SDSL payload channels every 125 μs.

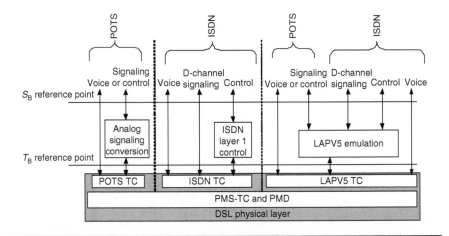

Figure 14.9   CVoDSL over Symmetric DSL (SDSL) protocols (CPE side).

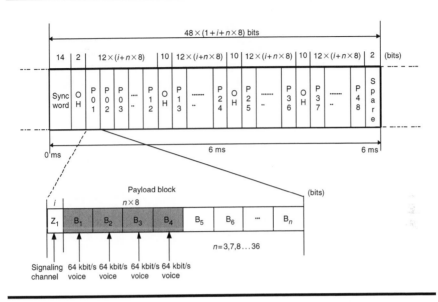

Figure 14.10   Mapping of the SDSL frame for channelized POTS.

2. The ISDN basic rate access (BRA) does not need a separate synchronization because the SDSL frames are synchronized to the same clock domain as the ISDN BRA. Therefore, the ISDN frame word is not needed, saving 12 kbits/s.

3. The ISDN M-channel is used for transporting ISDN line-status bits, transmission control information, as well as signaling to control the ISDN connection. Only the ISDN M-channel functions, which are needed to control the interface to the ISDN terminal equipment, are transported over a messaging channel (SDSL embedded overhead channel (eoc) or fast signaling channel).

The ISDN B- and D-channels are transported within the SDSL payload subblocks. The SDSL payload data is structured within the SDSL frames as illustrated in Figure 14.11.

The ISDN D-channel(s) are transparently forwarded by the CP-IWF from the S_B to the T_B reference point. Voice samples from the ISDN terminal(s) have to be mapped every 125 μs to the respective payload block B-channel of the SDSL frame. With respect to the IOM-2 (ISDN Oriented Modular) interface (where B-channels are transmitted before the D-channel) the ordering in the SDSL frame (where the D-channels are mapped first in the SDSL payload block) where the B-channels bits should be mapped in the Nth payload block and D-channel bits in the $(N+1)$th payload block.

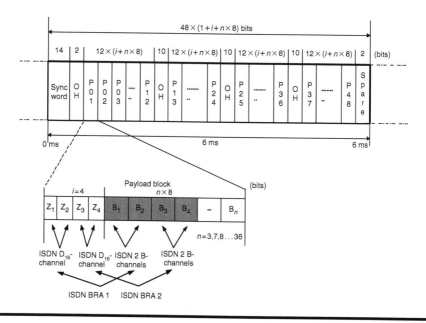

Figure 14.11 **Mapping of the SDSL frame for channelized Integrated Services Digital Network (ISDN).**

The signaling channel used for the ISDN M-channel is either the SDSL standard eoc channel (3.3 kbit/s) or the optional fast eoc channel. The optional 8 kbit/s fast signaling channel is always conveyed in the first z-bit of the SDSL payload block. If this fast signaling channel is used, up to 6 ISDN BRA can be transported over SDSL. To avoid unnecessary shifting of ISDN D- and B-bits, the respective D-bits are transmitted after their B-bits in the subsequent SDSL payload block. The ISDN S-buses, which connect the ISDN terminals with the NT, can be controlled independently with the defined ISDN message codes in SDSL. The main functions of these messages are as follows:

- Initiate activation form the LT side
- Initiate activation from the NT side
- Initiate deactivation from the LT side
- Initiate loopback functions from the LT side
- Allow control of the NT ISDN finite state machine from the LT side

The LAPV5 TC is based on mapping and time slot allocation of CVoDSL based, LAPV5 enveloped POTS or ISDN transport, which is for ISDN an alternative procedure to the simple use of D-channel messages as described in the ISDN TC (see Figure 14.12). Either CVoDSL based POTS or ISDN transport is possible at a time.

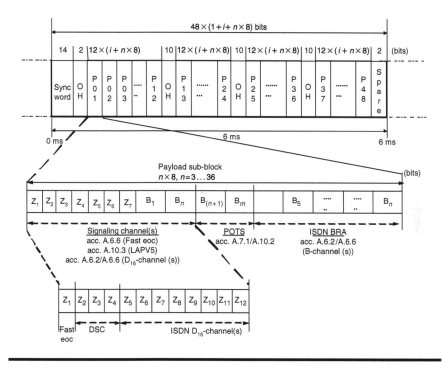

Figure 14.12   Example for channelized LAPV5 enveloped POTS and ISDN.

The existence of available signaling channels is negotiated at the SDSL start-up via handshake procedure.

The protocol architecture in Table 14.4 covers the LAPV5 enveloped simultaneous support for POTS and ISDN.

The LAPV5-EF (Envelope Function) address envelopes the frames for signaling of an individual ISDN access, or for POTS signaling or for POTS/ISDN port control.

Table 14.4   LAPV5 Protocol Architecture

POTS Signaling	POTS/ISDN Port Control	
EN 300 324-1, Clause 13	**EN 300 324-1, Clause 14**	**ISDN Signaling**
LAPV5-DL		LAPD
EN 300 324-1, clause 10		
LAPV5-EF Address		
EN 300 324-1, clause 9		
TPS-TC		
PMD-TC, PMS-TC		

For the reliable transport of POTS signaling and POTS/ISDN port control messages, the data link protocol LAPV5-DL (Data Link) is used which is a simplified version of LAPD (Link Access Protocol D). The LAPV5-DL protocol means that only one common instance of LAPV5-DL is used for both the POTS signaling and the POTS/ISDN port control, and that the LAPV5-DL address (ETSI EN 300 324-1, clause 10.3.2.3) takes the value of all zeros. POTS signaling messages and POTS/ISDN port control messages are distinguished by means of the message type information element (ETSI EN 300 324-1, clause 13.4.4). Operation of ISDN layer 2 links is defined by LAPD.

14.5.5   Differences Relevant to VDSL

VDSL does currently not provide a channelized voice path.

14.6   VoATM over DSL

The transport of voice services, voice-band data, and fax traffic over a broadband subscriber line connection (including ADSL, SDSL, and VDSL) between the customer premises and the service provider's switched telephone network is realized with the Broadband Loop Emulation Service (BLES). BLES is a subset of functions defined by the DSL Forum, based on the ATM Forum's Loop Emulation Service (LES) specification. BLES includes support for the VoDSL reference model for functional blocks and interfaces as defined in the reference model (see Figure 14.13). In addition, BLES includes support for the VoDSL-IWF reference model for both POTS and ISDN delivery methods, and the services supported by a service provider class 5 switch and CPE. BLES networks are expected to operate in

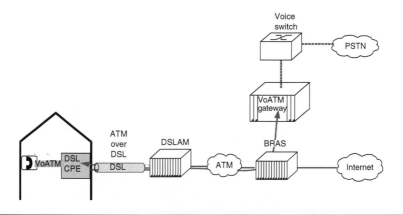

Figure 14.13   Voice over ATM (VoATM) over DSL network architecture.

environments that may inter-mix voice channels with data services at the DSL interface. BLES network topologies rely on IWF located in the access network and at the customer premises to derive narrowband services. A DSL interface provides a broadband access between the customer premises and a network capable of supporting these derived narrowband services. BLES includes the support for compressed voice and non-compressed voice together with or without silence removal. This section includes details of the architecture, a list of requirements, and recommended protocols and interfaces based on (where possible) existing standards that are specific to BLES.

14.6.1 VoATM Network Architecture

The Figure 14.13 illustrates the VoATM based network architecture and its elements.

The DSL CPE must provide functions to packetize voice samples in ATM cells, as well as to provide emulation functions for the call signaling. At the DSL link, the ATM layer used for data service is also used for voice service. The voice information transported over a dedicated ATM VC (virtual channel). In addition, the ATM layer provides a highly sophisticated QoS called traffic management. This service ensures a predefined priority for the voice cells. The cells are transported by the DSLAM and routed through the ATM network. At the destination of the ATM VC is a VoATM gateway. This gateway terminates the ATM VC and voice cells are de-packetized and converted to standard based format (GR-303, TR-08 or V5.x) for delivery to Class 5 voice switch. The Class 5 switch is a telephony switch providing dial tone, call routing, and other services. Class 5 switches are connected to the PSTN.

14.6.2 VoATM Functionality and Protocols

The VoATM protocol AAL2 makes use of features like compression, voice activity detection (VAD), and comfort noise generation (CNG). Along with this comes the requirement to implement an echo cancelation.

The functions mentioned in Figure 14.14 are characterized by the following capabilities.

The ATM layer is specified by the ITU-T I.356, I.361, and I.432. ATM includes other functions like Operation and Maintenance (OAM) (ITU-T I.610), traffic management (ATM Forum af-tm-0056.000), management (ILMI (af-ilmi-0065.000)), and signaling (ATM Forum af-sig-0061.000). All these functions are also required for the data service over DSL and therefore not discussed in this chapter.

For the above mentioned ATM layer function, the traffic management at the CPE side is performed in direction from CPE to network called traffic shaping.

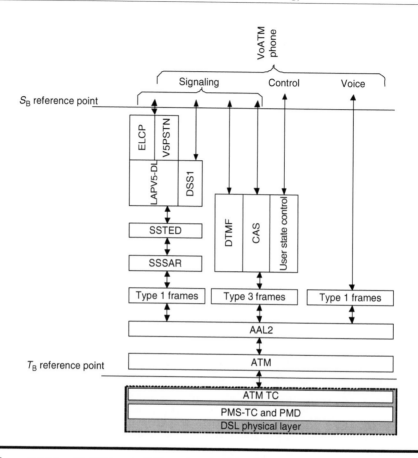

Figure 14.14   VoATM over DSL protocols (CPE side).

For the AAL2 functionality, the ATM Forum af-vmoa-0145.000 specification LES using AAL2 was adapted by the DSL Forum (TR-039: requirements for Voice over DSL access facilities to BLES and reduced to general requirements. Coming to the AAL2 protocol, it is divided into the common part sublayer (CPS) and the service specific convergence sublayer (SSCS) functions. The CPS function takes care of encapsulating the AAL2 mini-cells in ATM cells. Two options can be used for this function, CPS-lite which means that multiplexing of type 1 packets from several channels is not supported. Unused bytes of these cells are filled with "padding bytes." The CPS function offers support for several channels multiplexed in ATM cells. The minimum requirement is the support of the CPS-lite function.

The SSCS function is divided in type 1 and type 3 frames. Type 1 frames are used to transport voice samples, silence descriptors, and signaling or management information. The transport of voice samples is defined with several profiles covering G.711, G726 ADPCM, G279a, and others. The optional functions of AAL2 are voice compression which offers a significant reduction of required bandwidth in the access network but requires an echo compensation at both sides of the VoATM connection.

Silence descriptors are cells used to control the optional function of VAD and silence suppression. This cell includes the noise level of the sender.

Signaling and management related functions transported over type 1 frames are using two additional functions on top of it. The SSSAR (service specific segmentation and reassembly) function is responsible to segment frames larger than 44 bytes in pieces. At the receiving side, segmented frames can be distinguished with the UUI (User–User-Identification) field, which is a part of the AAL2 mini-cell. Above is the SSTED (service specific transmission error detection) function which is used to provide reliable transmission of the transported content. The transported content consists of call signaling or control for V5.x based networks and AAL2 management information. Only the call signaling or control functions are described in this chapter.

The DSS1 (Digital Subscriber Signaling System No.1) message relay function is used to tunnel the ISDN D-channel through the ATM network, as illustrated in Figure 14.15.

The ISDN message relay function gives an example of how VoATM works in terms of interworking. The similar concept is used for functions like ELCP (Emulated Loop Control Protocol) and PSTN (public switched telephone network) protocol. These parts are required to ensure complete interworking with existing V5.x networks. Functions transported over type 3 packets are used for GR-303 core networks. Parts of type 3 packets, namely DTMF (Dual-tone multi-frequency) and user state control, might also be in V5.x networks.

As conclusion, it should be noticed that VoATM has defined a complete protocol suite for voice services.

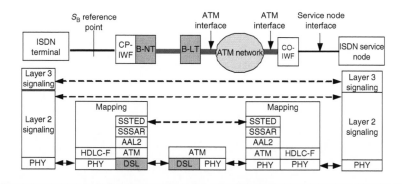

Figure 14.15 **DSS1 (Digital Subscriber Signaling System No.1) message relay function over ATM adapter layer 2 (AAL2) over DSL.**

14.6.3   Differences among ADSL, SDSL, and VDSL

Voice transport using ATM is possible for all DSL technologies. The only difference is the number of supported voice calls. For ADSL bandwidth, valid for ADSL and VDSL, the upstream bandwidth is the limitation. The symmetric bandwidth DSLs, SDSL and VDSL symmetric option, offer more voice calls if the upstream bandwidth is always higher than the ADSL upstream.

14.7   VoIP over DSL

VoIP is the most emerging voice application over DSL. Because of the only voice technology which allows merging the data access for Internet service with traditional voice service, it reduces the network infrastructure requirements. Key factors like sharing the infrastructure for broadband Internet access and reducing the maintenance cost are just some of them. From the DSL perspective, the required protocol layers and physical layer adaptation are already in place. DSL offers high bandwidth Internet data access which is based on IP.

14.7.1   VoIP Network Architecture

Figure 14.16 should illustrate the VoIP-based call connection transported over ATM network architecture and its elements.

The DSL CPE must provide functions to packetize voice samples into VoIP-frames and then into ATM cells. The call signaling information is

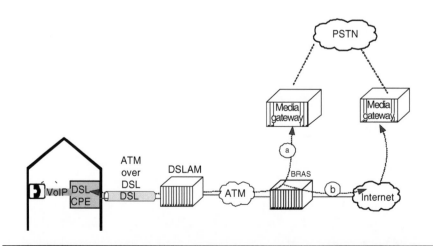

Figure 14.16   Voice-over-Internet Protocol (VoIP) using ATM over DSL network architecture.

translated into VoIP-based signaling and is also packetized into ATM cells. At the DSL link, the ATM layer used for data service is also used for voice service. The voice information is transported over a dedicated ATM VC or shared with other connections which are multiplexed at higher layer. In addition, the ATM layer provides a highly sophisticated QoS, called traffic management. This service allows ensuring a predefined priority for the voice cells, in the case of dedicated ATM VC. The cells are transported by the DSLAM and routed through the ATM network. At the BRAS (broadband remote access server), two paths for the VoIP packets are possible:

1. Destination of the VoIP packets is a media server which is part of the access network service provider's network. The BRAS will terminate the ATM VC and the VoIP connection is terminated by the media gateway. Voice cells are de-packetized and converted to standard based format (GR-303, TR-08 or V5.x). The Class 5 switch services like providing dial tone, call routing, and other services are part of the media gateway. The Media Gateway is connected to the PSTN.
2. VoIP packets are forwarded to the Internet, which means that the ATM connection must be terminated at the BRAS. From the Internet the VoIP connection is routed to the destination, which could be a media gateway. Voice cells are de-packetized and converted to standard based format (GR-303, TR-08 or V5.x) for delivery to the PSTN. Here the Class 5 voice switch functionality is part of the media gateway.

The IEEE 802.3ah Ethernet in the First Mile (EFM) task force has developed standards for Ethernet PHYs to be used on twisted pair copper wiring in the public access network (see Figure 14.17). The task force is basing these new PHYs on DSL transceivers, namely G.993.1 (VDSL1), G.993.2 (VDSL2), G.991.2 (SHDSL), and G.992.3/5 (ADSL2/2+). The DSL technology for short-reach EFM is VDSL. This leads to new network architecture for VoIP using Ethernet-based network architecture and its elements.

The DSL CPE must provide functions to packetize voice samples in VoIP frames and in addition to Ethernet frames. The call signaling information is translated into VoIP-based signaling and is also packetized into Ethernet frames. At the DSL link, the IEEE EFM TC layer is used for data service as well as for voice service. In this scenario, the IP must provide a QoS. This service must ensure a predefined priority for the voice cells. The EFM frames are transported by the DSLAM and routed through the Ethernet network. At the BRAS, two paths for the VoIP Ethernet frames are possible:

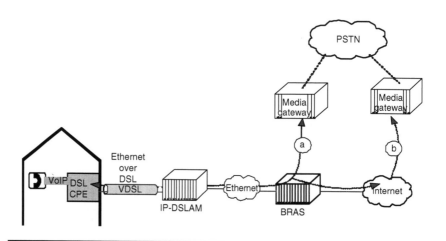

Figure 14.17   VoIP using IEEE (Institute of Electrical and Electronics Engineers) EFM (Ethernet in the First Mile) over DSL over Ethernet network architecture.

1. Destination of the Ethernet connection is a media gateway which belongs to the access network service provider. This gateway terminates the Ethernet protocol and the VoIP connection. Voice cells are de-packetized and converted to standard based format (GR-303, TR-08, or V5.x). The Class 5 switch function is part of the gateway, which is connected to the PSTN.
2. VoIP packets are forwarded to the Internet, which means that the media gateway belongs to an independent service provider. In this example, the VoIP connection is routed to the media gateway over the Internet. Voice cells are de-packetized and converted to standard based format (GR-303, TR-08, or V5.x) for delivery to the PSTN. Here the Class 5 voice switch functionality is part of the media gateway.

VoIP allows many other call scenarios, like calls to other DSL broadband users which avoid the connection to the PSTN.

14.7.2   VoIP Functionality and Protocols

The VoIPs make use of the IP, which is common for both voice and signaling or control information. At the higher-layer protocols, these streams are treated differently. The following example should describe the basics for VoIP.

The A/D converter works with at least 14-bit resolution. The 14-bit sample will be converted to μ-Law or A-Law resulting into 8-bit sample. μ-Law

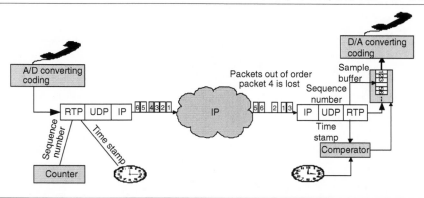

Figure 14.18   Example of real-time packets (RTP) application (one direction).

is used in North America and A-Law is used in Europe and Japan. The RTP are designed for real-time applications on IP. The most important features are the sequence number (SN) to allow the right re-assembly of particular packets, and the time stamp which allows synchronization and jitter calculation at the destination. The User Datagram Protocol (UDP) provides the primary mechanism that application programs use to send datagrams to other application programs. UDP provides protocol ports used to distinguish among multiple programs executing on a single machine. That is, in addition to the data sent, each UDP message contains both a destination port number and a source port number making it possible for the UDP software at the destination to deliver the message to the correct recipient and for the recipient to send a reply. The RTP/UDP encapsulated voice samples are then encapsulated in IP frames. These frames can be transmitted over DSL. Assuming that over the IP network the packets are out of order and one packet is lost (example in Figure 14.18), at the receiver side, the information of the RTP is used to re-order the packets and to insert noise for the lost packet. The signaling or control information is transported over the Transport Control Protocol (TCP) to achieve a reliable transport. The higher-layer application is typically H.323 or Session Initiation Protocol (SIP).

The protocol suite required for VoIP is in the case of H.323 based on existing voice network protocols. SIP is defined independent from the voice network protocols used. The choice of the protocol suite is independent of the DSL technology (see Figure 14.19).

14.7.3   Commonalities Relevant to All DSLs

With respect to the network architecture shown in Figure 14.16, the protocols required for VoIP using ATM as transport protocol is shown in Figure 14.20.

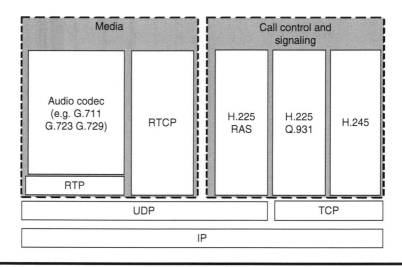

Figure 14.19   H.323-based VoIP application.

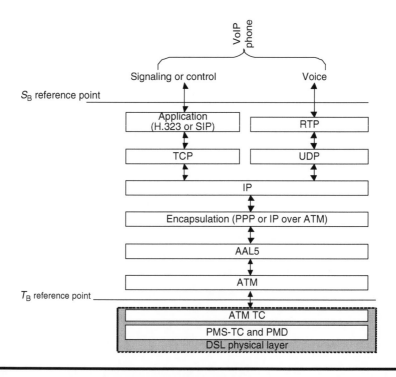

Figure 14.20   VoIP over DSL using ATM over DSL.

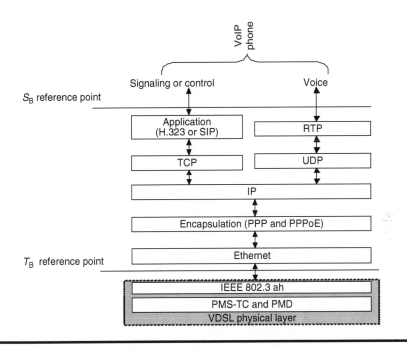

Figure 14.21   VoIP over Very High Bit-Rate DSL (VDSL) using IEEE Ethernet transmission convergence (TC).

As shown in Figures 14.20 and 14.21 , the transport of voice samples and signaling is processed over several protocol layers. For the transport over DSL, the ATM TC is used. The handling in this layer does not differ from the VoATM-based system. Main difference starts with the ATM adaptation layer. For VoIP, AAL5 is used, which is the same as for data services. This means, the ATM layer is typically configured for multiple VCCs (virtual channel connections), to ensure a higher QoS for the ATM cells containing VoIP samples. It should also be considered to implement a QoS on IP/Ethernet level to avoid unnecessary added delay at this level.

14.7.4   SHDSL- and VDSL-Specific Aspects

As already mentioned, for VDSL1 or 2 and SHDSL, it is possible to use another TPS-TC function. The IEEE 802.3 ah specifies how to transport the network protocol Ethernet over DSL. With this, the protocol architecture for VoIP below the IP protocol looks different to the one when ATM is used.

The amount of protocol layers in this approach is reduced compared to the VoIP over ATM-based solution. The packetization delay added by

Table 14.5   Comparison of Overhead Bandwidth Occupation

	Efficiency (percent)	
Ethernet Frame (Bytes)	**Ethernet over ATM (EoA) with LLC Encapsulation, no FCS**	**VDSL EFM-TC Mode (4-Byte Ethernet CRC, 2-Byte TC CRC)**
64	60.37	92.67
128	80.5	95.48
256	80.5	96.95
512	87.82	97.7
1024	87.82	98.08
1516	89.38	98.2

the ATM protocol is eliminated with this method. On the other hand, it requires a QoS support at IP/Ethernet level because of the not existing ATM QoS. What is more straight-forward is the reduced overhead bandwidth occupation. A comparison is given in Table 14.5.

14.7.5   Differences Relevant to ADSL

ADSL has currently not specified an EFM-TC layer function. ADSL2 and ADSL2plus have specified the use of EFM-TC under the definition PTM-TC (Packet transfer mode transmission convergence).

Chapter 15

Bonded DSL

J. Lane Moss and Matt Squire

CONTENTS

Abstract   Loop bonding is a method by which carriers can use two or more copper pairs to the customer to increase the throughput to an end user. Most traditional Digital Suscriber Line (DSL) services have been based on single pair technologies (e.g., residential asymmetric DSL [ADSL]) or two pair technologies (e.g., High Bit-Rate DSL4 [HDSL4]), and have been relatively rigid in their deployment options. Loop bonding allows multiple pairs to be used together in a very dynamic, flexible manner to achieve more bandwidth to the subscriber. These techniques are being applied for residential triple-play services as well as high-speed symmetric business services. Loop bonding applications have focused on supporting a particular type of payload, for example, ATM (Asynchronous Transfer Mode) or Ethernet, and offer very efficient mechanisms for using multiple pairs for bandwidth and resiliency improvements. This chapter discusses loop bonding technologies for supporting both ATM and Ethernet, and shows how these techniques are being used by carriers for a wide variety of applications.

15.1 Introduction

Because the bit rate that can be delivered by any Digital Subscriber Line (DSL) is limited by loop topology and crosstalk, a single copper pair may not provide adequate capacity for a desired service. In such cases, loop bonding can be used to combine the payloads of multiple DSLs into a single aggregate stream, which can be used to increase the rate at a given reach or the reach of a given rate. Because most DSL technologies are rate adaptive and can automatically connect at the maximum possible rate given the individual loop conditions, it is desirable that loop bonding solutions provide the capability to bond lines running at different data rates.

Using multiple pairs in a bonded application is not new. Inverse multiplexing over ATM (IMA), Multilink Point-to-Point Protocol (MLPPP), and 802.3ad link aggregation provide methods to enable the use of multiple lines in a bonded application. However, none of these solutions were specifically designed for operation over DSLs, and consequently each method has inherent drawbacks when used in the DSL environment.

To address limitations of the existing bonding technologies, ATIS NIPP-NAI and the ITU-T have standardized three distinct methods that can be used to bond DSLs running at different data rates:

1. ATM-based DSL bonding
2. Ethernet-based DSL bonding
3. Time-division inverse multiplexing (TDIM)

Each method is designed to pair with a particular transmission protocol specific transmission convergence layer (TPS-TC), i.e., Asynchronous Transfer Mode (ATM)-based bonding assumes that DSL transceivers use an ATM transmission convergence layer (ATM-TC, see chapter 17), and

Ethernet-based bonding assumes a packet transfer mode transmission convergence layer (PTM-TC) based on an encapsulation scheme known as 64/65B, which is used in the "Ethernet in the First Mile" (EFM) standard from the IEEE (see chapter 13). As such, the bonding functionality can reside outside the DSL transceivers and communicate over the defined TC layer.

To bond pair running at different data rates, four fundamental functions must be specified [DSL Anywhere]:

1. Segmentation—aggregate data stream is partitioned into fragments.
2. Framing—overhead bits used for fragment delineation.
3. Sequencing—a specific tag for each fragment used to reconstruct the aggregate stream.
4. Delineation—reception of the individual fragments.

In addition to these basic functions, the bonding standards also address operations and maintenance functions including initialization of the bonded group and removal of faulty pairs. This chapter discusses the ATM and Ethernet bonding standards in detail.

15.2 ATM-Based Bonding

Although operators have begun recently to convert to Internet Protocol (IP)-based networks, the majority of existing equipment is ATM-based. This is especially true for ADSL, because the DSL Forum Interoperability specifications TR-048 and TR-067 assume that ATM is used between the DSLAM and the customer premises equipment (CPE) [TR-048] and [TR-067]. Because ATM bonding only requires ATM to be used on the DSLs, ATM-bonding can be used even if other parts of the network are not ATM-based.

Loop bonding using ATM was first standardized by the ATM forum as IMA [ATM Forum AF-PHY]. This method forces each link to operate at the same nominal bit rate, and, by using a round-robin procedure to allocate cells to the various links in the group, explicit cell sequencing information is not required to reassemble the original stream. Also, special control cells, called ICP (IMA Control Protocol) cells, are used for operations, administration, and maintenance (OAM) messaging, and the timing of these ICP cells is defined so that the round-robin order is not disrupted in an IMA frame.

In 2004, ATIS committee NIPP-NAI (formerly T1E1.4) approved T1.427.01 [T1.427.01], which is a standard for ATM-layer bonding of pairs that may have disparate data rates.* In ATM, the segmentation, framing, and

---

* ITU-T Recommendation G.998.1 [G.998.1] is virtually identical to T1.427.01.

delineation functions are already defined in the ATM-TC, so only a sequencing function is needed to bond pairs with different rates. Each 53-byte ATM cell has a 5-byte header, and three of those bytes are used to identify the particular virtual path, virtual circuit (VP/VC) connection. Because only a very few of these are needed between the DSLAM and the CPE, the ATM-based bonding standards redefine the ATM header to use some of these bits to carry sequence information with no additional overhead. Because the existing cell format is reused in a way that is transparent to the ATM-TC, no additional framing protocol is necessary to maintain the other functions required for bonding.

15.2.1   Benefits of the New ATM-Based Bonding Approach

In addition to those benefits common to each of the DSL bonding standards, ATM-based bonding provides several additional advantages [DSL Anywhere]:

1. *Uses existing ATM infrastructure.* Allows reuse of existing ATM-based equipment.
2. *Works with existing DSL transceivers.* ATM-based bonding works with the standardized DSL ATM-TC. In contrast, IMA requires the ATM-TC not discard errored cells and not generate or terminate idle cells.
3. *Facilitates centralized implementation.* The sequence number (the SID, which is described below) provides the ability to reassemble the original cell stream even if the member links terminate on different line cards.
4. *Low bonding overhead.* ATM bonding uses existing bits in the ATM cell header to convey sequence information. The only overhead required for bonding is the status messages, which are nominally one cell per second per link.

15.2.2   Technical Overview

Figure 15.1 provides a high-level operational diagram of ATM-based bonding. The transmitting entity receives the aggregate ATM stream, inserts the sequence ID (SID) into each header, and distributes these modified cells to each of the bonded links in the group to be transported to the CPE modem. The receiving entity takes the individual streams from the member links and reconstructs the original stream from the SID values.

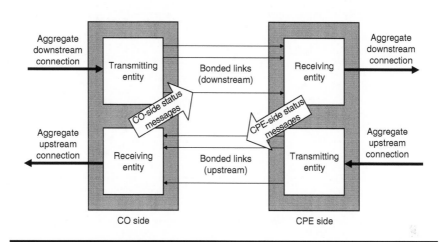

Figure 15.1   Operational diagram of Asynchronous Transfer Mode (ATM)-based bonding.

The central office (CO)-side and CPE-side status messages are used to communicate OAM information. During initialization, these messages are used for pair discovery to determine the member links of the group and determine which lines are capable of carrying bonded traffic. In data mode, the quality of each link is monitored and frequently communicated to the transmitter via this overhead channel. If the performance of a particular link deteriorates, these status messages are used to instruct the transmitter not to use this link.

15.2.2.1   Reference Model

The ATM bonding functionality resides above the ATM-TC of the DSL transceiver and below the ATM transport layer, as shown in Figure 15.2. The ATM bonding protocol takes a cell stream from the ATM transport

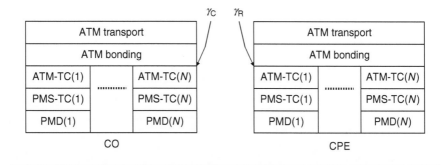

Figure 15.2   User plane protocol reference model of an ATM-bonded group.

GFC (4 bits)	VPI (8 bits)	VCI (16 bits)	PTI (3 bits)	CLP (1 bit)	HEC (8 bits)

(a) Standard ATM cell header

GFC (4 bits)	VPI (8 bits)	SID (8 bits)	VCI (8 bits)	PTI (3 bits)	CLP (1 bit)	HEC (8 bits)

(b) Modified ATM cell header for use with 8-bit SID

SID (4 bits)	VPI (8 bits)	SID (8 bits)	VCI (8 bits)	PTI (3 bits)	CLP (1 bit)	HEC (8 bits)

(c) Modified ATM cell header for use with 12-bit SID

Figure 15.3   Standard and modified ATM cell headers.

layer, splits it into lower rate cell streams, and passes the streams to the individual transceivers across the γ-interfaces.

15.2.2.2   Cell Header Format

In ATM-based bonding, the SID may be either 12 or 8 bits. Various ATM cell header formats are shown in Figure 15.3, with Figure 15.3a being the standard ATM cell format. With bonding, the upper 8 bits from the VCI field are used for the SID, and in the case of the 12-bit SID, the 4-bit GFC (generic flow control) field forms the upper four bits of the SID, as shown in Figures 15.3b and 15.3c, respectively. It should be noted that if the modified cells must have layer 2 transparency, the 8-bit format will likely be required, because ITU Recommendation I.361 and ATM Forum UNI 4.1 specify that the GFC bits shall be set to zero.

15.2.2.3   Autonomous Status Messages

ATM-bonding also defines a control protocol that is used to communicate status of the group and the constituent links. This information is conveyed via a single-cell message called the autonomous status message (ASM), which includes several fields that are described in Table 15.1. These messages are sent nominally once per second on each link, although ASMs may be sent at any time to communicate urgent information such as degradation of link quality.

The status of each link in the group in both the transmit and receive directions is communicated via the Tx link status and Rx link status fields of the ASM. Each is an eight-octet field, which carries two status bits for

Table 15.1 Select Fields of the Autonomous Status Message (ASM)

Field	Function
Group ID	Unique identifier for a bonded group
Tx link number	Identifies link that carries a particular ASM
Number of links	Total number of links in the group
Rx link status	Status of each link within the group at the receiver
Tx link status	Status of each link within the group at the transmitter
Insufficient buffers	Receiver has insufficient buffers to support bonding all links in the group
Group lost cells	Count of cells lost at the bonding layer

each of the potential 32 links in the group. These status bits are defined in Table 15.2. It should be noted that the upstream and downstream directions are independent, and the standards do not require link usage to be the same in both directions.

15.2.2.4 Operation

Under normal operation, the bonding transmitter takes the aggregate cell stream from the ATM transport layer, inserts a sequential SID into each cell header, and sends cells to the constituent links in any arbitrary order. At the receiver, each link passes the unmodified cells to the bonding receiver, which reassembles the original cell stream based on the SIDs and replaces the SIDs with zeros before passing this stream to the ATM transport layer.

The receiving entity is primarily responsible for determining which links can carry bonded traffic. Using ASM transactions, the status of each link is communicated on all links, so if a particular link drops or otherwise has insufficient performance, this information is communicated to the

Table 15.2 Tx Link Status and Rx Link Status

Status	Value	Link Status Description
Not provisioned	00	Link number not part of the provisioned group
Should not be used	01	Inform far end not to use a link that is a member of the group (e.g., quality is unacceptable at the receiver, not needed by the transmitter)
Acceptable to carry bonded traffic	10	Link under consideration to carry bonded traffic
Selected to carry bonded traffic	11	Bonded traffic may flow on this link

transmitter on all links. The receiver will continue to monitor links that are not being used to carry traffic, and when quality is restored on the troublesome link, the receiver will inform the transmitter by sending ASMs on all links. The content of the ASMs will indicate that the particular link in question can be used to carry bonded traffic again. The transmitter has final control over which links carry traffic, and it is not required to use a link the receiver has deemed acceptable. Although no link state information is formally communicated or required in ATM-based bonding, the standards do provide an informative link status state diagram based on Tx link status and Rx link status. This state diagram, shown in Figure 15.4, applies to both directions of a given link, although the two directions have independent state machines. Note that traffic can only flow on a link if the Tx link status and Rx link status are both set to "11," which means they are selected to carry bonded traffic.

Minimization of differential delay between the links in a group is important, because significant differences in delay between the links must be compensated by buffering, and it is also possible to exhaust the available number of SIDs. ATM bonding provides some methods to minimize these effects. One of the major factors influencing delay is the setting of the interleaver depth for DSL systems that use a combination of Reed–Solomon

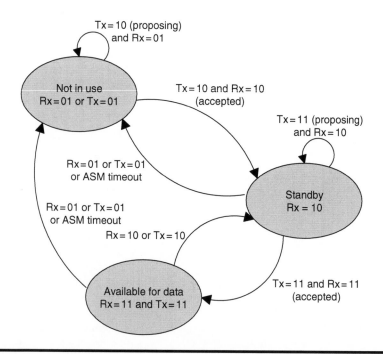

Figure 15.4   Informative link status state diagram.

coding and interleaving, such as ADSL and VDSL (Very High Bit-Rate DSL). Typically only a maximum delay is specified, and the modem may choose any interleaver depth that does not exceed the maximum delay. To force the differential delay between links in a bonded group to be as small as possible, ATM bonding can also specify a minimum delay tolerance via the G.994.1 handshake transactions, which informs the individual transceivers of a minimum delay that must be met as well.

The receiver is responsible for determining buffer requirements given the rates and delays of the constituent links, and the receiver must never allow the transmitter to use links that will exceed the buffering capabilities of the receiver. If the receiver cannot use all of the links which have been determined to be acceptable to carry bonded traffic, the receiver must set the insufficient buffers field in its outgoing ASMs to inform the transmitter of this situation.

Because network-side processing and memory are typically more expensive than at the CPE side, ATM bonding provides a mechanism to further reduce the upstream differential delay. Through ASM transactions, the bonding entity at the CO can specify additional delays to be added to each upstream link to minimize the differential delay at the upstream receiver.

15.2.2.5 Initialization

The initialization procedure is defined for links that are in data mode and have an ATM path present. The bonding transmitter at the CO begins by sending ASMs on each link with the correct group ID and link number so that the bonding receiver at the CPE side can identify the constituent links. After the bonding entity at the CPE side receives at least one ASM on each link in the group, the CPE side responds by sending ASMs to the bonding receiver at the CO. Once the links have been identified, the upstream and downstream directions operate independently: the bonding receiver informs the bonding transmitter which links are acceptable to carry traffic, but the transmitter is always in control of which links are actually used.

15.2.3 Summary

Because the vast majority of DSL deployments use ATM between the DSLAM and the subscriber, ATM-based bonding provides a means to dramatically improve data rates using multiple pairs served using existing equipment. Channel conditions such as loop attenuation and crosstalk are primary factors that determine the bit rate that can be supported by a DSL, and ATM-based bonding improves on traditional IMA in that the constituent links are allowed to operate at different data rates, thereby maximizing the aggregate rate. Because segmentation, framing, and delineation are already defined in ATM, only the sequencing function must be added to provide this functionality, although implementations will also need the ASM state machine, the initialization state machine, and a rate adaptive mechanism to distribute traffic to the pairs. By redefining bits in the ATM header, sequencing information is inserted without any additional overhead. With fewer constraints and complexity than IMA, ATM-based bonding is a rather simple, robust means by which to provide higher data rates using the existing ATM access network.

15.3   Ethernet-Based Bonding

In June 2004, IEEE 802.3 Working Group ratified a new version of the ever-evolving Ethernet standard [802.3]: IEEE 802.3ah [802.3ah], called Ethernet in the First Mile (EFM). This new standard adapts Ethernet—the best known and most widely used LAN technology in history—for deployments in carrier access networks. EFM can replace the complex and costly ATM and SONET (synchronous optical networking) access networks with simpler, more cost-effective Ethernet access networks, resulting in immediate savings in capital and operating expenditures, as well as increased bandwidth and service options to the subscriber.

EFM defined two new physical layer specifications for delivering Ethernet over plain old telephone lines: 2BASE-TL and 10PASS-TS. These deliver the simplicity and flexibility of Ethernet while still maintaining the spectral compatibility of xDSL.

The EFM specifications for Ethernet over copper defined two key enhancements to traditional xDSL technology. First, EFM defined a new, very efficient, encapsulation method for transporting Ethernet over lower-speed connections with a relatively high bit error ratio (the relative bit error ratio [BER] of 10^{-7} for xDSL is much higher than that of optical Ethernet). Second, EFM defined a method by which multiple pairs of xDSLs can be aggregated into a single Ethernet connection. These two enhancements represent a significant improvement in the efficiency and flexibility of Ethernet transport over xDSL.

The multipair aggregation strategy of EFM has also been standardized by ATIS as T1.427.02 [T1.427.02] and also by the ITU as G.998.2 [G.998.2], part of the G.bond initiative. The following sections describe the Ethernet bonding strategy and how it differs from other multipair aggregation technologies.

15.3.1   Benefits

The Ethernet transport and bonding mechanisms of IEEE 802.3ah, which are generalized in G.998.2, provide some key benefits for the data access market when compared to other multilink access technologies.

1. *Support of pairs running at different rates.* One of the key benefits first introduced by Ethernet bonding, and later capitalized on by other bonding approaches, is that the pairs in the aggregate group can be running at different bit rates. In bonding approaches prior to Ethernet bonding, all lines in an aggregate group had to be running at the same rate, even if one of them had to be run at an unnecessarily low rate to do so. This essentially wasted significant bandwidth (data rate). With Ethernet bonding, pairs running at different speeds can be fully utilized, thus providing optimal use of the limited bandwidth available in the access network.

2. *Low overhead.* Ethernet bonding uses a very small (2-octet) fragmentation header per fragment. This results in negligible overhead (e.g., on a 512-octet fragment, the overhead is $2/514 < 0.4$ percent), especially when compared to more traditional technologies such as ATM where the overhead generally averages over 20 percent. This high overhead comes from ATM cell encapsulation, in which 5 of every 53 octets are ATM cell headers, and where parts of ATM cells are padded when a frame does not partition evenly into ATM cells. For example, a 64-octet Ethernet frame has to be carried by two ATM cells (so only 64 of 106 octets are actually used).

3. *Dynamic handling of new and failed pairs.* With Ethernet bonding, new pairs can enter a group without causing any disruption. Failed lines can be quickly detected and removed from the aggregate group, with negligible loss and correction times well below 50 ms.

4. *Flexible transmit algorithms optimized for different applications.* As discussed later in this section, Ethernet bonding can be designed to minimize overhead or to minimize latency, or to find some balance between the two. This flexibility allows implementers to design bonding algorithms that are optimal for the intended applications.

5. *Pure Ethernet.* A standard rule in telecommunications is that more than 95 percent of all data transmitted either starts or ends as an Ethernet frame. Because of this reality, more and more carriers are moving to a pure Ethernet/IP network, eliminating the complexity of ATM and TDM (time-division multiplexing) infrastructures. Ethernet bonding allows carriers to build a pure Ethernet next-generation network and eliminate the hassles and complexity of ATM VCs and SONET circuits.

As access networks continue to evolve, the Ethernet bonding initiative is positioned to be a critical technology required to deliver the business and residential services of today and tomorrow. Already, carriers deploying video services over xDSL have realized that more than one xDSL may be required to support the bandwidth needs of complex IP video services. Ethernet bonding provides a simple and natural mechanism to deliver these services today.

15.3.2   Architecture

A key design goal of the Ethernet bonding initiative was to leverage existing xDSL and Ethernet standards as much as possible. In particular, the specification had to support the use of existing Ethernet Media Access Control (MAC) layers and existing xDSL technologies, all while providing the flexibility for new and different physical layers to be supported in the future.

As shown in Figure 15.5, the EFM specifications introduced three new layers into the Ethernet/xDSL hierarchy. The rate matching layer provides an adaptation that allows existing Ethernet MACs running at 100 Mbps to operate properly over Ethernet over copper interfaces running at variable rates that are less than 100 Mbps. The loop aggregation layer provides the mechanisms for fragmenting and reassembling so that multiple copper pairs provide a single Ethernet interface. The 64/65-octet encapsulation layer provides the framing and encapsulation so that Ethernet frames and fragments can be delivered properly over individual copper pairs.

The architecture of Figure 15.5 is important in that it is very flexible in the type of physical layer that can be supported, and the loop aggregation functionality is independent of the physical layer and framing mechanisms used to transport the data.

15.3.3   Ethernet Bonding—Overview

The loop aggregation techniques of IEEE 802.3ah are simple yet powerful. Frames are passed to the loop aggregation layer from the higher layer, where they are fragmented and distributed across the loops within the aggregate group. When transmitted across the individual loops,

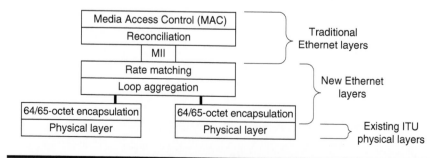

Figure 15.5   Ethernet in the First Mile (EFM) Ethernet over copper architecture.

a fragmentation header is added, which includes a sequence number and frame markers. This header is used by the receiver to resequence the fragments, and to reassemble them into complete frames.

The Ethernet bonding function can be partitioned into four components:

1. Group Definition—deciding if a set of lines can be aggregated into a single Ethernet port.
2. Fragmentation—the process of partitioning an Ethernet frame into multiple fragments for transmission over multiple lines.
3. Reassembly—the process of reassembling a stream of Ethernet fragments into Ethernet frames.
4. Initialization—the process of bringing up a line and joining an aggregate group.

Each of these functions is covered in the following sections.

15.3.3.1 Group Definition

In any process by which a stream of data is partitioned across multiple connections, it is important to control the differential latency between the connections. If there is a very high differential latency between two connections in an aggregate group, then reconstituting the original stream can require a great deal of memory at the receiver, and it can add latency to the traffic flow. This leads to a costly and poor overall solution.

The differential latency between two pairs is affected by a number of factors, including the following:

- *Speed of the pairs.* A higher bit rate line can transmit more bits per second than a lower bit rate line.
- *Effective distance.* Two lines with significant distance differences support a different number of bits "in transit."
- *Coding parameters.* Some technologies utilize complex coding techniques at the physical layer (e.g., Reed–Solomon coding with interleaving) that are parameterized and add latency that depends on the configuration.

Ethernet bonding places a restriction on the amount of differential latency that can be tolerated in an aggregate group. Specifically, the differential latency of two lines is defined to be the number of bits that can be sent down the faster of the two lines in the same amount of time that a maximum size fragment (which is 512 octets, as discussed in the next section) can be sent on the slower of two lines.

For any two pairs in an aggregate group, the differential latency guaranteed to be tolerated by Ethernet bonding is 15,000 bit times, where a bit time is the time required to send one bit down the slowest speed line. Two lines with a higher differential latency than this cannot be supported in the same aggregate group.

There are additional restrictions on which pairs can be in an aggregate group. Two pairs can be placed in the same aggregate group if:

- Differential latency is no larger than 15,000 bit times.
- Differential rate is no greater than four (i.e., one link is, at most, four times faster than the other).
- Number of pairs in the aggregate group is no greater than 32.

As long as these restrictions are met, any two pairs can be placed in the same bonded Ethernet connection.

15.3.3.2 Fragmentation

The fragmentation process for Ethernet bonding takes as input an Ethernet frame from a higher layer entity (an Ethernet MAC, for instance) and partitions that frame into some number of fragments for distribution over one or more lines in the aggregate group, see Figure 15.6.

Each fragment is encapsulated with a fragmentation header before being transmitted onto the line. The fragmentation header is used by the reassembly process on the other end of the connection to reconstitute the original stream. The fragmentation header consists of a sequence number that indicates the relative order of the fragment, as well as a flag field that indicates which fragments contain the first and last part of an Ethernet frame.

The Ethernet bonding fragmentation process is designed to be generic. The algorithm for partitioning the frames over the loops is not specified. Different vendors can develop different approaches and algorithms to solve

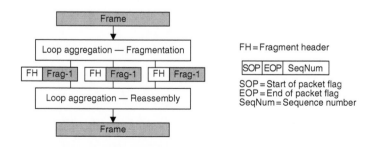

Figure 15.6   **Ethernet bonding fragmentation process.**

the problem. Some algorithms might yield a higher efficiency (i.e., a higher percentage of "user" data versus "overhead" data), whereas other algorithms might yield a lower latency. The designer of the transmitting equipment is free to make engineering trade-offs that best support the application at hand.

As an example, consider the problem of partitioning a 1200-octet Ethernet frame over an eight-pair aggregate group. A simple algorithm would split that frame into eight fragments, each of 150-octets, and transmit one fragment on each pair. A smarter algorithm might realize that one of those pairs is running at three times the speed of the other pairs and send 360-octets down the fastest pair and 120-octets down each of the other seven slower pairs. This approach would attempt to minimize the latency across the pairs by sending more bits down the pairs that can support higher bit rates. A different algorithm might send 400-octets on three pairs, and zero octets on the other five pairs. This would result in less overhead (three fragmentation headers versus eight) and thus potentially higher throughput. All of these algorithms are valid approaches, and implementers have the flexibility to design the algorithm that best fits their needs.

However, the fragmentation process must obey certain rules in that fragments must obey size constraints. Fragments, not including the fragment header, must be between 64 and 512 octets in length. As long as the loop aggregation algorithms obey these constraints and restrictions, any fragmentation algorithm can be handled by the reassembly process, yielding a flexible and interoperable solution.

15.3.3.3 Reassembly

The reassembly process for Ethernet bonding is also simple. The fragments received on each line are kept in a queue for that line. A receiver process for the aggregate group simply waits for the reception of the next expected sequence number over any pair in the aggregate group, as illustrated in Figure 15.7.

As fragments are reassembled into the proper order, the start and end of packet flags are used to determine packet boundaries. When an end of packet flag is found, that fragment is considered the last fragment in the frame. A checksum is then calculated on the frame (from the check sequence field of the Ethernet frame), and if the checksum is correct, the frame is handed off to the higher layer.

In addition to the basic processing, there are a number of error cases to consider (lost fragments, corrupt fragments, etc.). In general, the reassembly process will simply wait a specified amount of time for the next fragment based on the expected sequence number, and if it does not arrive in time, the assembly process is restarted using the next sequence number found.

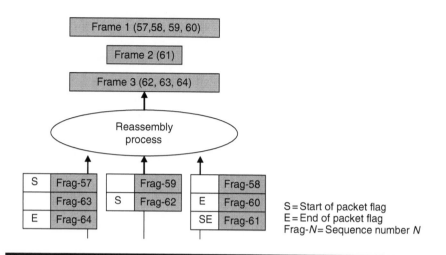

Figure 15.7   Bonding reassembly.

15.3.3.4   Initialization

The final part of the Ethernet bonding standard is the process by which two devices determine which pairs belong to the same aggregate group. This process is done during the handshaking phase (G.994.1) of the initialization sequence of an xDSL. (See Chapter 17 on Standardization.)

In this process, it is assumed that the CO system knows which pairs should be placed in the same aggregate group. During initialization, the CO system will attempt to "set" an identifier at the customer premises device to indicate which aggregate group this pair belongs to. If this pair already has an identifier and the set fails, the CO system will read the existing identifier and see if it matches any identifier defined for an existing group. If the identifier already exists, the new pair is added to the group indicated by the identifier. Otherwise, there is an incorrect configuration, which means some other CO system has already indicated this pair in some other group. However, if the set succeeds, then this is considered the first pair in the new aggregate group, and all other lines in the same group at the CPE device will store this identifier so they can indicate it to the CO device when the set is performed on other pairs.

15.3.4   *802.3ah Ethernet Bonding versus 802.3ad Link Aggregation*

There are often questions about the differences between Ethernet bonding ("loop aggregation") and 802.3ad link aggregation. 802.3ad link aggregation provides a way to use multiple Ethernet connections as a single aggregate

Ethernet connection. However, it has some limitations that make it a poor choice for the aggregation of xDSLs into a single Ethernet connection.

The primary limitation of using 802.3ad over multiple copper pairs is that link aggregation requires each link to be the same speed. This is a severe limitation in the outside copper plant, where different lines are capable of supporting different bit rates depending on the loop length and noise environment, and in some cases, the rate of a pair can change over time to adapt to a changing noise environment. Additionally, the distribution algorithm in link aggregation is often difficult to use in residential or small business environments, in which only a small number of computers are attached. In many cases, link aggregation might map all of the traffic onto the same line, which would not result in any increased bit rate.

Ethernet bonding, on the other hand, provides an excellent choice for xDSL aggregation, allowing flexible use of pairs supporting disparate bit rates, and offering flexible distribution algorithms that can better guarantee utilization of all available bandwidth.

15.3.5 Summary

The multipair aggregation work of the IEEE 802.3ah task force pioneered a new breed of bonding technologies capable of delivering higher bit rate services to the end user. This simple yet elegant solution is ideally suited to emerging access networks that are being built purely on Ethernet, IP, and MPLS infrastructures. By leveraging Ethernet bonding, carriers can deliver high-speed symmetric services to business customers, distancing their services from the common low-speed T1. Ethernet bonding can also be used to enable residential video services, in many cases providing 20–30 Mbit/s and more downstream bandwidth over long distances on just two pairs into the home. The flexibility and resiliency of Ethernet bonding make it ideally suited to meet the difficult requirements of video and business class services.

Bibliography

[ATM Forum AF-PHY] ATM Forum, *Inverse Multiplexing for ATM (IMA) Specification Version 1.1*, AF-PHY-0086.001, March 1999.

[DSL Anywhere] DSL Forum, *DSL Anywhere*, Issue 2, September 2004.

[G.998.1] ITU-T Recommendation G.998.1, *ATM-Based Multi-Pair Bonding*, 2005.

[G.998.2] ITU-T Recommendation G.998.2, *Ethernet-Based Multi-Pair Bonding*, 2006.

[IEEE 802.3] IEEE 802.3, *Carrier Sense Multiple Access with Collision Detection (CSMA/CD) Access Method and Physical Layer Specification*, 2002.

[IEEE 802.3ah]	IEEE 802.3ah, *Carrier Sense Multiple Access with Collision Detection (CSMA/CD) Access Method and Physical Layer Specification—Amendment: Media Access Control Parameters, Physical Layers, and Management Parameters for Subscriber Access Networks*, 2004.
[T1.427.01]	ATIS T1.427.01, *ATM-Based Multi-Pair Bonding*, 2004.
[T1.427.02]	ATIS T1.427.02, *Ethernet-Based Multi-Pair Bonding*, 2005.
[TR-048]	DSL Forum, *ADSL Interoperability Test Plan*, TR-048, April 2002.
[TR-067]	DSL Forum, *ADSL Interoperability Test Plan*, Issue 2, TR-067, December 2004.

Chapter 16

Multiline MIMO DSL System Architectures

Michail Tsatsanis and Thorkell Gudmundsson

CONTENTS

Abstract   Copper transmission technologies have made remarkable progress over the last 20 years and have produced mature, nearly optimal architectures for signaling over a single copper pair. As data rates and signal bandwidths increase, however, crosstalk interactions among the binder pairs become a major performance bottleneck. Next generation Digital Subscriber Line (DSL) systems are expected to provide substantial performance improvements by coordinating transmission across multiple copper pairs to mitigate crosstalk impairments. This chapter reviews various advanced MIMO (multiple-input–multiple-output) signal processing architectures for multiline transceivers and discusses their benefits and drawbacks. It explains how these techniques are able to mitigate (self and alien) NEXT (near-end crosstalk) and FEXT (far-end crosstalk), and discusses their performance under various loop and disturber conditions. Both point-to-point and point-to-multipoint methods are described and their performance and complexity tradeoffs are discussed.

16.1   Introduction

Transmission over copper pairs has attracted renewed attention from researchers, technologists, and businesses over the past several years. Long considered an old fashioned and antiquated network, the copper infrastructure has recently sprang into new life enabling broadband connectivity and being a key ingredient of the solution to the last mile access problem. At the time of this writing, copper provides broadband connectivity (Asymmetric Digital Subscriber Line [ADSL]) to more than 16 million residences in the United States, while servicing more than 1 million business establishments with high-speed (T1-type) connections.

The success of copper-based broadband services has been made possible by relentless advances in communications technology and ever increasing signal processing capabilities of silicon chips. Digital Subscriber Line (DSL) modems have come a long way from early Dataphone Data Service (DDS) and Alternative Mark Inversion T1 (AMI-T1) transceivers with simple alternating mark inversion signaling, spectrally inefficient square pulses, and crude spatial duplexing schemes (separate pairs for upstream and downstream transmission). Modern DSL modems use sophisticated modulation techniques, advanced coding, flexible spectral band plans, and sophisticated frequency division or echo canceled duplexing.

Several chapters in this book review the technological advances in the areas of modulation and coding, signal processing, analog front-end design, hardware and software that have enabled this progress. Although further improvements in most transceiver's subsystems are possible, it is debatable whether they will automatically translate into further significant performance improvements. Modern DSL modems operate in an environment of crosstalk interference that constitutes a major performance bottleneck.

For example, near-end crosstalk (NEXT) into an ADSL modem can easily be 20 or 30 dB higher than any other noise source induced by effects like A/D quantization noise, line driver dynamic range, thermal noise, etc. (see chapter 3 in [Golden 2006] for more details on crosstalk modeling). Controlling the detrimental effects of crosstalk in the network is one of the highest priorities in the effort to achieve a next generation of performance in copper modems.

It is well known to DSL engineers that the effects of NEXT can be controlled with appropriate frequency division duplexing (FDD) (or even time division duplexing [TDD]). A plethora of FDD spectral plans is constantly added to the ADSL standard (see all spectral annexes in [ITU G.992.3]) for various loop conditions and desired symmetry ratios. Further advanced studies have investigated the dynamic allocation of spectral bands or regulation of power distribution across frequencies depending on the interference conditions in the given binder. Those techniques come under the name of dynamic spectrum management (DSM) and are described in more detail in Chapter 8 of this book (see also [Song 2002] and references therein). They provide impressive performance gains when all modems in the binder cooperate and implement the DSM rules. The gains are smaller when legacy disturbers are present, which do not adhere to the DSM etiquette.

Even more impressive performance results are achieved if the crosstalking modems are designed to operate synchronously and coordinate their transmit signals at the waveform level. This type of coordinated multichannel signaling is often called vectored modulation (see [Verdu 1998] and [Ginis 2001]) and is well suited for multichannel media with strong interactions across the channels (crosstalk). Vectored transmission requires joint processing of the signals of all channels at the receiver or the transmitter to align amplitudes and phases in a way that counteracts the detrimental effects of the channel cross-couplings. These multichannel signal processing techniques are commonly referred to as Multiple-Input–Multiple-Output (MIMO) processing or space-time processing.

Vectored transmission techniques have a long history originating in phased array radar systems and then migrating to multiantenna wireless links (multiple-transmit-multiple-receive antennas). After more than ten years of intense study in the research community, vectored techniques are currently being incorporated into a host of standards for wireless LANs (IEEE 802.11n) and MANs (IEEE 802.16). In the wireline world, MIMO NEXT cancelers have been incorporated in 1 Gbps Ethernet transceivers (over four copper pairs), while more advanced vectored schemes are currently developed for the next generation 10 Gbps Ethernet modems.

Given the success of vectored signaling in so many diverse applications and the proven record of the technology, it seems strange that similar

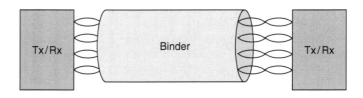

Figure 16.1   Point-to-point multipair link.

advanced multichannel architectures have not yet found their way into mainstream DSL applications. The reasons for that are both technical and operational. The copper plant was originally designed to transport low frequency signals (voice signals) with minimal crosstalk. It has always been thought that each individual pair is an individual channel, well-isolated from other channels. This is in contrast with the wireless medium which is by definition a broadcast medium where interference among users has always been of paramount importance. Of course, as the frequency bands used by copper modems keep expanding, crosstalk is more pronounced and electromagnetic pair separation less perfect.

Furthermore, MIMO techniques are usually associated with applications where multiple channels connect the source to the destination. For example, Gig-Ethernet copper connections utilize four pairs that are bonded together (see Figure 16.1). In contrast, the public copper network has a point-to-multipoint architecture and in the vast majority of cases, the service is provided over a single pair (see Figure 16.2).

Despite these difficulties, the potential of vectoring techniques in the copper network is significant and clever ways will be found to provide the associated benefits within the given network architecture. For example, there are several applications where high data rates are required, well beyond what a single copper pair can provide. These include high-end business access applications and DSL access multiplexer (DSLAM) and data

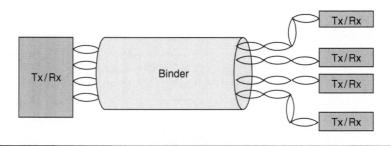

Figure 16.2   Point-to-multipoint access network.

link control (DLC) uplink connections. Standard solutions to this problem involve inverse multiplexing over ATM (IMA) hardware, where a high-rate bit stream is partitioned into multiple low rate streams, which are transported over copper pairs and then reassembled at the receiver. This configuration closely resembles the architecture of Figure 16.1 and is a prime candidate for the application of vector signaling technologies.

Even in the point-to-multipoint configuration of Figure 16.2, there are significant performance gains possible by coordinating the modems on the central office (CO) side. This type of technologies have received increased attention in the context of wireless cellular telephony and utilize techniques from multi-user communication theory [Ginis 2002]. Similar signal processing architectures have been proposed for Very High Bit-Rate DSL (VDSL) systems [Verdu 1998]. Simplified linear architectures have been proposed in [Cendrillon 2006a and b], while reduced complexity adaptation methods have been explored in [Louveaux 2005]. Finally, an overview of methods against alien crosstalk (see next section) is presented in [Ginis 2006].

MIMO processing technologies offer elegant technical solutions to crosstalk containment. They are however no panacea and their introduction into the network is far from straightforward. Currently deployed technologies, while adhering to minimal crosstalk interference protection guidelines (see spectral management standard [ANSI T1.417]), are not designed with interference mitigation in mind. The migration to a next generation of modems is a technical, operational, and financial challenge that will require innovative thinking from the best minds of all parties involved. The technical people in the organizations of the telecommunication carriers, system vendors, silicon suppliers, as well as the research community will certainly be counted to assess and recommend a path forward to embrace those new technologies where possible.

The goal of this chapter is to help technologists in the DSL space become familiar with these newer multiline transmission technologies. Our objective is not to describe a complete product or argue for a particular application. We rather focus on the fundamental concepts and expose basic principles and architectures. Our efforts will have been successful if readers are motivated by these concepts to take on the major challenges of implementation, deployment, and operation that may turn those concepts into advanced products and services.

The rest of the chapter is organized as follows: In Section 16.2, we briefly review crosstalk issues in the loop plant and illustrate the potential of vectored technologies. In Section 16.3, we review possible one-sided and two-sided multipair architectures; while in Section 16.4, we discuss extensions to the point-to-multipoint case. Finally, some performance examples are presented in Section 16.5 and some concluding thoughts in Section 16.6.

Figure 16.3   A single pair transmission setup.

16.2   The Challenges of Crosstalk for Single-Line and Multiline Systems

If each DSL in the network were operating in isolation, then the modem designer would only have to contend with the attenuation and dispersion of the copper loop. Imagine the situation in Figure 16.3, where the line in question was the only high-speed service in the binder. Then, the equivalent mathematical model for the received signal would be

$$y(\omega) = h(\omega)s(\omega) + v(\omega) \tag{16.1}$$

where

$h(\omega)$ is the copper loop's frequency response (including all transmit and receive filters)

$v(\omega)$ denotes the received noise

In the absence of crosstalk, this noise term can be reasonably modeled as white Gaussian with power spectral density $S_v(\omega)$. This frequency domain representation of the transmission system is also depicted in Figure 16.4.

Under this scenario, the goal of the transceiver designer is to develop an architecture that comes reasonably close to the channel capacity

$$C = \int_B \log_2 \left(1 + \frac{S_s(\omega)|h(\omega)|^2}{S_v(\omega)} \right) d\omega \tag{16.2}$$

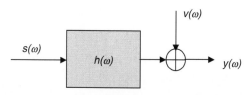

Figure 16.4   Mathematical model for single pair transmission.

where

    B is the available bandwidth

    $S_s(\omega)$ is the transmitter power spectral density (PSD)

This is generally achieved with a combination of appropriate processing to mitigate intersymbol interference (ISI) and advanced coding. When we consider the more general case of multiple high-speed modems transmitting through the same binder (see Figures 16.1 and 16.2), then the noise term in the capacity Equation 16.2 should be modified to account for crosstalk.

16.2.1 FEXT

We start our discussion on crosstalk by first examining the effects of far-end crosstalk (FEXT). Interference from the far-end aggressor signal is generated due to electromagnetic coupling between aggressor and victim pairs as explained in detail in chapter 3 in [Golden 2006] (see also Figure 16.5). Precise modeling of the interference term is possible from first principles using multiport network theory techniques (see [Joffe 2002] for a single aggressor and [Cioffi 2006] for multiple aggressors). Empirical models that predict the interference PSD from a number of FEXT aggressors have been developed and are extensively used in testing performance compliance (see [ANSI T1.417]). Further, several actual measurements of FEXT cross-coupling have been performed by several organizations and have been reported in the literature.

FEXT is generally attenuated as the loop length increases. If a 26-AWG loop has length in excess of 3 km, FEXT is generally attenuated under the noise floor. In contrast, FEXT presents an important impairment for relatively short loops. Figure 16.6 illustrates this degradation through a signal-to-noise ratio (SNR) plot. The SNR is plotted as a function of frequency for a 26 AWG, 1 km loop for a bandwidth of 1.1 MHz (a flat transmit PSD of −40 dBm/Hz is assumed and a noise floor of −140 dBm/Hz). The plain line depicts the case of no disturbers, where the effects of channel attenuation are evident. The dashed line shows the case where the 25 pair binder is full of similar disturbers injecting FEXT interference into the victim pair.

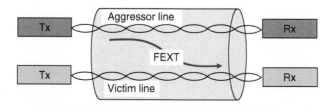

Figure 16.5 **Illustration of far-end crosstalk (FEXT) interference.**

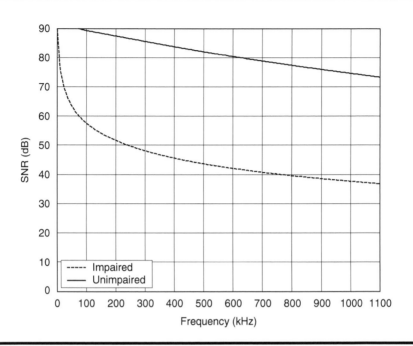

Figure 16.6   FEXT effects on signal-to-noise ratio (SNR).

The SNR loss due to crosstalk is evident and the corresponding loss of capacity is significant, or about half of the unimpaired capacity in this case.

If each line is studied in isolation, very little can be done to contain crosstalk. Single line, joint co-channel detection architectures have been studied for similar problems but have limited success in the presence of multiple interferers [Zeng 2002]. It is an interesting question therefore, whether multiline processing suffers from the same limitations.

To answer this question, we should shift our perspective from focusing on a single line to considering the whole binder as a joint transmission medium. There is of course a difference between the point-to-point transmission of Figure 16.1 and the point-to-multipoint setup of Figure 16.2. The intricacies of multipair signaling are better understood in the point-to-point setup of Figure 16.1, so we will cover that case first. We will revisit the point-to-multipoint architecture later in the chapter.

Let us consider a binder of M pairs and let us collect all received signals $y_k(\omega)$, $k = 1, \ldots, M$ in a vector $\mathbf{y}(\omega) = [y_1(\omega), \ldots, y_M(\omega)]^T$. Then, those signals can be described by a vector extension of Equation 16.1.

$$\mathbf{y}(\omega) = \mathbf{H}(\omega)\mathbf{s}(\omega) + \mathbf{v}(\omega) \qquad (16.3)$$

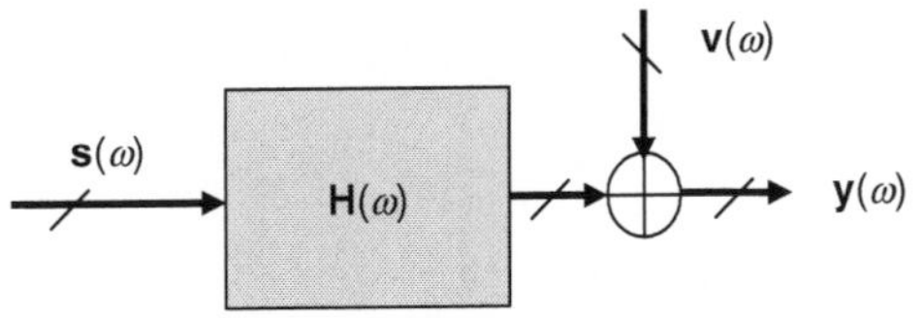

Figure 16.7 **Mathematical model of vector transmission system.**

where the vectors $\mathbf{s}(\omega)$ and $\mathbf{v}(\omega)$ are similarly defined and the $M \times M$ channel matrix models both the main loop channels (diagonal elements) and the FEXT coupling channels (off diagonal elements). Figure 16.7 illustrates this vector signal model.

Given this mathematical formulation, the benefits of vectored transmission can be quantified by examining the MIMO capacity of this vector channel. In the general vector case, the capacity can be calculated according to the equation (e.g., [Cover 1991])

$$C = \int_B \log_2 \left\{ \det \left(\mathbf{I} + \mathbf{S}_v^{-1/2}(\omega) \mathbf{H}(\omega) \mathbf{S}_s(\omega) \mathbf{H}^H(\omega) [\mathbf{S}_v^{-1/2}(\omega)]^H \right) \right\} d\omega \quad (16.4)$$

where

$\mathbf{S}_s(\omega)$ is the spectral matrix of the transmitted signals
$\mathbf{S}_v(\omega)$ is the spectral matrix of the additive noise
$\mathbf{I}$ is the identity matrix

In this particular case, some simplifications are possible without loss of generality: the transmitted signal matrix is assumed diagonal with identical power in each diagonal entry ($\mathbf{S}_s(\omega) = \sigma_s^2(\omega)\mathbf{I}$) and the noise is assumed spatially and temporally white ($\mathbf{S}_v(\omega) = \sigma_v^2\mathbf{I}$). With those simplifications, Equation 16.4 can be written as

$$C = \int_B \log_2 \left\{ \det \left(\mathbf{I} + \frac{\sigma_s^2(\omega)}{\sigma_v^2} \mathbf{H}(\omega) \mathbf{H}^H(\omega) \right) \right\} d\omega \quad (16.5)$$

An evaluation of this formula for the binder of Figure 16.6 reveals a per line capacity that is comparable to that of an unimpaired line. This result is remarkable as it shows the potential of vector transmission to completely eliminate the detrimental effects of FEXT in this setup.

Despite its predictive power, Equation 16.5 provides little insight on what the mechanism is through which the additional capacity is achieved.

To provide a connection between capacity and SNR, similar to Equation 16.2, let us rewrite Equation 16.5 utilizing the system's eigenvalues.

Note that $\mathbf{S}_x(\omega) = \sigma_s^2(\omega)\mathbf{H}(\omega)\mathbf{H}^H(\omega)$ is the received (noiseless) signal's spectral matrix and denote by λ_m, $m = 1, \ldots, M$ its eigenvalues. Then, using the fact that the determinant is the product of the eigenvalues we can write

$$C = \int_B \sum_{m=1}^{M} \log_2 \left(1 + \frac{\lambda_m(\omega)}{\sigma_v^2} \right) d\omega \tag{16.6}$$

Equation 16.6 is now similar in form to Equation 16.2. If we consider the eigenvalues λ_m as representing received signal power, then we maintain the notion of "per line" SNR for the "MIMO processed" lines. Figure 16.8 compares the individual and average per line MIMO SNR (plain and multiple dotted lines) to the "single-input–single-output (SISO)" SNR (dashed line).

It is now clear from Equation 16.6 why FEXT does not negatively affect the MIMO capacity. The received FEXT signal actually contributes to the numerator of the "MIMO SNR"; in other words, with appropriate processing

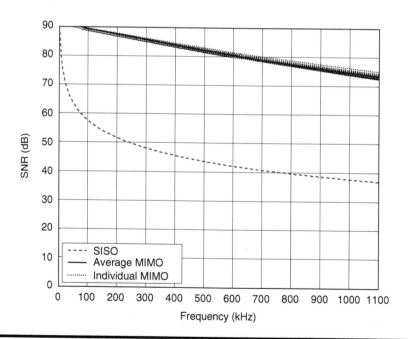

Figure 16.8 **SNR for multiple-input–multiple-output (MIMO)-processed lines and single-input–single-output (SISO) lines in the presence of FEXT.**

it can be utilized as useful signal as opposed to interference. We should caution the reader however, that these capacity gains are possible only if the vectored system includes all the modems in the binder. If the vectored system only includes a portion of the binder, then the remaining interferers will add "alien crosstalk" which cannot be completely eliminated by the vectored system. Similar situations arise in the presence of alien NEXT as well, and we will discuss them shortly.

16.2.2 NEXT

Interference from near-end transmitters can also couple into a victim pair (see Figure 16.9) and is generally stronger than FEXT interference, especially for long loops. The situation is worst when the upstream and downstream transmission bands overlap. Similarly to FEXT, both analytical and empirical models have been developed to assess the effects of NEXT interference (see [Joffe 2002] and [Cioffi 2006]).

If there is no coordination among pairs, NEXT can be detrimental on long loops. For the case of coordinated transmission however, one could argue that NEXT is more straightforward to mitigate than FEXT. The receiver has access to the interfering signals and can cancel them by utilizing MIMO NEXT cancelers. This approach is conceptually a straightforward generalization of the SISO echo canceler currently used in several modems with overlapping upstream and downstream spectra. It is however a computationally expensive approach, as a bank of $M \times M$ filters is now needed. Furthermore, the approach breaks down if there are legacy disturbers in the binder, not participating in the vector transmission group.

Traditionally, NEXT interference has been addressed by separating the upstream and downstream transmission band. Especially, for the higher frequency bands (over 500 kHz), FDM (frequency division multiplexing) is generally a better way to engineer the network. Examples include the ADSL and VDSL standards. Even FDM architectures suffer from NEXT interference and from legacy disturbers (e.g., ADSL modems on long loops impaired by HDSL [High Bit-Rate DSL] or SHDSL [Symmetric High Bit-Rate DSL] interferers). This is illustrated in Figure 16.10 for a system operating

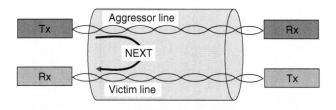

Figure 16.9   Illustration of near-end crosstalk (NEXT) interference.

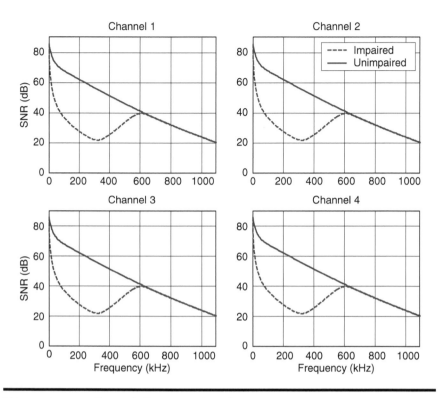

Figure 16.10   Effects of alien NEXT interference on SNR.

on a 3-km long loop. At that distance, FEXT is negligible and is ignored in this plot. In the absence of alien disturbers, each line operates in an undisturbed environment and enjoys the SNR depicted by the plain line. If one SHSDL disturber is present however, the SNR of each line degrades by an amount that depends on the coupling strength between the aggressor and the victim lines. The impaired SNR for four different lines in the binder is shown with the dashed lines in Figure 16.10.

To investigate the effect of that disturber to a vectored system utilizing those four victim pairs, we turn our attention back to the capacity Equation 16.4. Substituting a diagonal system matrix with identical diagonal entries, we obtain

$$C = \int_B \log_2 \left\{ \det \left(\mathbf{I} + \sigma_s^2 |h(\omega)|^2 \mathbf{S}_v^{-1}(\omega) \right) \right\} d\omega \qquad (16.7)$$

Notice that the noise or interference spectral matrix $\mathbf{S}_v(\omega)$ is now in general a full matrix because the interference is no longer spatially white. In particular, $\mathbf{S}_v(\omega)$ will have a signal subspace of dimension one and a noise subspace of dimension $M - 1$, as it contains a single interference source. Using the eigenvalues of $\mathbf{S}_v(\omega)$, we can rewrite Equation 16.7 as

$$C = \int_B \sum_{k=1}^{M} \log_2 \left(1 + \sigma_s^2 |h(\omega)|^2 / \lambda_{v,k}\right) \qquad (16.8)$$

which again maintains a notion of MIMO per line SNR.

Figure 16.11 plots the per line MIMO SNR $\sigma_s^2 |h(\omega)|^2 / \lambda_{v,k}$ for the four lines of Figure 16.10. Notice the distinct difference of the MIMO SNR plots compared to the SISO SNR plots of Figure 16.10. Although Figure 16.10 indicates that the alien disturber affects each of the four individual lines, Figure 16.11 shows that this is not true when those lines are coordinated in a vectored system. In particular, the plots of Figure 16.11 show that appropriate MIMO processing can capitalize on the predictability of interference across lines, and compact all that interference on one line. The performance gains materialize as SNR improvements on the remaining three lines which are now freed of alien crosstalk. In fact, the aggregate

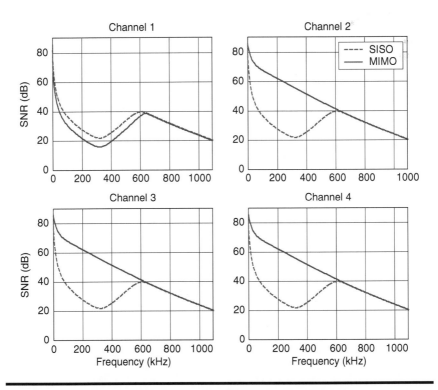

Figure 16.11   **Effects of alien NEXT interference on MIMO SNR.**

bit rate, reduced by about 40 percent with SISO, is recovered to within about 10 precent of the unimpaired rate.

The reader may be surprised by the fact that the SNR of channel 1 actually decreased with MIMO. But this is only because the above procedure only cares to maximize the total capacity. Other methods can be used that yield the same total capacity, and do not degrade the SNR of any channel, as we will see later.

The results developed in this section and the examples presented are intriguing. They show the potential of MIMO vectoring techniques to combat FEXT as well as self and alien NEXT interference. They are of course only illustrative examples and do not fully explore the system's envelope of performance, which generally depends on the number of vectored lines versus the number of alien crosstalkers. Further, they only articulate capacity arguments, and do not provide concrete methodologies and transceiver designs that achieve those gains.

In the next several sections, we investigate specific MIMO architectures for combating self and alien NEXT and FEXT. We discuss their effectiveness and assess their performance. We start with the case of FEXT dominated systems, and then move on to the more general case. We explain the main concepts in a point-to-point multiline setup, and then we comment on applicability in the star architecture case.

16.3 FEXT Mitigation Architectures

Let us consider the FEXT multichannel setup of Figure 16.7. Similarly, to the single line case, some kind of an equalization block is needed to remove the detrimental effects of the channel. In our case, we need to equalize a MIMO system, hence a MIMO equalization block would be appropriate.

16.3.1 Linear Architectures

Figure 16.12 shows a linear $M \times M$ equalization block $\mathbf{F}(\omega)$.

In equation form, the received signal is processed as

$$\hat{\mathbf{s}}(\omega) = \mathbf{F}^H(\omega)\mathbf{y}(\omega) = \mathbf{F}^H(\omega)\mathbf{H}(\omega)\mathbf{s}(\omega) + \mathbf{F}^H(\omega)\mathbf{v}(\omega) \qquad (16.9)$$

16.3.1.1 Linear Zero-Forcing Architecture

The primary purpose of this equalization block is the elimination of FEXT interference. This can be accomplished by selecting $\mathbf{F}^H(\omega) = \mathbf{H}^{-1}(\omega)$, in

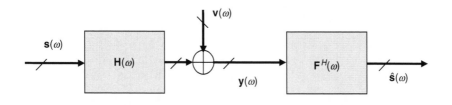

Figure 16.12   Linear MIMO receiver architecture.

which case all self FEXT components are forced to zero and the end-to-end signal model becomes

$$\hat{\mathbf{s}}(\omega) = \mathbf{s}(\omega) + \mathbf{H}^{-1}(\omega)\mathbf{v}(\omega) \tag{16.10}$$

The details of the MIMO filter implementation depend on the line code that the transceiver uses. If a single carrier pulse amplitude modulation (PAM) or quadrature amplitude modulation (QAM) line code is used, then the frequency domain description of Equation 16.9 should be realized through a finite impulse response (FIR) or a pole-zero implementation. If a discrete multitone modulation (DMT) line code is used, then Equation 16.9 can be applied directly in the frequency domain after the Fast Fourier Transform (FFT) processing at the receiver. In other words, the $M \times M$ equalization block $\mathbf{F}(\omega)$ can be applied on each tone independently and will replace the single tap Frequency domain Equalizer (FEQ) common in single line DMT modems (see chapter 8 in [Golden 2006] for more details on DMT modulation and chapter 7 in [Golden 2006] for more details on single carrier modulation).

Zero-forcing equalizers generally have a bad reputation because they tend to amplify the noise (see Equation 16.10). This is especially true in ill conditioned channels (e.g., single line channels with deep nulls). The difference here is that the equalization is across the spatial domain and not across the frequency domain (at least in the DMT modulation case). For a given frequency, the channel matrix $\mathbf{H}(\omega)$ is generally well conditioned because the off-diagonal (FEXT) components can be orders of magnitude smaller than the diagonal ones. Therefore, the inversion of this diagonally dominant matrix will not generally produce catastrophic amplification of the noise. In any event, such problems can be avoided by minimum mean-square error (MMSE) solutions presented next.

A special case arises if one channel uses one frequency and another channel does not use that frequency. The zero-forcing architecture then needs to take account of this case, and invert only the portion of the matrix that corresponds to the lines that are active at this given frequency.

16.3.1.2 Linear MMSE Architecture

MMSE designs appropriately balance FEXT reduction and noise increase by considering the correlation characteristics of the noise in the design. The solution is obtained by minimizing the MSE cost function

$$J_{\text{MSE}}(\omega) = E\{\|\mathbf{s}(\omega) - \mathbf{F}^H(\omega)\mathbf{y}(\omega)\|^2\} \tag{16.11}$$

Equation 16.11 provides the cost function in the frequency domain and is more suitable for a DMT design. A similar formulation could be developed in the time domain for single carrier modems.

The optimal MMSE solution is given by (see for example [Haykin 1996])

$$\mathbf{F}(\omega) = \left(\mathbf{H}(\omega)\mathbf{R}_s(\omega)\mathbf{H}^H(\omega) + \mathbf{R}_v(\omega)\right)^{-1}\mathbf{H}(\omega)\mathbf{R}_s(\omega) \tag{16.12}$$

where

$\mathbf{R}_v(\omega)$ is the noise spectral matrix

$\mathbf{R}_s(\omega)$ is the transmitted signal spectral matrix which is generally diagonal $\mathbf{R}_s(\omega) = \sigma_s^2(\omega)\mathbf{I}$

If the additive noise is also white $\mathbf{R}_v(\omega) = \sigma_v^2\mathbf{I}$, the solution reduces to

$$\mathbf{F}(\omega) = \left(\mathbf{H}(\omega)\mathbf{H}^H(\omega) + (\sigma_v^2/\sigma_s^2(\omega))\mathbf{I}\right)^{-1}\mathbf{H}(\omega) \tag{16.13}$$

Further performance improvements can be obtained using decision-directed architectures as explained next.

16.3.2 Decision Feedback Architectures

A general decision feedback (DF) structure for MIMO transmission is shown in Figure 16.13.

Under the assumption that all decisions are correct, the output of the slicer is equal to $\mathbf{s}(\omega)$, which is fed back to aid in other decisions. Mathematically, the receiver operation can be described as

$$\hat{\mathbf{s}}(\omega) = \mathbf{F}^H(\omega)\mathbf{y}(\omega) - \mathbf{B}^H(\omega)\mathbf{s}(\omega) \tag{16.14}$$

For this structure to be realizable, the decisions must be made sequentially in conjunction with the feedback in a causal manner, so that whenever a decision is fed back it must already have been made. This restricts the structure of $\mathbf{B}^H(\omega)$ to a strictly triangular matrix (or a permuted version thereof).

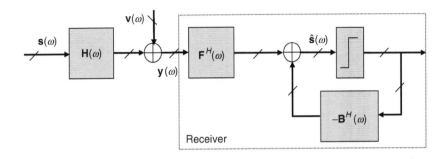

Figure 16.13   **General MIMO decision feedback (DF) architecture.**

16.3.2.1 *Zero-Forcing DF Architecture*

A generalized DF architecture is proposed for FEXT cancelation in coordinated DMT systems in [Ginis 2001] and expanded in [Verdu 1998]. The architecture is based on triangularizing the channel at each tone using the QR decomposition and generating suitable feedback and feed-forward matrices $\mathbf{B}^H(\omega)$ and $\mathbf{F}^H(\omega)$. This has been shown to be a zero-forcing version of the generalized decision feedback equalizer (GDFE).

Thus, consider the signal model from Equation 16.3 and assume that the noise $\mathbf{v}(\omega)$ is white. Note that this is without loss of generality, because coordination between the receivers allows a channel with arbitrary nonsingular noise covariance matrix $\mathbf{S}_v(\omega)$ to be turned into an equivalent AWGN channel though a noise-whitening filter at the receiver (see for example [Ginis 2002]). Let

$$\mathbf{H}(\omega) = \mathbf{Q}(\omega)\mathbf{R}(\omega) \tag{16.15}$$

where
  $\mathbf{Q}(\omega)$ is a unitary matrix
  $\mathbf{Q}(\omega)\mathbf{Q}^H(\omega) = \mathbf{I}$
  $\mathbf{R}(\omega)$ is upper triangular

As a first step in the receive chain, apply $\mathbf{Q}^H(\omega)$ to the input vector to produce a triangular aggregate channel

$$\mathbf{Q}^H(\omega)\mathbf{y}(\omega) = \mathbf{R}(\omega)\mathbf{s}(\omega) + \mathbf{Q}^H(\omega)\mathbf{v}(\omega) \tag{16.16}$$

and note that because $\mathbf{Q}^H(\omega)$ is unitary, the noise $\mathbf{Q}^H(\omega)\mathbf{v}(\omega)$ is still white.
  Next, decompose the triangular matrix into its diagonal and off-diagonal components

$$\mathbf{R}(\omega) = \mathbf{R}_{\text{diag}}(\omega) + \mathbf{R}_{\text{off}}(\omega) \tag{16.17}$$

where

$\mathbf{R}_{\text{diag}}(\omega)$ is strictly diagonal

$\mathbf{R}_{\text{off}}(\omega)$ is strictly upper triangular

Assuming that $\mathbf{R}_{\text{diag}}(\omega)$ is invertible* the received signal may be scaled further by its inverse, yielding the signal

$$\tilde{\mathbf{y}}(\omega) = \mathbf{R}_{\text{diag}}^{-1}(\omega)\mathbf{Q}^{H}(\omega)\mathbf{y}(\omega)$$

$$= \left(\mathbf{I} + \mathbf{R}_{\text{diag}}^{-1}(\omega)\mathbf{R}_{\text{off}}(\omega)\right)\mathbf{s}(\omega) + \mathbf{R}_{\text{diag}}^{-1}(\omega)\,\mathbf{Q}^{H}(\omega)\mathbf{v}(\omega) \quad (16.18)$$

Observe that the second term in the parenthesis is a strictly upper-triangular matrix. The transmitted signal contribution through that term may therefore be canceled via the feedback, giving a signal that is free of FEXT.

$$\hat{\mathbf{s}}(\omega) = \tilde{\mathbf{y}}(\omega) - \mathbf{R}_{\text{diag}}^{-1}(\omega)\,\mathbf{R}_{\text{off}}(\omega)\,\mathbf{s}(\omega) \quad (16.19)$$

Observe also that the noise gets amplified by the inverse of the diagonal elements of $\mathbf{R}(\omega)$.

Finally, the decisions $\mathbf{s}(\omega)$ are obtained by slicing $\hat{\mathbf{s}}(\omega)$. The full algorithm is outlined below.

Zero-Forcing DF Algorithm:

Initialization: For each tone ω, set

$$\mathbf{Q}(\omega)\,\mathbf{R}(\omega) = \mathbf{H}(\omega)$$
$$\mathbf{F}^{H}(\omega) = \mathbf{R}_{\text{diag}}^{-1}(\omega)\,\mathbf{Q}^{H}(\omega)$$
$$\mathbf{B}^{H}(\omega) = \mathbf{R}_{\text{diag}}^{-1}(\omega)\,\mathbf{R}_{\text{off}}(\omega)$$

Showtime: For each DMT symbol and each tone

1. Compute $\tilde{\mathbf{y}}(\omega) = \mathbf{F}^{H}(\omega)\mathbf{y}(\omega)$
2. For $k = M$ down to 1

    (1) Compute $\hat{s}_k(\omega) = \tilde{y}_k(\omega) - \sum_{j=k+1}^{M} b_{jk}(\omega)\, s_j(\omega)$

    (2) Set $s_k(\omega) = \left\lfloor \hat{s}_k(\omega) \right\rfloor$

where the subscripts denote matrix or vector elements and $\lfloor \cdot \rfloor$ is the slicing operation.

---

* If the channel matrix at this tone is rank-deficient by k, then k of the transmit signals should be eliminated, the corresponding rows and columns of $\mathbf{R}(\omega)$ removed, and the corresponding columns of $\mathbf{Q}(\omega)$ also removed, leaving the modified channel with full rank. In the DSL environment, this most likely means that k of the transmit channels is zero and should not be considered for transmission in the first place.

The algorithm makes the assumption that the channel is perfectly stationary so that the initial matrices do not need to change. In reality, the DSL environment varies slowly over time, so that adjustments must be made. In general, the changes are slow enough so that this can be done by applying the same initialization algorithm occasionally to an updated channel estimate. Other adaptation schemes may also be possible, but are outside the scope of this tutorial.

The primary benefits of this architecture over the linear one is that the noise tends to get amplified less. In particular, it avoids coloring the noise seen at the slicer, thereby yielding an optimal environment for detection.

The downside is the same as with any other DF architecture; an incorrect decision may lead to a propagation of incorrect decisions on the channels that use it for feedback and may thus be catastrophic in environments with high error probabilities. However, because DSL modems usually operate at very low error rates and use coding to further reduce detection errors, this problem is mitigated.

16.3.2.2 MMSE-DF Architecture

A decision feedback architecture may also be derived from the MMSE formulation in Section 16.3.1.2, as described in [Al-Dhahir 2000] and [Yu 2000]. We start with the MSE cost function similar to Equation 16.11

$$J_{\text{MSE}}(\omega) = E\left\{\|\mathbf{e}(\omega)\|^2\right\} = \text{trace}\left[E\left\{\mathbf{e}(\omega)\mathbf{e}^H(\omega)\right\}\right] \tag{16.20}$$

where

$$e(\omega) = \hat{\mathbf{s}}(\omega) - \mathbf{s}(\omega) = \mathbf{F}^H(\omega)\mathbf{y}(\omega) - \tilde{\mathbf{B}}^H(\omega)\mathbf{s}(\omega) \tag{16.21}$$

with $\tilde{\mathbf{B}}^H(\omega) = \mathbf{I} + \mathbf{B}^H(\omega)$. Substituting Equation 16.21 into Equation 16.20 we obtain

$$E\left\{\mathbf{e}(\omega)\mathbf{e}^H(\omega)\right\} = \mathbf{F}^H(\omega)\mathbf{R}_\mathbf{y}(\omega)\mathbf{F}(\omega) - \mathbf{F}^H(\omega)\mathbf{R}_{\text{ys}}(\omega)\tilde{\mathbf{B}}(\omega)$$

$$-\tilde{\mathbf{B}}^H(\omega)\mathbf{R}_{\text{ys}}^H(\omega)\mathbf{F}(\omega) + \tilde{\mathbf{B}}^H(\omega)\mathbf{R}_\mathbf{s}(\omega)\tilde{\mathbf{B}}(\omega) \tag{16.22}$$

where

$$\mathbf{R}_\mathbf{y}(\omega) = \mathbf{H}(\omega)\mathbf{R}_\mathbf{s}(\omega)\mathbf{H}^H(\omega) + \mathbf{R}_\mathbf{v}(\omega)$$

$$\tag{16.23}$$

$$\mathbf{R}_{\text{ys}}(\omega) = \mathbf{H}(\omega)\mathbf{R}_\mathbf{s}(\omega)$$

By manipulating Equation 16.22 we obtain

$$E\left\{\mathbf{e}(\omega)\mathbf{e}^H(\omega)\right\} = \tilde{\mathbf{B}}^H(\omega)(\mathbf{R}_s(\omega)$$

$$-\mathbf{R}_{ys}^H(\omega)\mathbf{R}_y^{-1}(\omega)\mathbf{R}_{ys}(\omega))\tilde{\mathbf{B}}(\omega)$$

$$+(\mathbf{F}^H(\omega) - \tilde{\mathbf{B}}^H(\omega)\mathbf{R}_{ys}^H(\omega)\mathbf{R}_y^{-1}(\omega))\mathbf{R}_y(\omega)\mathbf{T}(\omega) \quad (16.24)$$

where

$$\mathbf{T}(\omega) = \mathbf{F}^H(\omega) - \tilde{\mathbf{B}}^H(\omega)\mathbf{R}_{ys}^H(\omega)\mathbf{R}_y^{-1}(\omega)^H$$

Equation 16.24 indicates that the third term can be made equal to zero for the optimal choice of the feed-forward matrix

$$\mathbf{F}^H(\omega) = \tilde{\mathbf{B}}^H(\omega)\mathbf{R}_{ys}^H(\omega)\mathbf{R}_y^{-1}(\omega) \quad (16.25)$$

Then the optimal choice for the feedback matrix is the result of minimizing the quadratic form trace $\{\tilde{\mathbf{B}}^H(\omega)\Phi(\omega)\tilde{\mathbf{B}}(\omega)\}$ where

$$\Phi(\omega) = \mathbf{R}_s(\omega) - \mathbf{R}_{ys}^H(\omega)\mathbf{R}_y^{-1}(\omega)\mathbf{R}_{ys}(\omega) \quad (16.26)$$

subject to the special triangular form of the feedback matrix.
It can be shown (see [Al-Dhahir 2000]) that the optimal solution is

$$\tilde{\mathbf{B}}(\omega) = \mathbf{L}(\omega) \quad (16.27)$$

where $\mathbf{L}(\omega)$ is the Cholesky factor of the matrix

$$\Phi^{-1}(\omega) = \mathbf{L}(\omega)\mathbf{D}(\omega)\mathbf{L}^H(\omega) \quad (16.28)$$

with
$\mathbf{L}(\omega)$ a triangular matrix with a unit diagonal
$\mathbf{D}(\omega)$ a diagonal matrix

Further results on the performance of the Decision Feedback Equalizer (DFE) receiver in comparison to the MIMO capacity can be found in [Yu 2000]. The full algorithm is outlined below.

MMSE DF Algorithm:

Initialization: For each tone ω, set

$$\Phi(\omega) = R_s(\omega) - R_{ys}^H(\omega)R_y^{-1}(\omega)R_{ys}(\omega)$$
$$\Phi^{-1}(\omega) = L(\omega)D(\omega)L^H(\omega)$$
$$B(\omega) = L(\omega) - I$$
$$F^H(\omega) = L^H(\omega)R_{ys}^H(\omega)R_y^{-1}(\omega)$$

Showtime: For each DMT symbol and each tone

1. Compute $\tilde{y}(\omega) = F^H(\omega)y(\omega)$

2. For $k = M$ down to 1

    (a) Compute $\hat{s}_k(\omega) = \tilde{y}_k(\omega) - \sum\limits_{j=k+1}^{M} b_{jk}(\omega)s_j(\omega)$

    (b) Set $s_k(\omega) = \lfloor \hat{s}_k(\omega) \rfloor$

where the subscripts denote matrix or vector elements and $\lfloor \cdot \rfloor$ is the slicing operation.

This solution is in some ways better than the zero-forcing one, but somewhat more difficult to translate into a point-to-multipoint solution (see Section 16.4.2), because of the fact that the feed-forward matrix $\mathbf{F}^H(\omega)$ is not unitary, and therefore does not preserve signal power. However, in most DSL applications the SNR at most tones is very large, so that the difference between the two solutions is minimal.

16.3.3 *Two-Sided Architectures*

An alternative approach which can achieve the capacity of the channel is to precondition the signal before transmission, then transmit along "virtual" channels that have some desired property, and finally compensate for both the preconditioning and the channel at the receiver.

A general structure for a linear version of such "two-sided" transmission is shown in Figure 16.14.

16.3.3.1 *SVD Architecture*

A two-sided architecture for FEXT cancelation in DMT systems is proposed in [Tauböck 2000]. The architecture is based on decomposing the channel into completely independent channels at each tone via the singular value decomposition (SVD).

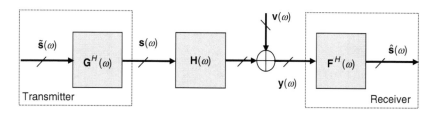

Figure 16.14   General linear two-sided structure.

Again, consider the signal model from Equation 16.3 and assume that the noise $\mathbf{v}(\omega)$ is white. Apply the SVD to the channel matrix at tone ω

$$\mathbf{H}(\omega) = \mathbf{Q}(\omega)\,\Lambda(\omega)\,\mathbf{P}^H(\omega) \tag{16.29}$$

where
$\mathbf{Q}(\omega)$ and $\mathbf{P}(\omega)$ are unitary matrices
$\Lambda(\omega)$ is diagonal with real and nonnegative values

The virtual channel represented by $\Lambda(\omega)$ is free of FEXT, so each sub-channel can be used independently for transmission.

To utilize this diagonal channel, define the transmitted signal as $\tilde{\mathbf{s}}(\omega)$, precondition it by $\mathbf{P}(\omega)$ to get the signal $\mathbf{s}(\omega) = \mathbf{P}(\omega)\tilde{\mathbf{s}}(\omega)$ on the physical channel and compensate by $\mathbf{Q}^H(\omega)$ at the receiver. Thus

$$\mathbf{Q}^H(\omega)\mathbf{y}(\omega) = \mathbf{Q}^H(\omega)\left(\mathbf{H}(\omega)\mathbf{s}(\omega) + \mathbf{v}(\omega)\right)$$

$$= \mathbf{Q}^H(\omega)\mathbf{H}(\omega)\mathbf{P}(\omega)\tilde{\mathbf{s}}(\omega) + \mathbf{Q}^H(\omega)\mathbf{v}(\omega) \tag{16.30}$$

$$= \Lambda(\omega)\tilde{\mathbf{s}}(\omega) + \mathbf{Q}^H(\omega)\mathbf{v}(\omega)$$

Note that as in the DF architecture, the noise $\mathbf{Q}^H(\omega)\mathbf{v}(\omega)$ is still white.

To retrieve the transmitted signal, the receiver next compensates for the attenuation of the virtual channels by inverting $\Lambda(\omega)$, assuming it is invertible,* yielding

$$\hat{\mathbf{s}}(\omega) = \Lambda^{-1}(\omega)\mathbf{Q}^H(\omega)\mathbf{y}(\omega)$$

$$= \tilde{\mathbf{s}}(\omega) + \Lambda^{-1}(\omega)\,\mathbf{Q}^H(\omega)\mathbf{v}(\omega) \tag{16.31}$$

* See previous footnote.

Observe that the noise gets amplified by the inverse of the virtual channel, but no FEXT is present.

As before, the decisions $\mathbf{s}(\omega)$ are obtained by slicing $\hat{\mathbf{s}}(\omega)$. The algorithm is outlined below.

SVD-Based Algorithm:

Initialization: For each tone ω, set

$$\mathbf{Q}(\omega)\,\Lambda(\omega)\mathbf{P}^H(\omega) = \mathbf{H}(\omega)$$
$$\mathbf{F}^H(\omega) = \Lambda^{-1}(\omega)\,\mathbf{Q}^H(\omega)$$
$$\mathbf{G}^H(\omega) = \mathbf{P}(\omega)$$

Showtime: For each DMT symbol and each tone ω

  1. Compute $\mathbf{s}(\omega) = \mathbf{G}^H(\omega)\,\tilde{\mathbf{s}}(\omega)$ at the transmitter

  2. Compute $\hat{\mathbf{s}}(\omega) = \mathbf{F}^H(\omega)\mathbf{y}(\omega)$ at the receiver

  3. Set $s_k(\omega) = \lfloor \hat{s}_k(\omega) \rfloor$ independently for each channel k

where $\lfloor \cdot \rfloor$ is the slicing operation.

As before, the algorithm makes the assumption that the channel is perfectly stationary so that the initial matrices do not need to change, but in practice, a good channel estimate should be maintained continuously and the matrices updated occasionally.

The primary benefit of this architecture over the DF one is that error propagation from channel to channel is avoided. However, the complexity associated with configuring the transmitter matrix for each tone detracts from its ease of implementation.

16.4 Point-to-Multipoint Architectures

The above architectures all utilize collocation of the receivers to remove or mitigate crosstalk as an interference source. Some loop configurations do not lend themselves to this, of course, or if they do so in one direction, they may not in the other. As an example, most consumer DSL installations are arranged in a point-to-multipoint configuration, where the downstream receivers are located in separate buildings, but the downstream transmitters in a single CO.

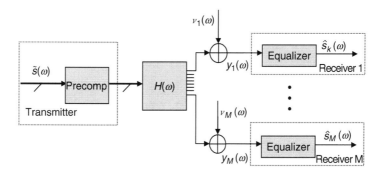

Figure 16.15   Point-to-multipoint architecture.

In these situations, some crosstalk can be addressed by pre-compensating at the transmitter (e.g., [Cendrillon 2005]). A general depiction of a transceiver architecture for doing this is shown in Figure 16.15, where the pre-compensator is selected to cancel crosstalk without unduly increasing the transmitted power.

16.4.1   Linear Architecture

Perhaps the simplest approach for canceling crosstalk at the transmitter is via a linear pre-compensator, see for example discussion in [Cendrillon 2006b].

Define the transmitted signal as

$$\mathbf{s}(\omega) = \frac{1}{\beta(\omega)}\mathbf{H}^{-T}(\omega)\,\mathbf{H}_{\text{diag}}(\omega)\,\tilde{\mathbf{s}}(\omega) \tag{16.32}$$

where
   $\tilde{\mathbf{s}}(\omega)$ is the desired transmit symbol
   $\mathbf{H}_{\text{diag}}(\omega)$ is the diagonal of the channel matrix
   $\mathbf{H}^{-T}$ is the inverse transpose of the channel matrix
   $\beta(\omega)$ is a scalar chosen so that the transmit power on each pair is limited

For example, this scalar may be chosen so that the power of $\mathbf{s}(\omega)$ is no larger than that of $\tilde{\mathbf{s}}(\omega)$ by setting

$$\beta(\omega) = \max_{n}\left\|\text{row n of}\,\mathbf{H}^{-T}(\omega)\,\mathbf{H}_{\text{diag}}(\omega)\right\| \tag{16.33}$$

Then the received signal,

$$\mathbf{y}(\omega) = \frac{1}{\beta(\omega)}\mathbf{H}_{\text{diag}}(\omega)\,\tilde{\mathbf{s}}(\omega) + \mathbf{v}(\omega) \tag{16.34}$$

is free of FEXT and can be equalized independently for each line. The resulting equalized signal for each receiver k is

$$\hat{s}_k(\omega) = \beta(\omega)\frac{y_k(\omega)}{h_{kk}(\omega)} \tag{16.35}$$

In general, this architecture is somewhat suboptimal, because the scalar $\beta(\omega)$ may have to be chosen conservatively to satisfy a power constraint on all transmit pairs. However, in most DSL environments, all rows in the channel matrix are strongly dominated by the diagonal element, so $\beta(\omega)$ is close to one and the architecture is close to optimal. Some care is needed when one channel uses a frequency, while another channel does not use that frequency; filtering should only consider the active lines for the given frequency. Further details may be found in [Cendrillon 2006b].

16.4.2   QR Architecture

The receiver-based QR architecture described in Section ´16.3.2.1 does, somewhat surprisingly, lead to a dual architecture that applies very well to FEXT cancelation in a point-to-multipoint configuration. This architecture is described for DMT systems in [Verdu 1998].

The first step in the derivation is to triangularize the transposed channel matrix at each tone using the QR decomposition,

$$\mathbf{H}^T(\omega) = \mathbf{Q}(\omega)\,\mathbf{R}(\omega) \tag{16.36}$$

where
$\mathbf{Q}(\omega)$ is a unitary matrix
$\mathbf{R}(\omega)$ is upper triangular

Then, break $\mathbf{R}(\omega)$ into its diagonal and off-diagonal parts

$$\mathbf{R}(\omega) = \mathbf{R}_{\text{diag}}(\omega) + \mathbf{R}_{\text{off}}(\omega) \tag{16.37}$$

It is easy to see that if the transmitted signal is defined as

$$\mathbf{s}(\omega) = \mathbf{Q}^*(\omega)\,\mathbf{R}^{-T}(\omega)\,\mathbf{R}_{\text{diag}}(\omega)\,\tilde{\mathbf{s}}(\omega) \tag{16.38}$$

where

$\tilde{\mathbf{s}}(\omega)$ is the desired transmit symbol

$\mathbf{Q}^*(\omega)$ denotes the element-wise conjugate of the matrix (without transposition)

$\mathbf{R}^{-T}(\omega)$ denotes the inverse transpose of $\mathbf{R}(\omega)$

Then the received signal $\mathbf{y}(\omega)$ is free of FEXT and can be equalized independently for each line. In fact, the received signal may be written as

$$\mathbf{y}(\omega) = \mathbf{R}^T(\omega)\,\mathbf{Q}^T(\omega)\,\mathbf{s}(\omega) + \mathbf{v}(\omega)$$

$$= \mathbf{R}_{\text{diag}}(\omega)\,\tilde{\mathbf{s}}(\omega) + \mathbf{v}(\omega) \tag{16.39}$$

so the appropriately equalized signal for each receiver k is

$$\hat{s}_k(\omega) = \frac{y_k(\omega)}{r_{kk}(\omega)} \tag{16.40}$$

A problem with the above derivation is that no means are taken to limit the power transmitted on the physical lines, while this may be strictly limited at each tone by a power mask. Fortunately, this is easily addressed by simply redefining the transmit symbol $\tilde{\mathbf{s}}(\omega)$ as follows. Let the range of valid transmit values for the constellation on tone ω and line k be $P_k(\omega)$ and define the true transmit symbol as $\tilde{s}_k^0(\omega)$. Then, both the real and imaginary values of $\tilde{s}_k^0(\omega)$ must lie in the range $[-P_k(\omega)/2, P_k(\omega)/2]$. Further, define

$$\mathbf{u}(\omega) = \mathbf{R}^{-T}(\omega)\mathbf{R}_{\text{diag}}(\omega)\,\tilde{\mathbf{s}}(\omega) \tag{16.41}$$

and

$$\tilde{s}_k(\omega) = \tilde{s}_k^0(\omega) + p_k(\omega) \tag{16.42}$$

with $p_k(\omega)$ as complex integer multiple $n_k(\omega)$ of the valid range

$$p_k(\omega) = n_k(\omega)P_k(\omega) \tag{16.43}$$

Then, choose the real and imaginary parts of $n_k(\omega)$ such that the corresponding parts of $u_k(\omega)$ are both within those same limits.

Note that, in practice, this can be done successively for each line, because of the triangular structure of the mapping from $\tilde{\mathbf{s}}(\omega)$ to $\mathbf{u}(\omega)$. In fact, because Equation 16.41 may be rewritten as

$$u_k(\omega) = \tilde{s}_k(\omega) - \sum_{j=1}^{k-1} \frac{r_{jk}(\omega)}{r_{kk}(\omega)} u_j(\omega) \qquad (16.44)$$

it may be accomplished by setting $\tilde{s}_k(\omega) = \tilde{s}_k^0(\omega)$ and following the computation of each $u_k(\omega)$ with a modulo operation on the range $P_k(\omega)$ (see [Verdu 1998] and [Yu 2001]). Note also, that once the real and imaginary parts of $\mathbf{u}(\omega)$ are all within allowable ranges, the power of the actual transmit signal $\mathbf{s}(\omega) = \mathbf{Q}^*(\omega)\,\mathbf{u}(\omega)$ is also within range, because the mapping between the two is unitary.

The equalized signal now becomes

$$\hat{s}_k(\omega) = \frac{y_k(\omega)}{r_{kk}(\omega)}$$

$$= \tilde{s}_k^0(\omega) + n_k(\omega)P_k(\omega) + \frac{v_k(\omega)}{r_{kk}(\omega)} \qquad (16.45)$$

Because $\tilde{s}_k^0(\omega)$ is in the valid range and any nonzero integer $n_k(\omega)$ yields a value outside of that range, the true transmit symbol (plus the noise) may be recovered with a simple modulo operation.

This is a very effective way of removing FEXT in the point-to-multipoint configuration, but may suffer somewhat from decoding errors owing to noise effects on the modulo operation at the receiver Equation 16.45. Similar to the discussion before, some care is needed when one channel uses a frequency, while another channel does not use that frequency; filtering should only consider the active lines for the given frequency.

16.5 Performance

All of the four receiver-based architectures described above cancel almost all FEXT in most situations and bring the average capacity to the capacity of an unimpaired loop shown in Figure 16.7. Moreover, the capacity of most of the individual pairs is also very close to the average.

598 ■ *Implementation and Applications of DSL Technology*

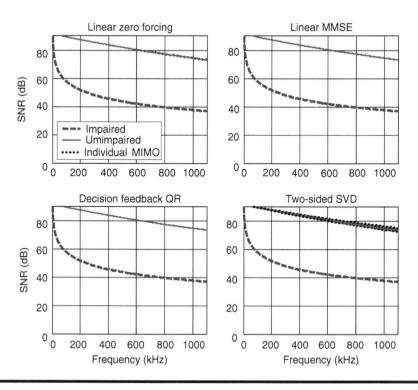

Figure 16.16   Comparison of performance for MIMO architectures.

This is depicted in Figure 16.16, where the capacity of each individual pair for each of the four architectures is compared to the unimpaired and impaired capacities. For the linear zero forcing, the linear MMSE and the decision feedback QR architecture, the individual capacities (multiple dotted lines are all indistinguishable from the unimpaired capacity [plain lines]), whereas the two-sided SVD architecture gives in a small spread around it.

A discussion on the near-optimality of linear crosstalk cancelation and pre-compensation schemes may be found in [Cendrillon 2006b]. For example, a bound on the capacity loss of the linear Zero-Forcing (ZF) architecture as a function of column-wise diagonal dominance of the upstream channel matrix is derived, which shows that in 99 percent of channels (using worst 1 percent case models) the capacity loss is less than 2 percent. A similar result holds for the downstream channel.

Because of the near-optimality of these simple architectures under normal conditions, there is typically little extra benefit in using MMSE and DF architectures. For other loop conditions, the performance comparison may be different. In particular, when the transmission is strongly impaired by alien noise, linear architectures are not sufficient for near optimal performance. Decision feedback and receiver-based architectures are then needed. When pre-compensation is considered however, linear architectures are likely to be close to optimal.

16.6   Conclusions

Vectored MIMO technologies have the potential to provide a step function increase in the performance of DSL systems. In this chapter, we have shown the applicability of several MIMO architectures to the DSL space and the copper network. We have also illustrated their crosstalk mitigation and performance enhancing properties. These concepts provide the basic tools that DSL engineers can use to design next generation DSL systems and solve the challenging problems in migrating the current infrastructure to the next generation one.

MIMO vectoring for DSL systems is an active research area and further advances are expected in several areas. Current challenges and research focus for vectoring systems lie in the following areas:

- Combining vectoring architectures with coding schemes.
- Investigating the challenges of training and adaptation, especially in the point-to-multipoint environment.

Glossary

ADSL: Asymmetric Digital Subscriber Line
CO: central office
DSM: dynamic spectral management
FDD: frequency division duplex
FEXT: far-end crosstalk
MIMO: multiple-input–multiple-output
NEXT: near-end crosstalk
PSD: power spectral density
SISO: single-input–single-output
SVD: singular value decomposition

Bibliography

[Al-Dhahir 2000]   N. Al-Dhahir and A.H. Sayed, *The Finite Length Multi-Input Multi-Output MMSE-DFE*, IEEE Transactions on Signal Processing, Vol. 48, No. 10, pp. 2921–2936, October 2000.

[Cendrillon 2004]   R. Cendrillon, G. Ginis, M. Moonen, and K. Van Acker, *Partial Crosstalk Precompensation in Downstream VDSL*, Signal Processing, Vol. 84, pp. 2005–2019, 2004.

[Cendrillon 2006a]   R. Cendrillon, G. Ginis, E. Van den Bogaert, and M. Moonen, *A Near-optimal Linear Crosstalk Canceler for VDSL*, accepted for IEEE Transactions on Signal Processing.

[Cendrillon 2006b]   R. Cendrillon, G. Ginis, E. Van den Bogaert, and M. Moonen, *A Near-optimal Linear Crosstalk Precoder for VDSL*, accepted for publication in IEEE Transactions on Communications.

[Cioffi 2006]   J.M. Cioffi, Ed., *Dynamic Spectrum Management*, NIPP-NAI Technical Report, 2006.

[Cover 1991]   T.M. Cover and J.A. Thomas, *Elements of Information Theory*, John Wiley & Sons, New York, New York. 1991.

[Ginis 2001]   G. Ginis and J.M. Cioffi, *Vectored-DMT: A FEXT Canceling Modulation Scheme for Coordinating Users*, Proceedings of IEEE International Conference on Communications, Vol. 1, pp. 305–309, June 2001.

[Ginis 2002]   G. Ginis and J.M. Cioffi, *Vectored Transmission for Digital Subscriber Line Systems*, IEEE Journal on Selected Areas in Communications, Vol. 20, No. 5, pp. 1085–1104, June 2002.

[Ginis 2006]   G. Ginis and C.-N. Peng, *Alien Crosstalk Cancellation for Multipair Digital Subscriber Line Systems*, EURASIP Journal on Applied Signal Processing, Vol. 2006, pp. 1–12, Article ID 16828, 2006.

[Golden 2006]   P. Golden, H. Dedieu, and K.S. Jacobsen, Ed., *Fundamentals of DSL Technology*, Auerbach Publications, Boca Raton, Florida, 2006, p. 457.

[Haykin 1996]   S. Haykin, *Adaptive Filter Theory*, 3rd Edition, Prentice Hall, New Jersey, 1996.

[Joffe 2002]   D. Joffe, *MIMO Cable Measurements and Models: An Intuitively Satisfying Approach*, T1E1 contribution T1E1.4/2002-239R1, November 2002.

[Louveaux 2005]   J. Louveaux and A.-J. van der Veen, *Downstream VDSL Channel Tracking Using Limited Feedback for Crosstalk Precompensated Schemes*, Proceedings of Acoustics, Speech, and Signal Processing (ICASSP '05), Vol. 3, pp. iii/337–iii/340, March 18–23, 2005.

[Song 2002]   K.B. Song, S.T. Chung, G. Ginis, and J.M. Cioffi, *Dynamic Spectrum Management for Next-Generation DSL Systems*, IEEE Communications Magazine, pp. 101–109, October 2002.

[Tauböck 2000]   G. Tauböck and W. Henkel, *MIMO Systems in the Subscriber-Line Network*, Proceedings of 5th International ODFM Workshop, Hamburg, Germany, September 2000.

[Verdu 1998]   S. Verdu, *Multiuser Detection*, Cambridge University Press, Cambridge, 1998.

[Yu 2000]   W. Yu and J.M. Cioffi, *Multiuser Detection in Vector Multiple Access Channels Using Generalized Decision Feedback Equalization*, 5th International Conference on Signal Processing, World Computer Congress, Beijing, August 2000.

[Yu 2001]   W. Yu and J.M. Cioffi, *Trellis Precoding for the Broadcast Channel*, Proceedings of Globecom, San Antonio, Texas, Vol. 2, pp. 1344–1348, 2001.

[Zeng 2002]   C. Zeng and J.M. Cioffi, *Near-End Crosstalk Mitigation in ADSL Systems*, IEEE Journal on Selected Areas in Communications, Vol. 20, No. 5, pp. 949–958, June 2002.

[ANSI T1.417]   *Spectrum Management for Loop Transmission Systems*, ANSI Standard T1.417.

[ITU G.992.3]   *Asymmetric Digital Subscriber Line Transceivers 2 (ADSL2)*, ITU Recommendation G.992.3.

Chapter 17

DSL Standardization

Les Brown, Angus Carrick, Krista S. Jacobsen, Ragnar H. Jonsson, and Sigurd Schelstraete

CONTENTS

The authors would like to thank the delegates to the T1E1.4, TM6, and SG15/Q4 meetings, whose work resulted in the standards and Recommendations that are the subject of this chapter. Needless to say, this chapter would not exist (nor would much of our gray hair) without their contributions to the industry. We also thank the reviewers of the chapter, who graciously committed significant amounts of time to improve the accuracy and clarity of this chapter.

Abstract   This chapter discusses the standardization of Digital Subscriber Line (DSL). It presents an overview of the groups responsible for DSL standardization and related activities and also provides some of the history of DSL. This chapter provides details of the Asymmetric DSL (ADSL), High Bit-Rate DSL (HDSL), Integrated Services Digital Network (ISDN), Symmetric High Bit-Rate DSL (SHDSL), and Very High Bit-Rate DSL (VDSL) standards and concludes with a description of handshaking, which is the common procedure used by all DSL modems at the beginning of the startup sequence.

17.1   Introduction

To enable connections between Digital Subscriber Line (DSL) modems from different manufacturers, there must be an agreement, in advance, of how the modems will behave. For example, to establish a connection, the modems on opposite ends of a telephone line must be aware of what sort of information the various initialization signals will contain, when those signals will be transmitted, and how the transmitted signals will be modulated. Furthermore, after a connection has been established, the modems may need to exchange information either to respond to changing channel and noise conditions or to optimize performance. Protocols and signal formats for this information must also be agreed in advance. The definition of those modem elements that are critical to ensure interoperability between any two DSL modems of the same type—whether Asymmetric DSL (ADSL), Very High Bit-Rate DSL (VDSL), or one of the symmetrical types such as Symmetric High Bit-Rate DSL (SHDSL)—is the role of the DSL standardization groups.

Several standards organizations address the standardization of DSL. The Telecommunications Sector of the International Telecommunication Union (known as the ITU-T) defines transceiver specifications for use internationally. A working group under the Alliance for Telecommunications Industry Solutions (ATIS) writes transceiver specifications for North America, and the European Telecommunications Standards Institute (ETSI) generates standards for European use. The Telecommunication Technology Committee (TTC) develops telecommunications standards for Japan, and the China Communications Standards Association (CCSA) does the same for China. Two additional organizations do not standardize DSL transceiver specifications but perform work that is relevant to the industry, and they reference the standards developed by the above mentioned groups as appropriate in their work. The first is the DSL Forum, which is an industry consortium focused primarily on initiatives, such as interoperability testing and management of modems, to accelerate DSL deployment. The second is the Institute of Electrical and Electronics Engineers (IEEE), which has based its Ethernet in the first mile (EFM) standard on two initial DSL transceiver specifications (namely, SHDSL and first generation VDSL; see Sections 17.3 and 17.5). This chapter provides a brief overview of each organization's role in the standardization of DSL.

They are the symmetric DSLs, which support the same bit rate in both the downstream and upstream directions; ADSL, which provides a higher bit rate in the downstream direction (toward the subscriber) than in the upstream direction (from the subscriber); and VDSL, which can support very high-speed symmetrical or asymmetrical services on shorter loops.

Also addressed is handshaking, the process by which two modems at opposite ends of a telephone line declare their basic capabilities (i.e., what type(s) of DSL they support, etc.), and negotiate operational parameters in preparation for a full initialization process specific to the selected DSL type.

17.2   The DSL Standards Organizations

Before delving into the details of DSL standards, it is useful to have an understanding of the organizations involved in DSL standardization and their relationships to each other. The groups that standardize or support standardization of DSL are the ITU-T, ATIS, ETSI, the DSL Forum, and the IEEE.

17.2.1 ITU

The ITU is a United Nations organization that relies on participation from both governments and the private sector to coordinate global telecommunication networks and services. The ITU-T generates standards called Recommendations for all fields of telecommunications. The ITU-T was created on March 1, 1993 to replace the International Telegraph and Telephone Consultative Committee (CCITT), which had been active in various forms since 1865.

Standardization work in the ITU-T is carried out by study groups. DSL standardization is carried out by Study Group 15, which is the lead study group on access network transport and optical technology. The focus of Study Group 15 is "the development of standards on optical and other transport network infrastructures, systems, equipment, optical fibres, and the corresponding control plane technologies to enable the evolution toward intelligent transport networks. This encompasses the development of related standards for the customer premises, access, metropolitan, and long haul sections of communication networks" [SG15 web]. The work of Study Group 15 is partitioned into work areas known as "Questions." Work on DSL takes place within Question 4 (abbreviated herein as SG15/Q4), which is entitled "Transceivers for customer access and in-premises phone line networking systems on metallic pairs." SG15/Q4 has been led since its inception by venerable rapporteur Richard "Dick" Stuart, and beginning in the late 1990s, SG15/Q4 became primarily responsible for DSL transceiver standards, including ADSL, VDSL, and SHDSL. In addition to generating transceiver specifications, SG15/Q4 generates other recommendations to support the transceiver recommendations. Table 17.1 provides a list of the DSL Recommendations that have been generated by SG15/Q4.

This chapter addresses the symmetric DSL Recommendations (G.991.x), the ADSL Recommendations (G.992.x), the VDSL Recommendations (G.993.x), and handshake (G.994.1). The bonding Recommendations (G.998.x) are addressed in Chapter 15.

17.2.2 ATIS

ATIS is a standardization organization based in the United States and accredited by the American National Standards Institute (ANSI). ATIS is "committed to rapidly developing and promoting technical and operations standards for the communications and related information technologies industry worldwide using a pragmatic, flexible, and open approach" [ATIS web]. A key objective of ATIS is to ensure interoperable end-to-end telecommunication solutions that can be implemented in a timely fashion.

Table 17.1 Digital Subscriber Line (DSL) Recommendations from the Telecommunications Sector of International Telecommunication Union (ITU-T)

Recommendation Number	Title	Comments
G.991.1	High Bit-Rate Digital Subscriber Line (HDSL) transceivers	Addressed in Section 17.3
G.991.2	Single-pair High Bit-Rate Digital Subscriber Line (SHDSL) transceivers	Addressed in Section 17.3
G.992.1	Asymmetric Digital Subscriber Line (ADSL) transceivers	Addressed in Section 17.4. Now referred to as "ADSL1"
G.992.2	Splitterless Asymmetric Digital Subscriber Line (ADSL) transceivers	Addressed in Section 17.4. Often referred to by its informal name, "G.lite" (pronounced "G dot lite")
G.992.3	Asymmetric Digital Subscriber Line transceivers 2 (ADSL2)	Addressed in Section 17.4
G.992.4	Splitterless Asymmetric Digital Subscriber Line transceivers 2 (splitterless ADSL2)	Addressed in Section 17.4
G.992.5	Asymmetric Digital Subscriber Line (ADSL) transceivers—Extended bandwidth ADSL2 (ADSL2plus)	Addressed in Section 17.4
G.993.1	Very High Bit-Rate Digital Subscriber Line transceivers	Addressed in Section 17.5. Now referred to as "VDSL1"
G.993.2	Very High Bit-Rate Digital Subscriber Line transceivers 2 (VDSL2)	Addressed in Section 17.5
G.994.1	Handshake procedures for Digital Subscriber Line (DSL) transceivers	Addressed in Section 17.6. Often referred to by its informal name, "G.hs" ("G dot h s")
G.995.1	Overview of Digital Subscriber Line (DSL) Recommendations	Often referred to by its informal name, "G.ref" ("G dot ref"). Not addressed in this book
G.996.1	Test procedures for Digital Subscriber Line (DSL) transceivers	Often referred to by its informal name, "G.test" ("G dot test"). Not addressed in this book
G.997.1	Physical layer management for Digital Subscriber Line (DSL) transceivers	Often referred to as "PLOAM." Not addressed in this book
G.998.1	ATM-based multi-pair bonding	Addressed in Chapter 15
G.998.2	Ethernet-based multi-pair bonding	Addressed in Chapter 15
G.998.3	Multi-pair bonding using time-division inverse multi-plexing	Not addressed in this book

ATIS committees work on various technical standards, both for wired and wireless networks. Among the work items are interconnection standards (i.e., from the customer to the network interface), including DSL, number portability, improved data transmission, Internet telephony, toll-free access, telecommunications fraud, and order and billing issues [ATIS web].

Work on DSL is carried out by the Network Access Interfaces (NAI) subcommittee of the Network, Interface, Power, and Protection (NIPP) committee, which is currently chaired by industry veteran Massimo Sorbara. (Prior to the restructuring of ATIS in 2005, the overarching committee was Committee T1, which contained the technical subcommittee called T1E1 and the working group with responsibility for DSL, which was called T1E1.4.* The reader can think of the NAI subcommittee as essentially the T1E1.4 working group with a new name. (As standards people are fond of saying, it is the same clowns, but a different circus.) The NIPP-NAI subcommittee generates and maintains standards and technical reports specifying systems and associated interfaces for high-speed bidirectional transmission on metallic interfaces and for access to telecommunications networks through optical and electrical interfaces [NIPP-NAI web]. The primary focus of the work in NIPP-NAI is on the physical layer of DSL transceiver specifications that are specific to North America.

In the early days of DSL, T1E1.4 (then chaired by Tom Starr, one of the earliest proponents of DSL) played an enormous role in the specification of both ADSL and HDSL, generating standards for both types of DSL before either the ITU or ETSI.

T1E1.4 was also the standards body that first broke the VDSL1 line code logjam by selecting discrete multitone (DMT) modulation in June of 2003. This decision was a key enabler of the work on second-generation VDSL (called VDSL2, see Section 17.5.4).

More recently, the focus of NIPP-NAI has shifted to spectrum management (both static and dynamic, see Chapters 7 and 8) and on identifying and communicating North American DSL requirements to SG15/Q4.

17.2.3   ETSI

"The European Telecommunications Standards Institute is an independent, nonprofit organization whose mission is to produce telecommunications standards for today and for the future" [ETSI web]. Within Europe, ETSI is responsible for standardization in telecommunications, broadcasting, and certain aspects of information technology, and is officially recognized by the European Commission and the European Free Trade Association as the

---

* Prior to 1987, the T1E1.4 group was known as T1D1.3.

region's competent body for standardization in these areas. Based in Sophia Antipolis in the south of France, ETSI has over 650 members from more than 50 countries inside and outside of Europe. These members include manufacturers, network operators and service providers, administrations, research bodies, and end users [ETSI web].

Work on DSL in ETSI began in the TM3 working group and later moved to its own working group, TM6, under the stewardship of Hans-Joerg Frizlen. The chairmanship was then assumed by Manfred Gindel. Today, work on DSL in ETSI continues to be carried out by the TM6 working group, which is now chaired by Peter Reusens. TM6 is essentially the European analog of NIPP-NAI, and, like NIPP-NAI, TM6 is now generally focused on European-specific issues such as spectrum management and transceiver requirements. European requirements are liaised by the group to SG15/Q4.

ETSI TM6 and ATIS NIPP-NAI have actively coordinated their work throughout their existence to promote convergence of solutions whenever possible.

17.2.4   DSL Forum

The DSL Forum is not an accredited standards development organization but rather is an international industry forum of approximately 200 companies. The DSL Forum generates technical reports to facilitate the testing and deployment of DSL. A key activity of the DSL Forum is the specification of interoperability performance requirements, i.e., the performance that must be achieved when the modems at the two ends of a telephone line are from different manufacturers. The DSL Forum also addresses home networks, network operations, and end-to-end network architecture. Although the technical reports generated by the DSL Forum are not standards per se, compliance with at least some of them is invariably a requirement of most service providers. Therefore, the DSL Forum commands a strong level of industry participation, and its work is highly regarded.

17.2.5   IEEE

The IEEE is the world's leading professional association for the advancement of technology, boasting over 365,000 members in over 150 countries. The IEEE is a leading authority on telecommunications, and its relationship to DSL stems from the work of one of its working groups on a last-mile Ethernet transport system.

The 802.3 working group of the IEEE is responsible for the standardization of Ethernet. In 2003, the 802.3ah subtask force completed the EFM standard. The objective was to transport Ethernet packets over the public telephone access network. Because DSL already addressed the physical layer needed for this application, 802.3ah decided to use the existing SHDSL and first-generation VDSL physical layers for EFM. However, 802.3 as a rule never specifies more than one physical layer for a standard, whereas the available first-generation VDSL standards (VDSL1) specified both single-carrier and multi-carrier modulation. As a result, in support of the activity in 802.3ah, T1E1.4 agreed to take on the task of selecting a single line code for the first generation of VDSL. Although T1E1.4 had always had a goal to choose a single line code, the time frame was vague, and the task often seemed insurmountable for various reasons. By formally encouraging and monitoring the line code selection activity in T1E1.4, 802.3ah worked a definitive plan for the development of the 802.3ah standard. The activity in T1E1.4 resulted in the selection of DMT modulation as the VDSL1 line code. IEEE 802.3ah shortly followed with selection of DMT as the line code for the VDSL portion of the 802.3ah specification. As a result of its need for a single line code for VDSL for use in the short-reach specification of EFM, IEEE 802.3ah was instrumental in expediting a VDSL line code decision.

17.3   Symmetric DSL

This section describes the symmetric DSLs, including ISDN (Integrated Services Digital Network), HDSL, and SHDSL.

17.3.1   ISDN—the First DSL

ISDN is often considered to be the first DSL. ISDN is a system that provides end-to-end digital connectivity via the public telephone network. There are two basic variants of ISDN: the basic rate interface (BRI) and the primary rate interface (PRI). Corresponding services are known as basic rate access (BRA) and primary rate access (PRA), respectively. The ISDN BRI provides two 64 kbit/s bearer channels (called "B channels") for voice or data at a data rate of 64 kbit/s. In addition, there is a 16 kbit/s signaling channel (called the "D channel"), and an additional 16 kbit/s of overhead for framing and signaling, bringing the total transmitted bit rate to 160 kbit/s. The line code for ISDN BRA is either 2B1Q (two binary, one quaternary) encoding or 4B3T. The 4B3T line code, also referred to as MMS43, is used in Germany. In the context of access networks, ISDN usually refers to basic rate access ISDN (ISDN-BA).

The primary rate ISDN is different in different regions of the world. In North America and Japan, the PRI consists of 23 B channels (64 kbit/s for each channel), one 64 kbit/s D channel, and an extra 8 kbit/s of overhead that results in an aggregate line rate of 1544 kbit/s; this is the same as

DS1 (Digital Signal Level 1) physical signal transport in the digital network hierarchy. In Europe and most of the rest of the world, the PRI consists of 30 B channels plus one 64 kbit/s D channel with an additional 64 kbit/s overhead channel for framing, signaling, and other overhead, bringing the total data payload to 2048 kbit/s [ETSI ETR 080]. B channels can also be aggregated to form H channels, for example, H0 = 384 kbit/s (6 B channels) and H12 = 1920 kbit/s (30 B channels).

A later innovation was a product called ISDN DSL (IDSL). IDSL is a proprietary arrangement whereby the entire ISDN data capacity (128 or 144 kbit/s) is combined to transfer packet data between a home user and an Internet access router. IDSL used the 2B1Q line code as the physical layer transport.

The concept of ISDN started to take shape in the late 1960s, and the name "ISDN" was first officially mentioned in a Japanese contribution to the CCITT in 1972 [Habara 1988]. The early work led to a CCITT Recommendation in 1980. Original Recommendations of ISDN were in CCITT Recommendation I.120 (1984 Red Books). However, because various countries had developed different versions of ISDN prior to this CCITT Recommendation, it was necessary to include country-specific information elements, using a code set mechanism allowing different use of different information elements within the data frames.

In the United States, the ISDN implementation proceeded slowly, even though ISDN has been under development since 1980. This was primarily due to discouraging regulatory policies and secondarily because of the divestiture of the Bell Operating Companies from AT&T, which removed Bellcore as a driver for deployment. Services were introduced in the United States on a trial basis in 1985, but these services never really took off in the same way as in Germany and the rest of Europe. In Canada in the early 1980s, Bell Canada conducted some early ISDN trials with equipment that used the Alternate Mark Inversion (AMI) line code with time-compression multiplexing (TCM).

The ISDN-BRI standard in the United States is T1.601, which was originally completed in 1987. The latest issue of T1.601 was released in 2004. The PRI standard is T1.605, which was developed in parallel with T1.601.

In Europe, the development of ISDN services proceeded at various rates, with Germany amongst the forerunners. Deutsche Telekom launched their ISDN service in 1989 at the CeBIT trade show in Hannover, with such success that by 2003 there were 10 million ISDN-BA connections in Germany (compared to 27 million plain old telephone service [POTS] lines) [T-Com]. In Japan, early introduction of ISDN began in 1984 and became widespread in the 1990s. In the rest of the world, ISDN modulation is based on echo canceled (EC) modulation, but the Japanese ISDN is based on a "ping-pong"

Time-Division Duplexing (TDD) system that uses TCM. The ISDN system in Japan is called TCM-ISDN. The use of TDD in ISDN would influence all future DSL development in Japan. In particular, the ADSL systems for Japan would need to account for the TDD nature of TCM-ISDN to minimize the effects of nonstationary crosstalk.

In the early 1990s, work began on specifications to improve interoperability of ISDN equipment in the United States, and a specific implementation for ISDN in the United States, National ISDN 1 (NI-1), was established in 1991. However, there were problems agreeing on this standard, and eventually a more comprehensive National ISDN 2 (NI-2) was adopted in 1992. These and number of subsequent clarifications of the ISDN operations greatly helped interoperability of ISDN equipment in North America.

In Europe, ETSI TM3 started work in 1990 on the specification of a basic rate ISDN digital transmission system on metallic local lines, later to be published as ETR 080 in July 1993. This ETSI technical report described the transmission properties of local networks suitable for transmission systems using metallic pairs of wires, with two line codes (2B1Q and 4B3T) described in appendices. This document was refined and improved over the years, with the latest published version, TS 102 080 Version 1.4.1, being generated by ETSI TM6 in 2003 [ETSI TS 102 080]. This latest version of TS 102 080 is intended to improve spectral compatibility with VDSL and ADSL by including a specification of the low-pass portion of a splitter to work on the same pair with a VDSL or an ADSL system.

The ITU-T ISDN-BRA Recommendations are I.411, I.412, and I.430. The I.430 Recommendation specifies the physical layer, with each of two line codes specified in annexes. The ITU-T Recommendation I.431 specifies the physical layer for ISDN-PRI.

17.3.2 HDSL

The vision of a single digital network capable of supporting a wide range of services—data, voice, and video—initiated by ISDN had a weak link. That weak link was the access network that connects customers to the local exchange via local lines, which are usually existing copper twisted-wire pairs. A typical connection consists of a cascade of cable sections of different diameters and lengths, bundled together with up to a hundred other such pairs, resulting in a low-bandwidth transmission medium that is prone to interference. This medium had been sufficient for 4 kHz bandwidth analog telephony, and ISDN-BA technology had been developed to allow the bidirectional transmission of 160 kbit/s over these lines to the end subscriber. However, the vision of an end-to-end connection for ISDN-PRA (1544 kbit/s in North America, 2048 kbit/s in Europe) required the installation of special coaxial or fiber-optic cable, or repeaters every few kilometers in carefully selected copper-pair cables.

In addition, there was a need for an alternative means to provision a T1 circuit in North America and an E1 circuit in Europe. Installation of T1 and E1 circuits, which usually required the installation of repeaters, could take as long as two or three months. Ideally, this new approach would eliminate the need for repeaters and thereby enable provisioning of a T1 or E1 circuit in two or three days.

In part in response to the need for a T1 replacement technology, in October of 1991 Bellcore released Technical Advisory TA-NWT-001210, generic requirements for HDSL [Bellcore 91-86]. The advisory called for a tenfold increase in the data rates to be transmitted over old-fashioned copper pairs, made possible by the improvements in processing speeds achievable with very large scale integration (VLSI), which would enable the high-speed digital signal processing that was required to compensate for the essentially unknown transmission characteristics of the cable plant. It was possible to implement the necessary adaptive echo cancelation, equalization, and filtering techniques in real-time. Early research had been carried out in the mid 1980s, but the laboratory prototypes had been large and unwieldy; but the possibility of a compact transceiver module using the current state of the art VDLSI was later becoming a reality.

The proposed HDSL was to provide repeaterless DSL rate access over nonloaded copper loops conforming to the North American carrier serving area (CSA)* design rules as a transparent replacement for the T1 repeatered lines then being used in the distribution networks. The proposed HDSL was to eliminate the need to remove bridged taps, separate binder groups, or install repeaters, resulting in considerable cost savings. In particular, eliminating the need for repeaters was seen as a significant improvement, because repeaters brought only problems: high-voltage power-feeding circuits were required, line-transformers were more bulky, the repeaters themselves were difficult to install and service, and the additional crosstalk from the boosted signals was a potential source of interference to adjacent, nonrepeatered systems.

The Technical Advisory (TA) defined overall system requirements to support the above goal and proposed a dual-duplex architecture in which two pairs carry full-duplex 784 kbit/s bit streams. The 784 kbit/s line rate was composed of 12 DS0 channels at 64 kbit/s each (for 768 kbit/s total) plus 8 kbit/s for the DS1 F bit (which was transmitted in each pair for redundancy) and an 8 kbit/s overhead channel. The two 784 kbit/s streams

---

* The "carrier serving area" (CSA) is defined as a 9-kft length of 26 AWG cabling or, equivalently, a 12-kft length of 24 AWG cabling.

were then combined to form a 1.544 Mbit/s DS1 bit stream conforming to the North American interface specifications. This release of the TA, which was largely based on the existing 2B1Q ISDN technology, defined the line code as 2B1Q and also proposed a framing format that included a scheme for synchronization and bit stuffing, a cyclic redundancy check (CRC) field, an HDSL embedded operations channel (EOC), and indicator bits for operations purposes, but not the protocol structure of the EOC or the start-up protocol.

The document reflected a close working relationship with vendors through the TlE1.4 working group. The first HDSL systems were placed into service in 1992. The T1E1.4 group then proceeded to work on a technical report specifying HDSL, referred to as TR-59 [TlE1.4/92-002R1].

At the same time work was ongoing on HDSL in T1E1.4, interest was stirred in Europe, where the same problems with high-speed access were being encountered. In the European access environment, there was a requirement for providing ISDN-PRA services (2048 kbit/s) to customers without having to lay new screened-copper or fiber-optic cables, or to provide repeaters. It was becoming clear that there was a gap between the demand for broadband data services and the rate of deployment of fiber-optic cables, which together with the economic pressures for reducing transmission costs provided an impetus to make the best use of the existing investment in copper cable.

In 1992, ETSI created a work item within sub-technical committee TM3 to study the application of the HDSL techniques developed in America to European requirements, in particular ISDN-PRA [ETSI STC-TM3]. In 1992, TM3 began work on a specification for HDSL transmission on metallic local lines [ETSI ETR 152]. The scope of this work was to specify HDSL transceivers capable of providing transparent capacity of 784 or 1168 kbit/s over one pair of the existing copper local line network. Target ranges were about 2.7 km (approximately 9 kft) on 0.4 mm wires (approximately equivalent to 26 AWG) and 3.7 km (approximately 12 kft) with 0.5 mm wires (approximately equivalent to 24 AWG), without repeaters. The original application was to implement the ISDN-PRA digital section using three pairs, although a two-pair option was very quickly added to the scope. The first draft of this technical report was completed in October of 1993 and described transmission techniques, performance objectives, architecture, and operations and maintenance. Two systems were described, both using the 2B1Q line code: triple-duplex operation transmitting 784 kbits/s per pair and dual-duplex operation transmitting 1168 kbit/s per pair. Further applications such as SDH (synchronous digital hierarchy) compatibility, fractional operation, and leased-line operation, as well as other line codes, were left for future editions. The triple-duplex method was chosen as an obvious extension of the American system, allowing the same transceivers to be used and with the added advantage of additional capacity for future compatibility with SDH TU-12 or VC-12 formats. (This additional capacity was, however, only implemented to any significant extent ten years later.) To maintain compatibility with HDSL standards in North America, the same basic HDSL frame structure was maintained, but with new mapping options for the European requirements. The HDSL frame on each pair has a nominal duration of 6 ms, and contains 48 payload blocks containing 12 data bytes for the three-pair system and 18 data bytes for the two-pair system, resulting in a payload rate of 2304 kbit/s (equivalent to 36×64 kbit/s channels). For 2048 kbit/s applications, only 32 of these 64 kbit/s channels are required, but future applications such as SDH TU-12 format data will use the full capacity. The report also described detailed start-up procedures, the embedded operations channel, operation and maintenance mechanisms, electrical characteristics, and performance requirements. The aim of this ETSI technical report was to facilitate the design of HDSL systems such that equipment from different vendors could work together. To simulate the local loop environment for test purposes, a set of test loops was also defined, and these, together with artificially generated test noise, allowed laboratory testing of HDSL equipment under controlled conditions.

The first version of ETR 152 was approved in October of 1993, but for various reasons publication did not follow until February of 1995. The second version of ETR 152 was completed in mid-1995 and included an annex describing the carrierless amplitude and phase modulation (CAP) line code (see chapter 6 of [Golden 2006] for details of CAP). This annex was included after a very interesting and heated debate. The third version of ETR 152 was published in late 1996 and included a one-pair system for 2320 kbit/s.

In mid-1997, the existing technical report was revised and issued as a technical specification, TS 101 135 [ETSI TS 101 135-a (1998)], which in turn was taken over by the ITU-T as the basis for G.991.1 [G991.1]. Finally, a specification for digital embedded ISDN and POTS was published in 1998 [ETSI TS 101 135-b (1998)].

The ETSI HDSL work was originally proposed as a "quick-fix" solution for the transport of 2 Mbit/s PRA services over three copper pairs using 2B1Q technology, but in the space of eight years it had matured into a complex system capable of supporting a wide range of services over one, two, or three wire pairs with a choice of 2B1Q or CAP line codes.

In 1997, ITU-T Study Group 15 established Question 4 to address DSL issues. This group became known as SG15/Q4 and would greatly influence all DSL development from then on. One of the first tasks for the new SG15/Q4 group was to create an international standard for HDSL. The ITU-T work was largely based on the ETSI TM6 work, which in turn was largely based on the latest ANSI T1E1.4 work [ATIS TR-28]. The HDSL ITU-T

Recommendation G.991.1 [G991.1] was published in 1998 and was mostly identical to ETSI TS 101 152, with the addition of support for 1544 kbit/s for North America. Figure 17.1 illustrates the timeline of HDSL standards development.

Despite HDSL being a leap in transmission capability, HDSL was in one sense a retrograde step. Whereas ISDN-BA integrated voice and data communications on the same phone line, HDSL required the voice to be separated out and transmitted over a completely separate network, to the extent that an additional copper pair was required if voice services were to be provided. It was not until later (see, for example, [ETSI TS 101 135-b (1998)] and [Habib 2002]) that the concept of voice and data integrated in the physical layer of the DSL was introduced, with voice-over-IP (VoIP) or voice over ATM (VoATM) having being introduced slightly earlier [DSL Forum TR-036].

A DSL designer must choose between carrying voice as a traditional voiceband signal (POTS) or providing DC (direct current) power via the line to power the customer-end equipment or mid-span repeaters. Because POTS uses DC signaling, it is not possible to do both on the same line.

17.3.3 SHDSL

In 1995, T1E1.4 started to study a second-generation HDSL system with the ambitious goal of replacing the two-pair HDSL systems with a new single-pair system covering the same carrier serving area [ATIS T1]. This second-generation HDSL was referred to as HDSL2, and the early work on HDSL2 became the basis for much of the SHDSL work in the ITU-T and ETSI.

Very precise requirements were formulated for the HDSL2 system design, in particular for the specific mix of disturbers that must be tolerated and the spectral compatibility that must be achieved with other systems. Various new line codes and modulation methods were evaluated, taking into consideration recent technological advances, including frequency-division duplexed (FDD) and EC line codes such as 3B1O and the single-carrier methods of quadrature amplitude modulation (QAM) and CAP. To achieve the performance targets with low latency while achieving spectral compatibility with specified services, it was found that different spectral shaping in the upstream and downstream directions was required, and various power spectral density (PSD) schemes were proposed, including POET (partially overlapped echo-canceled transmission), overCAPed (oversampled CAP/QAM), OPTIS (overlapped pulse amplitude modulation [PAM] transmission with interlocked spectra), and MONET (margin optimized interlocked extended-range transmission). Eventually, a solution based on the OPTIS PSD with the trellis coded PAM line code was agreed upon as

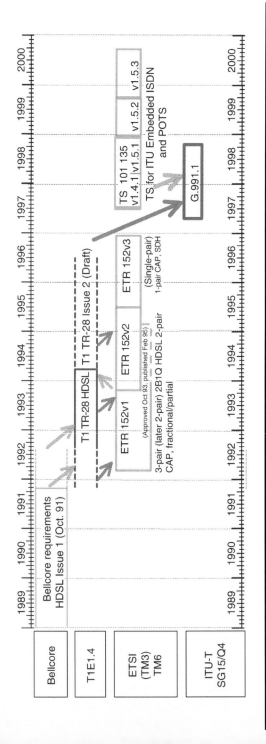

Figure 17.1 **Overview of HDSL timeline and relationships between standards bodies.**

being the best compromise between performance and spectral compatibility. A detailed description of the development of Committee T1 HDSL2 in North America can be found in [Starr 2003].

In ETSI TM6, the work on ETR 152 had come to a natural completion, and the group began to consider what potential there was for future symmetric DSL systems. The group decided to consider what synergy could be attained with the HDSL2 work in T1E1.4 [ETSI DTS/TM-06011]. A first proposal was made for requirements for the new system, including data-plus-ISDN/POTS, symmetrical services, negotiated data rates based on transmission distance, and support of the TCP/IP (Transmission Control Protocol/Internet Protocol). The acronym SDSL (to stand for single-pair DSL) was proposed for this new ETSI system [ETSI TM6].

Unfortunately, in North America the term SDSL is also used for proprietary 2B1Q multi-rate systems, based on the use of HDSL transceivers, that can run at lower-than-standard rates to provide Internet access to homes and small businesses. To avoid confusion between the ETSI specified SDSL system and the proprietary 2B1Q system, it is common to distinguish between the two systems by referring to the ETSI systems as ETSI SDSL and the HDSL transceiver based systems as 2B1Q SDSL. The 2B1Q SDSL systems were popular with North American competitive local exchange carriers (CLECs), and for a while the 2B1Q SDSL systems were probably the most common DSL systems for residential access. This early success of 2B1Q SDSL further increased the interest in developing standardized multi-rate SHDSL systems.

When it came to requirements, the situation in Europe was less clear cut than in North America: there was no real equivalent of the CSA, and the mix of potential disturbers was much greater. ETR 152 allowed for at least five variants of HDSL (one-, two-, and three-pair systems based on 2B1Q, and one- and two-pair based on CAP). Consideration was given to disturbance from HDB3 coded PRA as well as ADSL over POTS and ADSL over ISDN, which made the possibility of finding a "sweet spot" in the spectrum with partially interlocking spectra remote.

Therefore, the European telecommunications systems operators within TM6 spent some time collating their requirements for SDSL and agreed to the following goals, which are documented in Annex J of [ETSI TC.TM-a]:

- Enable cost-effective implementation.
- Support one embedded ISDN-BA channel or up to 3 (digital) POTS channels.
- Support life-line narrow-band service.
- Range was more important than the bit rate: a range equivalent to ISDN-BA (about 4.5 km on 0.4 mm) was required for low bit rates, and a range of about 3 km was needed for high bit rates such as 2 Mbit/s.

- Support payload bit rates of 384–2304 kbit/s, in 64 kbit/s steps (determined by operator).
- Provide a transfer delay of less than 1.25 ms for narrowband data and 5 ms for broadband data.
- Provide an auxiliary 64 kbit/s channel.

It can be seen that these requirements are completely different from the T1E1 HDSL2 requirement of achieving CSA range (9 kft of 26 AWG cabling) with exactly 1544 kbit/s in a precisely specified crosstalk environment [Starr 2003]. In particular, the mix of systems with which the new system would be expected to be spectrally compatible was much broader than for HDSL2; merely the possible presence of one-, two-, and three-pair ETSI HDSL systems made it very difficult to find an optimum solution (see Figure 17.2). The ETSI SDSL system could therefore not simply leverage the HDSL2 chipsets as had been done with three-pair HDSL, but rather would make use of the technical studies used in generating the OPTIS line code to devise a modulation scheme optimized for the European SDSL require-

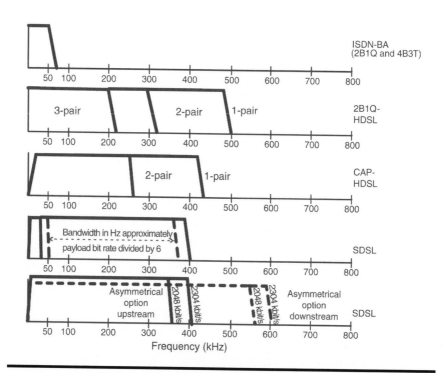

Figure 17.2   Comparison of European Telecommunications Standards Institute (ETSI) Symmetric Digital Subscriber Line (SDSL) bands.

ments. The manufacturers of HDSL2 chipsets cooperated fully with TM6, making various proposals as to how best to take advantage of the advances made in the United States. By May of 1999 [ETSI TC.TM-b], TM6 could agree that although solutions based on CAP, QAM, PAM, or DMT would provide equally good results in terms of spectral compatibility, reach, impairment tolerance, and interference immunity, the majority of technology providers preferred trellis coded PAM. ETSI TM6 therefore provisionally agreed that Ungerboeck-Coded baseband PAM would be used as the line code for ETSI SDSL. The term "Ungerboeck-Coded PAM (UC-PAM)" was adopted by ETSI TM6 to replace the term "Trellis-Coded PAM (TC-PAM)" to honor Gottfried Ungerboeck, the acknowledged inventor of the concept of trellis coded modulation [ETSI TC.TM-a]. (The reader can find trellis coding explained in Dr. Ungerboeck's own words in chapter 8 of [Golden 2006].)

After much hard work both during and between meetings, the first part of the ETSI SDSL specification (functional requirements) [ETSI TS 101 524-1] was approved and made available to SG15/Q4 as an input to the G.shdsl work, scheduled for determination in April of 2000. Part 2 of the ETSI SDSL specification (SDSL transceiver requirements) then received the group's attention and was approved for publication in May of 2000 [ETSI TS 101 524-2]. After that, work began on combining the two parts of the ETSI SDSL TS 101 524 into one document, which was published in June of 2001 [ETSI TS 101 524-2]. Further work followed, where new features were added and the document was better aligned with the ongoing SHDSL work in SG15/Q4. New revisions of ETSI SDSL TS 101 524 were published in 2003 [ETSI TS 101 524] and 2005. By the end of 2004, the SHDSL standards development was so mature that the ETSI and ITU-T standards were almost identical in content. Therefore, it was decided to rewrite ETSI SDSL TS 101 524 as a pointer document to ITU-T G.991.2. This pointer document was published in February of 2006 as ETSI TS 101 524 v1.4.1.

The SHDSL work in SG15/Q4 began in late 1998. The early ITU-T work was primarily focused on specifying requirements based on input from the regional bodies (mainly T1E1.4 and ETSI TM6). It quickly became clear that the ITU-T SHDSL requirements were the combined requirements for HDSL2 and ETSI SDSL. When ETSI decided in May of 1999 to adopt the HDSL2 line code for ETSI SDSL, it was decided only week later to use the same line code for ITU-T SHDSL. With a common line code and common requirements, it was clear that ITU-T SHDSL and ETSI SDSL had so much in common that the two standards would almost be identical. This resulted in close cooperation between SG15/Q4, ETSI TM6 and T1E1.4, where the three groups worked together to define the new SHDSL standard, and all along it was made sure that HDSL2 could be accommodated as a special case within ITU-T SHDSL.

There was also early effort to specify initialization procedures for ITU-T SHDSL based on the handshake procedures according to Recommendation

G.994.1 (see Section 17.6). The use of the ITU-T handshake procedure was new to the symmetric DSL standards, and initially there was some hesitation among some operators to accept the use of G.994.1 as part of the SHDSL initialization. European operators were particularly concerned that introducing G.994.1 handshake would not be appropriate, because it would not fit their deployment model of providing SHDSL services. As a result, there was some resistance in ETSI TM6 to using the G.994.1 handshake procedure for ETSI SDSL, even if it was to be used for ITU-T SHDSL. However, eventually both ETSI SDSL and ITU-T SHDSL adopted identical startup procedures based on G.994.1.

The first version of ITU-T SHDSL Recommendation G.991.2 was published in February of 2001. A revised version was published in December of 2003 and updated with amendments in 2004 and 2005. Figure 17.3 illustrates the timeline of SHDSL standardization.

It is interesting to note that a change in the standards environment had also taken place. During the development of the HDSL standard, cooperation between ITU-T, T1E1.4, and ETSI TM6 had been highly regimented and linear; written liaisons would be approved at the end of one meeting for formal consideration at the next meeting of the other body, leading to long delays before the simplest of changes could be coordinated. The SDSL project, on the other hand, was more of a frantic spiral, with delegates visiting all relevant standards meetings (see Figure 17.4), with a synergy developing that left those on the sidelines amazed at the speed of agreement on key issues.

17.3.4   The SHDSL Standards

The SHDSL technology has been standardized by the ITU-T in G.991.2 [G991.2] and ETSI in TS 101 524 (referred to simply as SDSL in ETSI [ETSI TS 101 524]). The two standards are essentially the same, and equipment that can support one should be able to support the other. The main difference between the two SHDSL standards is that ETSI TS 101 524 is European-specific while ITU-T G.991.2 also contains a North American variant. As the standards were emerging, there were some differences in exactly which optional feature was supported in which standard, but with time the features supported by the two standards mostly converged. In February of 2006, ETSI published TS 101 524 V1.4.1 [ETSI TS 101 524-1], which references ITU-T G.991.2 [G991.2] wherever possible and only specifies the differences from G.991.2.

The official name for the ITU-T's SHDSL is "single-pair high-speed digital subscriber line (SHDSL) transceivers," and during its development it was usually referred to as "G.shdsl." The official name for ETSI SDSL is "symmetric single pair High Bit-Rate Digital Subscriber Line (SDSL)." Although the two standards have different names and different acronyms, they are essen-

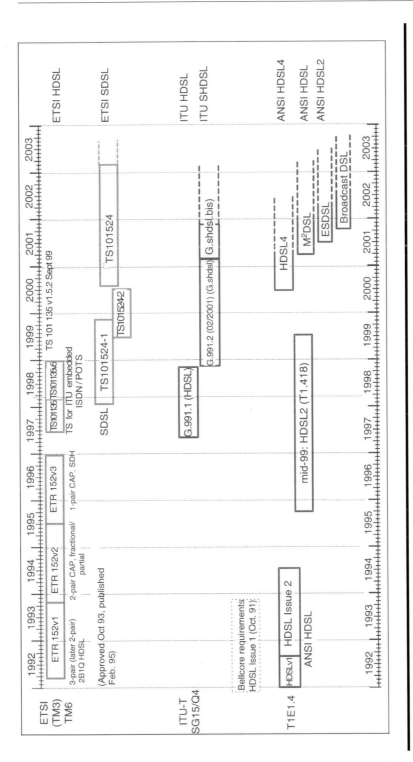

Figure 17.3 Timeline for symmetric DSL standardization.

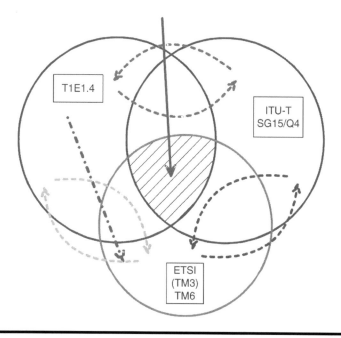

Figure 17.4   Overlap between standards committees.

tially identical in content. When referring to the two standards as a single type of technology, it is common to use the acronym SHDSL. This convention is followed in the remainder of this chapter. In addition, the HDSL2 and HDSL4 standards are considered to be special cases of SHDSL*, and most of the discussion about SHDSL in this chapter is directly applicable to HDSL2 and HDSL4.

The original motivation for developing the SHDSL standards was to develop a replacement for HDSL. The SHDSL systems were primarily targeted for the small and medium enterprise (SME) business environment, using the existing unshielded wire pairs in the local access network. However, the noise environment in the access environment had become more congested than in the early days of HDSL. Therefore, one of the design challenges for the SHDSL projects was to specify this noise environment and a modem offering optimum performance within it. There was also a need for supporting a wider range of bit rates, and not just the T1 or E1 rates that HDSL was originally specified to support.

The original SHDSL specifications provide for a bidirectional symmetrical channel with variable payload bit rates from 192 up to 2312 kbit/s.

---

* The development of HDSL2 preceded the development of SHDSL. SHDSL was designed to contain HDSL2 as a "special mode of operation," which means that an SHDSL transceiver can be used in the implementation of an HDSL2 system.

(The highest rate facilitates SDH TU-12 transport.) An option is provided for transporting an independent narrowband channel able to carry an ISDN-BA channel or analog telephone channels within this payload. A two-pair (four-wire) option is also specified, and regenerators are also included in the specification. Later revisions of the ITU-T G.991.2 Recommendation and the TS 101 524 standard added support for more pairs in multi-pair mode and specified support of extended data rates up to 5696 kbit/s on each pair.

17.3.4.1 The SHDSL Reference Model

The ITU-T G.991.2 and ETSI TS 101 524 specifications essentially share the same reference model, but they use different nomenclature. One of the more confusing things when comparing ITU-T and ETSI text is the naming of the transceivers or termination units. In the ETSI terminology, the equipment at the customer premises is known as the "network termination unit" (NTU), and that at the operator side of the line as the "line termination unit" (LTU). In the ITU-T terminology, the SHDSL equipment at the customer side is known as the STU-R (SHDSL transceiver unit at the remote end) and on the operator side as the STU-C (STU at the central office [CO]).

Figure 17.5 illustrates the ETSI SDSL reference configuration (adapted from Figure 4.1 of [ETSI TS 101 524]), with a description of the functional blocks superimposed. Although the diagram initially appears complicated, there are two main parts to a complete "termination unit." There is an application-invariant part responsible for converting data to a format suitable for transmitting over metallic pair cables, and there is an application-specific part that converts the data at the external interfaces into the format required for the application-invariant part. Additional maintenance functions are provided at each stage, and it is also interesting to note that the "common circuitry" block, inherited from the original three-pair HDSL reference configuration, is required for the multi-pair enhancement to the original single pair (two-wire) configuration.

Figure 17.6 illustrates the protocol model supported by this reference configuration, showing how the various functions correspond to the layers. This layered structure can be applied to all DSL technology (see [?]). The structure helps to ensure as much commonality as possible between all the DSL technologies and will facilitate future convergence between them.

The different processing layers are as follows:

- The Transmission Protocol Specific (TPS) layer corresponds to the user interfaces at the customer premises and network operator sides of the line, based on existing transport protocol stacks.

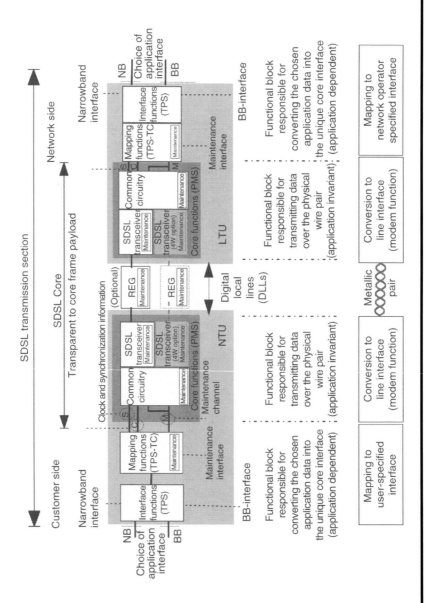

Figure 17.5   ETSI SDSL reference configuration.

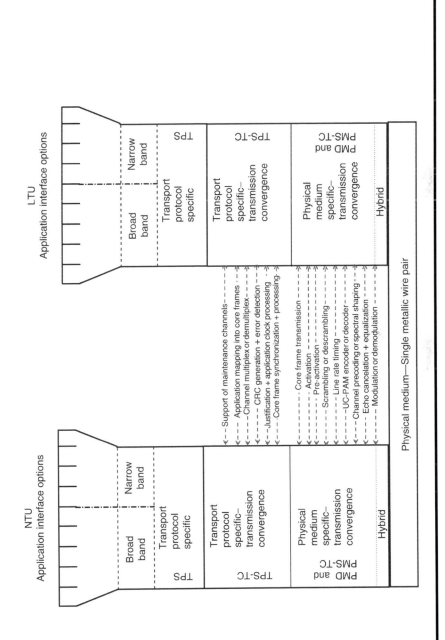

Figure 17.6   Protocol reference model.

The specification of these protocol stacks is outside the scope of the SHDSL technical specification.

- The transmission protocol specific–transmission convergence (TPS-TC) layer provides the interface between the user interface and the standard core.
- The physical medium specific–transmission convergence (PMS-TC) layer is responsible for mapping the application data into the core frames, multiplexing and demultiplexing application-specific channels, operations and maintenance channels, timing adaptation by means of justification procedures, and CRC functions.
- The physical medium dependent (PMD) layer provides the duplex connection to the physical transmission medium (metallic pair) by means of modulation and demodulation via a hybrid circuit, EC, equalization, coding and decoding, and bit level timing. This function is sometimes referred to as a "bit pump," because the PMD layer operates on the bit stream at its input without requiring any knowledge of the organization of the bits into frames containing various higher-order functions.

The choice of interface is deliberately left open-ended, with the possibility of adding TPS and TPS-TC layers to encompass future applications. Current applications are described in Annex E of ITU-T G.991.2 and Annex A of ETSI TS 101 524.

17.3.4.2   The SHDSL PMD Layer

The line code used for SHDSL is TC-PAM, also known as UC-PAM, with Tomlinson–Harashima precoding. This line code is essentially the same as used for HDSL2, and key components of the encoding were adopted directly from HDSL2. In particular, the scrambler, the programmable one-dimensional convolutional encoder, a generalized version of the PAM mapping, and the Tomlinson–Harashima channel precoding were all adopted from HDSL2.

The trellis coding improves the performance of the PAM line code by making it more resilient to noise. The trellis coding used for SHDSL can typically provide about 4.4–5.1 dB coding gain, depending on the specific convolutional codes used and how much latency can be tolerated. This means that the TC-PAM code can tolerate about 4.4–5.1 dB worse signal-to-noise ratio (SNR) than the equivalent uncoded PAM code.

The first generation of SHDSL standards only used 16-level TC-PAM, with three data bits and one redundant bit for each symbol. The 16-level TC-PAM was chosen for SHDSL after extensive study, because it was shown to provide an excellent trade-off between performance and bandwidth utilization. Using more than 16 levels would improve the bandwidth utilization but would in most cases degrade performance. Using fewer than 16 levels would in some cases provide slightly better performance but would mean less bandwidth utilization. Therefore, only 16-level TC-PAM was used in first generation of the SHDSL standards.

In the second generation of SHDSL, known as "enhanced SHDSL" or "ESHDSL," the maximum bit rate was raised from 2312 to 5696 kbit/s. The increase in bit rate prompted the introduction of 32-level TC-PAM for improved spectral compatibility at high data rates. In 32-level TC-PAM, there are four data bits per symbol, so for the same data rate the symbol rate can be 3/4 of the symbol rate needed for 16-level TC-PAM. This in turn implies that the bandwidth needed for the signal is 3/4 of the bandwidth needed for 16-level TC-PAM, which can result in less crosstalk interference into other systems, such as ADSL. The drawback is that 32-level TC-PAM does not have as good performance as 16-level TC-PAM at these high rates.

One drawback with trellis coding is that the use of decision feedback equalization (DFE) is impractical because error propagation introduces correlation in the error signals, which can eliminate the coding gain from the trellis coding. One solution to this problem would be to use maximum likelihood sequence estimation instead of DFE. However, this approach will typically increase the complexity of the decoder. An alternative approach is to use Tomlinson–Harashima precoding (see chapter 11 of [Golden 2006]), and this is the approach chosen for SHDSL.

Figure 17.7a shows a general block diagram of the transceiver (i.e., transmitter and receiver) functions belonging to the PMD layer. The transmitter and receiver paths are connected to the physical medium (the metallic line pair) by a hybrid circuit that provides electrical isolation between the two directions. The line interface block also contains the Analog Front-End (AFE) that performs the analog processing such as digital-to-analog and analog-to-digital conversion and contains analog filters, line drivers, and analog gain control functions.

The transmit data is first passed through a scrambler to remove data patterns such as long runs of zeros or ones, or bit patterns because of the overlying frame structure. Such patterns could otherwise adversely affect the performance of the receiver blocks, which work on the assumption that the received data is random. The data is then passed through a trellis encoder (Ungerboeck encoder), which introduces redundancy in the encoded symbols by adding one extra (redundant) bit for each symbol. A channel precoder (Tomlinson–Harashima precoder) then pre-equalizes the data, based on coefficients acquired during the activation stage, and a spectrum shaper ensures that the final signal on the line has the

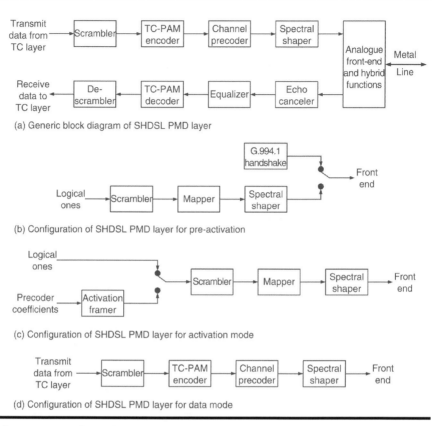

(a) Generic block diagram of SHDSL PMD layer

(b) Configuration of SHDSL PMD layer for pre-activation

(c) Configuration of SHDSL PMD layer for activation mode

(d) Configuration of SHDSL PMD layer for data mode

Figure 17.7   Physical medium dependent (PMD) layer modes.

required power spectral density (see Section 17.3.4.6). In the receiver path, an EC removes echoes of the transmitted signal, which are caused by imperfections in the hybrid circuit and reflections on the line. Following this, an adaptive equalizer minimizes any distortion due to intersymbol interference (ISI) and minimizes the effects of the highly colored crosstalk noise. The channel equalization is done jointly by the adaptive equalizer in the receiver and the Tomlinson–Harashima precoder in the transmitter at the other end of the line. The trellis decoder, typically implemented using the Viterbi algorithm, does a maximum likelihood estimation of the received symbol sequence. The self-synchronizing de-scrambler then performs the inverse of the function performed by the scrambler in the transmitter and passes the recovered data to the PMS-TC layer.

When initialized, the SHDSL transceivers go through three stages: pre-activation mode, core activation mode, and data mode. In pre-activation mode, the transceivers on either end of the line use G.994.1 handshake procedures to exchange information about their capabilities and agree on common mode of operation (see Section 17.6 for more details). During pre-activation, the transceivers can agree on whether to enter a line probe session. In line probe mode, the two transceivers transmit two-level probing signals at negotiated symbol rates to evaluate the characteristics of the transmission line. Following the line probe sequence, the transceivers enter into a second G.994.1 handshake session for final line rate and power back-off negotiations. Figure 17.7b shows the configuration of the PMD layer during pre-activation.

The core activation mode (see Figure 17.7c) begins immediately after the pre-activation phase. In the core activation phase, the two transceivers exchange training signals generated as two-level PAM encoding of scrambled ones. This commences with the STU-R* sending a training signal to the STU-C to enable the STU-C to choose gain values and possibly initialize equalizer functions, and the STU-R to possibly start training its EC. The STU-R then stops sending the training signal, and the STU-C starts sending its training signal to the STU-R to allow corresponding functions to be carried out at the other end of the line. Then the STU-R resumes sending its training signal so that duplex transmission takes place. This allows both sides to continue the optimization of their equalization and echo-cancelation functions. After a time specified to allow the STU-R timing extraction to be fully synchronized, the STU-C then passes the precoder coefficients to the STU-R, and the STU-R responds by returning precoder coefficients to the STU-C. Finally, the STU-C unit sends two activation frames with reversed synchronization words to indicate successful completion of the activation procedure, and the two units move to data transmission mode (see Figure 17.7d).

The basic data mode line code for SHDSL is 16-level TC-PAM, where a symbol consists of three data bits and one redundant bit that is used for the trellis coding. SHDSL supports data rates in the range of 192–2312 kbit/s. The core TPS-TC frame adds fixed overhead of 8 kbit/s, so the line rate for SHDSL is in the range 200–2320 kbit/s, respectively. More specifically, line rates of $(n \times 64 + i \times 8 + 8)$ kbit/s are supported, where n is an integer from 3 to 36, and i is an integer value from 0 to 7. The corresponding symbol rates are equal to the line rate divided by three.

Second generation SHDSL optionally supports bit rates up to 5696 kbit/s in increments of 8 kbit/s, which corresponds to a maximum line rate of 5704 kbit/s after including the 8 kbit/s of overhead. The 16-level TC-PAM can be used for data rates up to 3848 kbit/s in increments of 8 kbit/s. For spectral compatibility reasons, it was decided not to use 16-level TC-PAM at data rates above 3848 kbit/s, but rather to use 32-level TC-PAM

---

* The STU-R is also called the NTU, and the STU-C is also called the LTU.

modulation at these higher rates, given that its bandwidth is narrower than with 16-level TC-PAM and the resulting crosstalk to other systems is lower. 32-level TC-PM modulation can be used for data rates from 768 to 5696 kbit/s.

An alternative approach to increasing the data rate of SHDSL systems is to use multiple pairs, similar to what was done for HDSL. The original idea behind the development of SHDSL was to create a single-pair system that could replace multi-pair HDSL systems. However, as the DSL market matured during the early development of SHDSL, operators soon became interested in using multi-pair variants of SHDSL to deliver higher data rates or to provide a particular data rate over longer distances. Therefore, the "single-pair" SHDSL standards added support for multi-pair operation. Early versions of SHDSL only provided for aggregation over two pairs (four-wire mode), but second generation SHDSL standards provided for multi-pair operation on up to four pairs. Methods have also been devised to support multi-pair operation by doing the pair aggregation at higher layer protocols (see Chapter 15 for details of the bonding protocol), but these methods are independent of the physical layer transport.

17.3.4.3 The SHDSL PMS-TC Layer

The SHDSL core frame is based on the HDSL core frame but is more flexible to support a variable bit rate, and the definition of the overhead bits is not always identical to the corresponding bits in the HDSL frame. For values of $i = 1$ and $n = 36$, the SHDSL frame is essentially identical to the ETSI HDSL frame.

Figure 17.8 shows the data mode core frame structure. This frame structure provides the flexibility to transport variable payload bit rates from 192 up to 2312 kbit/s, based on different values of n and i. The data rate can be as high as 5696 kbit/s for ESHDSL. The framing can be done in either plesiochronous or synchronous mode.

Each core frame is subdivided into four blocks. The first group of the frame starts with a 14-bit synchronization word followed by two overhead bits and 12 payload blocks. Each of the three subsequent blocks consists of ten overhead bits and twelve payload blocks. The payload blocks can be configured as zero to eight individual Z-bits followed by three to thirty-six data bytes, and, therefore, depending on the payload bit rate, each payload block contains between 24 and 289 bits. As each of the 48 payload blocks is identically configured for a specific application, the SHDSL frame contains $48 \times (n \times 8 + i)$ payload bits plus 46 overhead bits and zero, two, or four (nominally two) stuffing bits within a 6 ms frame length, for a bit rate of $(n \times 64 + i \times 8 + 8)$ kbit/s. The 46 overhead bits are divided into the 14-bit synchronization word, 20 EOC message bits, 6 indicator bits, and 6 CRC bits.

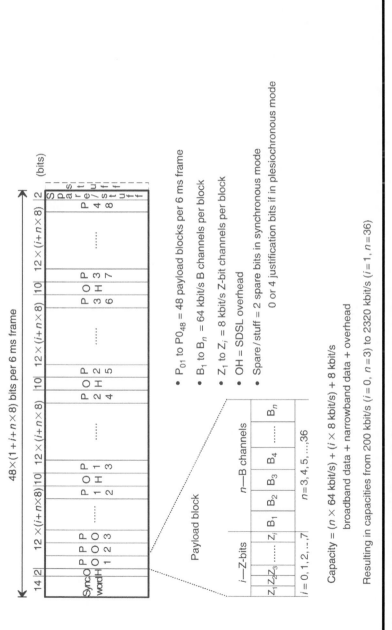

Figure 17.8   Symmetric High Bit-Rate DSL (SHDSL) core frame structure.

The allowed values for n are 3 through 36, and i can take values 0 through 7, but in the North American annex of G.991.2 (Annex A), the values of i are restricted to 0 or 1. For ESHDSL the maximum value of n is 89, corresponding to 5696 kbit/s.

In the optional multi-pair mode (M-pair mode), the payload data is interleaved between pairs. This is done by interleaving each payload sub-block among all pairs. An equal number of bits from each sub-block is carried on each pair, such that if the sub-blocks on each pair can carry k bits, then the aggregated sub-block can carry $M \times k$ bits on the M pairs. An M-pair SHDSL system can carry M times the data rate for each pair. This means that the maximum data rate for an M-pair system is $M \times 2312$ kbit/s for first generation SHDSL and $M \times 5696$ kbit/s for ESHDSL systems.

The STU-R symbol clock is always locked to the symbol clock of the STU-C, but the symbol clock may or may not be locked to the data clock. If the symbol clock is locked to the data clock, then the framing is synchronous and every frame is 6 ms long, including two stuffing bits at the end of each frame. In plesiochronous framing mode, the symbol clock is not locked to the data clock so the frame size cannot stay fixed. Therefore, in plesiochronous mode, the average frame size is maintained at 6 ms by adjusting the size of the individual frames, using either zero or four stuffing bits to adjust for relative offset of the data and symbol clocks.

17.3.4.4 The SHDSL Application-Specific TPS-TC Layer

Various application-specific TPS-TC mappings are specified in Annex E of ITU-T G.991.2 and Annex A of ETSI TS 101 524. As described above, the SHDSL frame structure is specified in such a way that a flexible number of 64 kbit/s time slots and 8 kbit/s time slots are available for the application data. By configuring the mapping of the application layer payload(s) into the core frame payload blocks, a wide range of applications can be seamlessly integrated into the SHDSL frame. Figure 17.9 summarizes the mapping options that allow the different application protocols to be mapped into the standard core frame payload block.

The simplest mapping is a clear channel unstructured mapping, in which application bits are mapped into the payload data blocks in the correct temporal order, but without regard to any overlying structure, as shown in Figure 17.9a. The next simplest mapping is a byte structured mapping, in which byte boundaries are conserved, as shown in Figure 17.9b. This can then be extended by also ensuring that a specific byte (for example, the first byte of a frame synchronization word) is mapped into the same position and payload block in each DSL frame, thus ensuring frame synchronization between the application layer and the SHDSL layer.

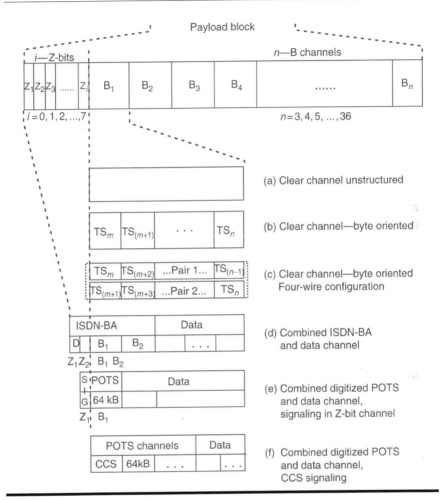

Figure 17.9   **SHDSL symbol synchronization options.**

These mappings are analogous to those in HDSL (ITU-T G.991.1 and ETSI TS 101 035), particularly when the four-wire (two-pair) option shown in Figure 17.9c is also considered. The main difference is that for the HDSL case, the Z-bits are "hardwired" with $i = 0$, whereas SHDSL can chose the most appropriate value for i.

The flexibility in assigning Z-bits can be used to map ISDN-BA 2B+D channels into the start of the SHDSL core frame, with two Z-bits being combined to give the capacity for the D-channel, and the B channels being mapped into the first two data bytes in the payload block, the remainder being free for data transmission (see Figure 17.9d). A similar mapping can be implemented for digitized POTS channels, with the signaling being carried either in a Z-bit channel (Figure 17.9e) or as a common channel signaling (CCS) channel (Figure 17.9f). In both these cases, more than one narrow band channel can be incorporated into the payload block.

It is also possible to split the payload mapping between two separate applications, for example, to support an additional broadband or ATM data path in parallel with one of the other applications mentioned so far (dual-bearer mode). A dynamic rate repartitioning algorithm has also been specified that allows time slots to be dynamically allocated between these dual bearers.

The TPS-TC annexes of ITU-T G.991.2 and ETSI TS 101 524 describe in detail the following application specific TPS-TC layers that make use of the principles outlined above:

- TPS-TC for clear channel data
- TPS-TC for clear channel byte-oriented data
- TPS-TC for DS1 transport
- TPS-TC for aligned DS1 or fractional DS1 transport
- TPS-TC for European 2048 kbit/s digital unstructured leased line (D2048U)
- TPS-TC for unaligned European 2048 kbit/s digital structured leased line (D2048S)
- TPS-TC for aligned European 2048 kbit/s digital structured leased line (D2048S) and fractional
- TPS-TC for synchronous ISDN-BA
- TPS-TC for mapping of 64 kbit/s POTS channels onto the SDSL frame
- TPS-TC for ATM transport
- TPS-TC for dual-bearer mode
- TPS-TC for LAPV5 enveloped POTS or ISDN
- TPS-TC for dynamic rate repartitioning
- TPS-TC for STM (Synchronous Transfer Mode) with dedicated signaling channel

17.3.4.5 Operations and Maintenance

There is an EOC specified for SHDSL, which allows the terminal units to maintain information about the SHDSL span. All SHDSL performance monitoring data is transported over the EOC, and the fixed indicator bits are used to indicate that an anomaly has occurred on a particular segment. Up to eight regenerators per span are supported by the EOC addressing.

The STU-C maintains a management information database for external access by network management or via craft terminal interfaces. The STU-R may also maintain a locally accessed management information database. A simple but robust packet protocol allows EOC messages to be "processed," "forwarded," or "ignored and terminated" at each node, thus ensuring that a message sent from any point of the SHDSL span will be correctly received at the appropriate destination.

Repeaters can be deployed to extend the range of SHDSL links. The deployment of regenerators slightly complicates the operation of the SHDSL link, but the SHDSL standards have special provisions to support the use of repeaters. The standards define how the SHDSL repeaters should forward EOC messages, how handshake is done by the repeaters, and the initialization sequence for links with repeaters.

17.3.4.6 SHDSL Power Spectral Density

The symbol rates for SHDSL vary with the data rate (refer to Section 17.3.4.2), and therefore the spectrum of the transmit signal also changes with the data rate. A family of symmetric (identical in the upstream and downstream direction) PSD masks is specified for SHDSL. Asymmetric PSD masks are also specified for improved performance at key data rates.

All the symmetric PSD masks for both North America and Europe share the same basic form. In the main band, the symmetric PSD masks all have the shape of a sixth-order Butterworth filter, but out of band the mask is specified such that it corresponds to constant near-end crosstalk (NEXT) across all frequencies. Most of the PSD masks have a nominal power of 13.5 dBm, and the 3-dB frequency is at half the symbol rate. The only exceptions to these general rules are that the North American masks for 1536 and 1544 kbit/s are slightly narrower for spectral compatibility reasons, and the European PSDs for data rates greater than or equal to 2048 kbit/s have 14.5 dBm nominal power for improved performance. Figure 17.10 shows examples of the SHDSL symmetric PSD masks. These are the European masks with increased transmit power for data rates above 2048 kbit/s.

There are two sets of asymmetric PSD masks specified for North America. There is one set of masks for 1536 and 1544 kbit/s, and there is one set of masks for 768 and 776 kbit/s. The asymmetric PSD masks for North America are taken directly from HDSL2 and HDSL4. Figure 17.11 shows the North American asymmetric masks for 1536 and 1544 kbit/s. The North American asymmetric PSD masks are carefully shaped to provide maximum performance with good spectral compatibility with ADSL.

There are also two sets of asymmetric PSD masks for Europe, for data rates of 2048 and 2314 kbit/s. The European asymmetric PSD masks

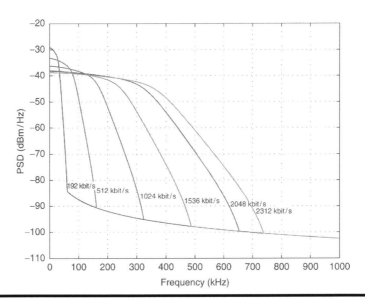

Figure 17.10   Symmetric power spectral density (PSD) masks for SHDSL at several data rates.

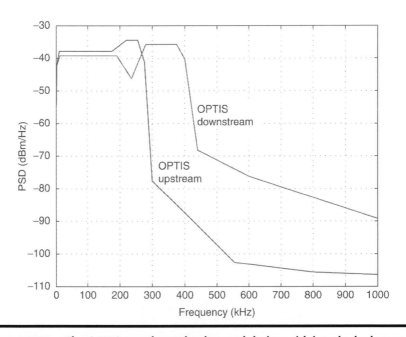

Figure 17.11   The OPTIS (overlapped pulse-modulation with interlocked spectra) PSD mask, originally conceived for HDSL2 and used as asymmetric mask for North American SHDSL at 1536 and 1544 kbit/s.

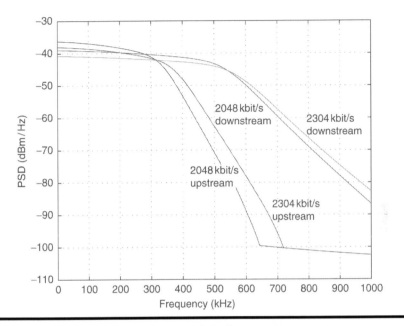

Figure 17.12   Asymmetric PSD masks for European SHDSL.

were also designed to improve SHDSL performance at these key rates. Figure 17.12 shows the European asymmetric PSD masks.

The extended rate SHDSL (ESHDSL) has a family of PSD masks that extend up to 3848 kbit/s for 16-level TC-PAM and up to 5696 kbit/s for 32-level TC-PAM. Figure 17.13 shows the European PSD masks for the highest ESHDSL rates for 16- and 32-level TC-PAM. For comparison, the figure also shows the European PSD mask for 2312 kbit/s.

17.3.4.7   SHDSL Performance Requirements

Both ITU-T G.991.2 and ETSI TS 101 524 specify transmission performance requirement tests that stress SHDSL transceivers in a way that is representative of a high-penetration scenario in operational access networks. These performance requirements are partially intended to enable operators to define deployment rules that apply to most operational situations.

There are separate performance requirements specified for North America and Europe, with different loop topology and noise environments for the two regions. The North American test loops have six different topologies, including loops with and without bridged taps. These test loops are based on a 26-AWG (American wire gauge) line of variable length. The European test loops have seven different topologies composed of loops with different diameters (and, therefore, different electrical characteristics). The key European test loop is a straight 0.4-mm loop of variable length.

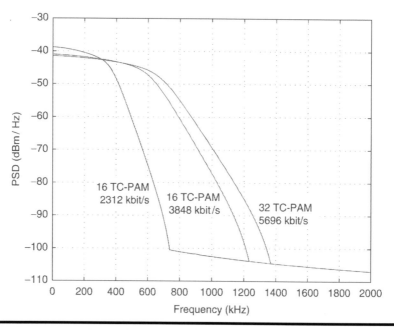

Figure 17.13   Example PSD masks for extended rate SHDSL (ESHDSL).

The North American 26-AWG wire is approximately 0.4-mm in diameter and has essentially the same characteristics as the European 0.4-mm test loop.

The main difference between the North American and European test cases is the noise profiles for the crosstalk noise. The North American noise profiles are based on using a composite of one or two types of disturbers (a total of 49 disturbers in each case), where the specific disturbers are different for each SHDSL bit rate being tested. The European performance testing, on the other hand, is based on four predefined noise profiles that have been constructed by combining many disturbers of different types. The European (ETSI) noise A assumes the very severe case of about 400 disturbers sharing the same cable, while noise B, C, and D assume 49 disturbers sharing the same cable. The 49-disturber noise profiles are representative of networks with very high penetrations of DSL services and therefore with high levels of crosstalk noise. The 400 disturbers in European (ETSI) noise model A generate extremely high-crosstalk levels, but they are considered to be representative of the crosstalk levels in the Netherlands.

Figure 17.14 shows representative examples of performance requirements for SHDSL transceivers operating in the European test environment. The circles and triangles in the figure show the specified minimum performance of an SHDSL transceiver in terms of the percents of European

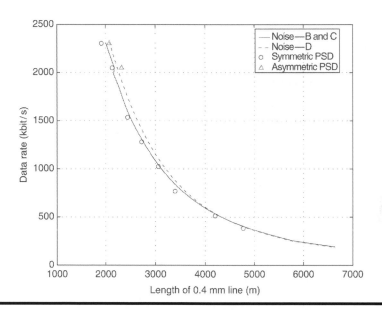

Figure 17.14 **European performance requirements for symmetric and asymmetric PSDs and simulated performance for SHDSL for European noise models B, C, and D.**

(ETSI) SHDSL test noise B, C, and D. The circles are the performance requirements when symmetric PSD masks are used (see Figure 17.10), and the triangles represent the performance requirements for the two asymmetric PSD masks (see Figure 17.12). The solid line in Figure 17.14 represents the simulation performance prediction for a very good SHDSL transceiver in the presence of European (ETSI) SHDSL test noises B and C. The dashed line represents the predicted SHDSL performance in the presence of test noise D. The predicted performance for test noises B and C shows a slight increase in performance for bit rates above 2048 kbit/s. This increase is because the European symmetric PSD masks allow a 1-dB increase in transmit power for data rates of 2048 kbit/s and above. The European performance requirements assume a 6-dB noise margin for a Bit-Error Ratio (BER) of 10^{-7}. This means that BER will be no higher than 10^{-7}, even if the noise level is increased by 6 dB. For the North American performance requirements, a 5-dB noise margin is assumed.

European (ETSI) test noise D is 49 self-NEXT, which means that it is the crosstalk from 49 identical SHDSL systems sharing the same cable binder. For low bit rates (and long loops), the performance with 49 self-NEXT (model D) is the same or slightly worse than with noise models B and C. This implies that at these rates, SHDSL is self-NEXT limited. For a self-NEXT limited system, an increase in transmit power does not lead to improved performance (assuming the 49 self-NEXT case) because the increased transmit power implies that all the other 49 SHDSL systems would increase their transmit power as well, resulting in an equivalent increase in noise and therefore no improvement in SNR.

For the higher rates, the performance with noise model D is better than with noise models B and C, which implies that at these rates the performance is limited by crosstalk from other kinds of DSL systems (mainly HDSL and ADSL). This performance limitation could have been avoided by increasing the SHDSL transmit power, but this would have resulted in performance degradations to other systems (e.g., HDSL and ADSL). These kinds of mutual degradation of various DSL systems are often referred to as "spectral compatibility issues." Addressing spectral compatibility is always a delicate issue and has been the subject of much debate in various standards bodies (see chapter 7). Choosing the transmit power for SHDSL was a balancing act between SHDSL performance and potential spectral compatibility issues with other types of DSL. A reasonable trade-off was found for SHDSL, allowing very good performance with limited spectral impact on other types of DSL services.

The SHDSL bit rates are symmetric, so the upstream and downstream data rates are the same. However, the upstream and downstream noise environments are not symmetric, because some of the disturbing systems (e.g., ADSL) use asymmetric PSD masks. Therefore, downstream transmission may be operating with a comfortable SNR margin while the upstream direction is struggling to maintain the minimum SNR margin (or vice versa). It is possible to improve SHDSL performance somewhat by using asymmetric PSD masks that take into account the asymmetric noise environment and asymmetry in spectral compatibility issues. Based on this, asymmetric PSD masks have been developed for key rates: 768 and 1544 kbit/s for North America and 2048 and 2304 kbit/s for Europe. These asymmetric PSD masks provide somewhat better performance than the symmetric PSD masks.

The performance requirements for the SHDSL standards were determined based on theoretical performance calculations (see chapter 4 in [Golden 2006]). In these calculations, it was assumed that SHDSL equipment would have an implementation loss of only about 1.6 dB. In other words, the imperfections in the real implementations would only degrade the performance margin by 1.6 dB from the theoretical performance margins of TC-PAM systems. It is a testimony to the quality of the SHDSL equipment on the market today that it is able to meet these very stringent requirements.

17.4 ADSL

Standardization of the Asymmetric Digital Subscriber Line began in 1989 in the T1E1.4 working group [Starr 1999]. Standardization of HDSL (see Section 17.3) was ongoing in both T1E1.4 and ETSI. For business customers, HDSL would address the need for a more efficient way to deliver T1 or E1 services. However, there was also a need for a system intended for residential customers. Unlike HDSL, this other system would need to preserve the operation of POTS on the same line as DSL. Furthermore, based on the belief at the time that Video-on-Demand (VoD) would be the so-called killer application, it would need to support a higher bit rate in the downstream direction (toward the subscriber) than in the upstream direction (away from the subscriber). It is because of the asymmetry of the bit rate in the two directions that asymmetric DSL was so named.

At the time ADSL standardization began, DMT modulation was still relatively new to the communications industry. Although the idea of partitioning an available channel bandwidth into a large number of subchannels and transmitting a distinct carrier on each was by no means new (see [Doelz 1957], for example), practical obstacles—primarily complexity and cost—had prevented widespread use of these sorts of techniques. Only with advances in digital signal processing technology did the use of multi-carrier modulation in products, rather than solely in laboratory prototypes, become feasible.

DMT is a type of multi-carrier modulation that uses an inverse discrete Fourier transform (IDFT) to partition a finite channel bandwidth into a large number of parallel subchannels (also called "tones") that are orthogonal (noninterfering) at the sub-carrier frequencies. (See chapter 7 of [Golden 2006] for details of multi-carrier modulation and DMT.) The key to the successful application of DMT in DSL is its use of the IDFT in the transmitter and the discrete Fourier transform (DFT) in the receiver. Because both the DFT and IDFT can be implemented efficiently using the fast Fourier transform (FFT), DMT emerged as an intriguing and attractive line code for digital communication applications soon after the arrival of practical FFT signal processing hardware.

However, at the time the definition of ADSL began, nearly all commercially available modems were still based on single-carrier modulation, such as PAM or QAM/CAP. (See chapter 6 of [Golden 2006] for details of single-carrier modulation.) Voiceband modems predominantly used single-carrier modulation (QAM, to be specific) and the emerging cable modem specification was also based on QAM. Furthermore, HDSL was based on the 2B1Q line code. As a result, momentum favored the selection of some form of single-carrier modulation for ADSL. But an enthusiastic and plucky young professor from Stanford University by the name of John Cioffi tirelessly promoted the merits of DMT (see [Cioffi 1991], for example), and eventually some of the more forward-thinking participants in T1E1.4 began to support his point of view.*

---

* In 2006, John Cioffi was awarded the Marconi Prize for his contributions to DSL. The prestigious Marconi Prize recognizes outstanding scientific contributions to communications science and the Internet.

In what would become an unfortunate recurring theme in DSL standardization, two groups of companies aligned, each supporting a different line code (modulation technique) for ADSL. One group believed single-carrier modulation was the right approach. The other group, led by Stanford's John Cioffi, believed DMT should be the line code. To settle the debate, the T1E1.4 working group defined a set of performance tests of prototype ADSL modems. The tests, commonly referred to as the "ADSL Olympics," were conducted in February of 1993 at Bellcore (now Telcordia). Two single-carrier modems (one using QAM and another using CAP) and one DMT modem (built by John Cioffi's own start-up company, Amati Communications Corporation) were tested. The performance of the DMT system was dramatically better than the performance of either single-carrier systems, and on the basis of the test results, DMT was chosen as the line code for ADSL.

Discussion continues to this day about why the DMT modems performed so much better than the CAP and QAM modems during the ADSL Olympics. Some people believe that DMT is simply a better line code for difficult channels, such as those encountered on telephone lines. Others believe that much of the difference was due to a phenomenon known as "market leader's handicap." At the time of the ADSL Olympics, CAP and QAM ADSL modems had a clear lead in the emerging market. The companies making CAP and QAM claimed to have focused most of their engineering resources on supporting their early customers and comparatively little effort on maximizing their performance for the ADSL Olympics. It has been claimed, but never verified, that the CAP and QAM vendors took stock modems from the factory and sent them to the testing laboratory. In contrast, Amati, which was a start-up company that had little market presence at the time, devoted the vast majority of its engineering resources to winning the ADSL Olympics, and did so handily. In spite of the hard lesson of the ADSL Olympics, a similar sequence of events appeared to occur again nearly ten years later at the VDSL Olympics (see Section 17.5).

One might wonder, therefore, whether T1E1.4 made the wrong decision or a premature decision. However, the success of ADSL in the years following the line code decision proves that right choice was made. Furthermore, because the DSL standards groups operate by consensus, it is clear that decisions that affect the most basic elements of modems become significantly more difficult to make when companies have invested resources to develop a product based on one of the alternatives. As the industry's experience in VDSL1 showed, the worst decision would have been to not make a choice between the candidate ADSL codes and to delay the decision until mature products were available for all line codes.

Today, all ADSL variants continue to use the DMT line code. ADSL is by far the most widely deployed type of DSL, with over 90 percent of all telephone lines carrying broadband services using ADSL. Even some business services are provided using ADSL, despite the fact that other DSLs were designed specifically with business user needs in mind.

The commercial success of ADSL has emphatically put to rest any questions about the feasibility of implementing a high-performance, cost-effective DMT modem. The success of ADSL undoubtedly also paved the way for the selection of DMT as the line code for VDSL (see Section 17.5).

17.4.1 Overview of the ADSL Standards

Several ADSL standards have evolved over the years. The first ADSL standard was finalized in 1993 for the United States as Committee T1 Standard T1.413 [T1.413]. This document specified the operation of ADSL on the same physical line as POTS. ETSI followed with TS 101 388 [TS 101 388 (1998)] in 1998, which specified ADSL operation on the same line as ISDN. The first international ADSL standard was ITU-T Recommendation G.992.1, which was published in 1999 [G.992.1]. The body of G.992.1 specified the major components of ADSL transceivers, whereas the annexes addressed various operational modes. Annex A of G.992.1, which specifies ADSL over POTS, was based heavily on T1.413, and Annex B, which specifies ADSL over ISDN, leveraged TS 101 388. After the completion of G.992.1, SG15/Q4 became the organization primarily responsible for continuing ADSL standardization, and gradually the original versions of T1.413 and TS 101 388 became obsolete. However, both T1E1.4 and TM6 generated "pointer" standards to G.992.1; these documents specify regional requirements and restrictions. For example, the North American pointer standard [T1.413 Issue 2] requires the support of nonoverlapped spectra, and the European pointer standard [TS 101 388 (2002)] specifies European performance requirements for ADSL modems, both over POTS and over ISDN.

Because newer versions of ADSL have now been defined, G.992.1 is sometimes referred to as the ADSL1 standard. ADSL1 defines both overlapped and nonoverlapped downstream and upstream spectra. In addition, a "category I" mode is defined, which supports downstream bit rates of up to 6.144 Mbit/s and upstream bit rates of up to 640 kbit/s. The G.992.1 Recommendation also defines "category II" mode, in which support of some optional parameter settings (including trellis coding) can be used to enable enhanced performance. In practice, virtually all ADSL1 modems operate in category II mode. The maximum loop length on which ADSL1 can be deployed (called the "reach") is dependent on the loop and noise conditions. On 24-AWG loops, the maximum reach on a loop with very low noise is approximately 18 kft, but because typical loops generally do not have such low noise, ADSL1 is seldom deployed on loops longer than 16 kft.

In parallel with G.992.1, a scaled-down version of ADSL was also developed. This effort resulted in the specification known as "G.lite" (pronounced "G dot lite"), which was published as ITU-T Recommendation G.992.2 [G.992.2]. Three major ideas drove G.lite: (1) ADSL modem cost would be reduced by halving the number of DMT subchannels and halving the bandwidth spanned, (2) the reduced digital signal processing requirements would enable the ADSL modem at the customer-end of the line to be implemented entirely as software running on the microprocessor in a personal computer, and (3) customers would install their own ADSL modems and eliminate the cost dispatching network technicians to install splitter filters at customer premises.

However, G.lite modems ended up costing the same as full-rate ADSL modems because the cost savings due to halving the bandwidth were insignificant relative to the total cost of a modem. Furthermore, high-performance modems implemented in software were not practical. Tests with a variety of telephones proved that it was necessary to place a simple in-line filter in series with most types of telephones, and once these filters were in place, a full-rate modem performed better than a G.lite modem.

G.lite seemed like a good idea at the time it was developed, but it never gained traction in the marketplace, and the authors are not aware of any statistically significant G.lite deployments. However, G.lite produced two benefits: its development helped produce a political dynamic that contributed to the unification of industry support for one line code, and G.lite introduced the concept of splitterless self-installation by the customer, which has become the predominant model for ADSL. However, because G.lite has not been deployed in volume, G.992.2 is not considered further in this chapter. Interested readers are referred to [G.992.2] for details of G.lite.

The ADSL2 specification, ITU-T Recommendation G.992.3 [G.992.3], was first published in 2002. It primarily improves the functionality of ADSL1 and also specifies a number of additional modes that expand the service variety of ADSL. Perhaps more importantly, however, G.992.3 serves as the foundation of ADSL2plus, which provides significantly higher bit rates on shorter loops. Some annexes of ADSL2 allow downstream bit rates as high as approximately 15 Mbit/s and upstream bit rates as high as 3.8 Mbit/s. In addition, a mode to extend the maximum reach of ADSL has been defined. ADSL2 also introduces loop diagnostic functions to assist service providers with characterization of service issues, as well as modes intended to reduce power consumption. Finally, ADSL2 defines all-digital modes of operation, which can be used when support of POTS or ISDN on the same physical loop is not necessary, such as in the provision of services to business customers.

In parallel with G.992.3, a second version of G.lite was also developed and documented in G.992.4 [G.992.4]. However, this second version was no more successful in the market than the original was and, like the original, is not considered further here.

At the time of writing, the most recent work in ADSL was in the specification known as ADSL2plus, which is captured in ITU-T Recommendation G.992.5 [G.992.5]. ADSL2plus defines additional subchannels in the downstream direction, thus increasing the maximum downstream bit rate to 24 Mbit/s with the mandatory settings and even higher when some optional settings are used. ADSL2plus also supports essentially the same upstream options as ADSL2. G.992.5 was completed in late 2003, although progress has continued through amendments to the Recommendation. The specification is a "delta" standard to G.992.3, meaning that an ADSL2plus modem by definition has all the functionality of an ADSL2 modem. It also means that ADSL2plus modems can revert to ADSL2 operation when loop conditions do not allow transmission at the higher frequencies available in ADSL2plus, such as when a loop is too long to allow use of the frequencies above the ADSL2 band.

Table 17.2 summarizes the ADSL standards and their approximate maximum bit rate capabilities. The maximum bit rates shown in this table are achievable only on short lines with very little noise, and thus are not typically achieved for service on real lines. Details of the individual standards and their modes of operation are given in the subsections that follow.

Table 17.2   The Asymmetric DSL (ADSL) Standards

ADSL Type	Relevant Standards	Maximum Bit Rate (Mbit/s)	
		Downstream	Upstream
ADSL1	ITU-T Recommendation G.992.1, ANSI T1.413, ETSI ETS 101 388	8	0.8
ADSL2	ITU-T Recommendation G.992.3	13[a]	3.5[b]
ADSL2plus	ITU-T Recommendation G.992.5	24[c]	3.5

[a] A maximum downstream rate of 13 Mbit/s assumes the use of nonoverlapped spectra, which is required in most ADSL deployments, and operation on the same line as POTS. The use of overlapped spectra would increase the achievable downstream bit rate by up to approximately 2 Mbit/s but would result in a decrease in the maximum upstream bit rate. If operation over ISDN is required, the maximum downstream bit rate is lower, regardless of whether overlapped or nonoverlapped spectra are used

[b] A maximum upstream rate of 3.5 Mbit/s assumes that operation over POTS is required, and nonoverlapped spectra are used. In the case that the system operates over ISDN, or when overlapped spectra are used, the maximum upstream bit rate is lower. In the case that all-digital operation is allowed, the upstream bit rate can be as high as about 3.8 Mbit/s with nonoverlapped spectra.

[c] A maximum downstream bit rate of 24 Mbit/s corresponds to use of the mandatory framing parameters. Use of the optional parameters allows a higher maximum bit rate.

The rest of this section refers either to a specific version of ADSL (i.e., ADSL1, ADSL2, ADSL2plus), to a specific operational mode of ADSL in general (i.e., ADSL over POTS, ADSL over ISDN, etc.), or to ADSL in general. The reader should interpret material about "ADSL over POTS" to apply to all ADSL over POTS systems, whether ADSL1, ADSL2, or ADSL2plus. Likewise, a discussion about ADSL that does not specify an operational mode or a version should be interpreted as applicable to all ADSL systems. However, the reader is cautioned that statements about "ADSL1 over POTS" in the section on ADSL1 do not necessarily apply only to ADSL1 over POTS systems; they may also apply to ADSL2 over POTS systems, for example.

17.4.2 ADSL1

The range of frequencies generated by an ADSL transmitter is dependent on the direction of transmission, i.e., upstream or downstream. Additionally, in the upstream direction, the range of frequencies generated by the transmitter depends on whether the ADSL system operates on the same physical loop as POTS or as ISDN.

In all ADSL1 operational modes, the downstream transmitter partitions the bandwidth from 0 to 1.104 MHz into 257 subchannels, 255 of which are 4.3125 kHz-wide passband subchannels, and two of which are baseband subchannels with bandwidths of 2.15625 kHz (see chapter 7 of [Golden 2006] for details). The subchannels are indexed in order of increasing frequency as subchannels 0 through 256. The passband subchannels (with indices 1 through 255) are centered at integer multiples of 4.3125 kHz (the sub-carrier frequencies). The two remaining subchannels (subchannel 0 at 0 Hz and subchannel 256 at 1.104 MHz) are not used in ADSL1, which means up to 255 of the subchannels generated by the downstream transmitter could be used to support transmission. (As will be discussed, however, fewer than 255 subchannels are generally used.) Clearly, the highest-frequency subchannel available for downstream transmission is centered at 1104 kHz − 4.3125 kHz. However, because this number is not particularly convenient, industry practice is to refer to the ADSL1 downstream bandwidth as 1.104 MHz, with the understanding that the subchannel with index 256 is not used. Figure 17.15 illustrates the downstream subchannels in ADSL1.

In the upstream direction, the range of frequencies generated by the transmitter depends on whether the ADSL system operates on the same loop as POTS or on the same loop as ISDN, and, in the case of ISDN, the method used to generate the ADSL signal. In the case of ADSL1 over POTS, the upstream transmitter generates subchannels from 0 to 138 kHz (subchannels 0 through 32). The first and last subchannels are not used, and the first several passband subchannels are disabled to allow a POTS signal

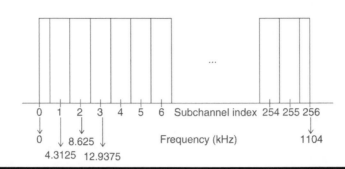

Figure 17.15   Downstream subchannels generated by an ADSL1 transmitter.

to share the line. (Although the POTS signal only spans 4 kHz, a generous guard band is imposed to ensure ADSL signals are sufficiently attenuated in the POTS band.) In the ADSL1 over ISDN case, at least the subchannels from 138 to 276 kHz (subchannels 33 through 64) are generated by the upstream transmitter. With one type of implementation, subchannels 0 through 32 are also generated. For either type of transmitter implementation, most or all subchannels below 138 kHz are not used so an ISDN signal can share the same telephone line. Figure 17.16 shows the upstream subchannels in ADSL1 over POTS and in ADSL1 over ISDN. In the ISDN figure, dashed

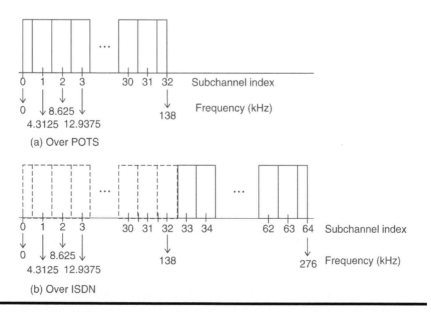

Figure 17.16   Subchannels generated by an upstream transmitter for (a) ADSL1 over POTS and (b) ADSL1 over ISDN.

lines are used to indicate subchannels that may or may not be generated, depending on the implementation.

Whether a particular subchannel generated by a transmitter is allowed to be used to support data in an ADSL link is dependent on the PSD masks defined for the selected operational mode, and, in the case of ADSL1 over ISDN, which type of upstream transmitter implementation is chosen. In ADSL1 over POTS, typically subchannels 7 through 31 are available for upstream transmission. If overlapped spectra are allowed, then the set of subchannels available for upstream transmission may also be used in the downstream direction. The use of overlapped spectra increases NEXT between lines in a cable binder. In ADSL, this increased NEXT degrades performance in the upstream direction (because, relative to when nonoverlapped spectra are used, noise levels on the upstream subchannels increase) but improves performance in the downstream direction (because more subchannels are available). Thus, in environments in which the reach of ADSL is limited by the failure of the downstream subchannels to provide a sufficient bit rate, overlapped spectra can be used to boost the downstream bit rate at the expense of the upstream bit rate. However, the use of overlapped spectra requires implementation of an echo canceller in both the downstream and upstream receivers, as well as appropriate transmit filters, so the enabling of overlapped spectra in a deployment to improve reach is entirely dependent on the implementations of the modems on both ends of the loop. Spectrum management requirements may also determine whether overlapped spectra are allowed (see Chapter 7). In the upcoming discussion of ADSL1 over ISDN, the trade-off between increased NEXT in the upstream direction and improved reach in the downstream direction when overlapped spectra are used is investigated.

In the case of ADSL over ISDN, a wider bandwidth at the low end of the spectrum is reserved to protect ISDN signals on the same physical loop. Depending on the ADSL implementation, the first subchannel available for upstream transmission may be subchannel 33, or it may be a lower-index subchannel. As with ADSL over POTS, if overlapped spectra are used, the subchannels allocated for upstream transmission are also available for downstream transmission. If nonoverlapped operation is required, typically subchannels 65 through 255 are available for use in the downstream direction.

ADSL1 was defined to allow either Synchronous Transfer Mode (STM traffic) or Asynchronous Transfer Mode (ATM traffic). Figure 17.17 illustrates the reference model for the ADSL1 transceiver at the central office (ATU-C) when the interfaces are STM, and Figure 17.18 shows the reference model for the ATU-C with ATM interfaces. In the STM reference model, seven bearer channels are defined: AS0, AS1, AS2, and AS3 are simplex bearer channels (downstream only), while LS0, LS1, and LS2 are

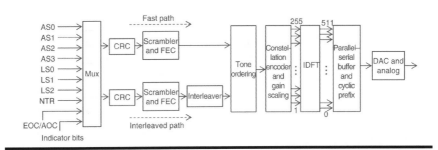

Figure 17.17   G.992.1 ATU-C reference model with Synchronous Transfer Mode (STM) interfaces.

duplex bearer channels (downstream and upstream). Support of at least AS0 and LS0 by the ATU-C is required. The network timing reference (NTR) can also be transmitted from the ATU-C to the subscriber modem (ATU-R). The bearer channels and, if present, the NTR are multiplexed together with the EOC, ADSL overhead control (AOC), and some indicator bits (Flags to indicate status and operational anomalies). In the ATM reference model, two ATM cell pipes are supported in lieu of the seven bearer channels, and support of LS0 is mandatory. Otherwise, the two reference models are identical.

Two latency paths are defined in ADSL1: the fast path and the interleaved path. Both paths provide a CRC and a scrambler, but the interleaved path also provides an interleaver. The purpose of the interleaver is to shuffle the bytes of several DMT symbols to improve the likelihood of correct decoding in the event strong impulse noise corrupts the channel during transmission. (See chapter 9 of [Golden 2006] for a detailed discussion of the value of interleaving.) The interleaver increases latency because the receiver must wait for all the bytes from a particular symbol to arrive before it can decode the symbol. Although some applications, such as one-way video transmission, can tolerate the increased latency, applications such as on-line video gaming and voice connections typically cannot.

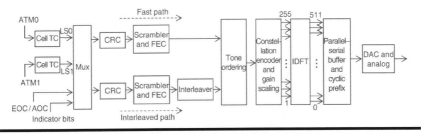

Figure 17.18   G.992.1 ATU-C reference model with Asynchronous Transfer Mode (ATM) interfaces.

When the data interfaces are STM, support of both latency paths (i.e., dual latency) is mandatory in the downstream direction, although support of two latency paths is optional in the upstream direction. When the data interfaces are ATM, support of one latency path is mandatory in both directions. Most implementations of ADSL1 modems have ATM interfaces and support only the interleaved path. In deployment scenarios requiring low latency, the interleaver can be turned off or set to its minimum depth to provide a true or approximate fast path.

Bits corresponding to user data are inserted, along with overhead bits, into data buffers for up to two latency paths, as described in Section 17.4.2.1. During each DMT symbol period, bits from the fast path appear first in the data buffer, and bits from the interleaved path follow. After Reed–Solomon encoding and scrambling to reduce the likelihood of a long sequence of zeros or ones, the data frame is routed to the tone ordering block, which determines the order in which subchannels will be assigned bits from the data frame. (The number of bits each subchannel will support is calculated during the initialization procedure. See chapter 7 of [Golden 2006] for details.) The process of assigning specific bits from the data frame to specific subchannels is called "tone ordering."

The tone ordering procedure depends on whether the ADSL1 modem supports dual latency and whether trellis coding, which is an optional capability in ADSL1, is in use. If neither dual latency nor trellis coding is supported, then there is effectively no tone ordering. Bits are simply assigned to the subchannels, starting with the lowest-index subchannel available, in the order in which they appear in the data buffer. To illustrate, if the downstream bit allocation is denoted as $(b_0, b_1, b_2, \ldots, b_{256})$, and the specific numbers of bits per subchannel are

$$b_0 \text{ through } b_{32} = 0$$
$$b_{33} = 8$$
$$b_{34} = 9$$
$$\cdot$$
$$\cdot$$
$$b_{205} = 2$$
$$b_{206} \text{ through } b_{256} = 0$$

then during each DMT symbol period, no bits would be assigned to subchannels 0 through 32, the first eight bits in the data buffer would be assigned to subchannel 33, the next 9 bits to subchannel 34, and so on until the end of the data buffer is reached after the final two bits in the buffer are assigned to subchannel 205.

If dual latency is supported but trellis coding is not used, then the tone ordering process is simple: the bits from the fast buffer are assigned to the subchannels with the smallest numbers of bits assigned to them, and then the bits from the interleaved buffer are assigned to the remaining subchannels. (In this case, it is possible that one of the subchannels may support a mixture of bits from the fast and interleaved buffers.) The ordered bit table is constructed by scanning the original bit table for all subchannels that support zero bits. These subchannels are then grouped together, in order, at the beginning of the ordered bit table. The original bit table is then scanned for subchannels that support two bits, which is the minimum number of nonzero bits a subchannel is allowed to support in ADSL1. These subchannels are then grouped together, in order, immediately following the group of subchannels that support zero bits. The scanning procedure continues for subchannels supporting successively higher-integer numbers of bits until the ordered bit table has been fully constructed. As an example, consider a simplified system with only 10 data-carrying subchannels with a bit allocation as follows:

$$(b_0, b_1, b_2, b_3, b_4, b_5, b_6, b_7, b_8, b_9, b_{10}) = (0, 8, 9, 7, 6, 7, 6, 5, 4, 2, 2)$$

The ordered bit table b' is

$$(b'_0, b'_1, b'_2, b'_3, b'_4, b'_5, b'_6, b'_7, b'_8, b'_9, b'_{10}) = (b_0, b_9, b_{10}, b_8, b_7, b_4, b_6, b_3, b_5, b_1, b_2)$$

In the example, each DMT symbol would support a total of 56 bits, so there would be exactly 56 bits (both user data and overhead) in the data buffer during each symbol period. The first two bits would be assigned to subchannel 9. The next two bits would be assigned to subchannel 10. The next four bits would be assigned to subchannel 8, and so on until the final 9 bits in the buffer are assigned to subchannel 2.

After the bits have been ordered by the tone ordering procedure, they are routed to the constellation encoder and gain scaling block. If trellis coding is not used, the bits are simply assigned directly to the subchannels according to the bit allocation computed during initialization. The complex number corresponding to each subchannel is a point from the QAM constellation that corresponds to the number of bits assigned to that subchannel and the values of the bits assigned to it during the symbol period being processed. For example, a subchannel supporting two bits has a standardized four-QAM constellation associated with it. The constellation point corresponding to the bits assigned to that subchannel during the symbol period being processed determines the complex number that is selected. Each complex number is then scaled so that the average power on all subchannels is the same. Each constellation point is also scaled by the gain scaling value of its corresponding subchannel. The gain scaling values are first computed during initialization and then are updated during a connection. They adjust the power on a subchannel to meet PSD requirements, to ensure sufficient SNR to support the number of bits transported on that subchannel, and to ensure the required noise margin is available. (See chapter 7 of [Golden 2006] for details.)

Both tone ordering and constellation encoding become more complicated when trellis encoding is used. The trellis code defined in ADSL1 is Wei's 16-state four-dimensional trellis code. (See chapter 8 of [Golden 2006] for details of trellis coding.) Because the trellis code is four-dimensional, the bits on sets of two subchannels (representing two dimensions each) are taken together as a single input to the trellis encoder. The constellation expansion of the code is one bit per four dimensions, which results in one-half bit of trellis coding overhead per subchannel. Because of the constellation expansion, the number of coded bits per subchannel must be between 2 and 15. When trellis coding is enabled, the subchannels are ordered in the same way as when trellis coding is not present; that is, the ordered table has all subchannels supporting zero bits grouped together at the beginning of the table, followed by all two-bit subchannels (in order), followed by all three-bit subchannels (in order), etc. However, because of the trellis code, bits are no longer assigned to individual subchannels but instead to pairs of subchannels, with the overhead of the trellis code taken into account in the allocation.

The example used previously can illustrate how the allocation of the bits to the subchannels changes when trellis coding is enabled. The ordered bit table is given as

$$(b'_0, b'_1, b'_2, b'_3, b'_4, b'_5, b'_6, b'_7, b'_8, b'_9, b'_{10})$$

$$= (b_0\{0\}, b_9\{2\}, b_{10}\{2\}, b_8\{4\}, b_7\{5\}, b_4\{6\}, b_6\{6\}, b_3\{7\}, b_5\{7\}, b_1\{8\}, b_2\{9\})$$

where, for convenience, the number of bits each subchannel can support has been incorporated in the representation in curly brackets $\{\cdot\}$. b'_0 is not used, not only because it supports no bits, but also because the number of data-bearing subchannels is even.* b'_0 would only play a role in the encoding of bits if the number of data-bearing subchannels were odd.[†]

---

[*] Furthermore, in a real ADSL system, b_0 would correspond to the subchannel at zero, which is never used. For convenience, the status of b_0 as always unused is ignored here.

[†] The fact that a zero-bit subchannel will have to be paired with a nonzero-bit subchannel if the number of data-bearing subchannels is odd is one reason why the minimum number of bits per subchannel is two in ADSL1. If one-bit subchannels were allowed in ADSL1, and a zero-bit subchannel happened to be paired with a one-bit subchannel, then it would not be

Therefore, b'_1 and b'_2, corresponding to subchannels 9 and 10, are assigned the first three bits from the data frame, because the other bit is reserved for trellis overhead. However, these three bits must pass through the trellis encoder before the constellation encoder can determine the correct QAM constellation points for transmission on subchannels 9 and 10. See chapter 8 of [Golden 2006] and [G.992.1] for details of the ADSL1 trellis encoder.

After the trellis encoder, two-coded bits emerge for subchannels 9 and 10. At this point, the joint processing of bits on subchannels 9 and 10 ends, and the bits for the two subchannels are mapped separately by the constellation encoder to the appropriate QAM constellation points.

Subchannel pairs (b'_3, b'_4) and (b'_5, b'_6) are processed in the same manner as (b'_1, b'_2). However, pairs (b'_7, b'_8) and (b'_9, b'_{10}) are handled slightly differently. To force the convolutional encoder in the trellis encoder to the zero state after every DMT symbol is processed (to eliminate the potential for error propagation from symbol to symbol), two more of the bits in the last two subchannel pairs are reserved. Therefore, for subchannel pair (b'_7, b'_8), only 11 data bits are sent to the trellis encoder, and for subchannel pair (b'_9, b'_{10}), only 14 bits are sent. (One can appreciate now why the tone ordering procedure places the subchannels supporting the most bits at the end of the ordered table.)

After all subchannels have been assigned constellation points corresponding to the trellis encoder outputs, average power scaling and gain scaling are applied in the same way as when trellis encoding is not used.

It is worth noting that when trellis coding is used, the number of user data and overhead bits in the data buffer must be calculated to account for the trellis overhead and reserved bits. Based on this observation, one might conclude that the use of trellis coding reduces the net data rate of a connection. However, the trellis code provides a coding gain, typically around 4 dB, which increases the total number of bits each DMT symbol can support. In general, a coding gain of 3 dB allows each subchannel to carry one additional bit, so a 4-dB coding gains more than compensates for the trellis code overhead. Therefore, although some bits in each symbol are indeed reserved for overhead, the total number of bits per symbol is generally significantly larger with trellis coding than without it, and the trellis overhead consumes a small percentage of the additional bits. The use of the same example to illustrate tone ordering with trellis coding as to illustrate tone ordering without trellis coding is somewhat misleading in this respect.

---

possible to have a bit for overhead as the trellis code requires and still have data transmitted on that subchannel pair. In ADSL2, one-bit subchannels are supported, and the trellis encoding procedure has been modified accordingly.

Regardless of whether trellis coding is used, after the constellation encoder has generated a complex-valued point for each subchannel, an Hermitian sequence is constructed using the complex values associated with the subchannels so that application of an oversized IDFT will result in an output that is real. (See chapter 7 of [Golden 2006].) As Figures 17.17 and 17.18 illustrate, the IDFT size is 512 points in the downstream direction. The output of the IDFT is converted to a serial stream of real-valued samples. The cyclic prefix, which ideally eliminates intersymbol interference between subsequent DMT symbols, is then prepended to the stream of 512 real values. A cyclic prefix length of 32 samples is used in the downstream direction in ADSL1. The prefixed signal is then converted to analog format, after which it is filtered to mitigate aliasing before being launched onto the telephone line.

To aid in timing recovery, one of the downstream subchannels in ADSL1 is allocated as a pilot tone. (See chapter 12 of [Golden 2006] for details about pilot tones.) The pilot tone does not support data transmission. Instead, the 4-QAM constellation point corresponding to the bits 00 is modulated onto the pilot. The subchannel that is used as the pilot tone depends on the operational mode (see Section 17.4.2.4).

Furthermore, to aid recovery of the DMT symbol boundary after certain types of micro-interruptions* that might otherwise cause modems to lose synchronization and retrain, ADSL1 defines a synchronization (abbreviated as "sync") symbol. The sync symbol consists of a pseudo-random sequence that is modulated onto the subchannels and transmitted after every 68 data symbols. After sync symbol insertion, a collection of 69 symbols, called a "superframe" (see Section 17.4.2.1), corresponds to 34,816 data samples, 2,176 cyclic prefix samples, and 544 sync symbol samples. With a subchannel spacing of 4.3125 kHz, data symbols (i.e., all symbols except the sync symbol) are transmitted at an average rate of

$$\frac{1}{T} = \frac{34{,}816}{34{,}816 + 2{,}176 + 544} \times 4.3125 \text{ kHz} = 4 \text{ kHz}$$

As one might expect, the operation of the ADSL1 transceiver unit at the customer end (the ATU-R) is nearly identical to the operation of the ATU-C. Figures 17.19 and 17.20 illustrate the transmitter reference models for STM and ATM transport, respectively. The primary difference between the functions of the ATU-R and the ATU-C is the number of subchannels generated. Because the upstream bandwidth is much smaller than the downstream

---

* A micro-interruption is a short disruption on the line, for example, because of impulse noise, that does not cause the modems to retrain but does cause a loss of data and possibly a loss of synchronization.

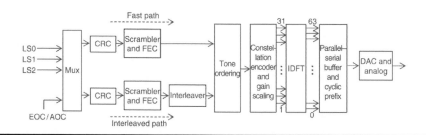

Figure 17.19   G.992.1 ATU-R reference model with STM interfaces.

bandwidth, fewer subchannels are needed in the upstream direction. (The subchannel spacing remains 4.3125 kHz.) Otherwise, most of the functionality of the ATU-C is also present in the ATU-R. One exception is the pilot tone, which is not used in the upstream direction because use of loop timing by the ATU-R makes a pilot tone unnecessary.

Because fewer tones are defined in the upstream direction, only 64 data samples are generated per symbol period.* A cyclic prefix of length 4 is defined, and, as in the downstream direction, a sync symbol is inserted after every 68 symbols. As a result, the average data symbol rate in the upstream direction is also 4 kHz.

17.4.2.1   Framing

Framing is the process by which the transmitter orders all bits that need to be sent over the link, including both data bits and overhead bits (except those from the trellis code). This section describes ADSL1 framing in the downstream direction; in the upstream direction, framing is somewhat simpler because there are fewer bearer channels to accommodate, and

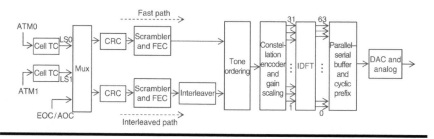

Figure 17.20   G.992.1 ATU-R reference model with ATM interfaces.

---

* It is possible for an ADSL1 over ISDN upstream transmitter to generate its subchannels using a 128-point IDFT, which would result in 64 subchannels. The reference model is only representative of ADSL1 over ISDN when a 64-point IDFT is used to generate the upstream signal.

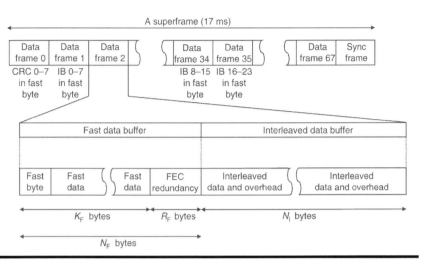

Figure 17.21   G.992.1 superframe.

only one latency path is defined. First, full-overhead framing for dual latency is described, and then the simplifications resulting from the use of reduced-overhead framing and a single latency path are presented.

Figure 17.21 illustrates the ADSL superframe, which is composed of 68 data frames followed by one sync frame. The 69 frames are mapped to 69 consecutive DMT symbols. Each superframe spans 17 ms. Distributed in selected data frames within the superframe are CRC bytes corresponding to the previous superframe and indicator bits for operations and maintenance functions.

Each data frame is composed of data from the fast buffer followed by data from the interleaved buffer. In addition to bearer channel data, the fast buffer contains a "fast byte" and forward error correction (FEC) redundancy bytes. The content of the fast byte depends on the index of the data frame within the superframe and the purpose the fast byte is serving. In the first data frame, with index 0, which is also the first frame of the superframe, the fast byte contains the CRC byte for the fast buffer in the previous superframe. In the second data frame, with index 1, the fast byte contains the indicator bits 0 through 7. (Indicator bits provide information about near-end and far-end defects, such as a loss-of-signal defect.) The other 16 indicator bits are transmitted in the fast byte of the data frames with indices 34 and 35. In the remaining data frames, the fast bytes of two consecutive frames may carry EOC messages or signals to control synchronization of the bearer channels assigned to the fast buffer. When a fast byte is used for synchronization control, bit 0 of the fast byte is set to 0. When the fast byte in two consecutive data frames is not needed for synchronization control, CRC, or indicator bits, the two bytes may be used to indicate "no

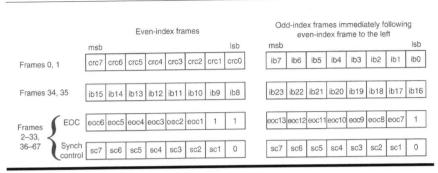

Figure 17.22   Contents of the fast byte in G.992.1.

synchronization action" or to transport a single 13-bit EOC message. (The no synchronization action setting is used when the bit timing base of the input user data streams is synchronous with the ADSL1 modem timing.) When an EOC message is transmitted, bit 0 of the fast byte is set to 1 in both data frames. Figure 17.22 illustrates the contents of the fast byte for the various frames within the superframe.

In the interleaved buffer, the first eight bits of each data frame are called the "sync byte" (which, the reader is cautioned, is entirely different from the sync frame). The first sync byte of the superframe (i.e., the first eight bits in the interleaved buffer for frame 0) is used to carry the CRC byte for the interleaved buffer from the previous superframe. In frames 1 through 67, the sync byte carries either information for synchronization control of the bearer channels assigned to the interleaved data buffer or messages from the AOC channel. The AOC channel transports, for example, bit swap messages (see Section 17.4.2.3). If no data has been allocated to the interleaved buffer, the sync byte carries the AOC channel data. When the interleaved buffer does contain data, the AOC channel is transported in a different byte of the frame, and the sync byte contains information about the content of this other byte (i.e., whether the byte contains AOC channel data or data from the bearer channel). Figure 17.23 illustrates the contents of the sync byte.

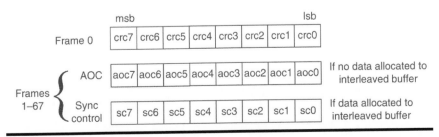

Figure 17.23   Contents of the sync byte in G.992.1.

During the initialization procedure, based on its latency and error protection requirements, each bearer channel is assigned either to the fast buffer or to the interleaved buffer. For each bearer channel X, the number of bytes allocated to the fast buffer is denoted as $B_F(X)$, and the number of bytes allocated to the interleaved buffer is denoted as $B_I(X)$. Because each bearer channel can be transmitted in only the fast or interleaved path, if $B_F(X) \neq 0$, then $B_I(X) = 0$.* The reader is referred to [G.992.1] for the details of how the bearer channels are ordered in the fast and interleaved buffers and other bytes that are also included in full-overhead framing mode.

For each data frame, the bits from the fast buffer are submitted to the Reed–Solomon encoder, which adds R_F redundancy bytes. The bits from the interleaved buffer must be treated differently, however, because the bits must be distributed among a set of data frames (and consequently among a set of data symbols) prior to Reed–Solomon encoding. Therefore, a "mux data frame" is defined for the interleaved path, as shown in Figure 17.24. Each mux data frame contains a total of K_I bytes, which include the sync byte, bearer channel data for the interleaved path (after the interleaver), and overhead bytes. A sequence of S mux data frames, which contain bits that will be allocated to a sequence of S DMT symbols, is then submitted to the Reed–Solomon encoder, which appends R_I redundancy bytes. The set of $S \times K_I + R_I$ bytes is then split into S sets of N_I bytes, where $N_I = (S \times K_I + R_I)/S$. Each set of N_I bytes is the interleaved buffer portion of a single data frame (refer to Figure 17.21).

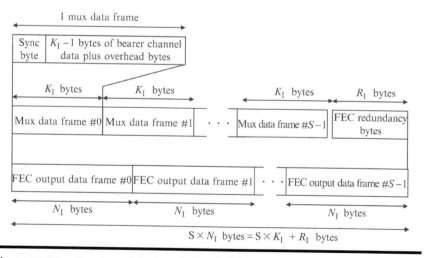

Figure 17.24   Framing of the interleaved buffer.

---

* There is an exception, which is the 16-kbit/s C-channel option. See [G.992.1] for details.

The value of S, which indicates the number of DMT symbols the Reed–Solomon codeword spans, is constrained to be a power of two in ADSL1. In particular, with mandatory settings, $S = 1$ for the fast buffer, and $S = 1$, 2, 4, 8, or 16 for the interleaved buffer. As a result of the power-of-two constraint, Reed–Solomon codewords are aligned with DMT symbols.

When $S = 1$, the maximum downstream data rate is limited to 8 Mbit/s. This bit rate is sometimes referred to as being "framing limited" because the maximum value is limited by the framing mechanism and not by the maximum number of bits each subchannel can support. Because the maximum ADSL1 bit rate could be much higher if all available subchannels were to support the maximum number of 15 bits, the value of $S = 1/2$ is an option that has been almost universally implemented by ADSL1 modem manufacturers. With this option, a single Reed–Solomon codeword spans two DMT symbols (and Figure 17.24 no longer applies), which reduces the overhead because of FEC. As a result, the maximum downstream bit rate increases (for example, to around 13 Mbit/s for a system operating over POTS using nonoverlapped spectra) because the framing with $S = 1/2$ does not restrict the achievable bit rate to an artificially low level.

Reduced-overhead framing ADSL1 was specified to enable support of synchronous user data streams. However, in practice, only ATM interfaces have been deployed. Therefore, the amount of overhead required for a connection can be reduced relative to the amount specified in full-overhead mode. Furthermore, although ADSL1 was specified to support dual latency, in practice only one latency path is usually implemented. As a result, all data is assigned either to the fast buffer or to the interleaved buffer.

The reduced-overhead framing mode is a simplification of the ADSL1 framing that can be used when synchronization control is unnecessary and only a single latency path is required. It is the framing mode most commonly deployed. In this mode, only one designated overhead byte is available— either the fast byte or the sync byte. The content of this byte varies based on the frame index. Table 17.3 describes the content of the fast byte or the sync byte as a function of frame index. The CRC byte remains in the fast byte or sync byte in the frame with index 0, and the indicator bits are still carried in frames 1, 34, and 35. The AOC and EOC bytes are then assigned to alternate pairs of frames, as indicated in the table.

17.4.2.2 Initialization

Before an ADSL1 connection can be established, the ATU-C and ATU-R must complete an initialization procedure. The two modems have exchanged basic capabilities during the handshake procedure (see Section 17.6), but they do not yet have any information about the line conditions or how to configure themselves to meet the service provider's objectives. The

Table 17.3   The Content of the Fast or Sync Byte in Reduced-Overhead Mode

Frame Index	Fast Byte Content (Valid When Fast Buffer Is Used)	Sync Byte Content (Valid When Interleaved Buffer Is Used)
0	Fast buffer CRC	Interleaved buffer CRC
1	Indicator bits 0–7	Indicator bits 0–7
34	Indicator bits 8–15	Indicator bits 8–15
35	Indicator bits 16–23	Indicator bits 16–23
$4n + 2, 4n + 3$	EOC or sync (see Note)	EOC or sync (see Note)
$4n, 4n + 1$	AOC	AOC

Note: In the reduced-overhead mode, only the "no synchronization action" code
        is used.

initialization procedure defines a set of signals and messages that allow the modems to prepare for the connection.

Initialization is a crucial component of ADSL1 transceiver operation. However, it is also a complicated component, and a thorough understanding of the signals and messages used during ADSL1 initialization requires a study of G.992.1. Chapter 7 of [Golden 2006] explains how modems can determine the channel characteristics and noise using the signals transmitted during initialization. Readers interested in details of the signals and messages defined for ADSL1 initialization are referred to [G.992.1].

17.4.2.3   Bit Swapping

After an ADSL connection has been established, the subchannel SNRs will probably not remain at the levels calculated during the initialization procedure. For example, the noise at the receiver may change as DSL modems on other lines activate or deactivate. Therefore, DMT-based standards define a protocol to allow bits to be moved from degrading subchannels to other subchannels without interrupting the connection. The total bit rate in each transmission direction remains the same, but the distribution of bits (and power) to subchannels changes. The process of moving bits and power among subchannels while maintaining a constant bit rate is called "bit swapping."

In either transmission direction, only the receiver can identify when the bits and gains on subchannels need to be modified to maintain the integrity of the connection. The transmitter, therefore, must be told when changes are necessary. When an ATU receiver detects that bits need to be moved from one subchannel to another, it initiates a bit swap request. In this chapter, this ATU is called the "initiating" ATU. The ATU on the other end of the line then makes the requested change to its transmitter. Here, this ATU is called the "responding" ATU. (In G.992.1, the initiating ATU is called the "receiver," and the responding ATU is called the "transmitter." This terminology leads to interesting statements such as "The receiver shall initiate a bit swap by sending a bit swap request…" Strictly speaking, receivers cannot send anything, so this chapter uses the alternative terminology.)

In G.992.1, bit swap messaging takes place on the AOC, which is composed of overhead bytes that are inserted in the framing structure, as described in Section 17.4.2.1. Each AOC message is preceded by a header that tells the responding ATU what type of message follows. All AOC messages are transmitted five consecutive times, and the responding ATU acts on a message if it receives three identical messages in a time period spanning five of that particular message. If the responding ATU does not detect three identical messages within the time period, or if it does not recognize the command in a message, it does not take any action.

The initiating ATU sends a bit swap request via the AOC, repeating it five times. The message is 9 bytes long. The first byte is the header, and the other 8 bytes contain instructions to modify the bits or gains of up to four subchannels, as shown in Figure 17.25. Each command can be to (a) increase or decrease the number of allocated bits by one; (b) increase the transmitted power by 1, 2, or 3 dB; (c) decrease the transmitted power by 1, 2, or 3 dB; or (d) do nothing. Typically, when a bit swap takes place, both the number of bits and the power allocated to each subchannel involved must change. The power usually has to be adjusted to maintain the target noise margin for the connection, which is dominated by the subchannel with the lowest individual noise margin. If a bit is swapped from a tone, usually the power must be decreased. If a bit is swapped to a tone, usually the power must be increased. Therefore, one can see that the bit swap message has been defined to cause a single bit to be moved from one subchannel to another and to communicate the associated power modifications necessary to maintain the noise margin following the swap.

The lowest number of bits allowed per data-carrying subchannel in ADSL1 is two. Clearly, the bit swap message defined above would not accommodate transitions from zero to two bits or from two to zero bits if power modifications are also needed. For cases when a subchannel supporting two bits needs to be turned off (allocated zero bits) or when a subchannel that is off needs to start supporting two bits, an extended bit swap message is used. This message is 13 bytes in length. The first byte is the header, which identifies the message as an extended bit swap request.

Header	Command	Subchannel index	Command	Subchannel index	Command	Subchannel index	Command	Subchannel index

← 1 byte →

Figure 17.25   Bit swap message format in G.992.1.

The remaining 12 bytes contain commands and subchannel indices in the same manner as shown in Figure 17.25. One can see that a transition from two to zero bits requires two commands to decrease the number of bits on a subchannel by one bit, followed by a command to reduce the power on that subchannel. Likewise, a transition from zero to two bits requires two commands to increase the number of bits on a subchannel by one bit, followed by a command to increase the power on that subchannel. Therefore, in total, a 2-to-0 or 0-to-2 bit swap requires 12 bytes of commands in addition to the header identifying the message as an extended swap message.

Bit swap transitions are coordinated using superframe counters. The ATU-C and ATU-R set their counters to zero immediately after they transition from the initialization procedure to steady-state operation, which is called the Showtime state. After every superframe (68 data symbols plus one sync symbol), the counters are incremented by one. The superframe counters are 8-bits long, and therefore they reset to zero after every 256 superframes.

When it receives a bit swap request, the responding ATU determines when the bit swap transition will take place. Within 400 ms of receiving the request, the responding ATU sends a bit swap acknowledge message to the initiating ATU. The acknowledgment is a three-byte message. The first byte is a header, the second byte is an acknowledge command, and the third byte is the value of the superframe counter at which the bit swap will occur. The value of the counter transmitted by the responding ATU is required to be at least 47 greater than the counter value when the bit swap request was received, which corresponds to a delay of over 800 ms. The reason for the long delay is to ensure that both the ATU-C and ATU-R have time to process all messages (the acknowledgment, for example) and ready themselves for the transition. Although it might seem that the modems should be able to prepare in far less time than 800 ms, when a single physical chip supports several ATU-C modems, as is almost always the case, the signal processing resources in that chip are shared. Imposing a limitation on the speed of bit swaps helps to ensure that the silicon at the central office has time to service the line that needs a bit swap. Because on any given line only one bit swap request is allowed to be outstanding in either transmission direction at any given time, ADSL1 bit swaps occur approximately once per second. The infrequency of bit swap opportunities in ADSL1 has been identified as a potential weakness, so ADSL2 defines a different mechanism that allows bit swaps at a much faster rate.

When the value of the superframe counters reaches the value selected by the responding ATU, the bit swap is implemented starting with the first symbol of the next superframe. The transmitter at the responding ATU changes the bit allocations on the affected subchannels and modifies their transmit powers as instructed. In addition, it reorders the subchannels using the appropriate tone ordering procedure. The receiver of the initiating ATU also updates the number of bits to be decoded on the affected subchannels and performs the tone reordering procedure. Furthermore, it updates other receiver parameters corresponding to the affected subchannels (such as the frequency-domain equalizer [FEQ] taps, see chapter 7 of [Golden 2006]) to account for the new power allocations and bit distributions.

17.4.2.4 Annexes of G.992.1

Details of the various ADSL1 operational modes are now described. The focus is on the ITU-T Recommendation G.992.1, which is the international ADSL1 specification. The annexes of G.992.1 define ADSL1 over POTS (Annex A), ADSL1 over ISDN (Annex B), and ADSL1 over POTS operating in the same binder as TCM-ISDN (Annex C).

17.4.2.4.1 Annex A: ADSL1 over POTS

The variant of ADSL that has been most widely deployed to date is that defined in Annex A of G.992.1, which specifies ADSL1 operation on the same loop as POTS. In over-POTS operation, the subchannels below 25 kHz are unused in both transmission directions. Although these subchannels are always generated by the downstream and upstream transmitters (because DMT is a baseband modulation), they are not allocated any bits or power. A highpass filter is used by ADSL transmitters to prevent interference from ADSL to the POTS signals.*

In ADSL1 over POTS, the bandwidth from 25.875 to 138 kHz is allocated for upstream transmission, and bandwidth above 138 kHz is allocated for downstream transmission. If overlapped spectra are used, the bandwidth from 25.875 to 138 kHz can also be used in the downstream direction.

Subchannel 64, centered at 276 kHz, is reserved as a pilot tone in Annex A.

In the downstream direction, ADSL1 is limited to a maximum of 20.4 dBm of power. In the upstream direction, the power is limited to 12.5 dBm. The distribution of the available power in each transmission direction within the passband is governed by a PSD mask and an average PSD requirement.

Three PSD masks are defined in Annex A of G.992.1. Figure 17.26 shows the PSD mask defined for upstream transmission. The mask extends from 25.875 to 138 kHz with a peak level of −34.5 dBm/Hz. However, the average PSD is required to be −38 dBm/Hz across the upstream band. The PSD mask lies 3.5 dB above the average PSD level to simplify bit allocation and to account for ripple in the transmitter filters. As discussed in chapter 7 of [Golden 2006], transmitter and receiver implementations are

---

* Similarly, a lowpass filter, called a splitter, is applied to the POTS signal to ensure it does not interfere with ADSL transmission. See Chapter 1 for details of splitters.

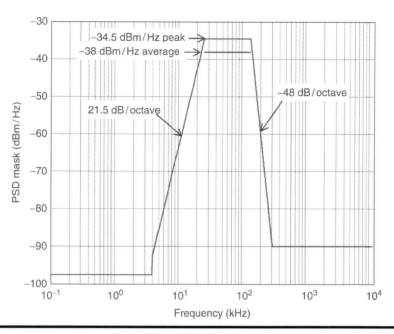

Figure 17.26   Upstream PSD mask for ADSL1 over POTS. Average PSD level also shown.

greatly simplified if the number of bits transmitted on each subchannel is an integer. However, because the SNR on any given subchannel is unlikely to correspond precisely to the SNR needed to support an integer number of bits, the number of bits must be rounded and the power adjusted accordingly to avoid a scenario in which the noise margin varies from subchannel to subchannel, which is undesirable from a performance perspective. If the PSD mask and average PSD level were coincident, then modems would always need to round down the number of bits on a subchannel. To allow the number of bits to be rounded up on some subchannels, a 2.5-dB gain is allowed. The remaining 1 dB of the 3.5 dB difference allows for ripple in the transmitter. Note that the integral of the upstream PSD mask over the passband results in a total power of 16 dBm. Clearly, the total power constraint ensures that modems cannot round up the number of bits on each subchannel.

Figure 17.27 illustrates the PSD mask defined for downstream transmission in the case that overlapped spectra are allowed. Note that the PSD mask extends from 25.875 to 1104 kHz, and that both the mask and average PSD levels are 2 dB lower than in the upstream direction. In the early days of pre-standard ADSL, both the upstream and downstream average PSD levels were −40 dBm/Hz. However, to provide a performance improvement in the upstream direction when overlapped spectra were used and to match

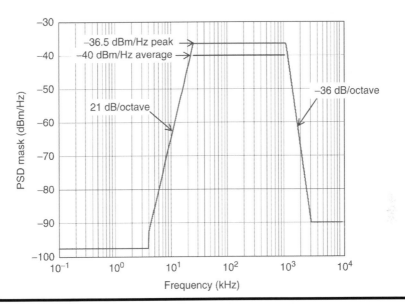

Figure 17.27   Downstream PSD mask for ADSL1 over POTS when overlapped spectra are allowed. Average PSD level also shown.

the PSD levels of HDSL transceivers, the upstream PSD was increased by 2 dB. This discrepancy in levels continues today, even though the vast majority of ADSL deployments use nonoverlapped spectra.

The downstream PSD mask defined for nonoverlapped ADSL over POTS operation is shown in Figure 17.28. The essential change from the overlapped mask is the sharp roll-off below 138 kHz, which at first appears from the plot to eliminate a large portion of the passband. However, because the plot is shown as a function of the logarithm of frequency, the amount of bandwidth actually eliminated by the use of nonoverlapped masks is not as significant as the figure might appear to indicate.

It should be noted that even though PSD masks are defined to allow transmissions within specific bandwidths, designers may choose to confine signals to a smaller range of frequencies. For example, when overlapped spectra are allowed, it may be advantageous to enable the use of only some of the subchannels between 25.875 and 138 kHz for downstream transmission to balance the improvement in the downstream bit rate and the degradation to the upstream bit rate. Alternatively, when nonoverlapped spectra are used, designers may choose not to use a small number of subchannels near the transition band or at the band edges to ease filter roll-off requirements.

The length of the telephone line also influences which frequencies are used for transmission in a particular connection. On long lines, for example, the highest frequencies are too attenuated to support transmission under

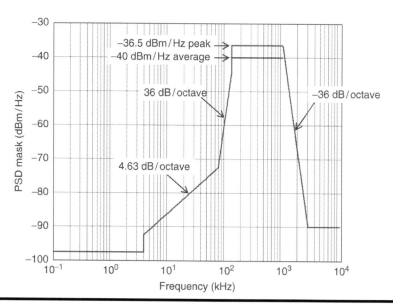

Figure 17.28 Downstream PSD mask for ADSL1 over POTS when nonoverlapped spectra are used. Average PSD level also shown.

the PSD constraints. Therefore, the uppermost subchannels are turned off on long lines, which is the reason why the bit rates achievable by ADSL are higher on shorter loops than on longer loops.

Furthermore, certain types of noise can cause some frequencies not to be used because the SNR on the corresponding subchannels is too low. For example, crosstalk from T1 lines can be severe enough to cause some subchannels in the lower part of the ADSL bandwidth to be turned off due to low SNR. Therefore, although the PSD masks define signal levels and bandwidths in which transmission is allowed to take place, other factors determine the frequencies that are actually used in a given connection.

Simulations can be used to illustrate ADSL1 performance as the bit rate that is achievable as a function of loop length. Furthermore, the DSL Forum has defined both North American and European performance requirements for ADSL1 modems in TR-067, which dictates the minimum performance required of implementations. The DSL Forum TR-067 performance requirements for North America are based on laboratory measurements of ADSL modems from several vendors. Table 17.4 documents the parameter assumptions used in a set of preliminary (theoretical) simulations of ADSL1. Figure 17.29 shows the theoretical performance of ADSL1 over POTS when it is alone in a cable binder; that is, there are no other DSL systems in the cable binder. Figure 17.30 shows the DSL Forum's North American performance requirements for ADSL1 over POTS when the only impairment

Table 17.4 Initial ADSL1 Simulation Settings

Parameter	Value
Noise margin	6 dB
Cable gauge	26 AWG
Number of disturbing lines	0
AWGN level[a]	−140 dBm/Hz
Minimum number of bits per subchannel	2
Maximum number of bits per subchannel	15
Trellis coding	On[b]

[a] AWGN simulates the noise floors of the downstream and upstream transceivers. A value of −140 dBm/Hz is commonly used in the industry, even though, as the DSL Forum performance requirements reveal, the actual noise floors are implementation dependent and could be higher or lower than −140 dBm/Hz.

[b] Although trellis coding is an option in ADSL1, support of the option is almost universally required by service providers. Therefore, because the objective of the simulations is to present a realistic set of performance results for ADSL1, trellis coding is presumed.

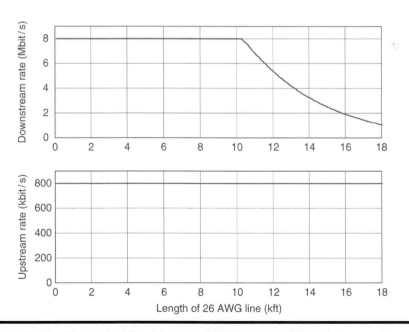

Figure 17.29 Theoretical line bit rates of ADSL1 over POTS with nonoverlapped PSD masks (−140 dBm/Hz AWGN [additive white Gaussian noise], 26 AWG [American wire gauge] cable).

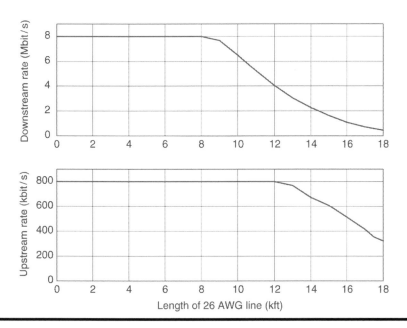

Figure 17.30   North American ADSL1 over POTS performance requirements in the AWGN case.

is additive white Gaussian noise (AWGN). A comparison of Figures 17.29 and 17.30 reveals that the DSL Forum's requirements are significantly below the performance level the theoretical simulations project. The difference is most striking in the upstream direction. The reason for the discrepancy is the AWGN level assumed in the simulations. Whereas a level of −140 dBm/Hz was assumed in the simulations (corresponding to the AWGN level injected in the DSL Forum tests), the noise floors of real modems can be significantly higher than this level, particularly in the upstream direction. The upstream receiver's noise floor is typically much higher than −140 dBm/Hz for a number of reasons, including the required density of ports on a line card and the constraints on component sizes these density requirements impose. Therefore, to determine the appropriate AWGN levels for the two directions of transmission, a second simulation was constructed. By trial and error, it was found that an AWGN level of −134 dBm/Hz in the downstream direction and −111 dBm/Hz in the upstream direction provides results very close to the DSL Forum requirements, as shown in Figure 17.31.

Using the downstream and upstream AWGN levels found to correspond to the DSL Forum's AWGN test requirements, it is possible to use simulations to predict the performance of real ADSL1 modems under various noise scenarios and using overlapped and nonoverlapped PSD masks. Figure 17.32 projects the achievable downstream and upstream line bit rates of ADSL1

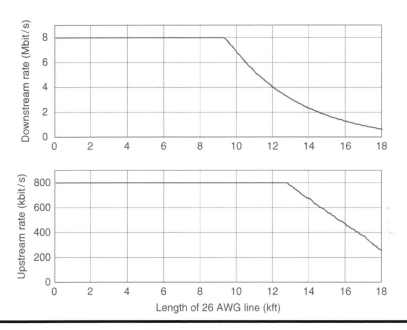

Figure 17.31   Line bit rates of ADSL1 over POTS with nonoverlapped PSD masks (−134 dBm/Hz AWGN in the downstream direction, −111 dBm/Hz AWGN in the upstream direction, 26 AWG cable).

when the nonoverlapped downstream PSD mask is used. The line bit rate includes overhead (AOC, EOC, indicator bits, and trellis overhead), so the user bit rate is lower than shown in the figure. To illustrate, in some sense, best case performance in a real environment, it was assumed in the simulation that all other systems in the cable binder are of the same type as the line being simulated; that is, all systems are also ADSL1 using the nonoverlapped downstream PSD mask. Subchannels 7 through 31 are enabled for upstream transmission, and subchannels 33 through 255 are enabled for downstream transmission. Table 17.5 documents the other settings used in the simulations.

Figure 17.32 shows clearly that the downstream line bit rate decreases a little more rapidly with increasing loop length than in the case of AWGN because of far-end crosstalk (FEXT) from other lines in the binder. However, the upstream bit rate curve is approximately the same as in the AWGN-only case, which indicates that FEXT does not significantly degrade the SNR in the upstream bandwidth on any of the loop lengths considered.

Figure 17.33 illustrates the achievable line bit rates of ADSL1 when overlapped spectra are used. In this case, subchannels 7 through 31 are used not only in the upstream direction, but also in the downstream direction.

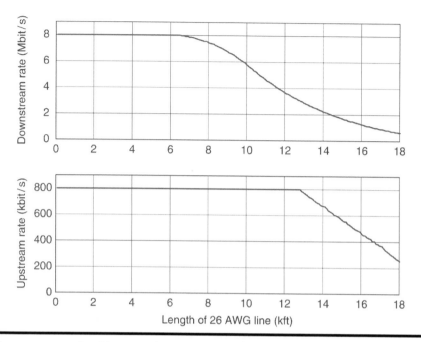

Figure 17.32   Line bit rates of ADSL1 over POTS with nonoverlapped PSD masks (49 ADSL1 over POTS disturbers, 26 AWG cable).

All other settings are the same as in Table 17.5. By comparing Figure 17.33 to Figure 17.32, one can see that the availability of additional subchannels in the downstream direction improves the downstream bit rate on loops longer than 6.5 kft. However, the upstream bit rate on loops longer than 7 kft is degraded significantly because of NEXT from downstream transmissions.

Table 17.5   ADSL1 Simulation Settings

Parameter	Value
Noise margin	6 dB
Cable gauge	26 AWG
Number of disturbing lines	49
AWGN level	−134 dBm/Hz downstream, −111 dBm/Hz upstream
Minimum number of bits per subchannel	2
Maximum number of bits per subchannel	15
Trellis coding	On

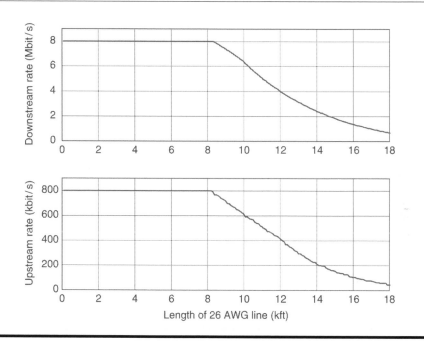

Figure 17.33   Line bit rates of ADSL1 over POTS with overlapped PSD masks (49 ADSL1 over POTS disturbers, 26 AWG cable).

The conclusions drawn about ADSL1 performance based on Figures 17.32 and 17.33 are a bit misleading because of the assumption that all lines in the cable binder support ADSL1. If instead of ADSL1 some of the other lines support another type of DSL, then the upstream bit rate of ADSL1 with nonoverlapped spectra will not be as high as Figure 17.32 indicates.

Figure 17.34 shows the results of a simulation of ADSL1 with nonoverlapped spectra when half of the lines in the cable binder support HDSL and the remaining lines support ADSL1 with nonoverlapped spectra. Other settings are as in Table 17.5. The upstream curve looks much more like that of Figure 17.33 than of Figure 17.32 because the lines of HDSL cause NEXT to ADSL1 in much the same way as overlapped ADSL1 spectra would. Note that because HDSL uses overlapped spectra that extend from 0 to at least 196 kHz, the ADSL1 downstream bit rate is also degraded by HDSL NEXT, particularly on longer lines. The downstream rate degrades to zero when the loop length approaches 16.5 kft. Because of the manner in which ADSL initializes and maintains a connection (such as through bit swaps), upstream transmissions cannot occur without at least some minimal downstream communication (and vice versa). Therefore, the ADSL link fails whenever the bit rate in either transmission direction drops below some level (assumed to be 4 kbit/s in the simulations).

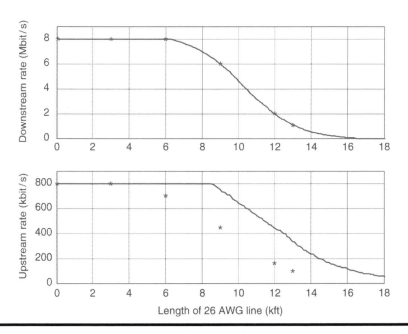

Figure 17.34   Line bit rates of ADSL1 over POTS with nonoverlapped PSD masks (24 ADSL1 over POTS disturbers plus 25 HDSL disturbers, 26 AWG cable). Asterisks are DSL Forum requirements.

The DSL Forum's performance requirements for a North American ADSL1 system operating in the presence of 24 HDSL are shown in Figure 17.34 by asterisks. In the downstream direction, there is a good match between the simulation projections and the performance requirements. However, in the upstream direction, the simulations are optimistic relative to the performance expected of real ADSL1 modems in this noise scenario, which suggests some factor unaccounted for in the simulations causes larger degradations in real ADSL1 modems.

17.4.2.4.2   Annex B: ADSL1 over ISDN

As discussed in Section 17.3, there are two types of duplexing used by ISDN systems. The variant of ISDN deployed exclusively in Japan uses time-compression multiplexing to separate downstream and upstream transmissions, and the variant deployed outside of Japan is a classic echo cancelled system. Annex B of G.992.1 defines ADSL1 for operation on the same line as the latter type of ISDN, whereas Annex C defines ADSL1 over POTS for operation in a TCM-ISDN environment.

When ADSL operates over ISDN, which spans a wider frequency band than POTS does, ADSL transmissions must be restricted to higher frequencies than in the over-POTS case to allow both ADSL and ISDN to share a loop. Therefore, both the downstream and the upstream lower band edges must be higher in frequency than in the case of Annex A. Frequencies from 276 to 1104 kHz are always allocated to the downstream direction. Subchannel 96 (centered at 414 kHz) is reserved for the pilot tone. The use of frequencies below 276 kHz is dependent on the implementation of the ATU-R and on whether overlapped spectra are allowed.

For upstream transmission, it is mandatory in Annex B of G.992.1 to generate subchannels in the band from 138 to 276 kHz. There are two ways in which these subchannels can be generated. The obvious way is to use a 128-point IDFT in the upstream transmitter to generate 64 subchannels that span from 0 to 276 kHz. The lower half of the subchannels are then potentially unused. In the early days of ADSL, it was desirable to use the same hardware for both ADSL over POTS and ADSL over ISDN. Therefore, Annex B describes a way to use a 64-point IDFT to generate the upstream spectrum in the band from 138 to 276 kHz. The transmitter performs "mirroring," which results in the appearance of the complex conjugates of subchannels 1 through 31 on subchannels 33 through 63 (subchannel 32 is unused). Essentially, mirroring uses aliasing to generate the desired signal, and the signal below 138 kHz is filtered out. Although mirroring was used successfully in early ADSL1 over ISDN modems, most implementations now use a 128-point IDFT to generate the upstream transmit signal. Generating 64 subchannels is inexpensive and straightforward with today's digital signal processing technology, and the same hardware can be used to support ADSL over POTS by turning off the upper 32 subchannels or by modifying the signal processing so that only 32 subchannels are generated. One benefit of generating 64 subchannels is that the Annex B PSD masks allow some of the subchannels below 138 kHz to be used for transmission. Therefore, generating the extra subchannels can result in improved performance relative to a system that uses mirroring with a 64-point IDFT.

The two non-TCM variants of ISDN have different bandwidths. The spectrum of 2B1Q ISDN is narrower than the spectrum of 4B3T ISDN. As a result, Annex B of G.992.1 allows some flexibility in how low in frequency ADSL1 transmissions may extend.

Figure 17.35 illustrates the upstream PSD mask defined in Annex B for ADSL1 operating on the same line as ISDN of the 2B1Q variety. Note that the PSD mask is undefined from 80 to 138 kHz, which allows some flexibility for implementations with 64 upstream subchannels to make use of the subchannels below 138 kHz. G.992.1 explains that the value of the transmitted PSD in the undefined region depends on the designs of the ISDN splitter lowpass filter and the ADSL1 highpass filter, but ADSL is expected not to degrade 2B1Q ISDN performance by more than 4.5 dB. Therefore, the utility of the subchannels below 138 kHz may be limited.

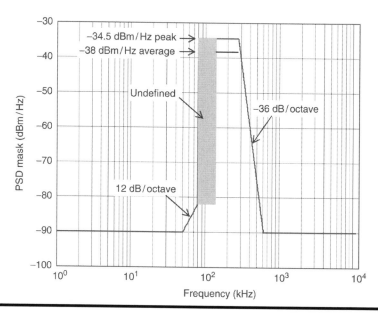

Figure 17.35   Upstream PSD mask for ADSL1 over ISDN of the 2B1Q variety. Average PSD level also shown.

Figure 17.36 shows the upstream PSD mask for use when the ISDN on the same line is of the 4B3T variety. The 4B3T mask is more restrictive than the 2B1Q mask, and the lower edge of the passband is at a higher frequency than in the 2B1Q case. Otherwise, the two masks are the same.

The aggregate upstream power for ADSL1 over ISDN is limited to 13.3 dBm, which corresponds to the integration of the PSD level of −38 dBm/Hz over the 31 available upstream subchannels between 138 and 276 kHz. If a modem transmits upstream using some subchannels below subchannel 33, it must transmit a lower PSD level on at least some of the subchannels in order not to exceed the allowed total power.

The downstream PSD mask for use on the same line as 2B1Q ISDN is shown in Figure 17.37, and Figure 17.38 shows the downstream PSD mask for use with 4B3T ISDN. Both masks look similar to the corresponding upstream masks, except they span a wider bandwidth, and the PSD mask and average PSD levels are lower than in the upstream direction. In both cases, as for ADSL1 over POTS, the average PSD level is −40 dBm/Hz within the passband, and the PSD mask level is −36.5 dBm/Hz within the passband. As in the upstream PSD mask definitions, the PSD masks are undefined over a small range, which allows some flexibility in filter design as well as in subchannel usage. Unlike in Annex A, there are no nonoverlapped PSD masks defined in Annex B of G.992.1, although nonoverlapped ADSL1 over ISDN operation is common.

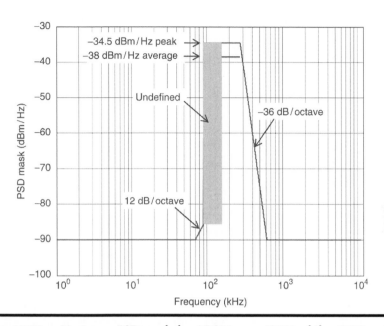

Figure 17.36 Upstream PSD mask for ADSL1 over ISDN of the 4B3T variety. Average PSD level also shown.

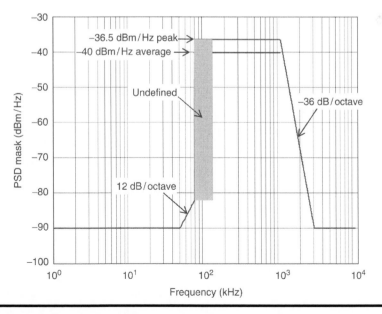

Figure 17.37 Downstream PSD mask for ADSL1 over ISDN of the 2B1Q variety. Average PSD level also shown.

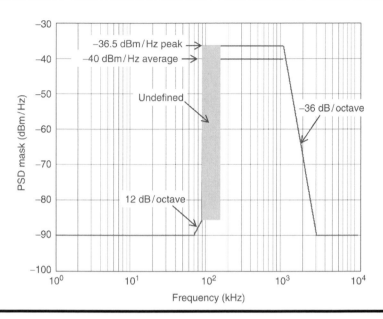

Figure 17.38   Downstream PSD mask for ADSL1 over ISDN of the 4B3T variety. Average PSD level also shown.

Because ADSL1 over ISDN transmissions are restricted to a higher range of frequencies than are ADSL1 over POTS transmissions, one would expect that the reach of ADSL1 over ISDN is shorter than the reach of ADSL1 over POTS, regardless of whether the ADSL1 over ISDN system operates using overlapped or nonoverlapped spectra. Figure 17.39 confirms this expectation under the conditions documented in Table 17.5. To eliminate the issue of whether the ISDN is of the 2B1Q variety or the 4B3T variety, the ADSL1 transmissions were restricted to frequencies above 138 kHz. The subchannels from 33 to 63 were used in the upstream direction, and subchannels 65 through 255 were used in the downstream direction. (Thus, nonoverlapped spectra were used.) A comparison of Figures 17.39 and 17.32 clearly shows that the downstream bit rate of ADSL1 over ISDN is lower than the downstream rate of ADSL1 over POTS on all but the shortest loop lengths. The degradation is solely because fewer subchannels are available to ADSL1 over ISDN in the downstream direction. The degraded downstream bit rate eventually limits the reach of ADSL1 over ISDN. Whereas the reach of nonoverlapped ADSL1 over POTS with self-crosstalk extends beyond 18 kft under the simulated conditions, Figure 17.39 shows that the reach of ADSL1 over ISDN under similar conditions is limited to about 17 kft. Note also that the upstream bit rate of ADSL1 over ISDN falls below the maximum rate on loops longer than 11 kft, whereas ADSL1 over POTS continues to support the maximum upstream rate on loops up to about 13 kft in length.

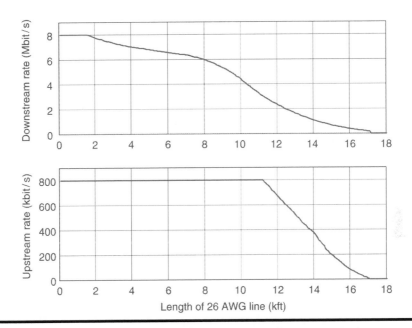

Figure 17.39   Line bit rates of ADSL1 over ISDN with nonoverlapped PSDs (49 ADSL1 over ISDN disturbers, 26 AWG cable).

Figure 17.40 illustrates the impact of allowing downstream transmissions to overlap the upstream band. In this case, subchannels 33 through 255 are also used in the downstream direction, and other settings are as given in Table 17.5. The figure shows that using overlapped spectra limits the reach of ADSL1 over ISDN to less than 12 kft. The failure of the upstream channel is clearly the problem, because the capacity of the downstream channel is still high at the length at which the upstream bit rate degrades to zero.

The premature failure of the upstream channel suggests that perhaps partial overlapping of the downstream and upstream transmissions would provide a means to extend the reach. For example, rather than extending its spectrum as low as subchannel 33, the downstream transmitter might be restricted to using only the subchannels above, say, subchannel 45. Figure 17.41 plots the downstream and upstream bit rates of ADSL1 over ISDN when subchannels 33 through 63 are used in the upstream direction, subchannels 46 through 255 are used in the downstream direction, and the simulation conditions are those documented in Table 17.5. The benefit of partially overlapping the downstream and upstream PSDs is obvious from the figure. The subchannels from 33 to 45 are free of NEXT, which provides a boost in upstream bit rate on longer loops. In the downstream direction, the bit rate decreases marginally relative to when full overlap of the spectra is allowed. However, this slight decrease in downstream bit rate results in

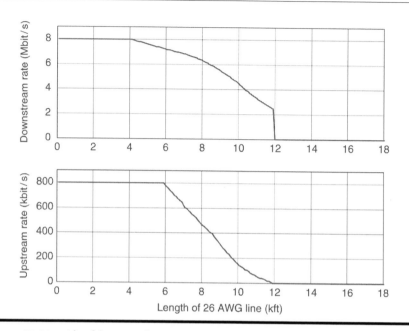

Figure 17.40   Line bit rates of ADSL1 over ISDN with fully overlapped PSD masks (49 ADSL1 over ISDN disturbers, 26 AWG cable).

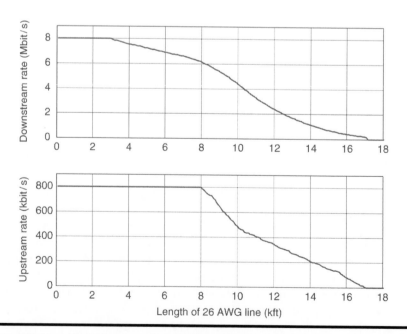

Figure 17.41   Line bit rates of ADSL1 over ISDN with partially overlapped PSD masks (49 ADSL1 over ISDN disturbers, 26 AWG cable).

a significant increase in the reach of the system, which is nearly 17 kft with partial overlap versus less than 12 kft with fully overlapped spectra.

17.4.2.4.3   Annex C: ADSL1 over POTS in the Same Binder as TCM-ISDN

ISDN service has been widely deployed in Japan. As discussed in Section 17.3, ISDN in Japan uses time-compression multiplexing to separate downstream and upstream transmissions. Therefore, the same channel bandwidth is time-shared by the two directions. All TCM-ISDN systems in a binder are synchronized so that they all transmit in the same direction (downstream or upstream) at the same time. As a symmetrical system, TCM-ISDN transmits approximately half the time in each direction.*

The spectrum of the TCM-ISDN signal has a null at 320 kHz and sidelobes above this first null. If the TCM-ISDN spectrum were confined primarily to the band below 320 kHz, it might be feasible to deploy ADSL on the same line as TCM-ISDN, albeit with transmissions at higher frequencies than in an Annex A or B system. However, the amount of bandwidth remaining would severely limit the reach of ADSL over TCM-ISDN. Therefore, there is no ADSL over TCM-ISDN system.

The energy in the TCM-ISDN mainlobe and sidelobes also presents challenges for ADSL over POTS systems operating in the same binder as TCM-ISDN lines. Because TCM-ISDN transmits approximately half of the time in one direction and half the time in the other direction, it causes time-varying crosstalk to ADSL systems that are using that bandwidth in a single direction. Roughly half of the time, an ADSL system will suffer from NEXT because of TCM-ISDN transmissions, and it will suffer from FEXT because of TCM-ISDN transmissions the other half of the time.

Annex C of G.992.1 defines the operation of ADSL1 systems in the same binder as TCM-ISDN lines. To cope with the two noise scenarios resulting from TCM-ISDN crosstalk, Annex C defines two types of ADSL1 data symbols: the FEXT symbol and the NEXT symbol. The FEXT symbol consists of the bit allocation that is appropriate when the TCM-ISDN systems are transmitting in the same direction as the ADSL1 system and the receiver performance thus suffers from FEXT. The NEXT symbol consists of the bit allocation that is appropriate when the TCM-ISDN systems are transmitting in the direction opposite to that of the ADSL1 system and the receiver performance suffers from NEXT. Figure 17.42 illustrates the relationship between TCM-ISDN transmissions and the type of crosstalk appearing at the ATU-C and ATU-R receivers.

---

* There is a small quiet period in TCM-ISDN when the transmission direction changes, which means the amount of time spent transmitting in either direction is slightly less than 50 percent.

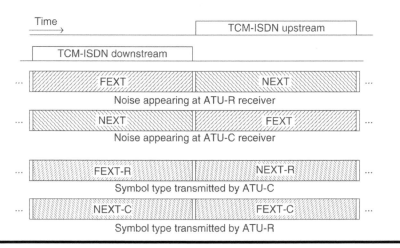

Figure 17.42   The relationship between TCM-ISDN transmissions and crosstalk appearing at ADSL receivers.

During the initialization procedure, the transceivers determine the appropriate allocation of bits during symbols that are corrupted by NEXT and during symbols corrupted by FEXT. Four bit allocations (symbol types) result, as shown in Table 17.6. Note that the "C" and "R" designations indicate the modem that will receive the symbol. As Figure 17.42 indicates, if propagation delays through the line are ignored, the ATU-C transmits FEXT-R when the ATU-R transmits NEXT-C. Likewise, when the ATU-R transmits FEXT-C, the ATU-C transmits NEXT-R.

The downstream and upstream PSDs in the original version of Annex C are the same as those defined in Annex A.

More recent work on Annex C of G.992.1 resulted in the addition of Annex I to G.992.1. Annex I specifies a wider-bandwidth downstream PSD mask for systems operating in the same binder as TCM-ISDN, and also doubles the number of subchannels in the downstream direction. Although SG15/Q4 was working at the time on ADSL2plus, which would also double the number of subchannels available in the downstream direction, the Japanese market required a solution before ADSL2plus was scheduled to be completed. Therefore, Annex I was written to accommodate the need of the Japanese market. Because ADSL2plus is addressed in the sequel,

Table 17.6   Symbol Types in Annex C of G.992.1

| Crosstalk Present | Symbol Used in Transmission Direction | |
	Downstream	Upstream
FEXT	FEXT-R	FEXT-C
NEXT	NEXT-R	NEXT-C

and Annex I is essentially an ADSL2plus system in the ADSL1 Recommendation, Annex I is not discussed here.

The reader is referred to [G.992.1] for additional details of Annex C and Annex I operation, including changes to the tone ordering and framing necessary to accommodate the different symbol types in Annex C.

17.4.3 ADSL2

ADSL2 is specified in ITU-T Recommendation G.993.2. As an evolution of ADSL1, ADSL2 generally uses the same bandwidth allocations as ADSL1. However, ADSL2 provides additional functionality and improves ADSL1 performance and utility in a number of ways, and ADSL2 also serves an important role as the basis for ADSL2plus.

To improve performance, the support of trellis encoding, which is an option in ADSL1, is made mandatory in ADSL2, although the receiver in either transmission direction may choose to disable trellis coding on a particular connection. (On very short loops, using trellis coding can actually result in a performance degradation because of the overhead it requires.) Technically, then, trellis coding is mandatory at the transmitter and optional at the receiver. However, the performance benefit of trellis coding is so significant on most loops that it is unlikely any receiver implementation does not incorporate trellis decoding. Although the gain of the trellis code is difficult to isolate when trellis coding is used in combination with FEC, the combination of the trellis code and FEC used in ADSL1 and ADSL2 provides about 5 dB of coding gain on most loops. (See chapter 9 of [Golden 2006] for details of how the coding gain is computed for concatenated codes.)

An improved initialization sequence in ADSL2 also helps to improve line rates. Some initialization signals found to be not sufficiently random in ADSL1 were made more random in ADSL2. Furthermore, an optional fast initialization sequence is defined to allow modems to resume a connection more quickly than they could by undergoing a full initialization.

A further performance improvement in ADSL2 is the result of allowing subchannels to support only one bit, both with and without trellis coding enabled, which provides a small performance benefit, particularly on longer loops. Annex L of G.992.3 defines a reach-extended version of ADSL2 that was specifically designed to provide even better performance on long loops.

ADSL2 defines other annexes that allow all-digital operation (i.e., without underlying POTS or ISDN), and others that enable higher upstream bit rates.

G.992.3 guarantees at least some performance benefit relative to a category I ADSL1 modems by requiring modems to support at least 8 Mbit/s in the downstream direction and at least 800 kbit/s in the upstream direction.

(The reader is reminded that most deployed ADSL1 modems provide features beyond those required by category I, such as trellis coding, which enables them to support 8 Mbit/s downstream and 800 kbit/s upstream.) Primarily because of market demand, commercial ADSL2 modems support significantly higher bit rates than ADSL1 modems in at least the downstream direction.

Furthermore, the restriction in ADSL1 that Reed–Solomon codewords are coupled with DMT symbols is eliminated in ADSL2. The absence of constraints on the value of S, which can be any value in ADSL2, results in slightly higher coding gain from the Reed–Solomon code. ADSL2 framing also allows flexibility in the overhead bit rate, which is fixed in ADSL1. Therefore, user data rates can be slightly higher if a lower-overhead bit rate is configured in ADSL2. Alternatively, if a higher-overhead bit rate is required for some reason, ADSL2 allows it.

Additional functionality is provided in ADSL2 through the definition of the options of dynamic rate repartitioning (DRR) and seamless rate adaptation (SRA). DRR is a mechanism that allows the bit rate to be reallocated among bearer channels when more than one bearer channel is used in a connection. For example, one bearer might support a videoconferencing signal, while another supports a data connection. DRR allows the total data rate of the connection to be repartitioned among the supported bearer channels without requiring the modems to retrain. Thus, in the example, if the videoconference ends, the data rate that had been previously allocated to it could be reallocated to the data channel.

SRA is a mechanism that provides a means to change the total bit rate of the connection without requiring the modems to retrain (which is the only way to modify the line bit rate in ADSL1). SRA provides additional robustness to changing loop impairments. If, for example, the noise on a line increases to the point that the target noise margin cannot be supported at the current bit rate, SRA can be used to reduce the bit rate to a level at which the target margin can be supported. Both DRR and SRA are options in ADSL2, which means that they are defined completely in G.992.3, but standard-compliant implementations are not required to support DRR or SRA.

An interface for packet transfer mode is also defined in ADSL2, in recognition of the surging popularity of Ethernet (rather than ATM) as the DSL transport mechanism. In addition, the specific packet transfer mode defined by IEEE 802.3 for (EFM) has been incorporated in ADSL2 (see Chapter 13).

To provide more visibility into failures and service difficulties, ADSL2 also defines a loop diagnostic procedure and various diagnostic parameters that can be retrieved by the management entity.

In terms of improving robustness, ADSL2 defines a new way to coordinate bit swap transitions. As discussed in Section 17.4.2.3, the transition time selected by an ADSL1 ATU responding to a bit swap request must be more than 800 ms in the future to ensure that both the initiating and the responding ATUs are fully prepared for the bit swap. To increase the speed and robustness of bit swaps and other on-line reconfiguration (OLR) transitions (such as DRR and SRA), the responding transceiver in ADSL2 signals timing by inverting the phase of the next transmitted synchronization symbol. Upon receiving a synchronization symbol with inverted phase relative to the last synchronization symbol, the initiating ATU prepares for the transition. In the upstream direction, the transition becomes effective five symbols after the inverted-phase synchronization symbol is received, and in the downstream direction, it becomes effective two symbols after the inverted-phase synchronization symbol is received. Thus, the ADSL2 OLR mechanism allows much faster bit swaps than the ADSL1 mechanism. Furthermore, because a synchronization symbol with inverted phase (relative to previous synchronization symbols) is nearly impossible for the initiating ATU to detect incorrectly, the ADSL2 OLR mechanism is thought to be more robust than the ADSL1 bit swapping mechanism.

Robustness is also improved by allowing the downstream receiver to decide whether it needs a dedicated pilot tone and, if so, which subchannel it designates as the pilot and whether it prefers to have a known sequence transmitted on that subchannel. If the receiver does not require a dedicated pilot tone, a modest bit rate improvement results because an additional subchannel is available for data transmission.

The modified tone ordering approach in ADSL2 can also potentially improve the robustness of ADSL2 modems. Unlike in ADSL1, in which the tone ordering is deterministic and based solely on the number of latency paths and whether trellis coding is in use, in ADSL2 both the downstream and upstream receivers determine the tone ordering for their respective directions. The receiver-determined tone ordering is thought to improve robustness to AM radio and similar interference. The tone ordering determined by the receiver is communicated to the transmitter during the initialization procedure.

Whereas in ADSL1 the minimum number of nonzero bits per subchannel is two, in ADSL2 one-bit subchannels are allowed, both with and without trellis coding enabled. As a result, the trellis encoding procedure has been modified. A receiver that allocates one-bit subchannels is required to do so in pairs, i.e., the total number of one-bit subchannels must be even. Sets of two one-bit subchannels are then treated as a single two-bit input ("a subchannel") for the purpose of trellis encoding.

Relative to G.992.1, G.992.3 defines several more operational modes in annexes. In this chapter, the emphasis is on the annexes that define the modes of operation. Those annexes not discussed include Annex D (ATU-C and ATU-R state diagrams), Annex E (splitters), Annex F (performance requirements for North America), Annex G (performance requirements for Europe), Annex H (SSDSL), Annex K (packet interface specification), and Annex N (64/65-octet PTM TC sub-layer functional specification). Readers interested in these other annexes are referred to [G.992.3]. In addition, chapter 13 of this text provides information about Annex N.

17.4.3.1 Annex A: ADSL2 over POTS

Like Annex A of G.992.1, Annex A of G.992.3 specifies the operation of ADSL2 over POTS. Many of the characteristics of ADSL2 over POTS are the same as in ADSL1. In particular, the downstream and upstream transmitter power limits are 20.4 dBm and 12.5 dBm, respectively, and the downstream and upstream PSD masks in the passbands are the same as in ADSL1 over POTS. However, the PSD masks are more restrictive in the out-of-band regions in Annex A of G.992.3 than in G.992.1.

In terms of performance, on shorter loops an ADSL2 over POTS system offers higher bit rates than an ADSL1 over POTS system because the framing in ADSL2 does not limit the maximum bit rate as it does in ADSL1. Instead, in ADSL2 the maximum number of bits per subchannel limits the maximum bit rate. On longer loops, ADSL2 over POTS provides slightly higher bit rates than an ADSL1 over POTS system that uses trellis coding. Figure 17.43 shows the downstream and upstream bit rates of ADSL2 over POTS under the conditions given in Table 17.7. Note that because the

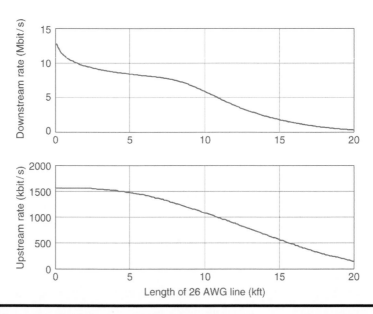

Figure 17.43   Line bit rates of ADSL2 over POTS with nonoverlapped PSD masks (49 ADSL2 over POTS disturbers, 26 AWG cable).

Table 17.7 ADSL2 Simulation Settings

Parameter	Value
Noise margin	6 dB
Cable gauge	26 AWG
Number of disturbing lines	49
AWGN level	−134 dBm/Hz downstream, −111 dBm/Hz upstream
Minimum number of bits per subchannel	1
Maximum number of bits per subchannel	15
Trellis coding	On

implementation constraints limiting ADSL1 performance are not expected to improve significantly in ADSL2 modem implementations, the AWGN levels used for the ADSL1 simulations are also used for the ADSL2 simulations.

By comparing Figure 17.43 to Figure 17.32, one can verify that ADSL2 provides higher bit rates than ADSL1 does. Simulations of other noise scenarios would also show better performance with ADSL2 than with ADSL1.

17.4.3.2 Annex B: ADSL2 over ISDN

Annex B of ADSL2 differs from Annex B of G.992.1 in several ways. First, separate masks are not defined for the 2B1Q and 4B3T variants of ISDN. In addition, the undefined regions of the downstream and upstream PSD masks are fully defined, and a mask for nonoverlapped operation is defined explicitly. In the case of the upstream mask, the slope of the PSD mask in the transition region above 276 kHz is steeper than in ADSL1. Figure 17.44 illustrates the G.992.3 Annex B upstream PSD. The downstream PSD mask for use with nonoverlapped spectra is shown in Figure 17.45, and the downstream mask for overlapped spectra is shown in Figure 17.46. The total power that may be transmitted in the downstream direction is 19.9 dBm when overlapped spectra are used and 19.3 dBm when nonoverlapped spectra are used; in the upstream direction, the total power constraint is 13.3 dBm.

In terms of performance, Annex B bit rates in G.992.3 are generally higher than in G.992.1 because of the mandatory trellis code, the availability of one-bit subchannels, and, on short loops, the absence of framing parameters that limit the bit rate. Figure 17.47 plots the downstream and upstream line bit rates of an ADSL2 Annex B system that uses nonoverlapped spectra under the conditions documented in Table 17.7. Subchannels 33 through 63 are used in the upstream direction, and subchannels 65 through 255 are available for downstream transmission. A comparison of Figures 17.47 and 17.39 reveals that the bit rate with ADSL2 is significantly higher than with ADSL1 on shorter loops and slightly higher on long loops.

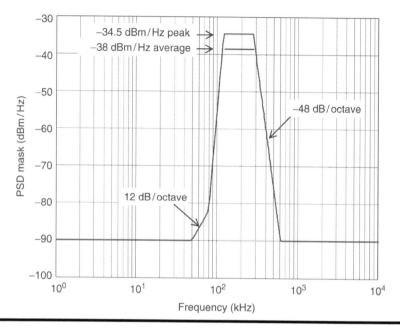

Figure 17.44 Upstream PSD mask for ADSL2 over ISDN. Average PSD level also shown.

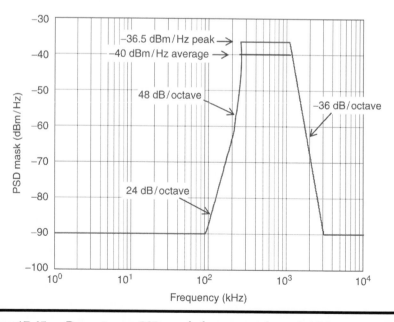

Figure 17.45 Downstream PSD mask for nonoverlapped ADSL2 over ISDN. Average PSD level also shown.

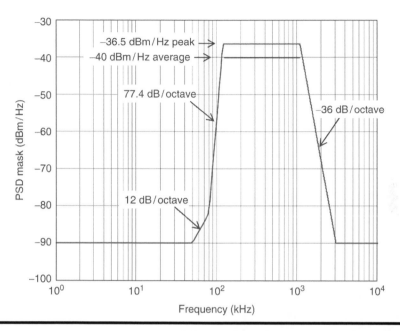

Figure 17.46   Downstream PSD mask for overlapped ADSL2 over ISDN. Average PSD level also shown.

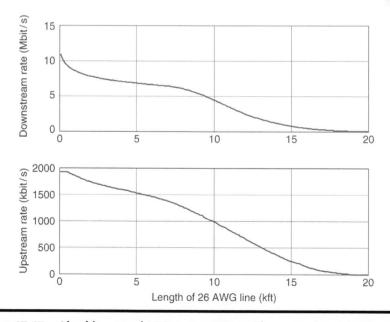

Figure 17.47   Line bit rates of ADSL2 over ISDN with nonoverlapped PSD masks (49 ADSL2 over ISDN disturbers, 26 AWG cable).

17.4.3.3 Annex C: ADSL2 over POTS in the Same Binder as TCM-ISDN

Annex C of G.992.3 operates in the same manner as Annex C of G.992.1, and the performance benefits due to the enhancements in ADSL2 provide performance improvements of the same order as in Annex A and Annex B.

17.4.3.4 Annex I: All-Digital Mode ADSL2 (Narrow Upstream)

One of the new operational modes introduced in G.992.3 is the "all-digital" mode. In all-digital mode, neither POTS nor ISDN is present on the same physical loop, and ADSL modems are allowed to use the frequencies that would otherwise be reserved for POTS or ISDN. In the first version of ADSL2, fully overlapped PSD masks were the only ones specified so that near-baseband transmission would be possible in both directions. Nonoverlapped PSD masks were added in a later amendment to ADSL2.

The PSD masks specified for all-digital operation in Annex I are based on the ADSL over POTS PSD masks. Figure 17.48 plots the upstream PSD mask defined in Annex I. Note that the mask is at the maximum level at frequencies as low as 3 kHz. The downstream PSD mask for overlapped operation is shown in Figure 17.49. The downstream PSD mask for nonoverlapped operation is the same as the nonoverlapped downstream PSD mask defined in Annex A of G.992.3.

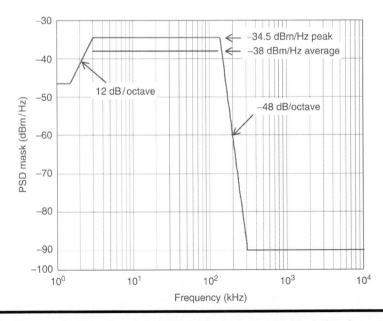

Figure 17.48   **Upstream Annex I PSD mask. Average PSD level also shown.**

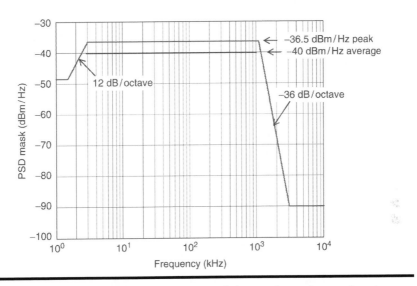

Figure 17.49   Downstream Annex I PSD mask for overlapped operation. Average PSD level also shown.

The total power constraint for an Annex I ATU-C is 20.4 dBm, as in Annex A, but for an ATU-R it is 13.3 dBm, as in Annex B.

Figure 17.50 shows the downstream and upstream bit rates achieved by a fully overlapped Annex I system under the conditions documented in

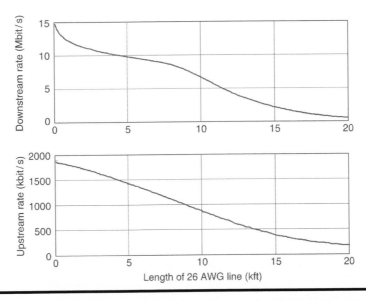

Figure 17.50   Downstream and upstream bit rates of a G.992.3 Annex I system using fully overlapped spectra.

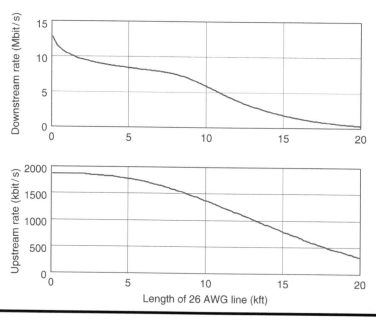

Figure 17.51   Downstream and upstream bit rates of a G.992.3 Annex I system using nonoverlapped spectra.

Table 17.7. Compared to an Annex A system using fully overlapped spectra, an Annex I system supports higher bit rates in both directions because of the availability of more subchannels. Figure 17.51 shows the bit rates achieved by a nonoverlapped Annex I system under the conditions documented in Table 17.7.

17.4.3.5   *Annex J: All-Digital Mode ADSL2 (Wider Upstream)*

Annex J of G.992.3 also defines all-digital operation for ADSL2, but the upstream bandwidth is wider. Two downstream PSD masks are defined, one for overlapped operation and one for nonoverlapped operation. The downstream PSD mask for overlapped operation is the same as in Annex I (shown in Figure 17.49). The downstream PSD mask for nonoverlapped operation is the same as the downstream nonoverlapped mask for Annex B (shown in Figure 17.45). In either case, the total power transmitted in the downstream direction is constrained to 20.4 dBm.

In contrast to Annex I, which defines a single upstream mask, Annex J defines a family of nine upstream PSD masks, as shown in Figure 17.52. The masks, which are indexed by the number of subchannels in their passbands, have successively wider passbands. However, the total upstream transmitter power is the same in all cases, and therefore the maximum and average PSD levels decrease as the passband width increases. To maximize Annex J

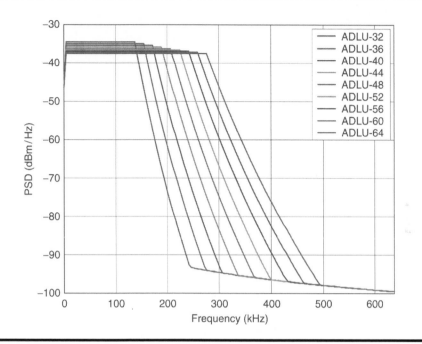

Figure 17.52   Upstream PSD masks defined in Annex J of G.992.3.

upstream performance, the total power in the upstream direction is as high as 13.4 dBm. The narrowest PSD mask spans 32 subchannels and has a maximum PSD mask level of -34.5 dBm/Hz, and the widest mask spans 64 subchannels and has a maximum PSD mask level of -37.5 dBm/Hz. The reason for specifying a family of upstream PSD masks is because the use of a wider spectrum in the upstream direction causes NEXT to any Annex A systems that might be operating in the same binder as the Annex J system. The availability of masks with different bandwidths allows a service provider to select a mask that results in a good trade-off between Annex J performance and the expected degradation to the downstream bit rate of any Annex A modems in the same cable binder.

Figure 17.53 plots the downstream and upstream bit rates achievable with a G.992.3 Annex J system that uses fully overlapped spectra, under the simulation conditions documented in Table 17.7. The upstream PSD mask in the simulation was selected based on the loop length and according to spectral compatibility requirements in the United States (see [T1.417-2003]). The widest PSD mask is used on loops up to 9.5 kft in length, and successively narrower masks are used as the loop length increases further. The results show that a fully overlapped Annex J system provides significantly higher upstream bit rates than an Annex A, B, or I system, even though the upstream spectrum is degraded by NEXT from the overlapped downstream spectrum.

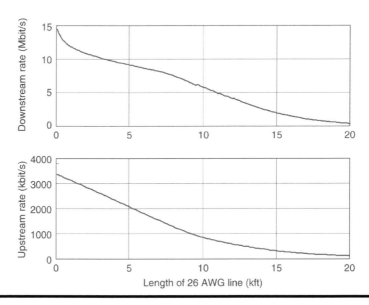

Figure 17.53   Downstream and upstream bit rates of a G.992.3 Annex J system using fully overlapped spectra.

Figure 17.54 shows the bit rates achievable with an Annex J system that uses nonoverlapped spectra under the simulation conditions given in Table 17.7. The upstream bit rate is significantly higher than when overlapped spectra are used. The downstream bit rate is the same as in the case of an Annex B system using nonoverlapped spectra, which can be verified by comparing the plot of the downstream bit rate with the corresponding plot in Figure 17.47.

With either nonoverlapped or overlapped spectra, the bit rates supported by an Annex J system are less asymmetrical than the bit rates of Annex A or Annex B systems. Furthermore, Annex J is generally more spectrally compatible with Annex A systems than is SHDSL or HDSL. For these reasons, when operation over POTS or ISDN is not necessary, Annex J is sometimes suggested as an alternative to SHDSL for providing symmetrical services.

17.4.3.6   Annex L: ADSL2 over POTS with Extended Reach

Annex L of G.992.3 defines a mode to extend the reach of an ADSL2 over POTS system by confining the downstream and upstream PSD masks to narrower, lower-frequency regions of the spectrum while maintaining the same transmitter power. The net result is a higher allowed PSD level in each direction, but less total bandwidth. On short loops, this approach would be detrimental to ADSL performance because the increased PSD level would

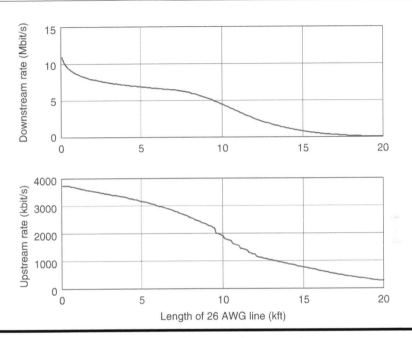

Figure 17.54 Downstream and upstream bit rates of a G.992.3 Annex J system using nonoverlapped spectra.

not compensate for the severe reduction in bandwidth. But on long loops, on which the higher frequencies are too attenuated to support transmission, this approach is very effective. To enable good performance on all loops, any modem that supports Annex L is required to support Annex A as well. The assumption is that Annex A will be used on most loops, whereas Annex L will be used only when it provides a benefit relative to Annex A.

Two upstream PSD masks are defined in Annex L. One mask, shown in Figure 17.55, has a passband that extends from approximately 25 to less than 60 kHz, whereas the other mask has a passband that resides between 25 and 103 kHz, as shown in Figure 17.56. The figures show clearly the PSD boost that results from maintaining the same upstream transmitter power as in Annex A.

Two PSD masks are also defined for the downstream direction: one is for nonoverlapped operation and the other is for overlapped operation. With either mask, the downstream signal is confined to the spectrum below 552 kHz, and the PSD level is higher than in the case of, for example, Annex A. Figure 17.57 illustrates the downstream PSD mask for nonoverlapped spectra, and Figure 17.58 shows the downstream PSD mask for overlapped operation. The curious notch in the PSD mask of Figure 17.58 is required to meet spectral compatibility requirements in the United States (see [T1.417-2003]).

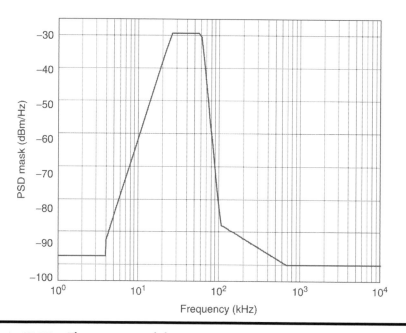

Figure 17.55 **The narrower of the two G.992.3 Annex L upstream PSD masks ("U1" in Figure 17.59).**

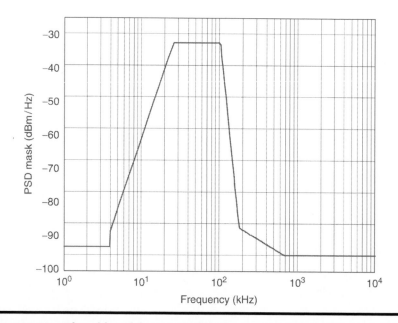

Figure 17.56 **The wider of the two G.992.3 Annex L upstream PSD masks ("U2" in Figure 17.59).**

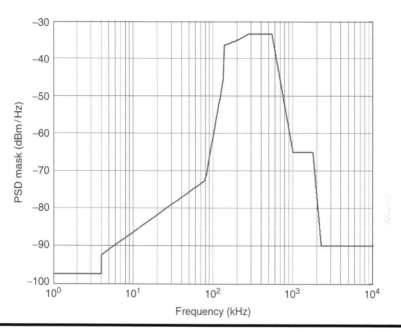

Figure 17.57   G.992.3 Annex L downstream PSD mask for nonoverlapped operation.

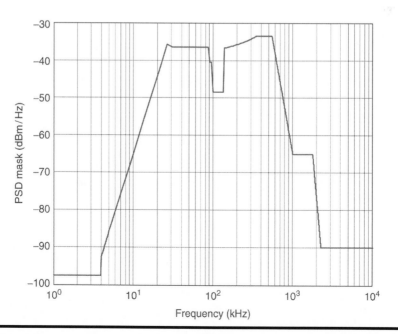

Figure 17.58   G.992.3 Annex L downstream PSD mask for overlapped operation.

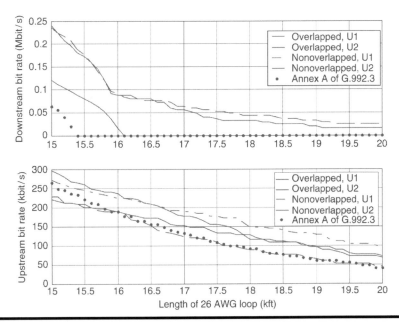

Figure 17.59 **Downstream and upstream bit rates with the PSD masks of G.992.3 Annex L. Annex A (nonoverlapped) performance is also shown for comparison purposes. In the plot of the downstream bit rate, the curves corresponding to the nonoverlapped Annex L masks are identical.**

Figure 17.59 illustrates the level of performance an Annex L system can provide under a harsh noise scenario that includes HDSL and T1 disturbers. The simulations represent the expected performance of an ADSL2 Annex L system under the conditions shown in Table 17.8. Also shown is the performance expected of a system that supports only Annex A of G.992.3.

Table 17.8   Annex L Simulation Settings

Parameter	Value
Noise margin	6 dB
Cable gauge	26 AWG
Noise	24 ADSL2 Annex L, 5 T1 (adjacent binder), 10 HDSL
AWGN level	−134 dBm/Hz downstream, −111 dBm/Hz upstream
Minimum number of bits per subchannel	1
Maximum number of bits per subchannel	15
Trellis coding	On

Because Annex L is intended to be used only on long loops, the plots show the bit rates on loops from 15 to 20 kft in length. The upper plot shows that whereas that downstream bit rate of an Annex A system degrades to zero on loops shorter than 15.5 kft, the Annex L system that is restricted to using nonoverlapped masks continues to operate on loops longer than 16 kft. When overlapped spectra are used, the Annex L downstream bit rate does not degrade to zero even on a loop as long as 20 kft. The upstream bit rate supported by Annex L is generally higher than that of Annex A on loops longer than 15 kft. Even when the upstream bandwidth is degraded by NEXT caused by the use of overlapped spectra, the upstream bit rate is not significantly lower than with Annex A because of the higher PSD level allowed by the Annex L PSD masks. Figure 17.59 clearly shows the performance benefit of Annex L on longer loops.

17.4.3.7 Annex M: ADSL2 over POTS with Extended Upstream Bandwidth

Annex M of G.992.3 defines ADSL2 over POTS operation with a wider upstream bandwidth. Annex M is like Annex J in that a family of nine upstream PSD masks is defined. However, the masks extend only as low as approximately 25 kHz so that POTS can share the loop with the Annex M system. Figure 17.60 shows the nine upstream masks defined in Annex M.

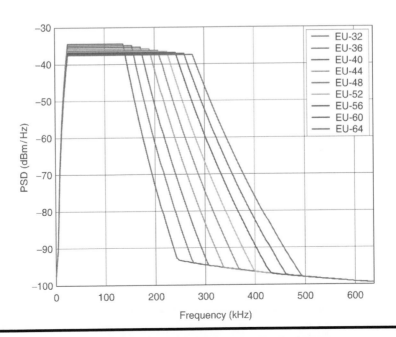

Figure 17.60   Upstream PSD masks defined in Annex M of G.992.3.

A comparison to Figure 17.52 reveals that the masks are similar in construction to those in Annex J. However, the total upstream power constraint in Annex M is 12.5 dBm, and the masks roll off rapidly below 25 kHz to accommodate POTS.

In the downstream direction, Annex M defines both nonoverlapped and overlapped downstream PSD mask. The mask for nonoverlapped operation is the same as the nonoverlapped downstream mask of Annex B (see Figure 17.45), and the overlapped PSD mask is the same as the overlapped mask in Annex A.

Figure 17.61 plots the expected bit rates of a nonoverlapped Annex M system under the conditions in Table 17.7. The upstream PSD mask in the simulation was selected based on the loop length and according to spectral compatibility requirements in the United States. The widest PSD mask was used on loops up to 9.5 kft in length, and then successively narrower masks were used as the loop length increased further. The benefit of the wider upstream PSD is immediately evident. On the shortest loops, an upstream bit rate as high as almost 3.5 Mbit/s is achievable. In the downstream direction, the bit rate is the same as in the case of a nonoverlapped Annex B system under the same conditions. (See Figure 17.47 to verify the rates are the same.)

Figure 17.62 shows the downstream and upstream bit rates of an Annex M system when fully overlapped spectra are used, and the simulation

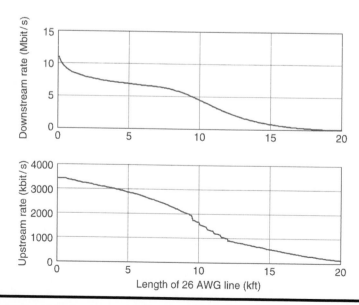

Figure 17.61 Downstream and upstream bit rates of a G.992.3 Annex M system using nonoverlapped spectra.

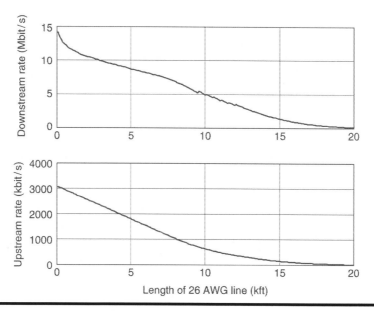

Figure 17.62 **Downstream and upstream bit rates of a G.992.3 Annex M system using fully overlapped spectra.**

conditions are those in Table 17.7. In this case, as expected, the downstream bit rate is higher than when nonoverlapped PSD masks are used. However, the upstream bit rate curve clearly exhibits degradations because of NEXT from downstream transmissions. The upstream bit rate is significantly lower than with nonoverlapped spectra, which suggests that nonoverlapped operation is likely to be preferred in practice, unless the upstream band is known to be degraded already by some other type of system in the binder that transmits downstream in the band overlapping the Annex M upstream band, such as SHDSL or HDSL.

17.4.4 ADSL2plus

ADSL was originally conceived as a means to provide broadband service over existing telephone lines. In most parts of the world, telephone lines are long. As a result, the maximum reach of ADSL, and not its maximum bit rate, was originally of paramount importance. When VDSL standardization began (see Section 17.5), a different deployment scenario was envisaged: VDSL would be deployed from street cabinets to subscribers, and therefore the maximum bit rate was of primary importance because the reach was not expected to be more than 4 or 5 kft in most applications. As work on these two disparate standards continued, however, and the VDSL1 line code logjam delayed progress on VDSL1, it became clear that there was a need for an ADSL system that would provide higher bit rates than ADSL1 and ADSL2 on short and mid-range loops without sacrificing the long reach of ADSL. This system would be suitable for deployment on loops with useful bandwidth above the 1.104 MHz maximum bandwidth of ADSL1 and ADSL2.

ADSL2plus was conceived to fill this gap in bit rates between ADSL and VDSL. To enable higher downstream bit rates, ADSL2plus doubles the downstream bandwidth to 2.208 MHz by defining 513 downstream subchannels. As in the other ADSL versions, the first and last subchannels (with indices 0 and 513) are not used, which means the maximum number of subchannels potentially available in the downstream direction is 511. In the upstream direction, ADSL2plus is identical to ADSL2.

ADSL2plus is defined in ITU-T Recommendation G.992.5. G.992.5 is a "delta" document to G.992.3, which means it does not stand alone but instead requires G.992.3 as a basis. The reliance on G.992.3 means that an ADSL2plus modem is also required to be capable of ADSL2 operation, which ensures both high bit rates on short loops and long reach.

Relative to G.992.3, G.992.5 defines some additional functionality. In particular, ADSL2plus specifies a method to allow shaping of the transmitted PSD. Shaping can be used, for example, to improve the spectral compatibility of ADSL2plus systems deployed at an intermediate point between the CO and subscribers and ADSL systems deployed from the CO. By shaping the ADSL2plus downstream PSD in the band below 1104 kHz, FEXT that ADSL2plus systems cause to ADSL1 or ADSL2 modems can be reduced.

G.992.5 also defines transmitter windowing as an optional feature. Windowing is useful to reduce out-of-band energy and to minimize the impact of out-of-band noise on the ADSL2plus signal. Chapter 13 of [Golden 2006] discusses windowing and its benefits in detail.

The annexes of G.992.5 define most of the same operational modes as are defined in G.992.3. One exception, however, is Annex L, which in ADSL2 provides a mode to extend the reach of ADSL. Because the intention of ADSL2plus is to increase the bit rate on short loops, Annex L is inconsistent with the objective of ADSL2plus and is therefore not defined.

17.4.4.1 Annex A: ADSL2plus over POTS

Annex A of G.992.5 defines ADSL2plus over POTS operation. The upstream PSD mask is nearly the same as in Annex A of G.992.3, except that the roll-off above the passband is steeper. Figure 17.63 illustrates the upstream PSD mask for Annex A of G.992.5, which is characterized by a roll-off of −72 dB/octave above 138 kHz. The upstream transmit power constraint

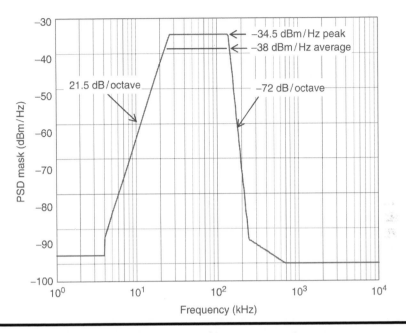

Figure 17.63   Upstream PSD mask defined in Annex A of G.992.5.

in Annex A of G.992.5 is 12.5 dBm, which is the same as in Annex A of G.992.3.

As in Annex A of both G.992.1 and G.992.3, two downstream PSD masks are defined. The PSD mask defined for nonoverlapped operation is shown in Figure 17.64, and Figure 17.65 shows the downstream PSD mask for overlapped operation. The total power constraint for either PSD mask is 20.4 dBm, as in Annex A of G.992.3.

Because the upstream PSD in Annex A of ADSL2plus is the same in the passband as the G.992.3 Annex A upstream PSD, the upstream bit rate will also be the same. However, on short and medium-length loops, the downstream bit rate with Annex A of ADSL2plus will be significantly higher than with G.992.3 Annex A, because of the availability of additional downstream subchannels above 1104 kHz. Figure 17.66 plots the downstream bit rate supported by a nonoverlapped G.992.5 Annex A system under the conditions documented in Table 17.7. Also shown for comparison purposes is the downstream bit rate supported by a nonoverlapped G.992.3 Annex A system. The figure clearly illustrates that ADSL2plus provides a higher downstream bit rate on loops shorter than about 9 kft and that, as expected, on longer loops the downstream bit rate provided by the ADSL2plus system is identical to the bit rate of the ADSL2 system.

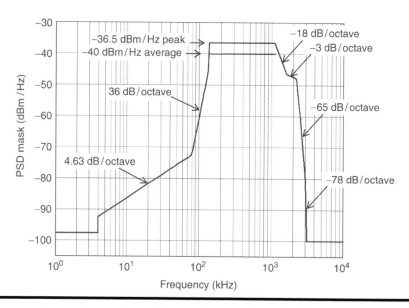

Figure 17.64 **Nonoverlapped downstream PSD mask defined in Annex A of G.992.5.**

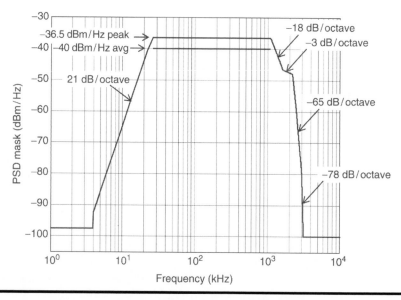

Figure 17.65 **Overlapped downstream PSD mask defined in Annex A of G.992.5.**

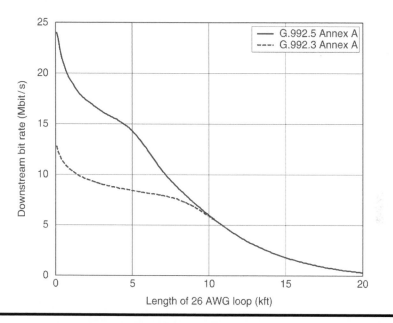

Figure 17.66   Downstream bit rate achieved by a nonoverlapped G.992.5 Annex A system. Also shown is the downstream bit rate of a nonoverlapped G.992.3 Annex A system under the same conditions.

17.4.4.2   Annex B: ADSL2plus over ISDN

As does Annex B of G.992.3, Annex B of ADSL2plus specifies operation over ISDN. The upstream PSD mask is the same as in ADSL2 in the passband but rolls off more steeply at frequencies above 276 kHz. Figure 17.67 illustrates the upstream PSD defined in Annex B of G.992.5. The total upstream transmitter power is allowed to be no higher than 13.3 dBm, as is the case in Annex B of G.992.3.

As in Annex B of G.992.3, two downstream PSD masks are defined in Annex B of G.992.5. The nonoverlapped mask is shown in Figure 17.68, and the overlapped PSD mask is shown in Figure 17.69. The masks look similar to their counterparts in G.992.3, except that the passband extends to 2208 kHz. As in Annex B of G.992.3, the total downstream transmitter power is constrained to 19.9 dBm with the overlapped PSD mask and 19.3 dBm with the nonoverlapped PSD mask.

Because the upstream PSD mask in Annex B of G.992.5 is the same in the passband as the upstream PSD mask in Annex B of G.992.3, the upstream bit rates will be the same. In the downstream direction, however, the availability of bandwidth above 1104 kHz significantly improves the downstream bit rate on short and medium-length loops. Figure 17.70 compares the downstream bit rate of a nonoverlapped G.992.5 Annex B

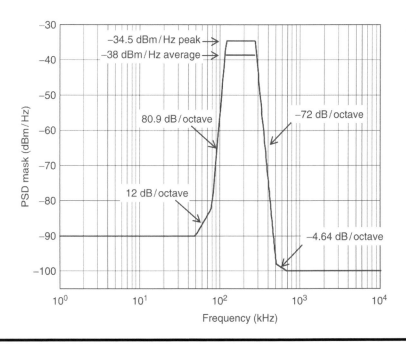

Figure 17.67   Upstream PSD mask defined in Annex B of G.992.5.

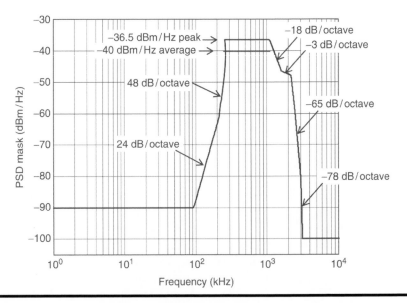

Figure 17.68   Nonoverlapped downstream PSD mask defined in Annex B of G.992.5.

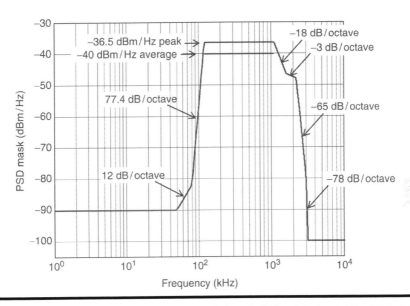

Figure 17.69   Overlapped downstream PSD mask defined in Annex B of G.992.5.

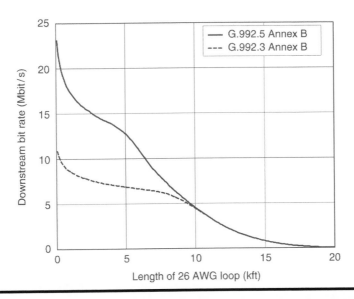

Figure 17.70   Downstream bit rate achieved by a nonoverlapped G.992.5 Annex B system. Also shown is the downstream bit rate of a nonoverlapped G.992.3 Annex B system under the same conditions.

system to that of a nonoverlapped G.992.3 Annex B system under the conditions documented in Table 17.7. The figure shows that ADSL2plus Annex B offers a bit rate improvement on loops shorter than about 9 kft, which is the same region in which Annex A performance improves.

17.4.4.3 Annex C: ADSL2plus in the Same Binder as TCM-ISDN

Annex C of G.992.5 is similar to Annex C of G.992.3, except that the downstream PSD masks are extended to 2.208 MHz (as in Annex I of G.992.1), and two sub-annexes define two upstream PSD mask bandwidth options. In Annex C.A, both nonoverlapped and overlapped downstream PSD masks are defined. They are identical to the corresponding masks in Annex A of G.992.5. Likewise, the upstream PSD mask in Annex C.A is identical to the upstream PSD mask in Annex A of G.992.5. In Annex C.B, the same two downstream PSD masks are defined, but the upstream PSD mask is EU-64 of Annex M of G.992.5 (which will be described in turn).

The availability of additional downstream bandwidth increases the downstream bit rate of G.992.5 Annex C relative to Annex C of G.992.3.

17.4.4.4 Annex I: All-Digital Mode ADSL2 (Narrow Upstream)

Annex I of G.992.5 is similar to Annex I of G.992.3, except the downstream PSD mask extends to 2208 kHz. Both nonoverlapped and overlapped PSD masks are defined for the downstream direction. The nonoverlapped PSD mask is the same as the nonoverlapped downstream PSD mask defined in Annex A of G.992.5. The overlapped PSD mask is shown in Figure 17.71. The total downstream power is 20.4 dBm.

The upstream PSD mask is similar to the upstream mask defined in Annex I of G.992.3, but the roll-off beyond 138 kHz is steeper, as illustrated in Figure 17.72. The upstream transmitter power is constrained to 13.3 dBm.

The upstream bit rate achieved by an Annex I ADSL2plus system will be identical to the bit rate of an Annex I ADSL2 system. However, the downstream bit rate will be higher on short loops because of the available additional bandwidth above 1104 kHz. Figure 17.73 illustrates the downstream bit rate that can be achieved by a fully overlapped G.992.5 Annex I system under the conditions documented in Table 17.7. The downstream bit rate curve for an overlapped G.992.3 Annex I system is also shown for comparison purposes. Annex I of G.992.5 achieves higher bit rates than Annex I of G.992.3 on loops up to about 9 kft in length.

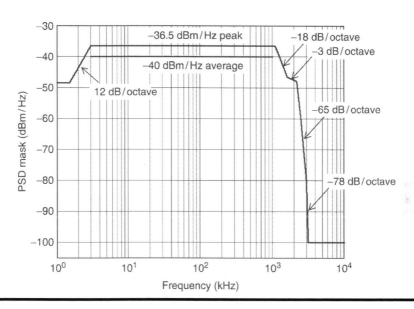

Figure 17.71   **Overlapped downstream PSD mask defined in Annex I of G.992.5.**

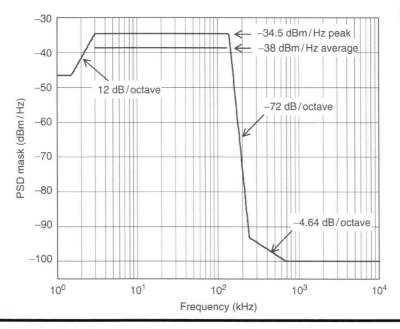

Figure 17.72   **Upstream PSD defined in Annex I of G.992.5.**

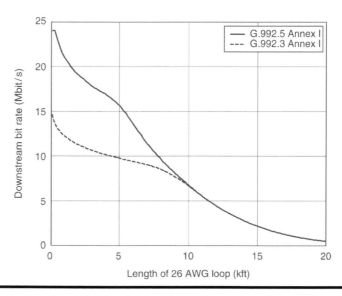

Figure 17.73   Downstream bit rate achievable with fully overlapped G.992.5 Annex I. Also shown for comparison is the downstream bit rate achievable with G.992.3 Annex I.

17.4.4.5   Annex J: All-Digital Mode ADSL2 (Wider Upstream)

As does Annex I, Annex J of G.992.5 defines an all-digital mode of operation. However, whereas the upstream bandwidth in Annex I is constrained to 138 kHz, Annex J defines the same family of upstream masks that Annex J of G.992.3 defines (see Figure 17.60). In addition, the total power allowed in the upstream direction is 13.4 dBm, as in Annex J of G.992.3.

Both overlapped and nonoverlapped PSD masks are defined for the downstream direction. The overlapped PSD mask is the same as the overlapped PSD mask defined in Annex I of G.992.5 and shown in Figure 17.71. The nonoverlapped PSD mask is the same as the nonoverlapped PSD mask defined in Annex B of G.992.5. With either mask, the downstream transmitter power is allowed to be no higher than 20.4 dBm.

In the upstream direction, the bit rate achieved by a G.992.5 Annex J system will be the same as the upstream bit rate of a G.992.3 Annex J system, assuming both systems use the same PSD mask on the same loop length, and assuming both operate with nonoverlapped spectra or both operate with overlapped spectra. In the downstream direction, the bit rate achieved by a nonoverlapped G.992.5 Annex J system is the same as the bit rate achieved by a nonoverlapped G.992.5 Annex B system, as shown in Figure 17.70. If fully overlapped spectra are used, the downstream bit rate will be higher than when nonoverlapped spectra are used. Figure 17.74

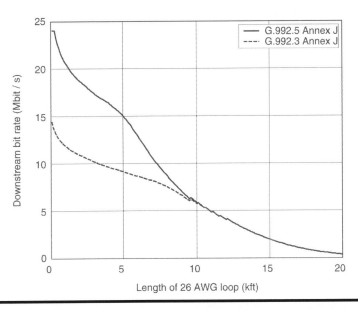

Figure 17.74   **Downstream bit rate achievable with fully overlapped G.992.5 Annex J. Also shown for comparison is the downstream bit rate achievable with an overlapped G.992.3 Annex J system.**

illustrates the achievable downstream bit rate of a G.992.5 Annex J system using the overlapped PSD mask. The simulation conditions are those given in Table 17.7.

17.4.4.6   *Annex M: ADSL2plus with Extended Upstream Bandwidth*

Annex M of G.992.5 is similar to Annex M of G.992.3, except that the downstream bandwidth extends to 2208 kHz. The upstream PSD masks are identical to the G.992.3 Annex M upstream PSD masks (shown in Figure 17.60), and the total allowed power is 12.5 dBm. In the downstream direction, the overlapped PSD mask is the same as the overlapped PSD mask in Annex A of G.992.5 (see Figure 17.65). The nonoverlapped PSD mask is identical to the nonoverlapped PSD mask defined in Annex B of G.992.5 and is shown in Figure 17.68. In either case, the downstream transmitter power is constrained to a level no higher than 20.4 dBm.

In the upstream direction, the bit rate achieved by an ADSL2plus Annex M system will be the same as the bit rate achieved by a G.992.3 Annex M system (under the same conditions). In the downstream direction, the bit rate achieved by a nonoverlapped G.992.5 Annex M system will be identical to the bit rate achieved by a nonoverlapped G.992.5 Annex B

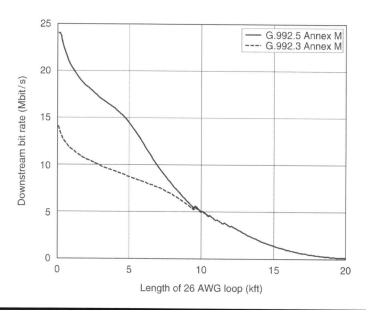

Figure 17.75 **Downstream bit rate achievable with fully overlapped G.992.5 Annex M. Also shown for comparison is the downstream bit rate achievable with an overlapped G.992.3 Annex M system.**

system. The downstream bit rate of an overlapped G.992.5 Annex M system will be slightly lower than that of an overlapped G.992.5 Annex J system. Figure 17.75 plots the downstream bit rate of an overlapped G.992.5 Annex M system under the conditions documented in Table 17.7. The downstream bit rate of an overlapped G.992.3 Annex M system is also shown. The ADSL2plus system provides higher downstream bit rates on loops shorter than about 9 kft.

17.4.5 Summary of ADSL Operational Modes

Table 17.9 summarizes the various ADSL operational modes. The values provided for the downstream and upstream passbands are the maximum allowed bands; modems may use less bandwidth in practice to accommodate band-splitting filters or because of excessive line attenuation or noise.

The simulations provided in this section show that ADSL2 provides significantly higher bit rates than ADSL1 on shorter loops and marginally higher bit rates on longer loops. ADSL2plus, in turn, provides significantly higher bit rates than ADSL2 on loops shorter than about 9 kft in length, assuming 26 AWG cabling.

Table 17.9   ADSL Operational Modes

Type of System	ITU-T Recommendation	Annex	Spectra	Upstream Passband (kHz)	Downstream Passband (kHz)
ADSL1 over POTS		A	Nonoverlapped	26–138	138–1104
			Overlapped	25–138	25–1104
ADSL1 over ISDN		B	Nonoverlapped	138–276	276–1104
			Overlapped	138–276	138–1104
ADSL1 over POTS in the same binder as TCM-ISDN	G.992.1	C	Nonoverlapped	25–138	138–1104
			Overlapped	25–138	25–1104
ADSL1 over POTS in the same binder as TCM-ISDN (extended downstream)		I	Nonoverlapped	25–138	138–2208
			Overlapped	25–138	25–2208
ADSL2 over POTS		A	Nonoverlapped	25–138	138–1104
			Overlapped	25–138	25–1104
ADSL2 over ISDN		B	Nonoverlapped	120–276	276–1104[a]
			Overlapped	120–276	120–1104
ADSL2 over POTS in the same binder as TCM-ISDN		C	Nonoverlapped	25–138	138–1104
			Overlapped	25–138	25–1104
ADSL2 all-digital mode 1	G.992.3	I	Nonoverlapped	3–138	138–1104
			Overlapped	3–138	3–1104
ADSL2 all digital mode 2		J	Nonoverlapped	3–276	276–1104
			Overlapped	3–276	3–1104
ADSL2 over POTS with extended reach		L	Nonoverlapped	25–104 or 25–60	25–552
			Overlapped	25–104 or 25–60	138–552
ADSL2 over POTS with extended upstream		M	Nonoverlapped	25–276	276–1104
			Overlapped	25–276	25–1104
ADSL2plus over POTS		A	Nonoverlapped	25–138	138–2208
			Overlapped	25–138	25–2208
ADSL2plus over ISDN		B	Nonoverlapped	120–276	276–2208
			Overlapped	120–276	120–2208
ADSL2plus over POTS in the same binder as TCM-ISDN	G.992.5	C	Nonoverlapped	25–138	138–2208
			Overlapped	25–138	25–2208

(Continued)

Table 17.9 (Continued)   ADSL Operational Modes

Type of System	ITU-T Recommendation	Annex	Spectra	Upstream Passband (kHz)	Downstream Passband (kHz)
ADSL2plus all-digital mode 1		I	Nonoverlapped	3–138	138–2208
			Overlapped	3–276	3–2208
ADSL2plus all-digital mode 2		J	Nonoverlapped	3–276	276–2208
			Overlapped	3–276	3–2208
ADSL2plus over POTS with extended upstream		M	Nonoverlapped	25–276	276–2208
			Overlapped	25–276	25–2208

[a] In ADSL2 over ISDN with nonoverlapped spectra, there is some flexibility in the crossover from upstream to downstream transmissions. The transition frequency can be as low as 254 kHz.

17.5   VDSL

At the time of writing, the most recent DSL standardization effort was VDSL, which supports higher bit rates than even ADSL2plus, is the highest bit-rate DSL that has been standardized to date.

17.5.1   *Relevant Standards Organizations*

Three standards organizations have been involved in VDSL standardization: T1E1.4 (now NIPP-NAI) for North America, ETSI TM6 for Europe, and ITU SG15/Q4 for international use. Each organization eventually published its own VDSL1 standard, with minor differences between them. VDSL2 was generated exclusively by SG15/Q4 with inputs from NIPP-NAI and TM6.

Like the standardization of ADSL before, progress in VDSL was initially limited by the inability of standards groups to settle on a line code (or modulation technique) for the new technology. Even with the issue decided in favor of DMT for ADSL, the standardization of VDSL re-opened the line code debate. Both multi-carrier and single-carrier modulation were considered, and each of them gained considerable traction in the various standardization bodies. Although the first "VDSL-like" proposals were made as early as 1995 (see [T1E1.4/94-087], [T1E1.4/94-088], and [T1E1.4/94-125], for example), a final decision on the line code was not reached until June of 2003. Since then, DMT has been considered the only line code for VDSL, and it was selected for the subsequent VDSL2 standard. Before that, standards that contain both line codes were published.

In 2002, T1E1.4 published a trial-use standard specifying both multi-carrier modulation (in particular, DMT) and single-carrier modulation (SCM)

[T1.424 Trial Use]. The final T1 VDSL1 standard was published in 2004 and was based on DMT modulation only [T1.424-2004]. At the same time as the publication of the VDSL1 standard, T1 also published a TRQ (Technical Requirements Specification) based on SCM [T1.TRQ.12-2004].

ETSI published its initial two-part VDSL1 standard in 2001. The first part [ETSI TS 101 270-1] is line-code-independent and lists the functional requirements such as operations, administration, and maintenance (OAM), spectral characteristics, and testing and performance requirements. The second part [ETSI TS 101 270-2] provides the transceiver specifications. The latter document specifies both DMT and SCM on equal footing. Both documents have undergone a number of revisions since their first publications.

The ITU-T published its first VDSL Recommendation G.993.1 [G.993.1-2001] in 2001 and an amendment in 2003, but neither document contained any transceiver requirements. A revision of G.993.1 that included the full transceiver specification (including line code) was published in 2004. This document, ITU-T Recommendation G.993.1-2004 [G.993.1-2004], contains a DMT-based VDSL1 specification in the main body, and a single-carrier based system is specified in an annex. After publication of G.993.1-2004, ITU started work on a new Recommendation called "VDSL2" that was exclusively based on DMT modulation. This Recommendation was sent out for consent in May of 2005 and was approved for publication in February of 2006. It was published in 2006 as ITU-T Recommendation G.993.2 [G.993.2].

17.5.2 VDSL: Origins and Challenges

17.5.2.1 Short Overview

VDSL was first proposed in the mid-1990s as the "next step" beyond ADSL [ETSI TM6 TD3R1]. The general objective was to specify a system that would provide higher bit rates than ADSL, albeit on shorter loops.

At the time VDSL standardization work began, the bandwidth of an ADSL system was 1.1 MHz. Later, an ADSL variant using up to 2.2 MHz in spectrum was specified (in ITU-T Recommendation G.992.5; see Section 17.4.4). However, the bandwidth of ADSL remains significantly less than the bandwidth used in VDSL. The crucial observation in VDSL was that on shorter loops, frequencies significantly higher than 1.1 MHz are useful for transmission of data, and therefore it is possible to support higher bit rates than with ADSL on these shorter loops. Typically, a maximum loop length of 1500 m (4.5 kft) was assumed in the early VDSL work. Proposals for ADSL-like transmission using higher bandwidths were presented at various standards meetings (see [T1E1.4/94-087], [T1E1.4/94-088], and

[T1E1.4/94-125], for example). Simulation results showed significant potential increases in data rate on short loops.

These early contributions led to the start, in about 1995, of a formal process to standardize VDSL. T1E1.4 and ETSI TM6 took the lead in VDSL standardization while SG15/Q4 was busy completing its Recommendations on ADSL.

The main appeal of VDSL was its promise to offer a new range of services over the existing copper infrastructure. ADSL was being primarily deployed to offer a high-speed internet (HSI) access service at relatively low rates. In most deployment scenarios, ADSL is offered from the CO. For a basic HSI service, this type of deployment has proved to be effective. The increased bandwidth and throughput of VDSL offered the possibility to significantly extend the current service offering of ADSL, but at the same time required deployment scenarios that are different from ADSL.

The desired VDSL services were investigated in detail by the FS-VDSL working group, under the umbrella of the operator collectively known as the Full Service Access Network (FSAN) operators. By and large operators expressed interest in offering the data rates shown in Table 17.10. These requirements became part of the various standard efforts on VDSL.

As can be seen from Table 17.10, VDSL was expected to offer both symmetrical and asymmetrical services. The need to support both types of services impacts the design of the system, and in particular the appropriate frequency plan for VDSL (see Section 17.5.3.2.4).

Roughly speaking, asymmetric data rates are targeted at residential customers, while symmetric data rates are intended for business customers. For residential customers, the ultimate goal of VDSL is offering a "triple-play" service, which is a combined service. This means a combined service offering that includes broadcast TV, HSI, and telephony. This service is typically asymmetric in nature because more data is sent from the network toward the user than from the user toward the network.

Business services, on the other hand, tend to be more symmetrical. A typical envisaged service is remote local area network (LAN) access,

Table 17.10 VDSL Service Requirements

Region	Asymmetric Requirements (Mbit/s)		Symmetric Requirements (Mbit/s)
	Downstream	Upstream	
North America	22	3	13
			10
			6
Europe	23	4	14
	14	3	6.4

which allows a business user to connect to the corporate LAN remotely. For this service, the upstream bit rate is more important than for Internet access, as the end user will not only download but also upload content. Also, high-quality videoconferencing requires the same bandwidth in the upstream and downstream directions. The 10 Mbit/s symmetrical service gained a lot of interest as a possible way to extend Ethernet into the access network, and in fact the EFM standard provides this functionality on short loops using VDSL.

Although the possible service offering with VDSL opens up a number of new possibilities for service providers, the inherent reach limitation of VDSL also requires a different type of deployment. For the data rates in Table 17.10, the reach will be much more limited than for the typical ADSL service because the higher rates depend on the availability of spectrum above 1.1 MHz. Whereas the ADSL service can be offered satisfactorily from the CO, deploying VDSL from the CO results in limited coverage (fewer than 30 percent of customers in some countries). To increase the coverage of the technology, VDSL needs to be installed closer to the end user, and therefore additional flexibility points are needed in the network.

From the start, the regional VDSL standards bodies recognized the need for different deployment scenarios. All standards specify scenarios for both deployment from the CO and the cabinet. Different PSDs and allowable transmit powers were defined (see Section 17.5.3.3), as were different noise models and performance requirements for the various deployment scenarios.

A third deployment scenario that was not explicitly considered in the ETSI or ANSI standards but that has proven very popular in Asia is MDU/MTU deployment. MDU stands for "Multi-Dwelling Unit" and refers to a residential building consisting of multiple homes, apartments, or suites. MTU stands for "Multi-Tenant Unit" and refers to office blocks occupied by many companies, usually small or medium-sized enterprises. In this type of deployment, the VDSL Digital Subscriber Line access multiplexer (DSLAM) is located in or close to the building and serves only that building. As such, the deployment uses only the very last part of the access network. This third scenario is explicitly addressed in the VDSL2 Recommendation.

17.5.2.2 Challenges

Compared to ADSL, VDSL presented some new technical challenges, many of which are due to the target loops being short and the availability of bandwidth at significantly higher frequencies than in ADSL. The particular topology in which VDSL systems are deployed also causes some issues that are unique to VDSL.

The following are some of the issues that required special consideration during the VDSL standardization process:

- Far-end crosstalk: Because VDSL loops are short, FEXT is a dominant noise source, particularly at higher frequencies where VDSL is likely to be the only occupant of the bandwidth. One consequence of a FEXT-dominated noise environment is that increasing the transmitted PSD does not improve bit rates because it must be assumed that all other VDSL systems transmit at the same PSD. This effect has an impact on the definition of VDSL PSDs.
- Power consumption: VDSL is to a large extent deployed from remote locations that tend to be very confined, because local regulations may limit the allowed installation space. As a result, the power consumption of VDSL modems needs to be low. This requirement, in part, motivated decisions on the total allowed transmit power and maximum PSD in VDSL, and particularly in VDSL1.
- Upstream power back-off: Even if two VDSL modems are deployed at the same service provider location, it is likely that a VDSL system deployed on a short loop shares a binder with a VDSL system deployed on a long loop.* Because FEXT is a dominant impairment for the typical VDSL loop lengths, the crosstalk caused by shorter loops will significantly reduce upstream performance on longer loops. To reduce the interference, VDSL standards specify a loop length-dependent PSD reduction called upstream power back-off (UPBO).
- Radio interference: Because VDSL1 uses transmit frequencies up to 12 MHz, and VDSL2 uses even higher frequencies, any radiated emission from the copper wire can interfere with a number of radio spectrum users. The VDSL PSD masks had to be designed specifically to avoid or at least mitigate this kind of interference. Conversely, VDSL can experience disturbances from radio users when a radiated signal is converted into a differential mode signal by unbalance in the loop or modem.
- Digital duplexing: After FDD was selected as the duplexing method, it was decided to use a novel scheme to separate the upstream and downstream transmissions. This new method allowed for unprecedented flexibility in the allocation of upstream and downstream frequencies and alleviated the need to separate the frequency bands with analog filters.

---

* "Short" and "long" loops need to be understood in the context of VDSL. The longest loop length that was considered in the first VDSL Recommendation was about 1500 m (about 4.5 kft).

These issues will be discussed in more detail in subsequent sections, and the solution adopted for each of them will be explained.

17.5.3   VDSL Fundamentals

17.5.3.1   Reference Model

Figure 17.76 shows the reference model of VDSL1. The VDSL1 standard is structured as a layered specification, with the different layers labeled as "TPS-TC," "PMS-TC," and "PMD."

The TPS-TC sub-layer interfaces with the transport protocol. The various VDSL standards specify systems that interface with ATM, STM, and packet modes (e.g., for Ethernet). The function of the transmit TPS-TC sub-layer is to transform the protocol-specific input format (e.g., ATM cells or Ethernet packets) into a uniform byte stream toward the PMS-TC sub-layer. The function of the transmit PMS-TC sub-layer is to perform any byte-oriented processing such as coding, interleaving, etc. The PMS-TC sub-layer is also responsible for framing the various data streams and overhead channels into a group of bytes to be modulated by the PMD sub-layer. The role of the PMD sub-layer is to modulate the incoming data bytes on a signal that is suitable for transmission over the physical medium (i.e., the copper wire pair). The same model applies to both SCM and multi-carrier modulation. Obviously, the main difference between the two resides in the PMD sub-layer, although there are some differences in the PMS-TC sub-layer as well.

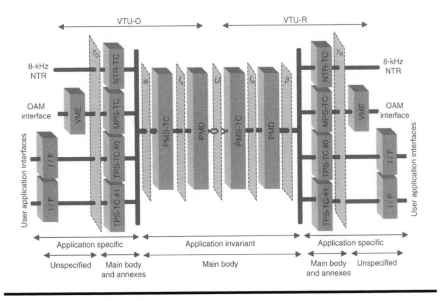

Figure 17.76   VDSL1 reference model.

At the receiver side, the receive TPS-TC, PMS-TC, and PMD sub-layers perform the inverse functions to complete the data transport from transmitter to receiver.

The VDSL2 Recommendation specifies two latency paths that allow different data streams with different latency and impulse noise protection requirements to be transported simultaneously.

17.5.3.2 Duplexing

Duplexing describes how the available channel bandwidth is used to accommodate bidirectional transmission. Two duplexing schemes—frequency-division duplexing and time-division duplexing—were considered for VDSL when it was first proposed. This section describes the two schemes and discusses their advantages and disadvantages in the context of VDSL. After a long debate, the standards organizations adopted FDD.

17.5.3.2.1 Frequency-Division Duplexing

FDD systems define disjoint frequency bands to support downstream and upstream transmissions. At least one band must be defined for each direction. Performance of an FDD system is closely tied to the bandwidths and placements of the downstream and upstream frequency bands.

The simplest FDD systems define a single downstream band and a single upstream band, as shown in Figure 17.77. As the figure shows, the downstream band could reside below or above the upstream band. Spectral compatibility considerations generally restrict flexibility in band placement.

Ideally, VDSL systems should operate on a wide variety of loop lengths, ranging from a few tens of meters to 1.5 km or longer. Because loop attenuation increases more rapidly with frequency as the loop length increases, the maximum useful frequency on a line decreases as the loop length increases. To support bidirectional transmission on a particular line, bandwidth must be available in both the downstream and upstream directions below the

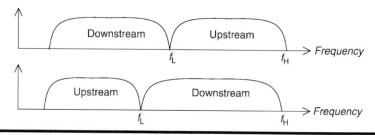

Figure 17.77   Possible arrangements of downstream and upstream bands in a frequency-division duplexing (FDD) system.

maximum useful frequency; otherwise, bidirectional data transmission is not possible.

The appropriate bandwidth choices for the downstream and upstream bands are dependent on the desired ratio of downstream and upstream bit rates. The appropriate allocation of bandwidth to support asymmetric 8:1 transmission differs significantly from the appropriate allocation to support symmetrical data transport. Furthermore, choosing the downstream and upstream channel bandwidths is complicated by the wide range of loop lengths on which a VDSL system must operate. Because the useful bandwidth of a loop decreases with increasing loop length, designing a frequency plan that provides downstream and upstream bands with the appropriate SNR to support the desired service is a significant challenge.

The final complication in determining the frequency plan for an FDD system is the number of bands into which the available bandwidth will be partitioned. One desirable characteristic often cited by operators is graceful degradation with increasing loop length, which means the system should not operate at a healthy bit rate on a loop of length L and then cease to operate altogether on a loop of length $L + 10$ meters. In the absence of spectral compatibility requirements, which generally make the design of a system with graceful degradation difficult, this objective implies that there must be several downstream and upstream bands so at least one band in each direction is available for all loop lengths of interest.

The optimal band plan depends on the targeted service and the specific loop and noise environment in which VDSL systems are deployed. A primarily asymmetric deployment will require a very different spectrum allocation from a primarily symmetric deployment. In addition to the service requirements, any frequency allocation must also be designed with the overall system in mind. Some frequency allocations may have a higher impact on the modem's overall complexity than others.

Because of the need for spectral compatibility, all modems in the same region must use the same band plan. The crosstalk coupling between the pairs in a binder group is significant. Especially at higher frequencies, any overlap in upstream and downstream bands would reduce the capacity of that overlapped region to the point of making it useless.

17.5.3.2.2 Time-Division Duplexing

In contrast to FDD solutions, which separate upstream and downstream transmissions by allocating disjoint frequency bands to the two directions, TDD systems transmit within the same, single band, but upstream and downstream transmissions occur during different time periods. (This method of duplexing is used in TCM-ISDN.) Use of the time-shared bandwidth is coordinated using a superframe, which is composed of a downstream transmission period, a guard time, an upstream transmission period,

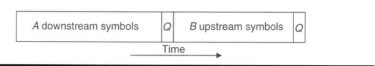

Figure 17.78   A superframe as defined in time-division duplexing (TDD).

and another guard time. Figure 17.78 illustrates the concept of a superframe. The durations of the downstream and upstream time slots are integer multiples of symbol periods. Superframes may be denoted as *A-Q-B-Q*, where *A* and *B* are the number of symbol periods allocated for downstream and upstream transmissions, respectively, and the *Q*s represent guard times (quiet or quiescent periods). The guard times, which need not be the same duration in the transmit and receive directions, are needed to account for the channel propagation delay (because the channel can only be used in one direction at a time) and to allow the channel impulse response to decay between transmit and receive periods.

To illustrate how TDD systems work, assume the use of a superframe of duration 20 symbol periods. Let the sum of *A* and *B* be 18 symbol periods, leaving a duration of 2 symbol periods for the guard times. The values of *A* and *B* can be selected by an operator to provide the desired downstream-to-upstream bit rate ratio. For example, if the transmit spectra and noise PSDs in the downstream and upstream directions are assumed to be equal, setting $A = 16$ and $B = 2$ results in a configuration that supports asymmetrical transmission with a bit rate ratio of 8:1. If $A = B = 9$, then symmetrical transmission is supported. If $A = 2B$, then a 2:1 bit rate ratio results. Figure 17.79 shows the superframes that support 8:1, 2:1, and symmetrical transmission.

The flexibility of the TDD superframe allows some level of compensation for differences in the downstream and upstream SNRs. For example, when UPBO is applied in the upstream direction, typically the upstream SNR is lower than the downstream SNR because the upstream power is reduced. A TDD system could compensate for this difference in SNR by allocating additional symbols for upstream transmission. If symmetrical transmission is desired, then $A = 8$ and $B = 10$ could be used instead of the nominal $A = B = 9$ superframe, which would increase the reach of a particular symmetrical bit rate (assuming the upstream bit rate limits performance with $A = B = 9$).

The use of TDD requires a certain level of synchronization of transmissions on lines within a single binder. Modems on lines in the same cable binder must transmit downstream at roughly the same time, and upstream at roughly the same time. If transmissions on lines do not occur in the same direction at approximately the same time, then NEXT from one TDD system to another can severely degrade performance.

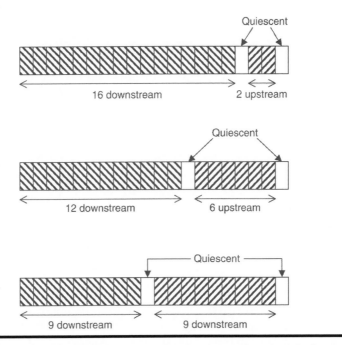

Figure 17.79 TDD superframes that support 8:1, 2:1, and symmetrical transmission (assuming equal signal-to-noise ratios (SNRs) in the downstream and upstream directions).

To ensure harmonious operation, all TDD modems synchronize to a common superframe clock. There are a number of methods by which this clock can be provided; for example, it can be derived from the 8 kHz network clock, sourced by one of the TDD modems, or derived using GPS (Global Positioning System) technology. In the case the clock is sourced by one of the modems, it must be assumed that all other modems operating the binder have access to the clock signal. Thus, in this case, coordination between modems is a requirement, which implies collocation.

17.5.3.2.3 TDD versus FDD

Both of the duplexing schemes proposed for VDSL1 offered advantages, but each also had disadvantages.

TDD systems can provide reduced complexity, both in digital signal processing and in analog components, if they use DMT. These complexity reductions result from sharing hardware common to both the transmitter and receiver. The hardware can be shared because the transmit and receive functions both require the computation of a(n) (I)DFT, which is usually accomplished using the FFT algorithm. Because TDD is used, a modem can either transmit or receive at any particular time, and only a single

FFT is required per modem. This FFT spans the entire system bandwidth and is active during the transmit and receive symbols. Additional analog hardware savings are possible in TDD modems because the same band is used to transmit and receive. Whichever path is not in use can be turned off, which reduces power consumption. In contrast, modems using FDD always must provide power to both the transmit and receive paths, because both are always active. Finally, TDD modems require no hybrid in the AFE (see Chapter 3).

The drawback of TDD solutions is, of course, the need to roughly synchronize downstream and upstream transmissions on lines in a binder so that performance is not compromised by NEXT from line to line. For this reason, a common superframe clock must be available to all modems at the line termination (service provider) end of the link. The common clock is easily provided; however, distribution of this common clock is a daunting task when unbundled loops (loops leased to competitive carriers) are considered.

FDD systems do not require any synchronization or coordination between different lines. On the other hand, the complexity reductions that can be obtained with a TDD system are not possible with FDD modems. The frequency division requires both the transmitter and receiver to be active all the time, so power savings cannot be realized, and a hybrid is required. In addition, some implementations may require additional filtering, involving either analog or digital filters.

At the start of the VDSL standardization effort, both TDD and FDD systems were proposed. Both proposals were developed in parallel, and an initial specification was drafted for each. The single-carrier proposal was based exclusively on FDD, possibly because a TDD single-carrier system would not benefit from a complexity reduction because its transmitter and receiver hardware are quite different. The SCM standard specified up to two carriers in each of the downstream and upstream directions. This arrangement resulted in a band plan with up to four bands in total. The separation of upstream and downstream bands typically required analog filtering to provide sufficient rejection of the out-of-band components of the spectrum.

The DMT specification was initially based on TDD. The PMD layer specified a 256-subchannel IDFT/DFT with a subchannel spacing of 43.125 kHz (i.e., ten times as wide as the subchannel spacing in ADSL). This configuration allowed the TDD system to use frequencies up to almost 11.04 MHz. In addition to the DMT-based TDD system, a DMT-based FDD system was proposed. This system introduced the concept of "digital duplexing." Digital duplexing uses a clever alignment of upstream and downstream transmissions that allows perfect separation of the upstream and downstream signals without the need for analog filtering to suppress echo. This capability, in turn, allows almost unlimited flexibility in the allocation of the spectrum.

Several companies designed VDSL systems according to the early proposals. Both TDD and FDD systems were demonstrated before a decision on duplexing was made.

In the end, the standards bodies chose FDD over TDD. This decision was driven by FSAN, the operator working group that was involved in the definition of the next-generation networks. After weighing the advantages and disadvantages of both FDD and TDD, the main driver in the decision appeared to be the need to synchronize all TDD modems deployed from a common node. With loop unbundling becoming a reality in many parts of the world, operators were uncomfortable with assuming the responsibility to provide a common, reliable clock to those who lease lines in their networks. The FSAN group officially communicated its position to TM6 and T1E1.4 in February and March of 1999, respectively ([TM6/991t10] and [T1E1.4/99-059]). As a consequence, all standards bodies adopted FDD as the only duplexing scheme for VDSL. All standardization activity on TDD systems ceased, and the system proposed for VDSL never became an official standard, although it did achieve commercial success in Japan, where synchronization of lines is commonplace because of the use of TCM-ISDN. The companies who had been supporting TDD began developing DMT systems based on FDD. Although the duplexing decision was somewhat of a set-back for the DMT-based VDSL developments, it did not change the position of DMT as a strong proposed line code for VDSL. However, the decision on the VDSL line code would not be made until several years after the duplexing decision.

17.5.3.2.4   Design of the VDSL Band Plans

After FDD was selected, the issue of the band plan was addressed. Operators expressed interest in both symmetric and asymmetric services with VDSL (see Table 17.10). However, the best FDD bandwidth allocations for symmetric and asymmetric services differ significantly. A band plan that is designed to optimally support asymmetric services will be sub-optimal for symmetric services and vice versa. On the other hand, it is not possible to "mix" different band plans in the same loop plant (for spectral compatibility reasons) to offer optimized symmetrical services to some customers and optimized asymmetrical services to others. At the frequencies used in VDSL, the strong NEXT would effectively render any region of overlap between different band plans useless. Therefore, a single VDSL band plan needed to be designed to offer both types of service. Inevitably, some compromises and trade-offs were required.

Initially, the number of bands in each direction was the subject of intense debate. At the frequencies used by VDSL, the available bandwidth quickly decreases as a function of loop length. Longer loops have much less spectrum available than shorter loops. Even so, it is desirable to offer the same symmetry or asymmetry ratio on both long loops and short loops. Therefore, at each loop length, an appropriate amount of upstream and downstream spectrum needs to be available. The number of bands in a frequency plan cannot be too low, or one direction may not have sufficient bandwidth while the opposite direction has excess bandwidth. The issue of band plan selection was also complicated by the discussion surrounding the line code decision. Single-carrier systems used a separate modulator (carrier) for each transmission band, with complexity constraints limiting the systems to two transmission bands in each direction. DMT-based systems, on the other hand, had large (I)DFTs and 4.3125 kHz bandwidth subchannels in each of the transmission directions. As a result, they were capable of supporting virtually any allocation of bandwidth, subject to 4.3125 kHz granularity. This difference clearly manifested itself in the positions of various companies. Single-carrier proponents argued for no more than four frequency bands in total (upstream plus downstream) from the beginning. DMT proponents, on the other hand, argued for the advantages of six or more bands. As a result, discussions on band plans stretched over many meetings, and opinions on the subject were divided along roughly the same lines as the line code preferences of the participants. Several official meeting reports appropriately but euphemistically refer to "lively discussions" on the topic.

In the end, to avoid a decision that would favor one line code over the other, it was decided that the VDSL standard(s) would specify a band plan for frequencies up to 12 MHz and that two bands would be specified in each transmission direction (upstream and downstream). The result is a four-band frequency plan. Use of the ADSL upstream band (called US0 in VDSL) was allowed as optional, although this band could be used in either the downstream or upstream direction in VDSL1. In May of 2000, T1E1.4 adopted a single band plan for North America (informally known as "plan 998") that was more biased toward asymmetrical services [T1E1.4/2000-083]. At about the same time, ETSI defined two band plans [TM6/001t13]. One of these is identical to "plan 998," and the second plan was designed as a better compromise between symmetrical and asymmetrical services. This second plan is known informally as "plan 997."

In addition to these regional band plans, the ITU-T defined a third band plan known as the "Fx-plan" [WP1/15/D.786]. This plan has one variable transition frequency that can be tuned to achieve the service requirements of an operator. However, within a loop plant, only one value of Fx can be used. This plan never received much support from operators, and it was

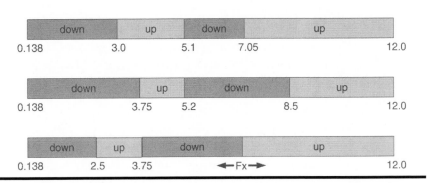

Figure 17.80   **VDSL band plans: "plan 997" (top), "plan 998" (middle), and the "Fx-plan" (bottom).**

removed from the subsequent VDSL2 Recommendation. The three VDSL1 band plans are shown in Figure 17.80.

Whereas VDSL1 considered frequencies up to 12 MHz, VDSL2 extended the available spectrum to as high as 30 MHz. The initial version of VDSL2 Recommendation G.993.2 contained one extended band plan intended for use in Japan. This band plan, which appears in Annex C of G.993.2, is an extension of "plan 998" and is shown in Figure 17.81. It is a six-band plan, with the spectrum between 12 and 30 MHz containing one downstream band and one upstream band, separated at 18.1 MHz.

The extension of VDSL to higher frequencies was initially driven by MDU/MTU requirements in Japan. Japanese operators advertized 100 Mbit/s symmetrical services, which required the use of bandwidth above 12 MHz. The TTC standardization body approved a 30-MHz band plan that was designed to offer this service over a distance of 200 m in low-noise conditions (specifically, with AWGN at a level of −130 dBm/Hz). It is expected that North American and European operators will eventually specify the use of higher frequencies as well. The need to use more bandwidth in VDSL2 and differences between the needs of disparate geographical regions may increase the number of band plans contained in future editions of the VDSL2 standard.

17.5.3.2.5   Digital Duplexing

Although DMT modulation was used in ADSL, VDSL1 for the first time introduced the principle of "digital duplexing" ([Sjöberg 1] and [Sjöberg 2]).

down	up	down	up	down	up
0.138	3.75 5.2		8.5	12.0	18.1	30

Figure 17.81   **Extended VDSL2 band plan.**

Digital duplexing eliminates the need for filters to separate the upstream and downstream bands. To do so, digital duplexing requires the use of a cyclic suffix in addition to the cyclic prefix used in DMT. Timing advance of the transmitted DMT symbols is used to reduce the overhead because of the use of the cyclic prefix.

In DMT VDSL, N orthogonal narrow-band channels (subchannels or tones) are modulated at the VDSL symbol rate with QAM subsymbols. During each symbol period, the N modulated subchannels are summed to form a time-domain block of $2N$ (real) samples. These blocks are cyclically extended to a total length of $2N + CE$ before transmission.

At the receiver, $2N$ samples are extracted from the time-domain signal during each symbol period. An FFT is used to demodulate the signal and recover the original QAM symbols on the N subchannels. The receiver must be symbol-aligned with the transmitter such that the $2N$ samples that are fed to the FFT correspond to the same transmitted DMT symbol. In this case, the N subchannels are orthogonal, which means that any given subchannel will not interfere with any of the other subchannels. Put differently, only the energy transmitted on a subchannel will contribute to the received energy on that subchannel. (See chapter 7 of [Golden 2006].)

In reality, of course, the received signal does not only consist of the desired signal. Near-end echo and NEXT from transmissions in the opposite direction will also contribute to the received signal and will therefore be present in the $2N$ samples that are processed by the FFT. In general, this crosstalk signal is not symbol-aligned with the transmitted signal. Usually, the $2N$ samples that enter the FFT contain contributions from two consecutive "crosstalk symbols" or "echo symbols." As a result, the energy from the crosstalk is not orthogonal to the useful received signal, which means that the crosstalk energy will also affect the desired received signal, even if the crosstalk was caused by a signal that uses subchannels in the other transmission direction.

One way of dealing with the NEXT and echo sidelobes is by filtering the transmitted signal such that the sidelobe energy of the signal (and hence the crosstalk) is reduced. However, the complexity of the filters may be high, which may result in an expensive solution. Digital duplexing provides a low-complexity alternative for separating the useful signal and the NEXT or near-end echo.

Sidelobe energy from crosstalk signals results because these "crosstalk symbols" are not aligned with the received symbols. If all transmit and receive signals are aligned, however, the NEXT and near-end echo will also be aligned with the received symbols. The crosstalk symbols will then be orthogonal to the useful signal, and there will be no interference on the "receive subchannels."

Aligning the two transmission directions is easily done at one side of the modem pair; in fact, such single-ended alignment is used in some ADSL implementations to simplify echo cancellation. In general, however, alignment at one side results in a misalignment at the other side of the modem pair. The essence of digital duplexing is that it provides a technique to achieve alignment on both sides simultaneously. The alignment is achieved by the use of the "cyclic suffix." A "timing advance" minimizes the required length of the cyclic suffix.

Because of the transmission delay of the line, it is normally not possible to align the transmit and receive symbols at both ends of a loop simultaneously. If the transmit and receive symbols are perfectly aligned at the line termination (LT) side, there will clearly be a misalignment at the network termination (NT) side. If the one-way delay of the line is denoted as Δ, the symbol at the NT must be transmitted at time $-\Delta$ for alignment at the LT to occur. On the other hand, the symbol from the LT will arrive at the NT at time Δ, which results in a misalignment of 2Δ between the transmit and receive symbols at the NT.

This problem can be solved by cyclically extending the transmitted symbols. The added samples are collectively called the cyclic suffix (CS). If, for example, the symbols are extended by 2Δ, it is possible at both sides of the link to choose $2N$ samples such that these samples contain contributions from only one received DMT symbol and only one crosstalk DMT symbol. At both the LT and the NT, the crosstalk will then be orthogonal to the received signal. No filtering is needed, because the receiver DFT will separate the useful signal and the near-end echo or NEXT onto different subchannels.

From this analysis, it follows that a choice of $CS = 2\Delta$ will keep the signal and the echo (or crosstalk) orthogonal. However, 2Δ is not the optimal choice for the CS length. In fact, the transmitted symbols at the LT can be advanced by Δ samples with respect to the received symbols. The advancement of the transmitted symbols at one end of the line is called "timing advance." With timing advance, it is possible to select $2N$ samples at both sides of the line such that signal and noise remain orthogonal when the CS length is just Δ. Furthermore, by setting the CS length to Δ (the one-way delay of the line), the overhead because of sending repeated samples is halved relative to if the CS were set to a length of 2Δ.

Conceptually, in a digitally duplexed system, the LT and NT start transmission of DMT symbols at different ends of the line at the same absolute moment in time. In a practical implementation, digital duplexing requires that the transmitter be capable of starting its transmission at a suitable time before the arrival of the DMT symbol from the opposite transmission direction.

Digital duplexing avoids the sidelobes caused by transmissions in the opposite direction by a suitable combination of cyclic extension and timing advance. The technique is called "digital duplexing" because, at least in theory, any band allocation can be accommodated solely through operations in the digital domain. No analog components are required. Note also that no (digital) filtering is required to obtain the desired orthogonality.

The reader is referred to chapter 7 of [Golden 2006] for more details of digital duplexing, including examples that illustrate the CS and timing advance.

17.5.3.3 Power Spectral Density

The power transmitted by any DSL system that is deployed in the copper network is subject to a number of constraints. One of the primary reasons to limit the allowable transmit power is because some of the transmitted power on one copper pair will couple as unwanted energy into other pairs in the same binder because of imperfect cable shielding and other effects. This energy will be perceived as crosstalk by the neighboring DSL modems and will inevitably degrade the performance of these systems. To guarantee the coexistence of different types of DSL in a cable binder, the PSD of each DSL system has to be chosen carefully. Spectral compatibility is one of the main reasons for imposing PSD limits (see Chapter 7).

The need to limit the transmitted power exists for VDSL. However, because VDSL uses frequencies that extend above the frequency band used by other DSL systems, the effect of crosstalk at higher frequencies on other DSL is small. Thus, there is no obvious limit that should be imposed on the VDSL PSDs based on crosstalk alone, at least not for frequencies above roughly 2.2 MHz. For VDSL, the main considerations that led to the definition of the PSDs are related to electromagnetic compatibility (EMC) concerns and allowable dissipated power in certain deployment scenarios.

Early VDSL proposals (also called BDSL, VADSL, NxDMT or HSAS at the time; see [T1E1.4/94-087], [T1E1.4/94-088], and T1E1.4/94-125) included PSD levels that were comparable to those used by ADSL systems. The levels proposed were as high as −40 dBm/Hz. It was soon realized, however, that power levels of this magnitude could create problems with licensed users of the radio spectrum. Essentially, the entire radio spectrum up to 12 MHz (the spectrum used by most current VDSL systems) has been licensed to various types of radio users. Although the VDSL signal is a differential signal on a copper wire pair, the unbalance of the loop results in a common mode component. The drop wire (or any type of exposed overhead wire) can act essentially as an antenna. Through this mechanism, power transmitted at VDSL frequencies over the copper pair can interfere with over-the-air radio services using the same frequencies.

The segments of phone lines closest to the customer (sometimes called drop wires, which can be aerial or buried) are usually unshielded. Balance on a phone line is often 60 dB or more in the voiceband but degrades with increasing frequency and can be as low as 10−30 dB at radio frequencies. To avoid that VDSL acts as an unintentional interferer, the PSD needs to be constrained appropriately.

Of all the radio spectrum users, amateur radio (HAM) users were singled out as the most likely victims of VDS1 egress interference. HAM radio receivers are very sensitive, and reception of a weak signal can be impaired easily by VDSL egress. Also, HAM users can be located virtually anywhere. Therefore, it must be assumed that VDSL and HAM systems can be located close to each other. For these two reasons, HAM is more likely to be disturbed by VDSL egress than any other over-the-air transmission system.

Both theoretical and experimental results show that VDSL transmission can have a real and measurable impact on HAM services. It was established during the early days of VDSL development that the disturbance could be limited to acceptable levels by reducing the VDSL PSD in the HAM bands to a level below −80 dBm/Hz. As a consequence, this restriction was imposed by standards as a mandatory requirement for VDSL PSDs. Similar requirements have been imposed on other systems deployed on copper wires (e.g., home phone line networks).

Although radio interference studies showed the need for a reduced PSD in certain parts of the spectrum, it was clear that reducing the entire VDSL PSD to this level would needlessly impair the performance of VDSL systems. Therefore, the requirement only applies within the HAM bands. The VDSL1 standards mandate the capability of VDSL systems to create "notches" at the frequencies corresponding to the HAM bands.* These frequency bands are slightly different in Europe and in North America. They are given in Table 17.11.

Table 17.11   Licensed HAM Bands

North American HAM Bands (MHz)	European HAM Bands (MHz)
1.81–2.0	1.81–2.0
3.5–4.0	3.5–3.8
7.0–7.3	7.0–7.1
10.1–10.15	10.1–10.15
18.068–18.168	18.068–18.168
21.0–21.45	21.0–21.45
24.89–24.99	24.89–24.99
28.0–29.7	28.0–29.1

---

* Taking the requirement even further, VDSL2 requires the ability to provide simultaneous notches in up to 16 arbitrary, operator-defined frequency bands, which may or may not correspond to the defined HAM bands.

The PSD notching capability is a feature that was introduced for the first time in VDSL1. Outside the HAM bands, the PSD level was initially fixed at −60 dBm/Hz in the first VDSL standards. This value was determined by the power requirements for VDSL systems. It was expected that VDSL systems would be deployed primarily from remote cabinets, where the allowable dissipated power must be limited due to uncontrolled climate conditions and limited space. Based on this requirement, a total power constraint of 11.5 dBm was imposed on VDSL1 systems. Distributing this power evenly over the VDSL spectrum yields the value of about −60 dBm/Hz.

Later, scenarios without EMC requirements were also considered, e.g., buried cables that pose less risk of interference to HAM. Depending on the considered noise scenario, the performance of VDSL can improve when higher PSD levels are used (although FEXT is generally a significant impairment because of the short lengths of VDSL lines). In situations where radio emissions or dissipation limitations are of less concern, it is not necessary to meet the PSD restrictions that have been designed for EMC. For this reason, the standards also define a second, higher* PSD level for use in environments with weak EMC coupling. For CO-deployed systems, the VDSL PSD is essentially allowed to be equal to the ADSL2plus PSD in the frequency range below 2.2 MHz. For frequencies above 2.2 MHz, the value of the PSD is also higher than −60 dBm/Hz. TM6 and T1E1.4 chose different values for these "boosted" PSD levels.

Later revisions of the European and North American standards made further refinements to the VDSL PSDs. Especially in Europe, the coexistence of cabinet-deployed VDSL with CO-deployed ADSL seemed to raise additional concerns. The issue arises when copper pairs carrying an ADSL signal from the CO share a binder with copper pairs that originate in the cabinet. When these pairs from the cabinet carry VDSL, the cabinet systems may generate significant FEXT in the downstream ADSL band. This FEXT may cause a degradation in the reach of the ADSL systems deployed from the CO. For this reason, the cabinet PSDs were reduced in the ADSL band (below 1.1 MHz). In T1E1.4, the PSD was limited to −60 dBm/Hz. In TM6, the use of the band below 1.1 MHz was precluded completely for VDSL systems that were deployed from the cabinet.

These considerations led to four different types of VDSL1 PSD. First, the standards distinguish between PSDs that can be used by CO-deployed systems and PSDs that can be used by cabinet-deployed systems. In addition, for both of these deployment scenarios there is a "low" PSD (typically called M1) that is to be used when radio emissions are of concern and a "high" PSD (typically called M2) that can be used when there are no concerns about radio emissions. Furthermore, the PSD requirements are (slightly)

---

* "Higher" in this case means above −60 dBm/Hz.

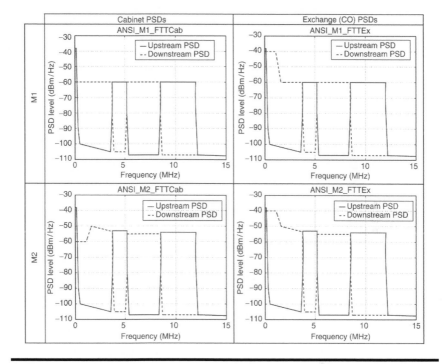

Figure 17.82   North American VDSL1 PSDs.

different in TM6 and T1E1.4. These PSDs are shown in Figure 17.82 and 17.83 for North America and Europe, respectively.

In addition to the PSD limits, VDSL systems are also constrained by the total amount of power they are allowed to transmit on the line. The power was limited primarily to reduce heat dissipation and to limit the complexity of the AFE. (See Chapter 3 for a discussion of the impact of transmitted power on complexity and performance of the AFE.)

The values of the total power allowed in VDSL1 are summarized in Table 17.12. For most cabinet PSDs, the PSD levels automatically constrain the total transmitted power to below the required value. However, for some types of PSD (notably the M2-type PSDs and the CO-based PSDs), the total power that could be transmitted if the PSD were completely "filled" is higher than the allowed total power. Therefore, a VDSL system must distribute its transmitted power in the available frequency band in a way that meets both the total power constraint and the applicable transmit PSD. As a result, it is not possible to transmit at a level that is equal to the applicable PSD at every frequency in the VDSL transmission band. In these cases, there is considerable freedom in allocating the power below the mask.

In VDSL2, the PSDs are similar to the PSDs that were defined for VDSL1, with some minor differences. The main difference between the VDSL1

734 ■ *Implementation and Applications of DSL Technology*

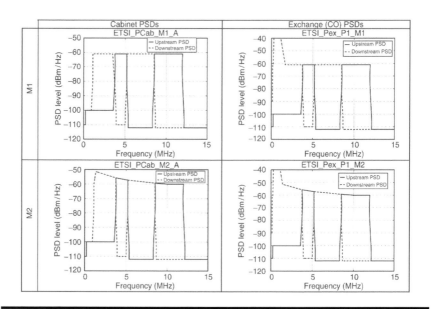

Figure 17.83   European VDSL1 PSDs.

and VDSL2 PSDs is that VDSL2 no longer explicitly defines cabinet PSDs. Instead, the VDSL2 standard includes a parameterized shaping mechanism similar to the one defined in ADSL2plus, which allows the PSD to be reduced within some constraints. This mechanism is used to mitigate the interference between cabinet-deployed and CO-deployed systems. In addition, VDSL2 allows transmit powers that are higher than 14.5 dBm in some cases. These higher transmit powers are part of the VDSL2 concept of "profiles" (see Section 17.5.4.1).

17.5.3.4   Upstream Power Back-Off

The loop lengths and frequency spectrum used by VDSL pose some unique problems for this technology. One of the main issues in VDSL standardization was nonreciprocal FEXT because of the disparity of loop lengths in a typical VDSL deployment scenario, such as the one illustrated in Figure 17.84. The LT VDSL system is placed in a central node (either at the

Table 17.12   Total Power Constraints for VDSL1

	Central Office (CO) (in dBm)		Cabinet (Cab) (in dBm)	
	Downstream	Upstream	Downstream	Upstream
North America	14.5	14.5	11.5	14.5
Europe	14.5	14.5	11.5	11.5

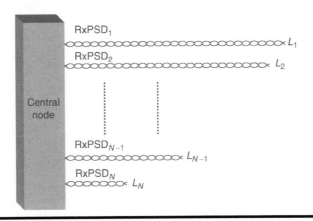

Figure 17.84 **Illustration of a distributed topology, whereby lines of different lengths share the same binder.**

CO or in a cabinet). From this central node, loops emanate to the various VDSL users. Because the users are located at different distances from the central node, the loop lengths over which the VDSL systems are deployed vary significantly in length. A crucial observation is that for the loop lengths in question (say, up to 1500 m), FEXT is a dominant noise source. As a result, the noise experienced by a VDSL system will be primarily caused by transmissions by other VDSL systems in the binder. Although this noise is unavoidable, it becomes problematic in the upstream direction when a long loop experiences FEXT because of upstream transmissions on a short loop. In that case, the (strong) transmit signals on the shorter lines couple into a (long) victim line over the last portion of the line, beginning from a location where the useful signal on the long line is already severely attenuated. The FEXT caused by this short loop is significantly stronger than the FEXT that would be caused by a (long) loop of the same length as the victim.

This effect is illustrated in Figure 17.85, which shows the performance of a VDSL system in the topology illustrated in Figure 17.84 for various loop lengths of the system under study (the victim loop). Two cases are considered. First, it is assumed that twenty disturbing systems are at the same distance from the LT as the system under test (called the "equal-length scenario"). Then the case where the disturbers are distributed at various distances, as illustrated in Figure 17.84 (called the "distributed topology"), is considered. The curves show that the performance in the two scenarios is significantly different. There is an enormous loss in performance for longer loop lengths in the case of a distributed environment. The relatively strong transmissions on the shorter loops practically exclude the use of longer loop lengths.

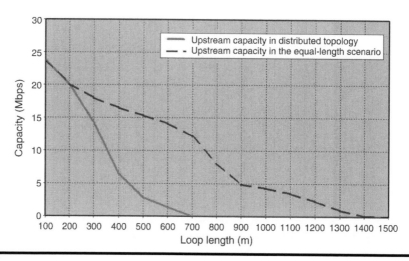

Figure 17.85   **Comparing the upstream bit rate in a distributed topology with the upstream bit rate in an equal-length scenario.**

It is clear from Figure 17.85 that a solidarity mechanism is needed to protect the upstream transmissions on longer VDSL loops. To remedy this so-called near–far problem, transmitters that are closer to the LT must reduce their upstream transmit PSDs in a systematic way to protect the capacities of longer lines. This reduction in transmit power reduces the relatively strong FEXT from shorter lines into the longer lines. The process of systematically reducing the power on shorter lines is called upstream power back-off (UPBO), and its use is crucial for the successful deployment of VDSL systems. The bit rates on shorter loops are necessarily somewhat reduced by the use of UPBO, but the overall performance of VDSL improves significantly.

The UPBO mechanism is parameterized by a "reference PSD," which roughly corresponds to the maximum transmit PSD that can be used on a loop of length zero. Longer loops are allowed to transmit at a level equal to the reference PSD plus a frequency-dependent factor that depends on the length of the loop.

The reference PSD depends on the deployment scenario, because the appropriate transmit levels are a function of the PSD mask and the assumed noise environment. The VDSL1 standards define the reference PSDs for the various applicable deployment environments. The ANSI standard defines four different reference PSDs that are linked to four different combinations of noise scenario and transmit PSD [T1E1.4/2001-157]. The ETSI standard defines five different reference PSDs for different noise environments [TM6/013w07].

To determine its allowed transmit PSD using the reference PSD, a modem autonomously determines the so-called electrical length kL_e of the loop. The electrical length is an estimate of the loop attenuation, assuming that all the sections of cable obey a $\sqrt{f}$ attenuation characteristic. Specifically, the electrical length provides an approximation of the loop insertion loss of the form

$$\text{Loss}(f) = kL_e\sqrt{f} \quad (\text{in dB})$$

The VTU-R estimates the electrical length during the first stages of initialization, based on the level of the received downstream signal. Once it has done so, it calculates its allowed transmit PSD as

$$\text{TxPSD}(f) = \text{PSD}_{\text{REF}}(f) + kL_e\sqrt{f} \quad (\text{in dBm/Hz})$$

This calculated PSD is, of course, not allowed to exceed the designated maximum PSD, which for VDSL1 is one of those shown in Figures 17.82 and 17.83.

It is clear from this expression that the UPBO mechanism provides the required solidarity mechanism between short loops and long loops. Transmissions on shorter loops will be at lower powers. As the loop length increases, higher upstream transmit PSDs can be used. Transmissions on the longest loops are at the maximum level allowed by the relevant standard (see Figures 17.82 and 17.83). Thus, the UPBO mechanism provides a trade-off between the bit rates on shorter loops and the bit rates on longer loops that mitigates the near–far problem. Figure 17.86 shows the upstream VDSL bit rates achievable when UPBO is applied. The results for two different noise scenarios (and, therefore, different reference PSDs) are shown. Compared to Figure 17.85, the bit rate on the long loops is significantly improved. At the same time, the bit rate on short loops has been reduced.

Note that a similar mechanism is not normally required in the downstream direction because the problem is primarily due to the fact that the transmitters are not collocated. In the downstream direction, the

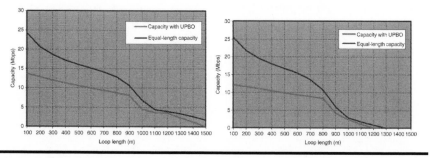

Figure 17.86   Upstream bit rate with standardized upstream power back-off (UPBO) mechanism.

transmitters are usually located in a single node, so the problem does not manifest itself for downstream transmission.*

17.5.3.5 Line Code

To transport data over the VDSL loop, the incoming bit stream must be modulated on a signal that can be sent over the copper wire pair. From the beginning of the VDSL standardization work, two distinct and noncompatible modulation techniques (line codes) were proposed for VDSL. Perhaps unsurprisingly in light of the work in ADSL, the line codes proposed were CAP/QAM (collectively denoted as "single carrier modulation" or SCM) and DMT (discrete multitone), which is the multi-carrier modulation technique standardized in ADSL. A detailed discussion of the two line codes can be found in chapters 6 and 7 of [Golden 2006]. Both line code alternatives enjoyed the support of a large group of companies. The "VDSL Alliance" represented a group of about 50 companies supporting the DMT line code. The "VDSL Coalition" represented about the same number of companies that were in support of SCM as the line code for VDSL.

The line code selection was one of the most long-standing unresolved issues in VDSL standardization. Initially, all standards bodies allowed both line codes to co-exist in their specifications, even while knowing that the specification of incompatible line codes was contrary to one of the primary goals of standardization, which is to enable interoperable solutions. Both the ETSI specification and the T1 trial-use standard described both SCM and multi-carrier technology, and the ITU-T's "Recommendation" avoided the issue altogether by not containing any transceiver specification. Although it was always considered desirable and necessary to select a single line code, all three standard bodies, all of which require consensus to make decisions, remained deadlocked in their attempts to decide the issue.

In March of 2002, T1E1.4 finalized its work on a three-part "trial-use" standard for VDSL1. This document contained a full transceiver specification for both line codes. T1E1.4 did maintain its goal to select a single line code, however. The trial-use standard was given a lifespan of only two years, at which point the line code issue would be reconsidered and a decision made on whether to extend the life of the trial-use standard or take some other course of action. The intention of the process was to allow both line codes an opportunity to demonstrate their respective strengths and somehow change a large number of minds during the trial-use period. The goal was to make a decision on the line code upon the expiration of the trial-use period. The subsequent revision of T1.424 would be a VDSL1 standard with a single line code.

---

* If there is a problem in the downstream direction, the PSD shaping mechanism of VDSL2 can be used to mitigate it.

ETSI published a transceiver specification with both line codes in 2001. This standard also contained a full specification of the two line codes. Unlike T1E1.4, however, TM6 never made a clear commitment to choose a single line code.

In August of 2002, prompted by the EFM group in the IEEE, T1E1.4 was the first standards group to start an effort to select a single VDSL line code. The selection methodology was similar to the one that was used almost ten years earlier in ADSL. Equipment makers of both "camps" were invited to submit their SCM or DMT modems for third-party testing by independent test labs. In keeping with tradition, this testing was informally referred to as the "VDSL Olympics."

From August of 2002 to February of 2003, through a number of additional "lively discussions," T1E1.4 developed a test plan that would be used to evaluate both technologies. The plan consisted of a series of 32 performance tests and a number of tests investigating the robustness and latency of the submitted systems [T1E1.4/2003-036]. Each of the 32 performance tests was characterized by a given service, loop type, noise type, and transmit PSD. The goal of each test was to achieve the maximum possible reach for the given service bit rate, which roughly corresponds to the maximum loop length over which the service can be offered with a guaranteed quality of service. Four systems were submitted for the VDSL Olympics: two DMT systems (one each from Ikanos Communications and Alcatel/STMicroelectronics) and two SCM systems (one each from Metalink and Infineon). The tests specified by the test plan were carried out by Telcordia and BTexact. Each of the labs recorded the results for each of the 32 performance tests for the four participating systems. Summaries of the results of the performance tests are shown in Figures 17.87 and 17.88. Overall, as the figures demonstrate, the DMT modems outperformed the SCM modems, as was the case ten years earlier in the ADSL Olympics. Detailed reports of the VDSL Olympics results can be found at [Telcordia Olympics] and [BT Olympics].

The results of the VDSL Olympics were analyzed and debated at an interim T1E1.4 meeting in June of 2003. After reviewing the results, T1E1.4 recommended the selection of DMT as the only line code for the T1 VDSL1 standard. The group did not completely discard the SCM technology, however; instead, SCM VDSL was documented as a TRQ. These two documents were published in June of 2004 as T1.424-2004 [T1.424-2004] and T1.TRQ.12-2004 [T1.TRQ.12-2004], respectively. Soon after T1E1.4 made its line code selection, the decision was adopted by the IEEE's 802.3ah group for the short-reach EFM standard. ETSI maintained a VDSL1 specification with both line codes but, in November of 2003, started a new VDSL2 project that was to be exclusively based on DMT modulation.

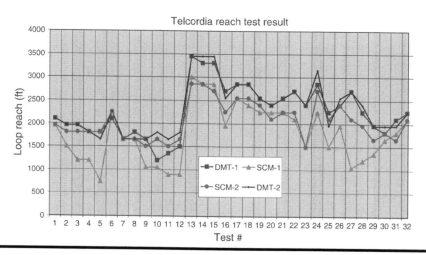

Figure 17.87   Summary of Telcordia VDSL Olympics results. This shows the loop reach that was achieved in each performance test (see [Telcordia Olympics]).

Last in line, SG15/Q4 resolved the line code issue in January of 2004. In spite of the T1E1.4 line code decision, the group was unable to reach consensus to publish a VDSL1 Recommendation that contained only DMT. After lengthy debates, a compromise was reached that would finish the work on VDSL1 and enable work on VDSL2. As per the compromise, the ITU-T VDSL Recommendation G.993.1 was updated so that it contained

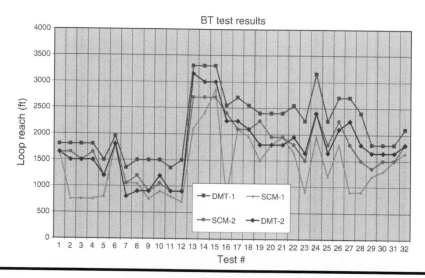

Figure 17.88   Summary of BT VDSL Olympics results. This shows the loop reach that was achieved in each performance test (see [BT Olympics]).

DMT in the main body of the standard. SCM, although fully specified, was put into an annex of the same document. The G.993.1 Recommendation is based on the regional VDSL specifications and as a result is very similar to the ETSI and ANSI standards. As part of the compromise, SG15/Q4 also agreed that a new Recommendation, called VDSL2, would be developed, and that this Recommendation would be based only on DMT. Any future enhancements to VDSL would be addressed in the VDSL2 Recommendation. As a result of the compromise, VDSL1 was effectively "frozen in time," and VDSL2, based on DMT modulation, was considered the way forward for the long term.

The line code decision was the last remaining hurdle to a complete VDSL1 standard. The decision generated renewed interest in VDSL and resulted in a major effort to develop the next-generation of VDSL: VDSL2.

17.5.4   VDSL2

Work on VDSL2 started in T1E1.4 in August of 2003. SG15/Q4 started its own version in January of 2004 with the goal to consent the document in May of 2005—a very short time in the world of standardization. Over the next 18 months, TM6, T1E1.4, and SG15/Q4 worked closely together in the development of VDSL2. Unlike in VDSL1, no regional versions of VDSL2 were developed. Both TM6 and T1E1.4 contributed their requirements directly to SG15/Q4 for incorporation in the emerging G.993.2 Recommendation.

In addition, showing great character, those who had passionately supported SCM for VDSL1 made significant and valuable contributions to the standardization of DMT for VDSL2. A new and somewhat surprising spirit of cooperation resulted, and in this healthy environment, a vastly improved VDSL Recommendation was generated. ITU-T Recommendation G.993.2 was officially approved in February of 2006.

One of the stated goals of VDSL2 was to facilitate multimode ADSL2/VDSL2 implementations. Facilitating multimode implementations does not mean that ADSL2 is an integral part of the VDSL2 standard, but rather that both technologies should be similar in terms of features, management, and the like so that they could potentially be implemented in a single platform. In the end, the VDSL2 work was an attempt to incorporate the best parts of the existing VDSL and ADSL standards into a single Recommendation.

17.5.4.1   Profiles

From the start of the VDSL2 work, it was clear that operators still had a wide and diverse range of applications in mind for the new technology, as they had for VDSL1. These applications were quite literally "across the spectrum." Some operators were targeting very wide-band, extremely high data rate services intended for deployment in MDU/MTU environments. These environments have short loops that can potentially sustain data rates up to 100 Mbit/s symmetric by using the spectrum up to 30 MHz. Other operators wanted a longer reach, lower bit rate service with VDSL2 and were interested in extending the reach of VDSL2 beyond that of VDSL1. Still other operators expressed interest in low-complexity systems that were suitable for confined enclosures such as cabinets and pole tops.

It was quickly realized that the requirements were too diverse to be incorporated in a single, practical, cost-effective system. High data rate applications require a very large bandwidth, which presents significant challenges to the designers of AFEs. In addition, the digital processing would have to run at higher sampling rates and, to keep memory requirements reasonable, wider subchannel spacing. On the other hand, systems that are designed for long reach do not need the high bandwidths, but need to be able to transmit in the ADSL upstream band (US0) and may benefit from higher downstream transmit power. This scenario results in a different set of analog and digital requirements. Although it is not impossible to build a single system that can work in all environments, such a system would likely be sub-optimal in all of its applications. For instance, in an MDU/MTU environment there is no need to implement the complexity required to support US0, because the capacity contributed by US0 is perhaps 1 Mbit/s, which is insignificant compared to 100 Mbit/s. For longer-reach applications, on the other hand, there is no need to design a complex AFE that can accommodate frequencies up to 30 MHz, but support of US0 is critical for providing the necessary reach. Therefore, a one-system-fits-all design would be burdened by some unnecessary complexity in all of its applications.

To handle the diversity of the envisaged applications, VDSL2 introduced the concept of "profiles." Essentially, a profile is a subset of the full set of parameter settings described in (allowed by) the VDSL2 standard. The specific set of parameter settings for a given profile is chosen to yield a system that is optimized for a specific deployment scenario (e.g., MDU/MTU, CO deployment with significant ADSL crosstalk, etc.). The profiles were formulated in terms of the system attributes that contribute most to system complexity. As such, each profile roughly corresponds to a system with a given expected complexity and a given expected capability. The definition of profiles allows manufacturers to limit the complexity of the system, and it also allows operators to select equipment that meets their needs with the expected features and performance, thereby avoiding unnecessary overhead.

The parameters that compose the definition of a profile are as follows:

- Transmit power (upstream and downstream): The maximum allowed transmit power affects the design and complexity of the line driver. Higher transmit power levels can also lead to higher-power consumption. Profiles allow designs that are either focused on long-reach/higher-power or short-reach/lower-power applications.

- Sub-carrier spacing: VDSL2 specifies both 4.3125 and 8.625 kHz sub-carrier spacing. The higher sub-carrier spacing requires more digital processing but the same amount of memory for processing of the (I)DFT. Profiles specify 4.3125 kHz sub-carrier spacing for CO/RT applications and 8.625 kHz spacing for MDU/MTU deployments.

- Support of ADSL upstream band (US0): Because of leakage in the lower part of the spectrum, support of the ADSL upstream band requires either echo cancellation or extra filters to separate US0 from the downstream spectrum. In addition, a time-domain equalizer (TEQ) may be needed to achieve optimal performance on long loops. (See chapter 11 of [Golden 2006].) Also, the POTS splitter can be simplified if US0 does not have to be supported. Some profiles specify the mandatory support of US0 for applications that are targeted for longer reach, whereas other profiles do not require support of US0.

- Data rate capabilities: The minimum bidirectional data rate capability is primarily a requirement for the external interfaces, although it also imposes a minimum level of processing that must be supported by the coding and framing mechanisms.

- Interleaver delay and depth: Interleaver memory is an important contribution to the chip area and is directly proportional to the interleaver delay. Different profiles allow manufacturers to tailor the interleaver memory—and hence complexity of the design—to the specific application.

- Number of codewords per DMT symbol: The number of codewords per symbol determines the complexity of the framing and coding.

- Highest supported sub-carrier in upstream and downstream direction (as a function of band plan): This parameter determines the highest frequency that the system is required to transmit. As such, it imposes requirements on both the digital and analog parts of the modem.

As the elements above indicate, profiles allow manufacturers to choose to build lower-complexity or higher-complexity systems within the framework of the VDSL2 Recommendation.

After considerable debate (because there is no such thing as a brief discussion in standardization), eight profiles were defined and included in the VDSL2 Recommendation. They are labeled as 8a, 8b, 8c, 8d, 12a, 12b, 17a, and 30a. The complete set of profile parameters and their values is given in Table 17.13.

The profiles 8a, 8b, 8c, 8d, 12a, and 12b have been defined with CO/cabinet deployment in mind. It should be noted, however, that the Recommendation does not place any restrictions on where a profile can

Table 17.13 VDSL2 Profile Parameters

Frequency Plan	Parameter	Parameter Value for Profile							
		8a	8b	8c	8d	12a	12b	17a	30a
All	Maximum aggregate downstream transmit power (dBm)	+17.5	+20.5	+11.5	+14.5	+14.5	+14.5	+14.5	+14.5
All	Minimum aggregate downstream transmit power (dBm)	For further study	For further study	For further study	For further study	For further study	For further study	For further study	For further study
All	Minimum aggregate upstream transmit power (dBm)	+14.5	+14.5	+14.5	+14.5	+14.5	+14.5	+14.5	+14.5
All	Minimum aggregate upstream transmit power (dBm)	For further study	For further study	For further study	For further study	For further study	For further study	For further study	For further study
All	Sub-carrier spacing (kHz)	4.3125	4.3125	4.3125	4.3125	4.3125	4.3125	4.3125	8.625
All	Support of upstream band zero (US0)	Required	Required	Required	Required	Required	Not Required	Not Required	Not Required
All	Minimum bidirectional net data rate capability (MBDC) in Mbit/s	50	50	50	50	68	68	100	200
All	Aggregate interleaver and de-interleaver delay (octets)	65,536	65,536	65,536	65,536	65,536	65,536	98,304	131,072
All	Maximum interleaving depth (D_{max})	2,048	2,048	2,048	2,048	2,048	2,048	3,072	4,096
All	Parameter $(1/S)_{max}$ downstream	24	24	24	24	24	24	48	28
All	Parameter $(1/S)_{max}$ upstream	12	12	12	12	24	24	24	28

	Row description								
Annex A, Annex B (998)	Index of highest supported downstream databearing sub-carrier (upper band edge frequency in MHz [informative])	1,971 (8.5)	1,971 (8.5)	1,971 (8.5)	1,971 (8.5)	1,971 (8.5)	1,971 (8.5)	N/A	N/A
	Index of highest supported upstream data-bearing sub-carrier (upper band edge frequency in MHz [informative])	1,205 (5.2)	1,205 (5.2)	1,205 (5.2)	1,205 (5.2)	2,782 (12)	2,782 (12)	N/A	N/A
Annex B (997)	Index of highest supported downstream sub-carrier (upper band edge frequency in MHz [informative])	1,634 (7.05)	1,634 (7.05)	1,634 (7.05)	1,634 (7.05)	1,634 (7.05)	1,634 (7.05)	N/A	N/A
	Index of highest supported upstream sub-carrier (upper band edge frequency in MHz [informative])	2,047 (8.832)	2,047 (8.832)	1,182 (5.1)	2,047 (8.832)	2,782 (12)	2,782 (12)	N/A	N/A
Annex C	Index of highest supported downstream sub-carrier (upper band edge frequency in MHz [informative])	1,971 (8.5)	1,971 (8.5)	1,971 (8.5)	1,971 (8.5)	1,971 (8.5)	1,971 (8.5)	4,095 (17.664)	2,098 (18.1)
	Index of highest supported upstream sub-carrier (upper band edge frequency in MHz [informative])	1,205 (5.2)	1,205 (5.2)	1,205 (5.2)	1,205 (5.2)	2,782 (12)	2,782 (12)	2,782 (12)	3,478 (30)

or should be deployed. Profile 17a could be useful in situations where the distance to the user is short enough to exploit the use of frequencies above 12 MHz. Profile 30a is essentially intended for use in MDU/MTU scenarios, where the distances are very short and the full spectrum up to 30 MHz can be used.

To be compliant with the VDSL2 Recommendation, a system must be compliant with at least one profile. There is no requirement that a system should support all profiles, or even more than one profile, to be a compliant VDSL2 system.

17.5.4.2 VDSL2 Initialization

Although VDSL2 initialization borrowed elements from both ADSL2 and VDSL1, the protocol is very similar to the one used in VDSL1. Both VDSL1 and VDSL2 use the Special Operations Channel (SOC) protocol. Initialization messages are encapsulated in HDLC (High-level Data Link Control) frames and communicated to the other side at a rate of one byte per DMT symbol. To improve robustness, this one byte is modulated on a multitude of subchannels to provide redundancy. In case the message is received in error, the protocol allows the receiving modem to ask for retransmission.

The VDSL2 initialization procedure consists of four main phases:

1. Handshake: Preliminary parameter settings are established during the handshake phase. See Section 17.6.
2. Channel Discovery: This phase is primarily intended to allow the modems to optimize the upstream and downstream PSDs for the specific loop and noise conditions. In addition, the modems perform timing recovery and determine appropriate values for a number of other modulation parameters (such as the cyclic extension length, timing advance, window lengths, etc.). At the end of Channel Discovery, the modems establish the transmit PSDs that will be used from that point on.
3. Training phase: During this phase, the modems can optionally train their TEQs and ECs. Depending on the need to train either the TEQ or EC, some of the stages during the Training phase can be skipped. The Training phase introduces a number of signals that are reminiscent of ADSL signals and that are not present in VDSL1 (such as the TEQ training signal). At the end of Training phase, the modem is ready to determine the bit allocation.
4. Channel Analysis and Exchange: At the beginning of this phase, the VTU-O communicates to the VTU-R what the desired service requirements are (such as bit rate, impulse noise protection, maximum interleaver delay, etc.). Both modems then determine the number of bits that can be allocated to each subchannel in

their receive bands and establish the appropriate modulation and framing parameters to provide a communication channel that is consistent with the service requirements. After these parameters have been exchanged, the modems transition to Showtime (steady-state transmission).

Like ADSL2, VDSL2 specifies a loop diagnostic mode. Loop diagnostic mode is essentially a more robust version of the initialization procedure wherein the redundancy has been increased significantly. It is intended to be used on loops on which the regular initialization procedure fails due to loop or noise conditions. Loop diagnostic mode does not lead to a regular Showtime session. Instead, the modems end their communications after exchanging a number of diagnostic parameters, such as loop attenuation and line noise.

17.5.4.3 Improvements and New Features

VDSL2 builds on the published VDSL1 and ADSL standards. During the line code decision process for VDSL1, essentially no new features were added to VDSL1. During that time, ADSL did progress, however, with SG15/Q4 developing the ADSL2 and ADSL2plus Recommendations (see Sections 17.4.3 and 17.4.4). Thus, many of the new features that were added to VDSL2 leveraged progress that was made in the various ADSL Recommendations. Others are entirely new and were included for the first time in any Recommendation in VDSL2. The most significant features of the VDSL2 Recommendation are as follows:

- Extended band plans, allowing the spectrum up to 30 MHz to be used for transmission. The maximum frequency was increased in some cases from 12 to as high as 30 MHz to allow aggregate bit rates on the order of 200 Mbit/s. The band plans and PSDs for VDSL2 are included in the annexes of G.993.2.
- PSD shaping. G.992.3 explicitly provides the means to define custom-shaped PSDs (within predefined limits). In addition, a downstream PSD shaping mechanism is included to facilitate the coexistence of CO- and cabinet-deployed systems.
- Mandatory support of US0 in profiles. Some profiles are required to support US0 as a mandatory capability. In VDSL1, the use of US0 was optional, and the band could be used in either the downstream or upstream direction. In VDSL2, the band can only be used in the upstream direction. In addition, the upper frequency of US0 can be extended to as high as 276 kHz using the same PSDs as are defined in Annex M of ADSL2 and ADSL2plus.

- Mandatory support of trellis coding. The requirement to support trellis coding allows a higher coding gain and therefore some improvement in data rate. Trellis coding was not specified, even as an option, in VDSL1.
- Improved coding parameters. VDSL2 specifies a larger set of Reed–Solomon codeword and interleaver parameter values than those specified in VDSL1. Also, unlike in VDSL1, the coding and interleaver parameters are determined by the receiver, which allows a more optimal selection of these parameters.
- Higher downstream transmit power in some profiles. The 8b profile allows a downstream transmit power as high as 20.5 dBm, which is on the order of the maximum power allowed for ADSL systems.
- Improved on-line reconfiguration mechanisms. OLR allows the modem to change some of its modulation and framing parameters to maintain the expected quality of service, even when line or noise conditions change. In the first version of G.993.2, bit swap is the only form of OLR specified. The mechanism for bit swapping is based on the mechanism used in ADSL2 (see Section 17.4.3). It is expected that more advanced types of OLR (such as DRR and SRA) will be added in the future.
- Improved framing. The framing in VDSL2 is based on the mechanism specified in ADSL2. VDSL2 supports two bearer channels and two latency paths. Some level of interleaving is required to be supported in each of the latency paths.
- Interleaver reconfiguration. This entirely new feature of VDSL2 allows the modems to seamlessly change the interleaving depth during steady-state transmission. A similar mechanism does not exist in any of the other standards. Note that the first edition of G.993.2 describes the method but does not provide a mechanism to activate the change. It is expected that the first amendment to VDSL2 will specify the mechanism.
- Improved initialization. The initialization protocol is largely based on VDSL1 principles, but new training states have been added to enable TEQ and EC training.
- Loop diagnostic mode. Like ADSL2, VDSL2 includes a special sequence that allows modems to obtain information about the loop and noise conditions, even when completion of the regular initialization procedure is not possible.

17.5.5 *Important Dates in VDSL Standardization*

Table 17.14 lists the dates that mark major events in the development of the VDSL standard(s) and Recommendations.

Table 17.14 Important Dates in VDSL Standardization

Date	Event
1995	First efforts in VDSL standardization
June 1998	ETSI publishes first VDSL document (functional requirements)
March 1999	FDD selected as duplexing method for VDSL
February 2000	T1E1.4 decides to develop three-part trial-use standard for VDSL, documenting both DMT and SCM
April 2000	Band plan selection for North America
November 2000	ANSI trial-use standard sent out for letter ballot
February 2001	ETSI publishes VDSL transceiver specification containing both line codes
October 2001	ITU publishes G.993.1
March 2002	Publication of ANSI trial-use standard (T1.424)
February 2003	Publication of "VDSL Olympics" test plan
February–June 2003	Lab testing of DMT and SCM systems ("VDSL Olympics")
June 2003	DMT selected as the only line code for VDSL in North America in T1E1.4
July 2003	ETSI publishes revised VDSL transceiver specification containing both line codes
August 2003	Start of VDSL2 project in T1E1.4
January 2004	Approval of G.993.1 in ITU Start of G.993.2
June 2004	Publication of T1.424-2004 (DMT) Publication of T1.TRQ.12-2004 (SCM)
May 2005	Consent of G.993.2 (VDSL2)
February 2006	Approval of G.993.2

17.6 Handshake

17.6.1 Handshake's Roots

To understand the handshake Recommendation, it is helpful to understand the origins and development of the concept. Handshake began as a mere twinkle in the eyes of the infamous "British Bulldog," the late John Brownlie. (May he rest in peace; he was loved by all.) During the standardization of Recommendation V.34 [V.34] in ITU-T SGXVII (1989–1992), it was recognized that a new method of interworking between the various voiceband modem Recommendations was needed. For voiceband modems, it is not known which type of modem is used at each end of the connection, and therefore it is a general market requirement that any new modem Recommendation must be able to interoperate with the older modem Recommendations. Prior to 1989, auto-mode operation between modem Recommendations relied upon being able to detect the initial tone transmissions from each modem and being able to respond appropriately within the time constraints defined for each modem Recommendation. An example of this type of auto-mode operation can be found in Annex A of Recommendation V.32 *bis* [V.32bis]. As the number of modem Recommendations increased, it became more and more difficult to define tone-based interworking procedures.

The twinkle in John's eye came in late 1991 in the form of Delayed Contribution D141 [BT 1991] to the October 1991 ITU-T SG XVII meeting in Geneva, in which he first proposed the concept of a "call menu" and "answer menu" carried over four-phase modulated signals. Handshake became official in late 1994 with the approval of Recommendation V.8 [V.8]. Recommendation V.34, which relies on V.8, was approved at the same time.

In early 1994, the voiceband modem industry started to focus on new applications such as voice, data or videoconferencing, and began standardization efforts in TIA TR-30.1 [Data Race 1994] and ITU-T (both SG15 [GSTN] and SG14—formerly SGXVII [SAVD]). One of the identified requirements was to be able to switch between analog voice and modem transmission (for data, video, and fax) during a connection. To meet this requirement, it was recognized that a more user-friendly signaling protocol was needed: "one that would not upset grandma when she picked up her telephone." This requirement ruled out the usual modem-like exchanges, including V.8. In August of 1994, the "White Knight" (standards elder Larry Smith) came to Handshake's rescue* with his proposal [AT&T 1994] for a new protocol for the exchange of communication capabilities and selection of an operating mode. These were formidable years in the development of Handshake, which was eventually adapted to accommodate emerging standards. The result was the approval of Recommendation V.8 *bis* [V.8bis] in late 1995, which was used with ITU-T Recommendations V.70 [V.70] and H.324 [H.324].

With Recommendation V.8 *bis* came a rich set of messages and transactions to allow both modems to exchange their capabilities, and to allow either modem to select an operating mode, and the concept of "future proofing" with a hierarchical tree structure for code points.

Having very nearly reached the channel capacity with Recommendation V.34, many members of SG14 felt that it was the end of the road for voiceband modem standardization.[†] They started to look around to see what new modem standards they could develop. They found

---

* John Brownlie, the father of V.8, viewed this "rescue" more as "attempted murder" of his baby.
† Little did they know that the network model was about to change with the advent of the pulse-coded modulation (PCM) modem, resulting in the channel capacity increasing toward 64 kbit/s.

some being developed in regional standards bodies like ANSI T1E1.4 and ETSI TM6. In March 1996 (the last meeting of ITU-T SG14 for the 1993–1996 study period), the United States proposed a new question for the next study period of SG14 entitled "DCEs for high speed digital access," which included the study of HDSL, ADSL, and VDSL [USA 1996]. This new question was approved by the WTSA in 1996 but was assigned to ITU-T SG15, where it became the new Question 4 (SG15/Q4).

As standardization work on voiceband modems started to wind down, many of the players began to participate in SG15/Q4. Naturally, they brought with them much of the experience garnered during the development of the voiceband modem Recommendations. This included the concept of Handshake between different types of modems. Initially, Handshake was not welcomed (ETSI TM6 and ATIS T1E1.4 saw nothing wrong with their primitive tone-based negotiation), but as time passed Handshake became accepted by the DSL community and grew strong and healthy.* In SG15/Q4, Handshake was given a new nickname, "G.hs." G.hs carried forward many characteristics from its early days as Recommendation V.8 *bis*, including some of the basic messages and transactions, along with the hierarchical tree structure for code points. (The details will be described in more detail in the following sections.) The first version of G.hs was approved as Recommendation G.994.1 in June of 1999 to support G.992.1 (ADSL), G.992.2 (ADSL lite), and G.991.2 (SHDSL). Means to support legacy ANSI T1.413 ADSL were included in an appendix.

G.994.1 is an evolving Recommendation, being revised as each DSL Recommendation is developed and approved. The latest revision of G.994.1 was approved in 2007 [G.994.1]. This is an integrated version incorporating all of the amendments up to and including Amendment 5 (the latest amendment as of the writing of this chapter), which supports VDSL2 Amendment 1, G.992.3 Amendment 3 and G.992.5 Amendment 3.

17.6.2   Overview of G.994.1

Recommendation G.994.1, "Handshake procedures for Digital Subscriber Line (DSL) transceivers" by name, provides a flexible mechanism for DSL transceivers to exchange capabilities and to select a common mode of operation prior to the transmission of initialization signals that are specific to a particular DSL Recommendation. It includes parameters relating to service and application requirements as well as parameters pertinent to various DSL transceivers. Recommendation G.994.1 is the first phase of the start-up procedure for almost all ITU-T DSL transceiver Recommendations.

---

* Some might argue that Handshake has become quite obese, with a girth far exceeding that of any of the other DSL Recommendations.

G.994.1 has two principal components:

1. A physical layer (signals, modulation, and initialization or cleardown procedures) and
2. A protocol layer (messages and transactions)

These components, which for the most part are independent of each other, will now be described in more detail.

The following description is intended to give the reader a basic understanding of Recommendation G.994.1. It is not a complete treatise on the subject. For example, the retransmission mechanism (in case of errors in reception) and message segmentation are not described. For additional details, the reader is encouraged to read the Recommendation.

17.6.3 Detailed Description of G.994.1

17.6.3.1 Signals, Modulation, and Initialization Procedures

G.994.1 defines sets of carriers that are modulated to transport the protocol's various messages. There are two signaling families, one based on $N \times 4$ kHz carrier spacing (to support G.991.2), and one based on $N \times 4.3125$ kHz carrier spacing (to support all other DSL Recommendations), where N is a positive integer.

Within the 4.3125 kHz signaling family, each carrier set consists of either two or three carriers both upstream and downstream, with the carriers for each transmission direction chosen to be far enough apart to work well in the presence of bridged taps that may create narrow band notches. The 4 kHz signaling family currently only defines a single carrier set with a single carrier frequency for each transmission direction.

Each DSL mode of operation has a mandatory carrier set associated with it, but use of as many carriers as possible is encouraged. To maximize robustness, all transmitted carrier frequencies are simultaneously modulated with the same data bits using Differentially encoded binary phase shift keying (DPSK) with a symbol rate of either $4312.5/8 = 539.0625$ symbols per second or $4000/5 = 800$ symbols per second, depending on the signaling family.

Both duplex and half-duplex transmission modes are defined for use within the G.994.1 Recommendation. The transmission mode supported is a function of the carrier set. Currently, duplex mode is used with all carrier sets within the 4.3125 kHz signaling family, and half-duplex mode is only used with the carrier set defined for the 4 kHz signaling family.

Table 17.15 shows the carrier sets that have been defined for the 4.3125 kHz signaling family, and Table 17.16 shows the mandatory carrier set for each of the associated DSL Recommendations.

Table 17.15   Carrier Sets for the 4.3125 kHz Signaling Family

| Carrier Set Designation | Upstream Carrier Sets | | Downstream Carrier Sets | | Transmission Mode |
	Frequency Indices (*N*)	Maximum Power Level/ Carrier (dBm)	Frequency Indices (*N*)	Maximum Power Level/ Carrier (dBm)	
A43	9 17 25	−1.65	40 56 64	−3.65	Duplex only
A43c	9 17 25	−1.65	257 293 337	−3.65	Duplex only
B43	37 45 53	−1.65	72 88 96	−3.65	Duplex only
B43c	37 45 53	−1.65	257 293 337	−3.65	Duplex only
C43	7 9	−1.65	12 14 64	−3.65	Duplex only
J43	9 17 25	−1.65	72 88 96	−3.65	Duplex only
V43	944 972 999	−16.65	257 383 511	−3.65	Duplex only
V43P	9 17 25	−1.65	257 383 511	−3.65	Duplex only
V43I	37 45 53	−1.65	257 383 511	−3.65	Duplex only
V43-S	944 999	−16.65	257 383	−3.65	Duplex only
V43P-S	17 25	−1.65	257 383	−3.65	Duplex only
V43I-S	45 53	−1.65	257 383	−3.65	Duplex only

Either the handshake transceiver unit at the customer end of the loop (HSTU-R) or the one at the service provider end of the loop (HSTU-C) may initiate a handshake session.

Figure 17.89 illustrates the timing for the duplex start-up procedure initiated by an HSTU-R. A similar procedure is defined for HSTU-C initiated duplex start-up, and procedures are defined for both HSTU-R and HSTU-C initiated half-duplex start-up. Readers interested in the details of these protocols are referred to [G.994.1].

Initially, the HSTU-R is in state R-Silent0 (not transmitting), and the HSTU-C is in state C-Silent1 (not transmitting). The HSTU-R initiates the start-up procedure by transmitting signals from one or both of its signaling families, with phase reversals every 16 ms (R-Tones-Req). When this has been detected by the HSTU-C, it responds by transmitting signals from one or both of its signaling families (C-Tones). When this has been detected by the HSTU-R, it remains silent (R-Silent1) for 50 to 500 ms and then transmits signals from only one signaling family (R-Tone1). When the HSTU-C has detected R-Tone1, it responds by transmitting Galfs* on modulated carriers (C-Galf1). When the HSTU-R has detected Galfs, it responds by transmitting Flags on modulated carriers (R-Flag1). When the HSTU-C has detected Flags, it responds by transmitting Flags (C-Flag1). When the HSTU-R has detected Flags, it begins the first transaction.

---

* A galf is an octet of value 81_{16}, i.e., the ones complement of an HDLC flag. "GALF" is "FLAG" spelled backward.

Table 17.16 Mandatory Carrier Sets

xDSL Recommendation(s)	Carrier Set Designation
G.992.1 – Annex A, G.992.2 – Annex A/B, G.992.3 – Annex A/I/L, G.992.4 – Annex A/I G.992.5 – Annex A/I G.993.2 where support of a profile requiring US0 (Note 4)	A43
G.992.5 – Annex A/I (Note 1) G.992.5 – Annex J/M (Note 2) G.993.2 where support of a profile requiring US0 (Note 1 and 4)	A43c
G.992.1 – Annex B, G.992.3 – Annex B G.992.5 – Annex B G.993.2 where support of a profile requiring US0 (Note 4)	B43
G.992.5 – Annex B (Note 3)	B43c
G.992.1 – Annex C/H/I, G.992.2 – Annex C G.992.3 – Annex C, G.992.5 – Annex C	C43
G.992.3 – Annex J/M, G.992.5 – Annex J/M	J43
G.993.1 – Using multi-carrier modulation (except Annex C) G.993.2 where support of a profile not requiring US0	V43
G.993.1 – Annex C using multi-carrier modulation, over POTS	V43P
G.993.1 – Annex C using multi-carrier modulation, over ISDN-BA	V43I
G.993.1 – Using single-carrier modulation, over POTS	V43P-S
G.993.1 – Using single-carrier modulation, over ISDN-BA	V43I-S
G.993.1 – Using single-carrier modulation, over TCM-ISDN	V43-S

Note 1: To be used where spectrum management forbids use of the downstream carrier set A43, typically where G.992.5 or G.993.2 is deployed from a cabinet.

Note 2: To be used where spectrum management forbids use of the downstream carrier set J43, typically where G.992.5 is deployed from a cabinet.

Note 3: To be used where spectrum management forbids use of the downstream carrier set B43, typically where G.992.5 is deployed from a cabinet.

Note 4: At least one of the tone sets A43 and B43 shall be transmitted, depending on the US0 band supported.

Note 5: If multimode operation is supported, the HSTU shall transmit the carrier sets corresponding to all enabled modes simultaneously.

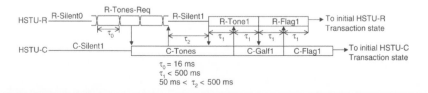

Figure 17.89   HSTU-R (handshake transceiver unit at the customer end of the loop) initiated duplex start-up procedure.

17.6.3.2 *Protocol*

G.994.1 defines several types of messages that are incorporated into a defined set of transactions.

17.6.3.2.1 Messages

Table 17.17 shows the messages defined in G.994.1.

17.6.3.2.2 Message Structure

G.994.1 messages are encapsulated in an HDLC-like frame structure as depicted in Figure 17.90.

17.6.3.2.3 Message Coding Format

The message information field consists of three components: an identification field (I) followed by a standard information field (S) and an optional nonstandard information field (NS). This general structure is shown in Figure 17.91.

17.6.3.2.4 Identification Field (I)

The identification field consists of four components: a one-octet message type field followed by a one-octet revision number field, an additional information field and a bit-encoded parameter field. This general structure is shown in Figure 17.92.

The purpose of the message type field is to identify the message type (i.e., MS, CL, etc.) of the frame. The field is one octet in length and occupies the first octet in the identification field.

The revision number field identifies the revision number of G.994.1 to which the equipment conforms. The field is one octet in length. The revision number of G.994.1 is updated when a change is made to the core content of the general handshake protocol. To ensure backward compatibility, newer revisions are proper supersets (include all aspects, e.g., message types, transactions, etc.) of previous revisions. The revision number of G.994.1 is not updated with the addition of new parameters (i.e., code points).

The encoding of the message type field is shown in Table 17.18, along with the revision of G.994.1 that supports each one. As can be seen, the MP message was added to revision 2 of G.994.1, and the REQ-RTX message was added to revision 3 to support the retransmission mechanism.

The encoding of the revision number field is shown in Table 17.19. As the table indicates, at the time of writing there were three revisions of G.hs.

The additional information field consists of either a vendor ID information block or a retransmission information block. See [G.994.1] for the encoding of this block.

Table 17.17   G.994.1 Messages

Message	Description
CLR	Capabilities List + Request: Sent by an HSTU-R, conveys a list of possible modes of operation of the xTU-R and requests the transmission of a CL message by the HSTU-C
CL	Capabilities List: Sent by an HSTU-C in response to the reception of either a complete CLR message, or an intermediate frame of a segmented CLR message [G.994.1]
MR	Mode Request: Sent by an HSTU-R, requests the transmission of an MS message by the HSTU-C
MS	Mode Select: Sent by an HSTU-C or an HSTU-R, requests the initiation of a particular mode of operation
MP	Mode Proposal: Sent by an HSTU-R, proposes a particular mode of operation and requests the transmission of an MS message by the HSTU-C
ACK(1)	Acknowledge, Type 1: Either (1) acknowledges receipt of a complete CL message or an intermediate frame of a segmented CL message [G.994.1] and ends a G.994.1 transaction, or (2) acknowledges receipt of a complete MS message or an intermediate frame of a segmented MS message [G.994.1] and initiates the G.994.1 session cleardown procedure (see Section 17.6.3.3)
ACK(2)	Acknowledge, Type 2: Acknowledges receipt of an intermediate frame of a segmented CL, CLR, MP, or MS message and requests the transmission of the next frame of the message. See [G.994.1] for details on message segmentation
NAK-EF	Negative Acknowledge, Errored Frame: See [G.994.1] for details
NAK-NR	Negative Acknowledge, Not Ready: See [G.994.1] for details
NAK-NS	Negative Acknowledge, Not Supported: See [G.994.1] for details
NAK-CD	Negative Acknowledge, Clear Down: See [G.994.1] for details
REQ-MS	Request MS Message: Sent by an HSTU-C in response to the reception of an MR message, requests the transmission of an MS message by the HSTU-R. It indicates that the HSTU-C does not wish to select a mode and is deferring the mode selection to the HSTU-R
REQ-MR	Request MR Message: Sent by an HSTU-C in response to the reception of an MS message, requests the transmission of an MR message by the HSTU-R. It indicates that the HSTU-C wishes to select the mode
REQ-CLR	Request CLR Message: Sent by an HSTU-C in response to the reception of either an MR, MS, or MP message, requests the transmission of a CLR message by the HSTU-R. It indicates that the HSTU-C wishes to perform a capabilities exchange
REQ-RTX	Retransmission Message: See [G.994.1] for details

8	7	6	5	4	3	2	1	Octet
Flag								1
Flag								2
Flag								
Flag (optional)								
Flag (optional)								
Message information field								
FCS (first octet)								
FCS (second octet)								
Flag								N−2
Flag								N−1
Flag (optional)								N

Figure 17.90   Frame structure.

The bit-encoded parameter field contains parameters that are independent of the mode to be selected and are typically either service or application related. Examples of these parameters are data rate and latency requirements, splitter information, and relative power level "per" carrier for the various G.994.1 carrier sets. This field is encoded in accordance with the rules for the hierarchical tree structure described below.

17.6.3.2.5   Standard Information (S) Field

In the standard information field, the parameters represent modes of working or capabilities relating to the xTU-R or xTU-C. The standard information field of CL, CLR, MP, and MS messages is encoded in accordance with the rules for the hierarchical tree structure described below. For MR, ACK, NAK, and REQ messages, the standard information field is not used and is therefore of zero length.

The standard information field consists of a set of octets in which each capability is either assigned a unique bit position, where a binary ONE in the assigned bit position indicates that the capability is valid, or a value, which may span more than one octet.

For CL and CLR messages, the validity of multiple capabilities may be conveyed by transmitting a binary ONE in each bit position corresponding to a valid capability (or each field corresponding to a range of values). For

Identification (I) field	Standard information (S) field	Nonstandard information (NS) field

Figure 17.91   Information field structure.

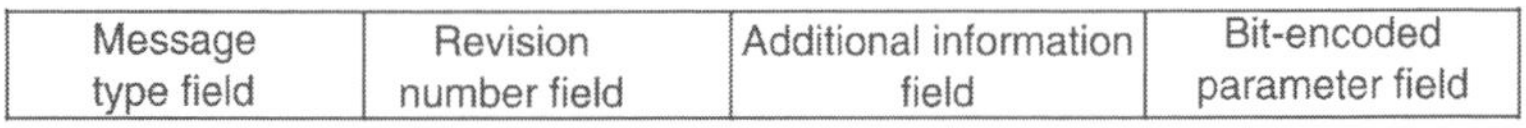

Message type field	Revision number field	Additional information field	Bit-encoded parameter field

Figure 17.92   Identification field structure.

messages MP and MS, multiple capabilities may be selected only if they can all be supported simultaneously within the xTU concerned.

Support for a specific DSL Recommendation is normally indicated at the highest level in the tree, with further details in the underlying branches. There are literally thousands of parameters currently defined in G.994.1. Needless to say, they are not described here. However, to give the reader a flavor for them, a small sampling is provided in the sequel.

17.6.3.2.6   Nonstandard (NS) Information Field

MP, MS, CL, and CLR messages may optionally contain a nonstandard information field to convey information beyond that defined in the Recommendation. See [G.994.1] for further details.

17.6.3.2.7   Hierarchical Tree Structure

In both the I and S fields, most of the information to be conveyed consists of parameters relating to particular modes, features, or capabilities associated with the two transceivers.

Table 17.18   Message Type Field Format

8	7	6	5	4	3	2	1	Message Type	G.994.1 Revision 1 Support	G.994.1 Revision 2 Support	G.994.1 Revision 3 Support
0	0	0	0	0	0	0	0	MS	X	X	X
0	0	0	0	0	0	0	1	MR	X	X	X
0	0	0	0	0	0	1	0	CL	X	X	X
0	0	0	0	0	0	1	1	CLR	X	X	X
0	0	0	0	0	1	0	0	MP	–	X	X
0	0	0	1	0	0	0	0	ACK(1)	X	X	X
0	0	0	1	0	0	0	1	ACK(2)	X	X	X
0	0	1	0	0	0	0	0	NAK-EF	X	X	X
0	0	1	0	0	0	0	1	NAK-NR	X	X	X
0	0	1	0	0	0	1	0	NAK-NS	X	X	X
0	0	1	0	0	0	1	1	NAK-CD	X	X	X
0	0	1	1	0	1	0	0	REQ-MS	X	X	X
0	0	1	1	0	1	0	1	REQ-MR	X	X	X
0	0	1	1	0	1	1	1	REQ-CLR	X	X	X
0	0	1	1	1	0	0	0	REQ-RTX	–	–	X

Table 17.19   Revision Number Field Format

Bits								Revision Number
8	7	6	5	4	3	2	1	
0	0	0	0	0	0	0	1	1
0	0	0	0	0	0	1	0	2
0	0	0	0	0	0	1	1	3

To encode these parameters in accordance with a consistent set of rules, and allow future extension to the parameter list is a way that permits present and future G.994.1 implementations to parse the information field correctly, the parameters are linked together in a predefined tree structure. The order in which the parameters in the tree are transmitted and the use of delimiting bits that enable the tree to be reconstructed at the receiver are described in the rules set out below.

Parameters (Pars) are classified as NPars, which are parameters that have no subparameters associated with them, and SPars, which are parameters that have subparameters associated with them. The general structure of this tree is as shown in Figure 17.93.

At level 1, the highest level of the tree, each SPar has associated with it a series of Pars (NPars and possibly SPars) at level 2 in the tree. At level 2 in the tree, each SPar has associated with it a series of NPars at level 3 in the tree. Level 3 is the lowest level in the tree. Therefore, there are no SPars at this level.

The order of transmission of NPars and SPars is shown in Figure 17.94. Transmission of parameters begins with the first octet of NPar(1) and ends with the last octet of $Par(2)_N$.

$Par(2)_n$ indicates a set of level 2 parameters associated with the nth level 1 SPar, and consists of $NPar(2)_n$ parameters and possibly $SPar(2)_n$

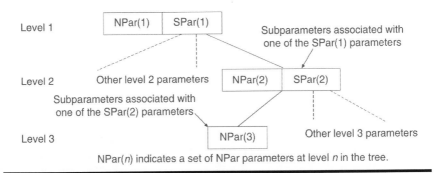

Figure 17.93   Tree structure linking parameters in the identification (I) and standard information (S) fields.

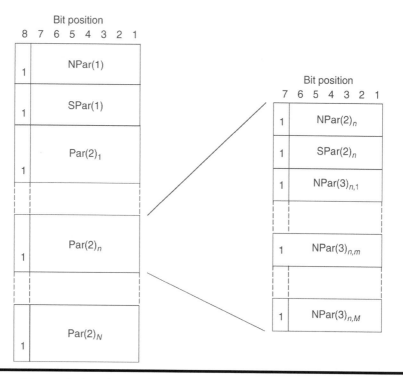

Figure 17.94   Order of transmission of NPars and SPars.

parameters. NPar(3)$_{n,m}$ indicates a set of level 3 NPars associated with the mth level 2 SPar which in turn is associated with the nth level 1 SPar.

The order of transmission of the Par(2) blocks is the same as the order of transmission of the corresponding SPar(1) bits. Similarly, the order of transmission of the NPar(3)$_n$ blocks is the same as the order of transmission of the corresponding SPar(2)$_n$ bits.

The use of delimiting bits is illustrated in Figure 17.94. Within each octet of a parameter block, at least one bit is defined as a delimiting bit. This is used to define the last octet in the block to be transmitted. A binary ZERO in this bit position indicates that there is at least one additional octet in the block to be transmitted. A binary ONE in this bit position indicates it is the last octet in the block to be transmitted.

Bit 8 is used to delimit the NPar(1) block, the SPar(1) block, and each of the Par(2) blocks. There are N of these Par(2) blocks, one for each of the capabilities in the SPar(1) block that is enabled (set to binary ONE).

For this parsing rule to function correctly, both the identification field (I) and the standard information field (S) must include at least one octet of NPar(1) and at least one octet of SPar(1).

Bit 7 is used to delimit each NPar(2) block, each SPar(2) block, and each of the associated NPar(3) blocks. Figure 17.94 indicates that there are M of these NPar(3) blocks, one for each of the capabilities in the SPar(2)$_n$ block that is enabled (set to binary ONE). M may be different for each of the Par(2) blocks.

A Par(2) block may either contain both NPar(2) and SPar(2) octets, or NPar(2) octets alone. To indicate that a Par(2) block contains only NPar(2) octets, bits 7 and 8 are both set to binary ONE in the last NPar(2) octet to be transmitted.

Bits 1 through 7 in each octet at level 1 of the tree and bits 1 through 6 in each octet at levels 2 and 3 of the tree may be used to encode parameters.

To allow for compatibility with future revisions of the G.994.1 Recommendation, receivers parse all parameter blocks and ignore any information that is not understood. However, to be able to correctly parse the parameter blocks, it is necessary to pay attention to the number of SPar(1) and SPar(2) bits that are set, even if the meaning of one or more of these bits is not understood.

17.6.3.2.8 Example Message Content For ADSL1

Tables 17.20 through 17.27 provide the octets (but not specific bit settings) for a single branch of the code tree that might be used for indicating some of the capabilities of a G.992.1 Annex A modem. The order of transmission is in ascending numerical order of the G.994.1 table numbering, which is provided in the table captions.

17.6.3.2.9 Transactions

Basic transactions may be classified as one of two types: those that exchange and negotiate capabilities between the HSTU-C and the HSTU-R, and those that select a mode of operation.

Table 17.28 shows the set of basic transactions specified in G.994.1. An "X" indicates that the basic transaction is supported for the stated revision number, and a dash indicates that it is not. Each transaction is initiated by the HSTU-R, and ends with an ACK(1). In basic transactions, the HSTU-R controls the negotiation procedure. At the end of a basic transaction, the transceivers either terminate the G.994.1 session (applies to transactions A, B, C, and D), or go to the Initial HSTU-x Transaction state (applies to transaction C only).

In transaction A, the HSTU-R selects a mode of operation and requests that the HSTU-C transition to the selected mode. When the HSTU-C responds with an ACK(1) message, both stations transition to the selected mode.

Table 17.20   Table 10 of G.994.1—Standard Information Field—NPar(1) Coding

			Bits					NPar(1)s
8	7	6	5	4	3	2	1	
x	x	x	x	x	x	x	1	Voiceband: V.8 (Note 1)
x	x	x	x	x	x	1	x	Voiceband: V.8 *bis* (Note 1)
x	x	x	x	x	1	x	x	Silent period (Note 2)
x	x	x	x	1	x	x	x	G.997.1 (Note 3)
x	x	x	1	x	x	x	x	Reserved for allocation by the ITU-T
x	x	1	x	x	x	x	x	Reserved for allocation by the ITU-T
x	1	x	x	x	x	x	x	Reserved for allocation by the ITU-T
x	0	0	0	0	0	0	0	No parameters in this octet

Note 1: Setting this bit to binary ONE in an MS message initiates the G.994.1 session cleardown procedure, and requests a V.8 or V.8 *bis* handshake in the voiceband, with the xTU-R taking on the role of a calling station and the xTU-C taking on the role of an answering station.

Note 2: This bit shall be set to binary ONE in a CLR or CL message. Setting this bit to binary ONE in an MS message initiates the G.994.1 session cleardown procedure, and requests a silence period at the other transmitter of approximately 1 min. The station that invoked the silent period by transmitting MS may terminate the silent period prior to the 1 min by restarting a G.994.1 session.

Note 3: The use of this bit is for further study and shall be set to binary ZERO in CLR, CL, and MS.

Table 17.21   Table 11 of G.994.1—Standard Information Field—SPar(1) Coding—Octet 1

			Bits					SPar(1)s—Octet 1
8	7	6	5	4	3	2	1	
x	x	x	x	x	x	x	1	G.992.1—Annex A (Note 1)
x	x	x	x	x	x	1	x	G.992.1—Annex B (Note 1)
x	x	x	x	x	1	x	x	G.992.1—Annex C (Note 1)
x	x	x	x	1	x	x	x	G.992.2—Annex A/B (Note 1)
x	x	x	1	x	x	x	x	G.992.2—Annex C (Note 1)
x	x	1	x	x	x	x	x	G.992.1—Annex H (Note 1)
x	1	x	x	x	x	x	x	G.992.1—Annex I (Note 1)
x	0	0	0	0	0	0	0	No parameters in this octet

Note 1: The spectrum information indicated in the NPar(3) fields associated with these Recommendations is of informative nature and does not imply any requirements on the transmit spectrum used during initialization and data mode. Regardless of the spectrum information, the transmit spectrum shall comply with their respective Recommendations. Spectrum information may only be included in a CLR or CL message, not in an MP or MS message. The spectrum information is coded in eight bits (across two octets) as a binary representation of the sub-carrier index.
Maximum frequencies: up to and including the sub-carrier index.

Table 17.22   Table 11.1 of G.994.1—Standard Information Field—G.992.1 Annex A NPar(2) Coding

			Bits					
8	7	6	5	4	3	2	1	G.992.1 Annex A NPar(2)s
x	x	x	x	x	x	x	1	R-ACK1
x	x	x	x	x	x	1	x	R-ACK2
x	x	x	x	x	1	x	x	Reserved for allocation by the ITU-T
x	x	x	x	1	x	x	x	STM
x	x	x	1	x	x	x	x	ATM
x	x	1	x	x	x	x	x	G.997.1—clear EOC OAM
x	x	0	0	0	0	0	0	No parameters in this octet

Table 17.23   Table 11.2 of G.994.1—Standard Information Field–G.992.1 Annex A SPar(2) Coding

			Bits					
8	7	6	5	4	3	2	1	G.992.1 Annex A SPar(2)s
x	x	x	x	x	x	x	1	Sub-channel information
x	x	x	x	x	x	1	x	Spectrum frequency upstream
x	x	x	x	x	1	x	x	Spectrum frequency downstream
x	x	x	x	1	x	x	x	Reserved for allocation by the ITU-T
x	x	x	1	x	x	x	x	Reserved for allocation by the ITU-T
x	x	1	x	x	x	x	x	Reserved for allocation by the ITU-T
x	x	0	0	0	0	0	0	No parameters in this octet

Table 17.24   Table 11.2.2 of G.994.1—Standard Information Field—G.992.1 Annex A—Spectrum Frequency Upstream NPar(3) Coding—Octet 1

			Bits					G.992.1 Annex A Spectrum Frequency
8	7	6	5	4	3	2	1	Upstream NPar(3)s—Octet 1
x	x	0	0	0	0	x	x	Spectrum minimum frequency upstream (bits 7 and 8)

Table 17.25   Table 11.2.2.1 of G.994.1—Standard Information Field—G.992.1 Annex A, Spectrum Frequency Upstream NPar(3) Coding—Octet 2

			Bits					G.992.1 Annex A Spectrum Frequency
8	7	6	5	4	3	2	1	Upstream NPar(3)s—Octet 2
x	x	x	x	x	x	x	x	Spectrum minimum frequency upstream (bits 1 to 6)

Table 17.26 Table 11.2.2.2 of G.994.1—Standard Information Field—G.992.1 Annex A, Spectrum Frequency Upstream NPar(3) Coding—Octet 3

Bits								G.992.1 Annex A Spectrum Frequency upstream NPar(3)s—Octet 3
8	7	6	5	4	3	2	1	
x	x	0	0	0	0	x	x	Spectrum maximum frequency upstream (bits 7 and 8)

Table 17.27 Table 11.2.2.3 of G.994.1—Standard Information Field—G.992.1 Annex A, Spectrum Frequency Upstream NPar(3) coding—Octet 4

Bits								G.992.1 Annex A Spectrum Frequency upstream NPar(3)s—Octet 4
8	7	6	5	4	3	2	1	
x	x	x	x	x	x	x	x	Spectrum maximum frequency upstream (bits 1 to 6)

Table 17.28 Basic G.994.1 Transactions

Transaction Identifier	HSTU-R	HSTU-C	HSTU-R	G.994.1 Revision 1 Support	G.994.1 Revision 2 and 3 Support
A	MS →	ACK(1)		x	x
B	MR →	MS →	ACK(1)	x	x
C	CLR →	CL →	ACK(1)	x	x
D	MP →	MS →	ACK(1)	–	x

In transaction B, the HSTU-R requests that the HSTU-C selects the mode of operation. The HSTU-C selects the mode by transmitting an MS message. When the HSTU-R responds with an ACK(1) message, both stations transition to the selected mode.

In transaction C, capabilities are exchanged and negotiated by the two stations. Transaction C is normally followed by either transaction A, B, or D during the same session to select a common mode of operation identified during the capabilities exchange.

In transaction D, the HSTU-R proposes a mode of operation and requests that the HSTU-C selects the mode of operation. The HSTU-C selects the mode by transmitting an MS message. When the HSTU-R responds with an ACK(1) message, both stations transition to the selected mode.

G.994.1 also defines a set of extended transactions, as shown in Table 17.29. Extended transactions are derived from a concatenation of two

Table 17.29   Extended G.994.1 Transactions

Trans-action Identifier	HSTU-R	HSTU-C	HSTU-R	HSTU-C	HSTU-R	G.994.1 Revision 1 Support	G.994.1 Revision 2 and 3 Support
A:B	MS→	REQ-MR→	MR→	MS→	ACK(1)	×	×
B:A	MR→	REQ-MS→	MS→	ACK(1)		×	×
A:C	MS→	REQ-CLR→	CLR→	CL→	ACK(1)	×	×
B:C	MR→	REQ-CLR→	CLR→	CL→	ACK(1)	×	×
D:C	MP→	REQ-CLR→	CLR→	CL→	ACK(1)	–	×

basic transactions. They are used when the HSTU-C wishes to control the negotiation procedure. At the end of an extended transaction, the stations either terminate the G.994.1 session or go to the Initial HSTU-x Transaction state.

In transaction A:B, the HSTU-R selects a mode of operation and requests that the HSTU-C transition to the selected mode. However, rather than responding to the MS message with an ACK(1) message as is the case for basic transaction A, the HSTU-C responds to the MS message with a REQ-MR message requesting the HSTU-R to proceed directly into basic transaction B without returning to the Initial Transaction state.

In transaction B:A, the HSTU-R requests that the HSTU-C selects the mode of operation. However, rather than responding to the MR message with an MS message as is the case for basic transaction B, the HSTU-C responds to the MR message with a REQ-MS message requesting the HSTU-R to proceed directly into basic transaction A without returning to the Initial Transaction state.

In transaction A:C, the HSTU-R selects a mode of operation and requests that the HSTU-C transition to the selected mode. However, rather than responding to the MS message with an ACK(1) message as is the case for basic transaction A, the HSTU-C responds to the MS message with a REQ-CLR message requesting the HSTU-R to proceed directly into basic transaction C without returning to the Initial Transaction state.

In transaction B:C, the HSTU-R requests that the HSTU-C selects the mode of operation. However, rather than responding to the MR message with an MS message as is the case for basic transaction B, the HSTU-C responds to the MR message with a REQ-CLR message requesting the HSTU-R to proceed directly into a transaction C without returning to the Initial Transaction state.

In transaction D:C, the HSTU-R proposes a mode of operation and requests that the HSTU-C selects the mode of operation. However, rather than responding to the MP message with an MS message as is the case for basic transaction D, the HSTU-C responds to the MP message with a

REQ-CLR message requesting the HSTU-R to proceed directly into basic transaction C without returning to the Initial Transaction state.

G.994.1 also defines Retransmission transactions for use when an HSTU receives an errored frame. See Recommendation G.994.1 [G.994.1] for further information on this mechanism.

17.6.3.3 Cleardown Procedures

After all transactions have been completed, a cleardown procedure follows. Figure 17.95 illustrates the timing for cleardown of a duplex session (by either the HSTU-R or the HSTU-C).

When an HSTU-R (HSTU-C) receives the last message of the last transaction of a session, usually an ACK(1) acknowledgment message, it initiates the cleardown procedure. After receiving the last message, the HSTU-R (HSTU-C) continues to transmit Flags for a period of ≤ 0.5 s. It then transmits four octets of Galf (referred to as R-GALF2 for an HSTU-R, C-GALF2 for an HSTU-C) and then stops transmitting, which terminates the G.994.1 session. When the HSTU-C (HSTU-R) detects either Galfs or silence, it continues to transmit Flags (referred to as C-FLAG2 for an HSTU-C, R-FLAG2 for an HSTU-R) for a period of ≤ 0.5 s and then stops transmitting, which terminates the G.994.1 session.

If a received MS message indicates a common operating mode, both stations transition to the selected mode upon termination of the G.994.1 session, and initialization continues using the procedures defined in the applicable DSL Recommendation.

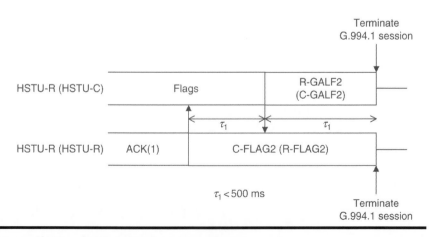

Figure 17.95   Duplex cleardown procedure.

17.7   Annex: History of Standardization

One of the earliest descriptions of standardization of digital transmission systems is recorded in the *Histories of Polybius* [Polybius 1922]. Polybius was born in Greece in about 208 BC, and the project of writing a history of the age probably suggested itself when he was detained in Italy for alleged opposition to the sovereignty of Rome. His assistance in the diplomatic discussions that preceded the last Punic War was appreciated by the Romans, whom he advised on questions of siege operations, exploration, and administration.

In Book X of his Histories (47:12), he emphasized the importance of standardization: "In offering these observations I am acting up to the promise I originally made at the outset of this work. For I stated that in our time all arts and sciences have so much advanced the knowledge of most of them may be said to have been reduced to a system. This is, then, one of the most useful parts of a history properly written." Earlier in the same book (45:6), he describes an early symmetric digital transmission method, which relies on an optical transmission medium. However, the methods of data compression, trellis coding, and the two-dimensional constellation scheme described could be considered a precursor of the methods used in DSL technology more than 2000 years later. Polybius wrote, "The most recent method, devised by Cleoxenus and Democleitus and perfected by myself, is quite definite and capable of dispatching with accuracy every kind of urgent messages, but in practice it requires care and exact attention. It is as follows: We take the alphabet and divide it into five parts, each consisting of five letters. There is one letter less in the last division, but this makes no practical difference. Each of the two parties who are about signal to each other must now get ready five tablets and write one division of the alphabet on each tablet, and then come to an agreement that the man who is going to signal is in the first place to raise two torches and wait until the other replies by doing the same. This is for the purpose of conveying to each other that they are both at attention. These torches having been lowered the dispatcher of the message will now raise the first set of torches on the left side indicating which tablet is to be consulted, i.e., one torch if it is the first, two if it is the second, and so on. Next he will raise the second set on the right on the same principle to indicate what letter of the tablet the receiver should write down. Upon their separating after coming to this understanding each of them must first have on the spot two tubes, so that with the one he can observe the space on the right of the man who is going to signal back and with the other that on the left. The tablets must be set straight up in order next the tubes, and there must be a screen before both spaces, as well the right as the left, ten feet in length and of the height of a man so that by this means the torches may be seen distinctly when raised and disappear when lowered. When all has been thus got ready on both sides, if the signaler wants to convey, for instance, that about a hundred of the soldiers have deserted to the enemy, he must first of all choose words which will convey what he means in the smallest number of letters, e.g., instead of the above 'Cretans a hundred deserted us,' for thus the letters are less than one half in number, but the same sense is conveyed. Having jotted this down on a writing-tablet he will communicate it by the torches as follows: The first letter is kappa. This being in the second division is on tablet number two, and, therefore, he must raise two torches on the left, so that the receiver may know that he had to consult the second tablet. He will now raise five torches on the right, to indicate that it is kappa, this being the fifth letter in the second division, and the receiver of the signal will note this down on his writing tablet. The dispatcher will then raise four torches on the left as rho belongs to the fourth division, and then two on the right, rho being the second letter in this division. The receiver writes down rho and so forth."

Nearly 2000 years after Polybius, in 1753, an unknown Scottish inventor, writing in the *Scots' Magazine* and identified by the initials CM, proposed a telegraph system using 26 wires, each representing a letter of the alphabet, which could carry an electric charge and attract a piece of paper with the appropriate letter to an electrified ball at the receiver. Fifty years later, in 1804, Francisco Salva demonstrated a system based on the principles outlined by CM. The system used 26 wires that were immersed in acidified water. Electric current passing through a wire would result in a stream of bubbles. A specific letter was indicated by bubbles emerging from the wire associated with that letter.

As the speed of progress accelerated, Baron Schilling developed the first working electromagnetic telegraph in St. Petersburg in 1830, based on Oersted's electromagnetic ideas. Signals were sent between one and six wires, causing movement in compass needles suspended over coils that indicated the letters. Czar Nicholas I saw this demonstration of the first telegraph system and saw in it an instrument of subversion, and therefore forbade publication of it by the press or in the scientific literature, resulting in Russia being the last major country to build a telegraph system (in 1853).

In 1835, Joseph Henry demonstrated, as had Faraday, that magnetism and electricity can be converted into each other. Henry went on to invent the first practical electric motor, the electric relay, and a simple early version of the telegraph. However, he did not patent any of his inventions, allegedly saying, "I did not then find it compatible with the dignity of science to confine benefits which might be derived from it to the exclusive use of any one individual."

Meanwhile, Sir William Fothergill Cooke, a medical student, attended a lecture in Heidelberg, Germany, at which one of Baron Schilling's electric telegraphs was demonstrated. He at once abandoned his anatomical studies and applied his energies to inventing a practical electric telegraph, joining forces with Sir Charles Wheatstone, a Cambridge physics professor. Together they worked out a telegraph system in which feeble currents of electricity deflected needles that point to letters of the alphabet laid out on a grid. Their invention involved 5 magnetic needles and a lattice grid of 20 letters (C, J, Q, U, X, and Z were omitted). Five wires are needed to send the signals, which deflect two of the five needles at the other end that point to the appropriate letter. One needle was used to point to a numeral. (They eventually refined their unit to use only one needle and a signal code.) On July 10, 1837, Wheatstone and Cooke patented the first British electric telegraph and became very rich men.

In the same year, the American inventor Samuel Finley Breese Morse, who had trained as a painter, developed the first American telegraph. This telegraph transmitted simple patterns of "dots" and "dashes," which were called Morse Code, over a single wire (the duration of a "dash" is three times the duration of a "dot"). The Morse Code included data compression: for example, the letter "e," which is frequently used, is a short "dot," whereas the infrequently used letter "y" is a much longer "dash dot dash dash." Morse's system was eventually adopted as the standard technique for telegraphs, because it was easier to construct and more reliable than Cooke and Wheatstone's, illustrating the truism that telecommunication standards must be at once technically elegant and practical to implement.

By 1850, telegraph services were operating in Britain between London, Dover, Birmingham, and Edinburgh, and the first telegraph lines had been completed in Belgium, Austria, Italy, and Germany, as well as from Washington to Baltimore in the United States. It was from the Capitol building in Washington, that Morse sent his famous Biblical quotation as the first formal telegraph message on the line to Baltimore, a message that revealed his own sense of wonder in God's creation: "What God Hath Wrought!" Digital subscriber lines had arrived. However, there was to be a short interruption in the development of DSL.

On March 3, 1847, Alexander Graham Bell was born in Edinburgh, Scotland. As an adult, he moved to the United States and was appointed professor of vocal physiology at Boston University, where, in 1874, he developed the basic ideas for an analog telephone. On March 10, 1876, he transmitted the first complete sentence over the telephone: "Watson, come here; I want you." In 1878, he was to say "It is conceivable that cables of telephone wires could be laid underground or suspended overhead, communicating with private dwellings... Not only so, but I believe in the future, wires will unite different cities, and a man in one part of the country may communicate with another in a distant place." This vision came to pass, eclipsing DSLs for nearly a hundred years. Voiceband modems, introduced in the late 1950s, transmitted digital data (masquerading as analog voice) over subscriber lines. T1 lines, which were introduced in 1970, used AMI modulation and were truly digital lines; some of them connected to subscribers. It was not until 1976 that the concept of an ISDN began to gain ground, and not until the mid-1980s that ISDN systems began to be standardized. The term "digital subscriber line" (or loop) was first coined to describe the metallic connection between the subscriber and the local telephone exchange that was capable of supporting the 160 kbit/s data transmission provided by the ISDN.

17.8 ACRONYMS

17.8.1 *Standards Organizations*

ATIS	Alliance for Telecommunications Industry Solutions
CCITT	International Telegraph and Telephone Consultative Committee
CCSA	China Communications Standards Association
ECSA	Exchange Carriers Standards Association (now ATIS)
ETSI	European Telecommunications Standards Institute
IEEE	Institute of Electrical and Electronics Engineers
ITU	International Telecommunication Union
ITU-T	ITU – Telecommunications
NAI	Network Access Interfaces (working group of NIPP)
NIPP	Network, Interface, Power, and Protection (committee in ATIS)
SG15/Q4	Study Group 15, Question 4 (of ITU-T)
TTC	Telecommunication Technology Committee

17.8.2 *DSL Types*

ADSL	Asymmetric Digital Subscriber Line
HDSL	High Bit-Rate Digital Subscriber Line
ISDL	ISDN DSL
ISDN	Integrated Services Digital Network
SDSL	Single-pair (or Symmetric) Digital Subscriber Line
SHDSL	Symmetric High Bit-Rate Digital Subscriber Line
VDSL	Very High Bit-Rate Digital Subscriber Line

17.8.3 Line Codes (Modulation Methods)

CAP	Carrierless amplitude and phase modulation
DMT	Discrete multitone
DPSK	Differentially encoded binary phase shift keying
MONET	Margin optimized interlocked extended-range transmission
OPTIS	Overlapped PAM transmission with interlocked spectra
PAM	Pulse amplitude modulation
QAM	Quadrature amplitude modulation
SCM	Single-carrier modulation

17.8.4 Duplexing

EC	Echo canceled (or canceler)
FDD	Frequency-division duplexing (or duplexed)
TCM	Time-compression multiplexing (or multiplexed)
TDD	Time-division duplexing (or duplexed)

17.8.5 Others

AOC	ADSL overhead control
ATM	Asynchronous Transfer Mode
ATU	ADSL transceiver unit
AWG	American wire gauge
AWGN	Additive white Gaussian noise
BRA	Basic rate access
BRI	Basic rate interface
CCS	Common channel signaling
CLEC	Competitive local exchange carrier
CO	Central office
CRC	Cyclic redundancy check
CS	Cyclic suffix
CSA	Carrier serving area
DA	Distribution area
DFE	Decision feedback equalizer (or equalization)
DFT	Discrete Fourier transform
DRR	Dynamic rate repartitioning
DSLAM	Digital subscriber line access multiplexer
ECSA	Extended carrier serving area

EFM	Ethernet in the first mile
EMC	Electromagnetic compatibility
EOC	Embedded operations channel
FEC	Forward error correcting (or correction)
FEQ	Frequency-domain equalizer
FEXT	Far-end crosstalk
FFT	Fast Fourier transform
FSAN	Full Service Access Network
HAM	Amateur radio
HSI	High-speed internet
IDFT	Inverse discrete Fourier transform
ISDN-BA	ISDN – basic access
ISI	Intersymbol interference
LAN	Local area network
LTU	Line termination unit
MDU	Multi-dwelling unit
MTU	Multi-tenant unit
NEXT	Near-end crosstalk
NTR	Network timing reference
NTU	Network termination unit
OAM	Operations, administration, and maintenance
OLR	On-line reconfiguration
PBO	Power back-off
PCM	Pulse-coded modulation
PMD	Physical medium dependent
PMS-TC	Physical medium specific–transmission convergence
POTS	Plain old telephone service
PRA	Primary rate access
PRI	Primary rate interface
PSD	Power spectral density
SDH	Synchronous digital hierarchy
SME	Small and medium enterprise
SNR	Signal-to-noise ratio
SRA	Seamless rate adaptation
STM	Synchronous Transfer Mode
STU	Symmetrical transceiver unit
TA	Timing advance
TEQ	Time-domain equalizer
TPS	Transport protocol specific
TPS-TC	Transport protocol specific–transmission convergence
TRQ	Technical requirements
UPBO	Upstream power back-off
VLSI	Very large scale integration

References

General

[ATIS web]   *ATIS Web site.* http://www.atis.org/about.shtml.

[ETSI web]   *ETSI Web site.* http://www.etsi.org/about_etsi/5_minutes/ home.htm.

[NIPP-NAI web]   *NIPP-NAI Web site.* http://www.atis.org/0050/nai.asp.

[Polybius 1922]   *The Histories of Polybius (200-118 BCE)*, Book X. 45.1–47.13 published in the Loeb Classical Library, 1922–1927 Harvard University Press.

[SG15 web]   *Study Group 15 Web site.* http://www.itu.int/ITU-T/ studygroups/com15/area.html.

Symmetric DSL

[ATIS TR-28]   *Committee T1, A Technical Report on High-Bit-Rate Digital Subscriber Line (HDSL)*, Technical Report TR-28, February 1994.

[ATIS T1]   *Committee T1, Draft Technical Report on High-Bit-Rate Digital Subscriber Line (HDSL)*, T1E1.4/96-006, April 1996.

[Bellcore 91-86]   *Generic Requirements for High-Bit-Rate Digital Subscriber Lines*, Bellcore 9l-86 Technical Advisory TA-NWT-001210, issue 1, October 1991.

[DSL Forum TR-036]   *DSL Forum TR-036*, Requirements for Voice over DSL v1.0 August 28, 2000.

[ETSI DTS/TM-06011]   *ETSI Work Item DTS/TM-06011*, Transmission and Multiplexing (TM) Single-pair Symmetric High bitrate Digital Subscriber Line (SDSL) transmission system on metallic local lines; SDSL applications with optimized line code for combined ISDN-BA (or POTS) and rate adaptive transmission up to 2048 kbit/s.

[ETSI ETR 080]   *ETSI ETR 080*, Transmission and Multiplexing (TM); Integrated Services Digital Network (ISDN) basic rate access; Digital transmission system on metallic local lines.

[ETSI ETR 152]   *High Bit-Rate Digital Subscriber Line Transmission System on Metallic Local lines*, European Telecommunications Standards Institute, Draft ETR 152, October 29, 1993.

[ETSI STC-TM3]   *ETSI STC-TM3, ETSI Sub-Technical Committee TM3 Architecture*, Functional Requirements and Interfaces for the Transmission Network Report of Meeting No. 8 Held in Bristol, United Kingdom, October 12–16, 1992.

[ETSI TC.TM-a]   *ETSI TC/TM WG TM6(99)1, Report of Meeting No. 13* Held in Villach, Austria, February 22–26, 1999.

[ETSI TC.TM-b]   *ETSI TC/TM WG TM6(99)2, Report of Meeting No. 14* Held in Grenoble, France, May 3–7, 1999.

[ETSI TM6]   *ETSI TM6, Temporary Document 962t13*, reproduced in Annex F of ETSI TC TM WG TM6(96)12 Report of Meeting No. 2 Held in Berne, Switzerlan, September 9–13, 1996).

[ETSI TS 101 135-a (1998)] *ETSI TS 101 135 Ver. 1.4.1* (1998-02-02), High bit-rate Digital Subscriber Line (HDSL) transmission system on metallic local lines; HDSL core specification and applications for 2048 kbit/s based access digital sections.

[ETSI TS 101 135-b (1998)] *ETSI TS 101 135 Ver. 1.5.1* (1998-11-19), High bit-rate Digital Subscriber Line (HDSL) transmission systems on metallic local lines; HDSL core specification and applications for combined ISDN-BA and 2048 kbit/s transmission.

[ETSI TS 101 135 (2000)] *ETSI Deliverable: TS 101 135 Ver. 1.5.3* (2000-09-05), High bit-rate Digital Subscriber Line (HDSL) transmission systems on metallic local lines; HDSL core specification and applications for combined ISDN-BA and 2048 kbit/s transmission.

[ETSI TS 101 524] *ETSI TS 101 524 Ver. 1.2.1*, Transmission and Multiplexing (TM); Access transmission system on metallic access cables; Symmetric single pair high bitrate Digital Subscriber Line (SDSL) (2003-03-28).

[ETSI TS 101 524-1] *ETSI TS 101 524-1 Ver. 1.1.1*, Transmission and Multiplexing (TM); Access transmission system on metallic access cables; Symmetrical single pair high bitrate Digital Subscriber Line (SDSL); Part 1: Functional requirements (2000-04-04).

[ETSI TS 101 524-2] *ETSI TS 101 524-2 Ver. 1.1.1*, Transmission and Multiplexing (TM); Access transmission system on metallic access cables; Symmetrical single pair high bit rate Digital Subscriber Line (SDSL); Part 2: Transceiver requirements (2000-06-23).

[ETSI TS 102 080] *ETSI TS 102 080 Ver. 1.3.2*, Ref. Transmission and Multiplexing (TM); Integrated Services Digital Network (ISDN) basic rate access; Digital transmission system on metallic local lines publication (2000-05-31).

[Habara 1988] K. Habara, ISDN: A look at the future through the past, *IEEE Communication Magazine*, November 1988.

[Habib 2002] A. Habib and H. Saiedian, *Channelized Voice over Digital Subscriber Line*, IEEE Communications Magazine, October 2002.

[G991.1] ITU-T Telecommunication Standardization Sector of ITU G.991.1 (10/98) Series G: Transmission Systems and Media, Digital Systems and Networks Digital transmission systems—Digital sections and digital line system—Access networks High bit rate Digital Subscriber Line (HDSL) transceivers.

[G991.2] *ITU-T G.991.2 (02/2001)*, Digital sections and digital line system—Access networks, Single-pair high-speed digital subscriber line (SHDSL) transceivers.

[G997.1] *ITU-T G.997.1 (1999)*, Digital sections and digital line system—Access networks, Physical layer management for digital subscriber line (DSL) transceivers.

[Starr 1999] T. Starr, J.M. Cioffi and P.J. Silverman, *Understanding Digital Line Technology*, Prentice Hall, 1999, ISBN 0-13-780545-4.

[Starr 2003] T. Starr, M. Sorbara, J.M. Cioffi, P.J. Silverman, *DSL Advances*, Prentice Hall 2003, ISBN 0-13-093810-6.

[T-Com]   http://www.telekom.de/untern/aktuell/1999/03189923.htm
                        (Stand 06.02.2002).
[TlE1.4/92-002R1]   *A Technical Report on High-bit-rate Digital Subscriber Lines
                        (HDSL)*, Exchange Carrier Standards Association Technical
                        Committee TlE1.4/92-002R1, February 14,1992.

ADSL

[Cioffi 1991]   J.M. Cioffi, *A Multicarrier Primer*, T1E1.4 contribution 91-
                        157, November 1991.
[Doelz 1957]   M.L. Doelz, E.T. Heald, and D.L. Martin, *Binary Data Trans-
                        mission Techniques for Linear Systems*, Proceedings of IRE,
                        pp. 656–661, May 1957.
[TS 101 388 (1998)]   ETSI TS 101 388, *Transmission and Multiplexing (TM); Ac-
                        cess transmission systems on metallic access cables; Asym-
                        metric Digital Subscriber Line (ADSL)*, 1998.
[TS 101 388 (2002)]   ETSI TS 101 388: *Transmission and Multiplexing (TM); Ac-
                        cess transmission systems on metallic access cables; Asym-
                        metric Digital Subscriber Line (ADSL)—European specific re-
                        quirements*, 2002.
[Golden 2006]   P. Golden, H. Dedieu, and K.S. Jacobsen, Eds. *Fundamen-
                        tals of DSL Technology*. Auerbach Publications, Boca Raton,
                        Florida, 2006.
[G.992.1]   ITU-T Recommendation G.992.1: *Asymmetric Digital Sub-
                        scriber Line (ADSL) Transceivers*, July 1999, available at
                        http://www.itu.int/rec/T-REC-G.992.1/en.
[G.992.2]   ITU-T Recommendation G.992.2: *Splitterless Asymmetric
                        Digital Subscriber Line (ADSL) Transceivers*, July 1999, avail-
                        able at http://www.itu.int/rec/T-REC-G.992.2/en.
[G.992.3]   ITU-T Recommendation G.992.3: *Asymmetric Digital Sub-
                        scriber Line Transceivers 2 (ADSL2)*, January 2005, available
                        at http://www.itu.int/rec/T-REC-G.992.3/en.
[G.992.4]   ITU-T Recommendation G.992.4, *Splitterless Asymmet-
                        ric Digital Subscriber Line Transceivers 2 (Splitterless
                        ADSL2)*, July 2002, available at http://www.itu.int/rec/T-
                        REC-G.992.4/en.
[G.992.5]   ITU-T Recommendation G.992.5, *Asymmetric Digital
                        Subscriber Line (ADSL) Transceivers—Extended band-
                        width ADSL2 (ADSL2plus)*, Jan 2005, available at
                        http://www.itu.int/rec/T-REC-G.992.5/en
[T1.413]   ANSI Standard T1.413. *Network to Customer Installa-
                        tion Interfaces—Asymmetric Digital Subscriber Line (ADSL)
                        Metallic Interface*, 1995.
[T1.413 Issue 2]   ANSI Standard T1.413, Issue 2. *Network and Customer In-
                        stallation Interfaces—Asymmetric Digital Subscriber Line
                        (ADSL) Metallic Interface*, 1998.
[T1.417-2003]   ANSI Standard T1.417-2003, issue 2. *Spectrum Management
                        for Loop Transmission Systems*, 2003.

VDSL

[T1E1.4/94-087]   J.M. Cioffi and K. Jacobsen. *A Solution to the T1 Problem for CSA-Range 6 Mbps Transmission*, T1E1.4 contribution 94-087, April 1994.

[T1E1.4/94-088]   J.M. Cioffi and K. Jacobsen. *15 Mbps on the Mid-CSA*, T1E1.4 contribution 94-088, April 1994.

[T1E1.4/94-125]   J.M. Cioffi and K. Jacobsen. *Range/Rate Projections for NxDMT*, T1E1.4 contribution 94-125, June 1994.

[TM6/001t13]   ETSI TM6 contribution 001t13, *Proposed VDSL Band Plans*, FSAN VDSL Working Group, Montreux, Switzerland, March 2000.

[ETSI TS 101 270-1] *Access transmission systems on metallic access cables; Very high speed Digital Subscriber Line (VDSL); Part 1: Functional requirements*, ETSI Technical Standard, TS 101 270-1 V1.3.1, July 2003.

[ETSI TS 101 270-2] *Access Transmission Systems on Metallic Access Cables; Very High Speed Digital Subscriber Line (VDSL);Part 2: Transceiver specification*. ETSI Technical Standard, TS 101 270-2 V1.2.1, July 2003.

[ETSI TM6 TD3R1]   *Draft ETSI Technical Report: Transmission and Multiplexing (TM); Very high bit-rate digital transmission on metallic local lines (VDSL)*, ETSI TM6 contribution TD3R1, December 1995.

[WP1/15/D.786]   ITU WP1/15 contribution D.786, *Frequency planning proposal*, Alcatel Bell, Geneva, Switzerland, April 2000.

[G.993.1-2001]   *Very High Speed Digital Subscriber Line Foundation*, ITU Recommendation G.993.1, November 2001; *Very High Speed Digital Subscriber Line*, ITU Recommendation G.993.1 (2001) amendment 1, March 2003.

[G.993.1-2004]   *Very High Speed Digital Subscriber Line Transceivers*, ITU Recommendation G.993.1, June 2004.

[G.993.2]   *Very High Speed Digital Subscriber Line 2*, ITU Recommendation G.993.2, approved February 2006, to be published.

[BT Olympics]   T1E1.4 contribution T1E1.4/2003-600, *VDSL Line Code Analysis of Ikanos—Mandatory Tests*, Anaheim, California, June 2003;

                     T1E1.4 contribution T1E1.4/2003-602, *VDSL Line Code Analysis of Infineon—Mandatory Tests*, Anaheim, California, June 2003;

                     T1E1.4 contribution T1E1.4/2003-604, *VDSL Line Code Analysis of Metalink—Mandatory Tests*, Anaheim, California, June 2003;

                     T1E1.4 contribution T1E1.4/2003-606, *VDSL Line Code Analysis of ST Microelectronics—Mandatory Tests*, Anaheim, California, June 2003.

[Telcordia Olympics] T1E1.4 contribution T1E1.4/2003-608, *Mandatory VDSL Transceiver Test Results for Infineon*, Anaheim, California, June 2003;

T1E1.4 contribution T1E1.4/2003-609, *Mandatory VDSL Transceiver Test Results for STMicroelectronics*, Anaheim, California, June 2003;
T1E1.4 contribution T1E1.4/2003-610, *Mandatory VDSL Transceiver Test Results for Metalink*, Anaheim, California, June 2003;
T1E1.4 contribution T1E1.4/2003-611, *Mandatory VDSL Transceiver Test Results for Ikanos*, Anaheim, California, June 2003.

[T1E1.4/2001-157] T1E1.4 contribution T1E1.4/2001-157, *Proposal for the North American Reference PSD*, Alcatel & Broadcom, Tampa, Florida, May 2001;
T1E1.4 contribution T1E1.4/2001-242, *Reference PSD for ANSI Noise F*, Alcatel & Broadcom, Greensboro, North California, November 2001.

[TM6/013w07] ETSI TM6 contribution 013w07, *Proposed Reference PSDs for Europe*, Alcatel & Broadcom, Stockholm, Sweden, September 2001.

[T1E1.4/2003-036] T1E1.4 contribution T1E1.4/2003-036R4, *Draft VDSL test specification*, Newport Beach, California, February 2003..

[T1E1.4/99-059] T1E1.4 contribution T1E1.4/99-059, *Recommendation for VDSL Duplexing Method*, FSAN VDSL Working Group, Orlando, Florida, February 1999.

[T1.424-2004] *Very-High-Bit-Rate Digital Subscriber Lines (VDSL) Metallic Interface (DMT Based)*, American National Standard T1.424-2004.

[T1.TRQ.12-2004] *Very-High-Bit-Rate Digital Subscriber Lines (VDSL) Metallic Interface (QAM Based)*, T1 Technical Requirement T1.TRQ.12-2004.

[T1.424 Trial Use] *Interface between Networks and Customer Installations—Very-High-Bit-Rate Digital Subscriber Lines (VDSL) Metallic Interface*, Trial Use standard, T1.424-Trial-Use.

[T1E1.4/2000-083] T1E1.4 contribution T1E1.4/2000-083, *VDSL Band Plan for North America*, Maui, Hawaii, February 2000.

[Sjöberg 1] F. Sjöberg et al., *Zipper: A Duplex Method for VDSL Based on DMT*, IEEE Transaction Communication, Vol.47, pp. 1245–1252, August 1999.

[Sjöberg 2] F. Sjöberg et al., *Asynchronous Zipper*, IEEE Proceedings of International Conference on Communications, Vancouver, Canada, June 1999.

[TM6/991t10] ETSI TM6 contribution 991t10, *Proposal for VDSL duplexing Method*, FSAN VDSL Working Group, Villach, Austria, February 1999.

Handshake

[AT&T 1994] AT&T, *Signals for proposed V.mcp protocol*, contribution TIA TR-30.1/94-08-041, Bellevue, Washington, 1–5 August, 1994.

[BT 1991] British Telecom, *An outline proposal for V.fast call-*

	ing/answering which can allow demod/remod by DCME/ PCME and assist automoding, Delayed contribution D.141, CCITT Study Group XVII, Geneva, 29 October–6 November, 1991.
[Data Race 1994]	Data Race, *Proposed modifications to TIA-578 for voice-data conferencing*, contribution TIA TR-30.1/94-02-003, Fort Lauderdale, Florida, 7-11 February, 1994.
[H.324]	*ITU-T Recommendation H.324.*
[GSTN]	ITU-T SG15 *Question 2 – GSTN Videophone.*
[G.994.1]	*ITU-T Recommendation G.994.1 – 2003.*
[G.994.1 Amd 4]	*ITU-T Recommendation G.994.1 Amendment 4 – 2005.*
[SAVD]	ITU-T SG14 *Question 1 – Simultaneous and alternating voice and data (SAVD).*
[USA 1996]	United States of America, *Proposed new questions for Study Group 14*, Delayed contribution D.164, ITU-T SG14, Geneva, 19 – 27 March, 1996.
[V.8]	*ITU-T Recommendation V.8.*
[V.8bis]	*ITU-T Recommendation V.8 bis.*
[V.32bis]	*ITU-T Recommendation V.32 bis.*
[V.34]	*ITU-T Recommendation V.34.*
[V.70]	*ITU-T REcommendation V.70.*

Index

Printed and bound by CPI Group (UK) Ltd, Croydon, CR0 4YY

17/10/2024

01775690-0020